NUCLEAR PHYSICS

Problem-based Approach Including MATLAB®

HARI M. AGRAWAL

Professor and Head
Department of Physics
G. B. Pant University of Agriculture and Technology
Pantnagar

PHI Learning Private Limited

Delhi-110092
2016

₹ 595.00

NUCLEAR PHYSICS—Problem-based Approach Including MATLAB® (with CD-ROM)
Hari M. Agrawal

ISBN-978-81-203-5252-0

Published by Asoke K. Ghosh, PHI Learning Private Limited, Rimjhim House, 111, Patparganj Industrial Estate, Delhi-110092 and Printed by Raj Press, New Delhi-110012.

In Memory of My Parents

Madan Mohan Agrawal and Sushila Agrawal

Contents

4. Nuclear Force 157–191

5. Nuclear Reactions 192–253

6. Interaction of Radiations with Matter 254–290

CD CONTENTS

Preface

A nuclear physics course is inevitable in the M.Sc. (Physics) program of all Indian universities and institutes. Recently, one of the discipline specific elective—"Nuclear and Particle Physics" has been outlined by the University Grants Commission (UGC) under choice-based credit system for B.Sc. (Hons.) Physics. Author felt an urgent and genuine need for one quality book that students could use as a standard textbook, which covers all the topics on nuclear physics with updated notions and current viewpoints. This book has evolved from the author's lecture notes for a nuclear physics course that he has been teaching for more than thirty years to the M.Sc. and the Ph.D. students at the G. B. Pant University, Pantnagar.

The book has been divided into ten chapters. The first chapter introduces the concept of the nucleus, which is followed by a chapter discussing the theoretical models that describe the nuclear structure. The phenomenon of radioactivity, including its applications, is covered in the next chapter. Chapter 4 is dedicated to the nuclear force, which is rather complex and can be understood to some extent by studying the two nucleon bound (deuteron) and unbound (*n-p* scattering) systems. Different reaction mechanisms are discussed in Chapter 5. The next two chapters cover, interaction of radiations with matter and detectors, including particle accelerators, from a practical point of view. A complete chapter is devoted to the sub-nuclear physics including the latest developments, the production and the properties of the elementary particles and their interactions. An overview of the applications of nuclear physics is presented in Chapter 9. Introductory nuclear astrophysics is covered in the last chapter.

While shaping this book, the scope has been sufficiently widened, at the same time retaining enough depth, so that the undergraduate students majoring in physics and the postgraduates may use it flexibly, according to their requirements. No prior knowledge of nuclear physics has been assumed on part of the students and almost every topic is started from the basic level. The mathematics has been kept as simple as possible and the subject has been presented in a manner that is both interesting and easily accessible to the student. The main text is, therefore, interspersed with numerous worked out exercises along with end chapter problems, helping the students grasp the key concepts and gradually build up a satisfactory approach to problem solving. This should make the book valuable also for the students wishing to study the subject on their own or for preparing for any competitive examinations like NET, SET, GATE, JEST, etc.

In today's information technology era, it is extremely important for scientists and engineers to be familiar with various computing and analyzing tools that are available and can be used to our advantage. Keeping this in mind, in many of the solved exercises in this book, MATLAB® and EXCEL have been used to not only exhibit the utility of these tools but to also demonstrate how these tools can make problem solving, which is an integral part of learning and understanding physics, fun.

Author shall be grateful if any errors, which might have crept in inadvertently, are pointed out along with any ideas to improve the book.

Hari M. Agrawal

Acknowledgements

This book became a reality because of the support of several people. First of all I should like to acknowledge Rajeev M. Agrawal, who helped in writing this book. He did his B.Tech. in Chemical Engineering and M.Tech. in Computer Applications in Chemical Engineering from I.I.T. Delhi in 2007. He has not only written and edited the text along with checking the correctness and arrangement of the numerical problems; but also encouraged me to organize my class notes and various other resources necessary for writing this book. The MATLAB® programs and the EXCEL charts in this book are prepared exclusively by him. He also drafted the complete layout of the book.

I convey my heartfelt thanks to the authors and the publishers of the various books, review articles and online resources that have been consulted extensively, while preparing this manuscript. I thank all my present and previous students whose contributions, in terms of their queries and suggestions, helped enrich my understanding of the subject. I also thank my wife, Kiran Agrawal and daughter, Rakhi Agrawal, for their continuous encouragement and patience during the course of working on this book.

Finally, I thank the editorial and production team of PHI Learning, in particular Ms. Shivani Garg, Senior Editor, for her wholehearted cooperation during the publication of this book.

Hari M. Agrawal

1

The Nucleus

"It was quite the most incredible event that has ever happened to me in my life. It was almost as incredible as if you fired a 15-inch shell at a piece of tissue paper and it came back and hit you.

[Recalling in 1936, the discovery of the nucleus in 1909, when some alpha particles instead of travelling through a very thin gold foil were seen to rebound backward, as if striking something much more massive than the particles themselves.]"

—Ernest Rutherford

The earliest prediction of the size of a nucleus was made by Rutherford on the basis of his famous α-scattering experiments by his team in Cavendish laboratory. On the basis of pure long-range Coulombian interaction, Rutherford confirmed the presence of a positively charged heavy nucleus.

1.1 RUTHERFORD'S NUCLEAR MODEL OF ATOM

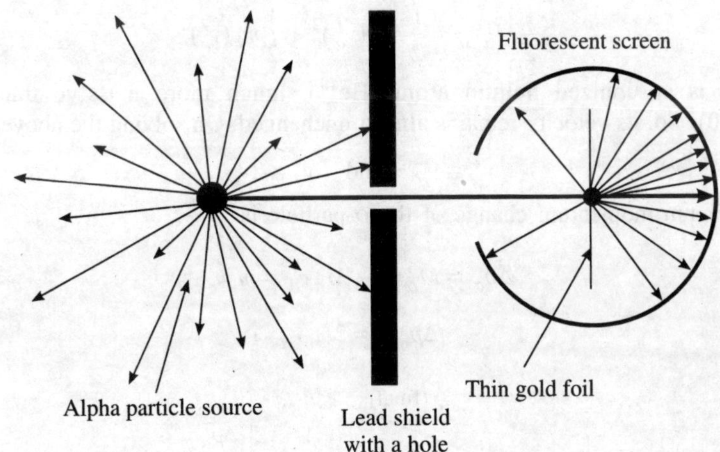

FIGURE 1.1 Schematic diagram of the Rutherford's scattering experiment.

Rutherford's famous α-scattering experiment is represented in Figure 1.1. Rutherford and his students (H. Geiger and E. Marsden) bombarded very thin gold foil (~600 nm) with a stream of high energy α-particles from a radioactive source. A circular ZnS screen, which gives off a visible flash of light when struck by an α-particle, was set up with a microscope to see the flashes. Based on the Thomson's plum-pudding model that was prevalent at that time, in which the uniform distribution of the electric charge throughout the volume of an atom is assumed, it was expected that the α-particles would pass through the foil with hardly any deflection.

EXERCISE 1.1: Assuming an α-particle scatters from an electron in the 600 nm gold foil, estimate the maximum scattering angle [Given: Au mass number is 197 and its density is 19.3 g/cc]

(a) in a single encounter,
(b) in multiple scattering events.

Solution: Considering elastic collision between α-particle and e^- in the gold foil, the maximum momentum transfer occurs when the α-particle hits the e^- (at rest) head-on as shown in Figure 1.2

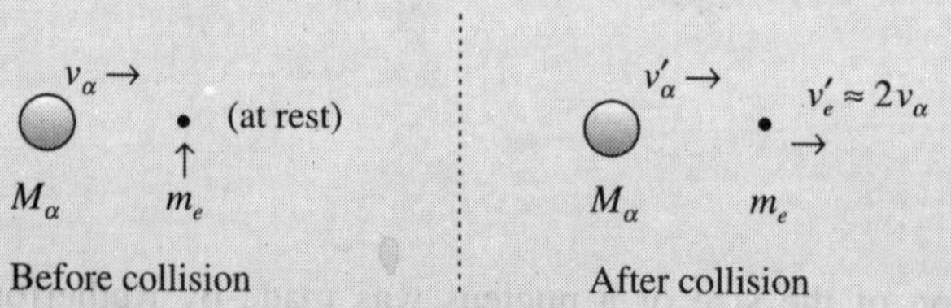

FIGURE 1.2 Schematic diagram of an α-particle making a head-on collision with an electron initially at rest.

According to the conservation of linear momentum

$$M_\alpha \overrightarrow{v_\alpha} = M_\alpha \overrightarrow{v'_\alpha} + m_e \overrightarrow{v'_e}$$

Conservation of kinetic energy gives

$$\frac{1}{2} M_\alpha v_\alpha^2 = \frac{1}{2} M_\alpha (v'_\alpha)^2 + \frac{1}{2} m_e (v'_e)^2$$

An α-particle is an ionized helium atom (He^{++}), much more massive than the electron ($M_\alpha / m_e \approx 7000$). So, its velocity remains almost unchanged. On solving the above two equations, we get

$$v'_\alpha \cong v_\alpha \quad \text{and} \quad v'_e \cong 2v_\alpha$$

Thus, the maximum momentum change of the α-particle is

$$\Delta p_\alpha = M_\alpha v_\alpha - M_\alpha v'_\alpha = m_e v'_e$$

or

$$\Delta p_{\text{max}} = 2 m_e v_\alpha$$

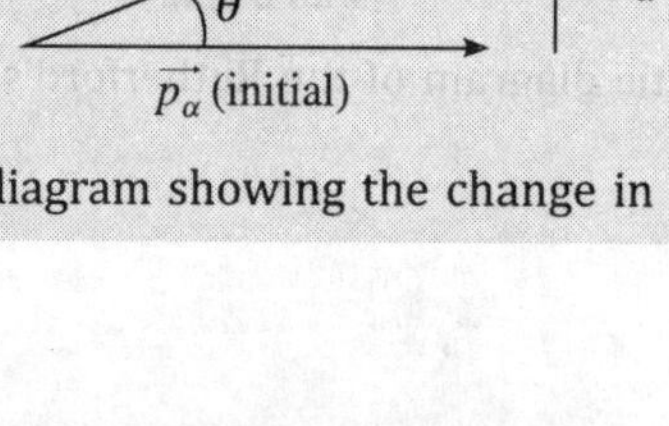

FIGURE 1.3 Vector diagram showing the change in α-particle momentum.

(a) Based on the above vector diagram (Figure 1.3) for α-particle momentum

$$\theta_{\max} = \frac{\Delta p_\alpha}{p_\alpha} = \frac{\text{length of arc}}{\text{radius}}$$

$$= \frac{2m_e v_\alpha}{M_\alpha v_\alpha} = \frac{2m_e}{M_\alpha} = 2.7 \times 10^{-4} \text{ rad} = 0.016°$$

(b) Multiple scattering is possible, and a calculation for random multiple scattering from N electrons yields an average scattering angle of

$$<\theta>_{\text{total}} \approx \sqrt{N} \times \theta$$

Number of atoms per cm^3 in the target, which is 600 nm thick gold foil is given by

$$\left(6.023 \times 10^{23} \frac{\text{atoms}}{\text{mol}}\right)\left(\frac{1\,\text{mol}}{197\,\text{g}}\right)\left(19.3 \frac{\text{g}}{\text{cm}^3}\right) = 5.9 \times 10^{28} \text{ atoms/m}^3$$

If there are 5.9×10^{28} atoms/m^3, then each atom occupies $(5.9 \times 10^{28})^{-1}$ m^3 of space. Assuming the atoms are equidistant, the distance, d between the centres is

$$d = (5.9 \times 10^{28})^{-1/3} \text{ m} = 2.6 \times 10^{-10} \text{ m}$$

In the foil, there are

$$N = \frac{600 \text{ nm}}{2.6 \times 10^{-10} \text{ m}} = 2300 \text{ atoms}$$

along the α-particle's path. If we assume that the α-particle interacts with one electron from each gold atom, then

$$<\theta>_{\text{total}} \approx \sqrt{2300} \times (0.016°) = 0.77°$$

If the α-particle interacts with all 79 electrons of each gold atom, then

$$<\theta>_{\text{total}} \approx \sqrt{79} \times \theta = 6.8°$$

The results of the scattering experiment were quite unexpected. It was observed that:

(i) Most of the α-particles passed through the gold foil undeflected.

(ii) A small fraction of the α-particles was deflected by small angles in agreement with the above solved example.

(iii) Few α-particles were scattered through large angles and a very few α-particles (one in twenty thousand) bounced back, that is, they were deflected by nearly 180°.

On the basis of the above observations, Rutherford drew the following conclusions regarding the structure of an atom:

(i) Most of the space in the atom is empty as most of the α-particles passed through the foil undeflected.

(ii) A few positively charged α-particles were deflected by very large angles. The deflection must be because of enormous repulsive force due to the concentrated positive charge in a very small volume of the atom.

(iii) Calculations by Rutherford showed that the radius of positively charged core is about 10^{-15} m.

After the discovery of protons, Rutherford proposed the nuclear model of atom. According to this model:

(i) The positive charge and most of the mass of the atom is densely concentrated in extremely small region, called the nucleus, a word first coined by Rutherford.

(ii) The nucleus is surrounded by electrons moving at high speeds around the nucleus in definite orbits.

(iii) Electrons and the nucleus are held together by electrostatic forces of attraction.

Rutherford's analysis of the observed α-scattering in terms of his nuclear model of the atom was based on the assumption that both nuclei, i.e., the target nucleus and the α-particle, could be considered as point charges. The force was thus, taken to be Coulombic:

$$F = \frac{1}{4\pi\varepsilon_0}\frac{(2e)(Ze)}{r^2} \tag{1.1}$$

where Ze is the charge on the nucleus and $2e$ is the charge of the α-particle.

Owing to the $1/r^2$ type fundamental form of the interaction, where r is the instantaneous separation between the α-particle and the nucleus, the trajectory of the α-particle is a hyperbola with the nucleus at the outer focus (Figure 1.4).

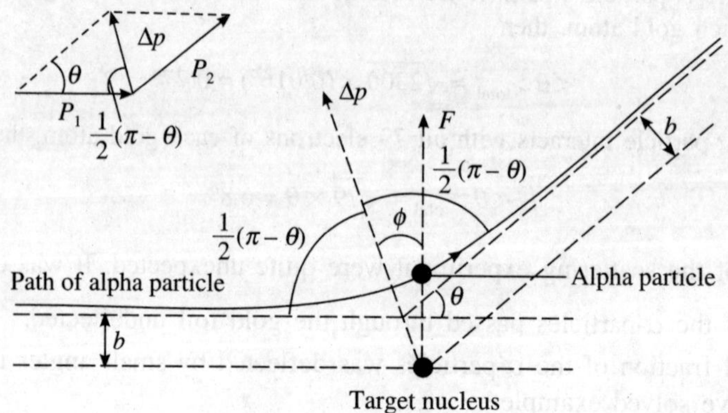

FIGURE 1.4 Geometrical relationships in Rutherford alpha scattering.

The impact parameter b is the minimum distance at which the α-particle would approach the nucleus if there were no force between them. φ is the instantaneous angle between $\vec{F}$ and $\overrightarrow{\Delta p}$ along the trajectory of the α-particle. The scattering angle θ is the angle between the asymptotic incident direction of the α-particle and the asymptotic direction of the deflected α-particle. Normally, detectors are placed at one or more scattering angles to count the particles scattered into small cones of solid angle subtended by the detectors.

1.1.1 Relationship between Scattering Angle and Impact Parameter

In the following discussion, we assume that the nucleus is sufficiently massive to be not displaced by the encounter with the α-particle. Therefore, the initial linear momentum $|p_1|$ is equal to the final linear momentum $|p_2|$, i.e., $|p_1| = |p_2| = mv$. However, $\overrightarrow{p_1}$ gets changed to $\overrightarrow{p_2}$, as a result of the change in direction, as shown in Figure 1.4.

From the law of sines, we have

$$\frac{\Delta p}{\sin\theta} = \frac{mv}{\sin\left(\dfrac{\pi-\theta}{2}\right)}$$

or

$$\Delta p = \frac{(mv)\left(2\sin\dfrac{\theta}{2}\cos\dfrac{\theta}{2}\right)}{\left(\cos\dfrac{\theta}{2}\right)} = 2mv\sin\frac{\theta}{2}$$

The change in momentum is equal to the impulse given by the nucleus,

$$\overrightarrow{\Delta p} = \int \vec{F}.dt$$

or

$$2mv\sin\frac{\theta}{2} = \int_{-\infty}^{+\infty} F\cos\varphi dt$$

Changing the variable from t to φ, the limits become:

$$\varphi = -\left(\frac{\pi-\theta}{2}\right) \text{ and } \varphi = +\left(\frac{\pi-\theta}{2}\right)$$

corresponding to $t = -\infty$ and $t = +\infty$, respectively. Therefore,

$$2mv\sin\frac{\theta}{2} = \int_{-\left(\frac{\pi-\theta}{2}\right)}^{+\left(\frac{\pi-\theta}{2}\right)} F\cos\varphi\left(\frac{dt}{d\varphi}\right)d\varphi \tag{1.2}$$

where $d\varphi/dt = \omega$, is the angular velocity of the α-particle about the nucleus.

The Coulomb force exerted by the nucleus on the α-particle is along the radius vector $\vec{r}$, basically a central force, therefore the angular momentum of the system must be conserved. So,

$$mr^2\frac{d\varphi}{dt} = mvb \tag{1.3}$$

$$\frac{dt}{d\varphi} = \frac{r^2}{vb} \tag{1.4}$$

Substituting equation (1.4) into equation (1.2), we get

$$2mv^2 b\sin\frac{\theta}{2} = \int_{-\left(\frac{\pi-\theta}{2}\right)}^{+\left(\frac{\pi-\theta}{2}\right)} Fr^2 \cos\varphi\, d\varphi$$

On substituting the expression for the electrostatic force F,

$$2mv^2 b\sin\frac{\theta}{2} = \frac{Ze^2}{4\pi \epsilon_0}\int_{-\left(\frac{\pi-\theta}{2}\right)}^{+\left(\frac{\pi-\theta}{2}\right)} \cos\varphi\, d\varphi = \frac{Ze^2}{4\pi \epsilon_0}\left[2\cos\frac{\theta}{2}\right]$$

or

$$b = \frac{Ze^2}{2\pi \epsilon_0 mv^2}\cot\frac{\theta}{2}$$

Thus, the equation becomes

$$b = \frac{Ze^2}{4\pi \epsilon_0 K}\cot\frac{\theta}{2} \tag{1.5}$$

where $K = mv^2/2$, is the kinetic energy of the bombarding particle. This is the fundamental relation between the impact parameter b and the scattering angle θ, which shows that the scattering angle decreases as the impact factor increases, as depicted in Figure 1.5.

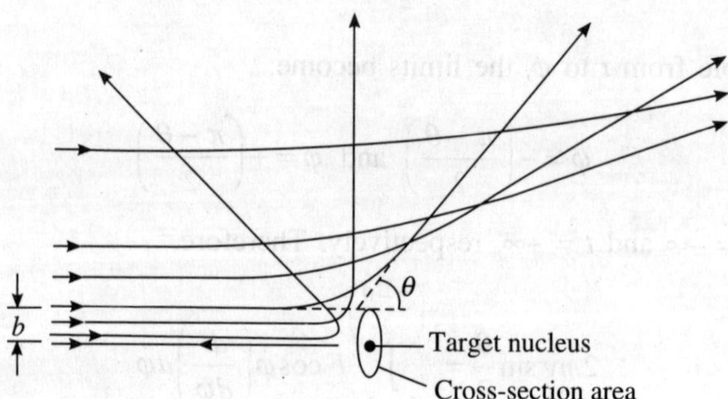

FIGURE 1.5 Scattering angle as a function of the impact parameter.

1.1.2 Angular Scattering Distribution

In an α-scattering experiment, a detector is placed at a particular angle θ, covering a finite $\Delta\theta$, which would correspond to a range of impact parameters Δb. The bombarding α-particles are incident at various impact parameters around the scatterer nucleus as shown in Figure 1.5. An α-particle that is initially directed anywhere within the area πb^2 around a nucleus will be scattered through an angle θ or more. For that reason, the area around a nucleus (πb^2) is called the cross-section, σ for the interaction, which is related to the probability for a particle being scattered by a nucleus.

Now, we consider a foil of thickness t containing n atoms per unit volume. The number of target nuclei per unit area is $n \times t$. An α-particle beam incident upon an area A, therefore, encounters ntA nuclei. If σ is the cross-section for each nucleon, then $ntA\sigma$ is the total area exposed by the target nuclei, and the fraction of incident particles scattered by an angle of θ or greater is

$$f = \frac{\text{Target area exposed by the nuclei}}{\text{Total target area}} = \frac{ntA\sigma}{A}$$

or

$$f = nt\sigma = nt\pi b^2$$

Substituting for b from equation (1.5), we get

$$f = nt\pi \left(\frac{Ze^2}{4\pi \epsilon_0 K} \right)^2 \cot^2 \frac{\theta}{2} \tag{1.6}$$

In an actual experiment, however, a detector is placed over a range of angles from θ to $\theta + \Delta\theta$. Thus, we need to find the number of particles scattered between θ and $\theta + d\theta$ that corresponds to the incident particles with impact parameters between b and $b + db$ as defined in Figure 1.6.

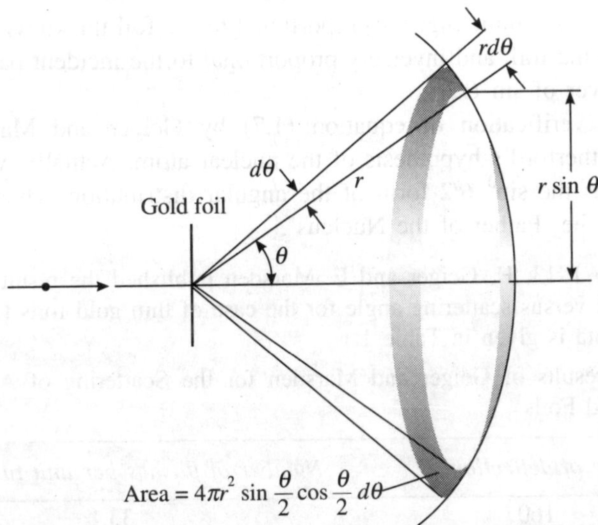

FIGURE 1.6 Detection of alpha particles scattered between θ and $\theta + d\theta$ in Rutherford experiment.

The fraction of the incident particles scattered between θ and $\theta + d\theta$ is found by differentiating equation (1.6) with respect to θ, which gives

$$df = -nt\pi \left(\frac{Ze^2}{4\pi \epsilon_0 K} \right)^2 \cot \frac{\theta}{2} \operatorname{cosec}^2 \frac{\theta}{2} d\theta$$

The minus sign indicates that f decreases with increasing θ. Geiger and Marsden placed a fluorescent screen at a distance r from the foil and the scattered α-particles were detected by means of scintillations they produced (Figure 1.6). The α-particles scattered between θ and $\theta + d\theta$ reached a zone of a sphere of radius r whose width is $rd\theta$. The area dA of the screen struck by these particles is

$$dA = (2\pi r \sin\theta)(rd\theta) = 2\pi r^2 \sin\theta\, d\theta$$

If the total number of incident particles is N_i, the number of particles scattered into $d\theta$ at θ is $N_i df$. Therefore, the number of particles scattered per unit area, $N(\theta)$, into the ring at scattering angle θ is

$$N(\theta) = \frac{N_i\,|df|}{dA} = \frac{N_i nt\pi\left(\dfrac{Ze^2}{4\pi \epsilon_0 K}\right)^2 \cot\dfrac{\theta}{2}\,\text{cosec}^2\dfrac{\theta}{2}\,d\theta}{4\pi r^2 \sin\dfrac{\theta}{2}\cos\dfrac{\theta}{2}\,d\theta}$$

or

$$N(\theta) = \frac{N_i ntZ^2 e^4}{(8\pi \epsilon_0)^2 r^2 K^2 \sin^4\dfrac{\theta}{2}} \tag{1.7}$$

This is the famous Rutherford's scattering formula. According to this formula, the number of particles scattered at a certain angle is proportional to the foil thickness and to the square of the nuclear charge of the foil, and inversely proportional to the incident particle kinetic energy and to the fourth power of sin $\theta/2$.

The quantitative verification of equation (1.7) by Geiger and Marsden constituted a conclusive test of Rutherford's hypothesis of the nuclear atom. Actually, what was verified in these experiments was the $\sin^4 \theta/2$ form of the angular distribution. Thus, Ernest Rutherford can rightly be called the 'Father of the Nucleus'.

EXERCISE 1.2: In 1913, H. Geiger and E. Marsden published the results on the number of scintillations obtained versus scattering angle for the case of thin gold foils [Phil. Mag., Vol. 25, 605 (1913)]. Their data is given in Table 1.1.

Table 1.1 Typical results of Geiger and Marsden for the Scattering of Alpha Particles from Thin Gold Foils

Angle of deflection, θ	*Number of flashes per unit time, N(θ)*
160	33.1
135	43.0
120	51.9
105	69.5
75	211
60	477
45	1435
37.5	3300
30	7800

Obtain a comparison between Rutherford's theory and the experiment.

Solution:

$$N(\theta) = \frac{N_i ntZ^2 e^4}{(8\pi \epsilon_0)^2 r^2 K^2 \sin^4 \frac{\theta}{2}}$$

The actual value of initial energy, the detector distance, the number of α-particles, and the number of nuclei needed in the Rutherford's scattering formula, are not provided in the experimental data. Therefore, we rewrite the above formula in terms of a constant (C), whose value is chosen so as to have minimum discrepancy between the theoretical curve and the actual experimental data.

$$N(\theta) = \frac{C}{\sin^4 \frac{\theta}{2}}$$

The comparison between theory and experiment is shown in Figure 1.7. y-axis is on logarithmic scale.

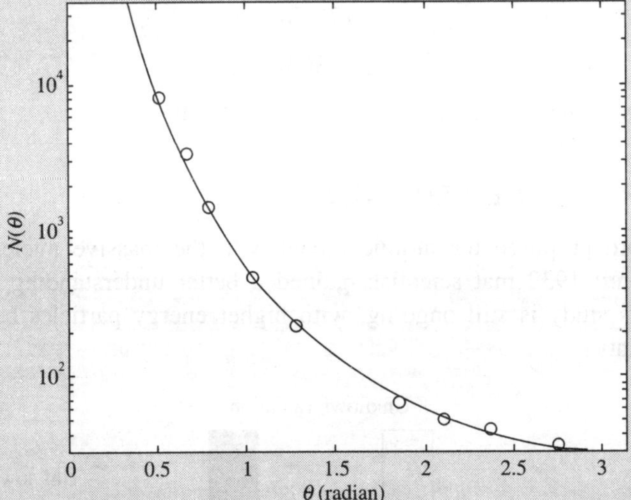

FIGURE 1.7 Comparison between Rutherford theory and actual experimental data.

The following Matlab code was used to generate the above plot.

```
clear all
t=[160 135 120 105 75 60 45 37.5 30]; %Theta in degrees
tr=t*3.14/180; %Theta in radians
N=[33.1 43.0 51.9 69.5 211 477 1435 3300 7800]; %N(theta)
C=30; %Constant (value chosen to get the closest match)

syms x real positive;
f=C/(sin(x/2))^4;
ezplot(f,[0,3.14])
hold on
plot(tr,N,'ro');
xlabel('theta (radian)')
ylabel('N (theta)')
set(gca,'yscale','log'); %Setting the y-axis on log scale
title('Comparison between Rutherford theory & actual experimental
data')
```

1.1.3 Estimate of the Nucleus Size

In the derivation of equation (1.7), Rutherford assumed that the target nucleus is small as compared with the distance of closest approach (= R), where the potential energy may be set equal to the initial kinetic energy (7.7 MeV) of the α-particle before it enters the electric field of the nucleus.

Thus, we have

$$K_{\text{initial}} = \frac{1}{4\pi \epsilon_0} \frac{(2e)(Ze)}{R}$$

or

$$R = \frac{2Ze^2}{4\pi \epsilon_0 K_{\text{initial}}} \tag{1.8}$$

For gold foil, $Z = 79$,

$$R = \frac{2 \times (79 \times 1.44 \text{ MeV fm})}{(7.7 \text{ MeV})}$$

$$R \approx 30 \text{ fm}$$

The radius of the gold nucleus is therefore less than 30 fm.

1.2 DISCOVERY OF THE NEUTRON

Although, Rutherford proposed the atomic model with the massive nucleus at the centre in 1911, it was not until 1932 that scientists gained a better understanding of the constituents of the nucleus. This study is still ongoing, with higher energy particles being utilized in the scattering experiments.

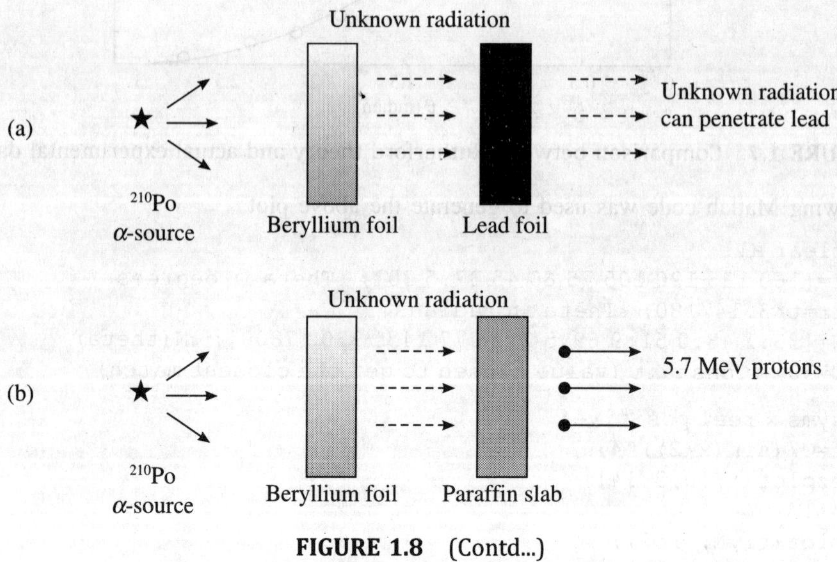

FIGURE 1.8 (Contd...)

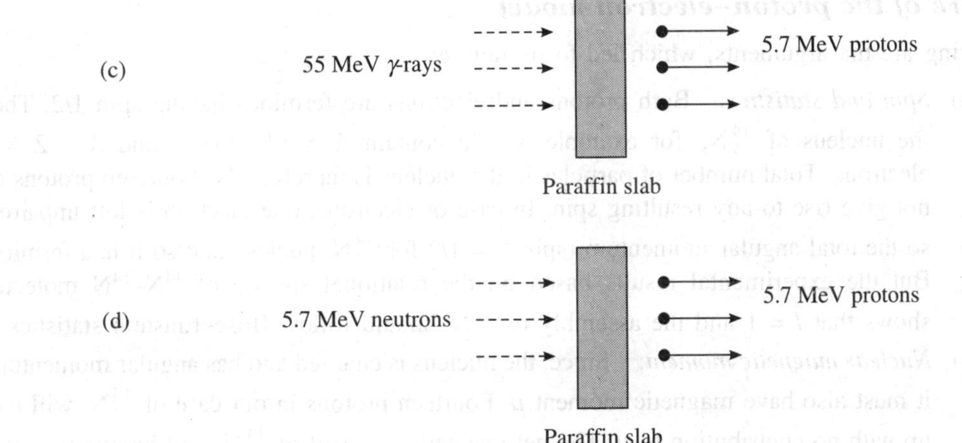

FIGURE 1.8 (a) Alpha particles from Po-210 radioactive source when fall on a Be foil cause the emission of an unknown, very penetrating radiation.
(b) Protons up to 5.7 MeV are produced when the radiation strikes a paraffin slab.
(c) If the unknown radiation are gamma rays, their energies should have been at least 55 MeV.
(d) If the radiation consists of neutral particles of mass approximately equal to proton mass, their energies must be less than or equal to 5.7 MeV.

The discovery of the neutron is the outcome of classic experiments carried out during 1930–32 (Figure 1.8). In 1930, German physicists W. Bothe and H. Beeker found that an uncharged, very penetrating radiation emitted when α-particles from ^{210}Po source were bombarded on Be foil. Subsequently in 1932, I. Curie and F. Joliot discovered that this mysterious radiation could knock protons with energies up to 5.7 MeV out of a paraffin slab. They assumed that γ-rays of nuclear origin are responsible for knocking out the protons in Compton scattering process from the hydrogen-rich paraffin. However, the hypothesis of Compton scattering requires γ-ray energies of at least 55 MeV to produce 5.7 MeV protons. Energies as large as 50 MeV were not known at that time.

In 1932, James Chadwick proposed that neutral particles with about the same mass as that of proton were responsible, in which case their energy needed to be only 5.7 MeV, because in an elastic head-on collision with another particle of the same mass, all of its kinetic energy gets transferred to the latter. Other experiments confirmed his hypothesis and he received the Nobel Prize in 1935 for his discovery of the neutron.

1.3 CONSTITUENTS OF THE NUCLEUS

1.3.1 Non-existence of Nuclear Electrons

When α and β-rays were first identified as $^4_2$He nuclei and electrons, the presumption was that both were contained in the nuclei because both were expelled from the nuclei. This led to the incorrect proton–electron model.

Failure of the proton–electron model

Following are the arguments, which led to its failure:

(a) *Spin and statistics*: Both protons and electrons are fermions having spin 1/2. Then the nucleus of ${}^{14}_{7}\text{N}$, for example, would contain A = 14 protons and $A - Z = 7$ electrons. Total number of particles in the nucleus is therefore 21. Fourteen protons do not give rise to any resulting spin. In case of electrons, one electron is left unpaired, so the total angular momentum (spin) I = 1/2 for ${}^{14}_{7}\text{N}$ nucleus and so it is a fermion. But the experimental results based on the rotational spectra of ${}^{14}\text{N}$–${}^{14}\text{N}$ molecule shows that I = 1 and the assembly of ${}^{14}_{7}\text{N}$ should follow Bose–Einstein statistics.

(b) *Nuclear magnetic moment*: Since, the nucleus is charged and has angular momentum, it must also have magnetic moment μ. Fourteen protons in our case of ${}^{14}_{7}\text{N}$ will pair up with no contribution to μ. The net magnetic moment of ${}^{14}_{7}\text{N}$ will be due to single unpaired electron. Thus,

$$\mu = \mu_B \cdot \text{where } \mu_B = \frac{e\hbar}{2m_e}$$

or

$$\mu = \frac{e\hbar}{2(m_p/1837)} \cong 1837\mu_N$$

Experimentally,

$$\mu({}^{14}_{7}\text{N}) = 0.4\mu_N$$

(c) $\beta^{\pm}$ *decay mode*: Dual β-decay is exhibited in few nuclei, e.g. ${}^{64}_{29}\text{Cu}$. If both electrons and positrons were in nuclei, they should annihilate each other making nuclei unstable but as such, we know it is relatively stable.

(d) *Binding energy*: The electron is a lepton (as discussed in Chapter 8) and cannot take part in strong interaction, which binds the nucleus together. If electrons were in the nucleus, it would be bound by Coulomb interaction and the binding energy will be of the order of

$$E \approx \frac{Ze^2}{R}$$

where R is the Coulomb radius of the nucleus, $R = 1.2A^{1/3}$ fm. Thus,

$$E = -Z\left(\frac{e^2}{\hbar c}\right)\left(\frac{\hbar c}{R}\right) = -\frac{(197\text{ MeV fm})Z}{137 \times 1.2A^{1/3}\text{ fm}}$$

or

$$E = -1.20\frac{Z}{A^{1/3}}\text{MeV}$$

Suppose, A = 140, Z = 58

$$E = -1.20\frac{58}{140^{1/3}} = -13.4 \text{ MeV}$$

The de Broglie wavelength of the electrons having energy 13.4 MeV would be

$$\lambda = \frac{h}{p}$$

$$= \frac{2\pi\hbar c}{pc} \approx \frac{2\pi\hbar c}{E}$$

$$\therefore \qquad \lambda = \frac{2 \times 3.14 \times 197.329 \text{ MeV fm}}{13.4 \text{ MeV}} \approx 92 \text{ fm}$$

This value is greater than the size of the nucleus and therefore, electrons cannot be bound in the nucleus.

1.3.2 Proton–Neutron Model

The experimental discovery of the neutron by Chadwick in 1932 led Heisenberg to suggest that the nuclei might be composed of protons and neutrons. This hypothesis supports the following facts:

(a) Spin and statistics: Both protons and neutrons are 1/2 spin particles. For the case of ${}^{14}_{7}N$, now the resultant spin would be 0 or $1\hbar$. Even number of fermions, obey Bose–Einstein statistics. This observation is in agreement with the experimental findings.

(b) *Magnetic moment*: If the nucleus consists of protons and neutrons, the nuclear magnetic moment is the vector sum of the contributions of the two kinds of nucleons. The resultant value of the magnetic moment $\approx \mu_N$. This is in agreement with the experimental values.

(c) *Quantum mechanical consideration*: In order for the nucleon to be inside the nucleus, it must be confined in a nucleus of size ~10^{-15} m. According to Heisenberg's uncertainty principle:

$$\Delta x \, \Delta p \sim \hbar$$

$$\Delta p \sim \frac{\hbar}{\Delta x}$$

Therefore, the kinetic energy of the nucleon will be given by

$$E = \frac{p^2}{2m_n} \sim \frac{\hbar^2}{2m_n(\Delta x)^2}$$

or $$E \sim \frac{\hbar^2 c^2}{2m_n c^2 (\Delta x)^2} = \frac{(197.329\,\text{MeV fm})^2}{2 \times 938 \times (1\text{ fm})^2}$$

or $$E \sim 20\text{ MeV}$$

Since, the nucleon is bound inside the nucleus, the average of the potential energy V must be negative and greater in magnitude than the kinetic energy. Therefore,

$$-(V) \geq 20\text{ MeV}$$

which indeed gives the correct value of the potential energy.

Another argument against the nuclear electron emerges from a simple quantum mechanical estimate of the nuclear energies. If an electron were confined in a region of nuclear size ($2R \sim 10$ fm), the energy difference between adjacent energy eigen states would be of the order of

$$\Delta E \cong \frac{\pi^2 \hbar^2}{2m_e L^2} = \frac{\pi^2 \hbar^2 c^2}{2m_e c^2 L^2}$$

$$= \frac{(3.14)^2 \times (197.329\text{ MeV fm})^2}{2(0.511\text{ MeV}) \times (10\text{ fm})^2} \approx 2 \times 10^9\text{ eV}$$

This is larger than the observed energy difference between nuclear states, typically 1 to 10 MeV.

For a proton, the estimate on the basis of the above equation gives

$$\Delta E = \frac{\pi^2 \hbar^2 c^2}{2m_e c^2 L^2} \approx 4\text{ MeV}$$

which is the right order of magnitude.

1.4 SIZE AND SHAPES OF THE NUCLEUS

In order to study the nuclear size, shape and density distribution, one employs electrons, protons and neutrons as probes. The basic criterion for selecting probes is that the de Broglie wavelength of the probe should be less than or equal to the size of the object being investigated. Thus,

$$\lambda = \frac{h}{p} \leq 2R$$

$$p \geq \frac{h}{2R}$$

where R is the radius of the nucleus. For an effective study of the nuclear density distribution, we require $\lambda \leq 2R$.

Electrons are the suitable probe to study the distribution of charge in a nucleus, because electrons interact with a nucleus only through electromagnetic interaction. For this purpose, high energy electron beam is required.

Interactions between nuclei and protons can be used to study the nuclear structure, shape and distribution of nuclear matter. Proton beams of high flux and suitable parameters are available using accelerators. However, proton–nucleus scattering results are difficult to analyze because both electromagnetic and strong interactions are present in the data.

Neutrons as probe are better than protons because no Coulomb interaction takes place in the scattering. But high flux and high energy neutron beams are difficult to obtain. Also, detection and measurements are more difficult for neutrons as compared to protons.

Thus, electron scattering measures the nuclear charge radius, while nucleon scattering measures the nuclear potential radius. Practically, the charge radius can be taken as equal to the potential radius, because the centres of charge distribution and nucleon (mass) distribution are assumed to coincide.

Since Rutherford's time, many comprehensive scattering experiments have been performed. Unlike the atom, the nucleus does not have a sharp boundary. Experiments have shown that the nucleus is shaped like a sphere with a radius that depends on the atomic mass number of the nuclide. The relationship between the mass number (*A*) and the nuclear radius (*R*) is given by the following equation:

$$R = r_0 A^{1/3} \tag{1.9}$$

where $r_0 \approx 1.2$ fm.

Nuclear radii for some light, intermediate and heavy nuclides are given in Table 1.2.

Table 1.2 Nuclear radii for some light, intermediate and heavy nuclides

Nuclide	*R* (fm)
$^{1}_{1}H$	1.25
$^{56}_{26}Fe$	4.78
$^{178}_{72}Hf$	7.01
$^{238}_{92}U$	7.74

EXERCISE 1.3: Show that nuclear density is a constant, same for all nuclei. Compare the nuclear density with the density of ordinary matter.

Solution: Equation (1.9) implies that the volume of the nucleus[1] is proportional to *A*.

$$V = \frac{4\pi}{3} R^3 = \frac{4\pi}{3} r_0^{\,3} A$$

Since, the mass of a nucleus is roughly proportional to *A*, the density $\rho = M/V$ of nucleons (nuclear density) is a constant and same for all nuclei.

The nuclear density ρ is of the order of 3×10^{17} kg/m³, which is 14 orders of magnitude greater than the density of ordinary matter like solids or liquids. For example, the density of water at NTP is 1000 kg/m³.

1. Not all nuclei are spherical, some being oblate and others prolate around the axis of rotation.

In general, the charge radius of nuclei is measured by the following four methods:

(i) Elastic scattering of fast electrons
(ii) Isotope shift
(iii) Mirror nuclei
(iv) Muonic X-rays

We will discuss the first method in detail because of high precision of fast electron scattering. In 1950s, Robert Hofstadter (Nobel Prize, Physics, 1961) and his colleagues at Stanford University performed electron scattering measurements of the nuclear charge distribution using electrons of energies from 100–500 MeV.

EXERCISE 1.4: What should be the energy of electrons and protons/neutrons to probe the size of $^{40}_{20}\text{Ca}$?

Solution:

$$R = r_0 A^{1/3} = 1.2 \times 40^{1/3} \approx 4 \text{ fm}$$

In order to probe $^{40}_{20}\text{Ca}$ nucleus, we need a de Broglie wavelength, $\lambda = h/p$, of 8 fm, which is equal to $2R$. From energy conservation, we have

$$E^2 = p^2c^2 + m_e^2c^4$$

$$= \left(\frac{h}{\lambda}\right)^2 c^2 + (m_ec^2)^2$$

$$= \left(\frac{1237 \text{ MeV fm}}{8 \text{ fm}}\right)^2 + (0.511 \text{ MeV})^2 = 23915 \text{ MeV}^2 + 0.26 \text{ MeV}^2$$

or

$$E \cong 155 \text{ MeV}$$

Therefore, the relativistic kinetic energy of electron needed for probe is

$$K = E - m_ec^2 \approx 154 \text{ MeV}$$

Proton/neutron energy is

$$E = \sqrt{(23915 \text{ MeV}^2) + (938 \text{ MeV})^2} \cong 950 \text{ MeV}$$

Therefore, the kinetic energy of proton/neutron needed for probe is

$$K = 950 \text{ MeV} - 938 \text{ MeV} = 12 \text{ MeV}$$

The experiments consist of shooting collimated high energy electrons at a thin target of the material under study and observing the probability of various angular deflections by a movable spectrometer. Typical results of this measurement are shown in Figure 1.9.

Assuming that the nucleus behaves as a circular disk and electrons undergo diffraction from the disk while scattering, one expects the first minimum at an angle of

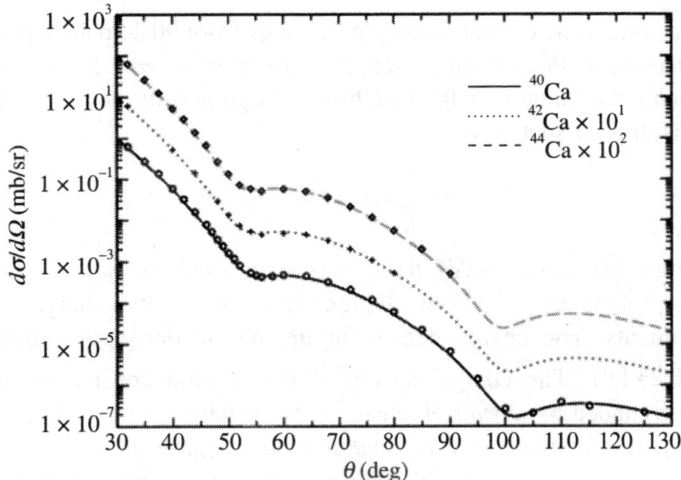

FIGURE 1.9 Scattering angle vs. differential cross-section for calcium isotopes Ca-40, Ca-42 and Ca-44 targets being hit by high energy electron beam at an energy of 250 MeV. [Data from R. F. Frosch et al., "Electron Scattering Studies of Calcium and Titanium Isotopes," Phys. Rev., 174., 1380 (1968)]

$$\sin\theta = 1.22\frac{\lambda}{R}$$

where R is the charge radius and λ is the de Broglie wavelength of the electron. From the above figure, the first minimum occurs at an angle of about 45°. However, the minima as seen in the figure do not occur with zero intensity, quite unlike what we observe in optical diffraction pattern. This is because the nuclear surface is not sharp and there is diffuseness in the density distribution near the edge. This is incorporated in the charge density distribution described by the Woods–Saxon form given by

$$\rho_{ch}(r) = \frac{\rho_0}{1 + \exp\left[\dfrac{r - R}{a}\right]}$$

where a is the diffusiveness parameter and ρ_0 is the constant density in the core. This distribution is shown in Figure 1.10.

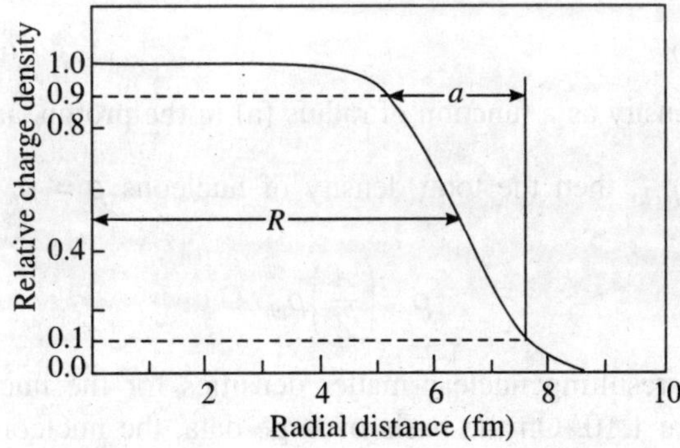

FIGURE 1.10 Nuclear charge density profile as a function of radial distance.

The distance at which the central density $\rho(0)$ varies from 90% to 10% is called the surface thickness or diffusiveness. For the entire range of nuclides, $a \approx 0.5$ fm. The significance of R is that it represents the value of r for which the charge density has fallen to one half of its central value. ρ_0 is chosen such that

$$4\pi \int_0^\infty \rho_{ch}(r) r^2\, dr = Z$$

Incidentally, fast electrons (GeV) have also been used to explore the charge density within individual protons and neutrons. Figure 1.11, shows the charge densities deduced from these experiments. The charge density of the proton decreases roughly exponentially ($\sim e^{-r/a}$ with $a \approx 0.23$ fm). The charge density of the neutron consists of an inner layer of positive charge, surrounded by a layer of negative charge which is again surrounded by a feeble layer of positive charge; however, the net charge of the neutron is zero.

To measure the nuclear potential radius, one could use neutrons as the probe. Neutrons suffer scattering via the nuclear force, which is not known as precisely as we know the Coulomb force.

Since, the strong nuclear forces which bind the nucleons together are of short range and are charge independent, we could assume to a reasonable approximation, the ratio of neutron density ρ_n to proton density ρ_p is the same at all points in the nucleus, i.e.,

$$\frac{\rho_n(r)}{\rho_p(r)} = \frac{N}{Z}$$

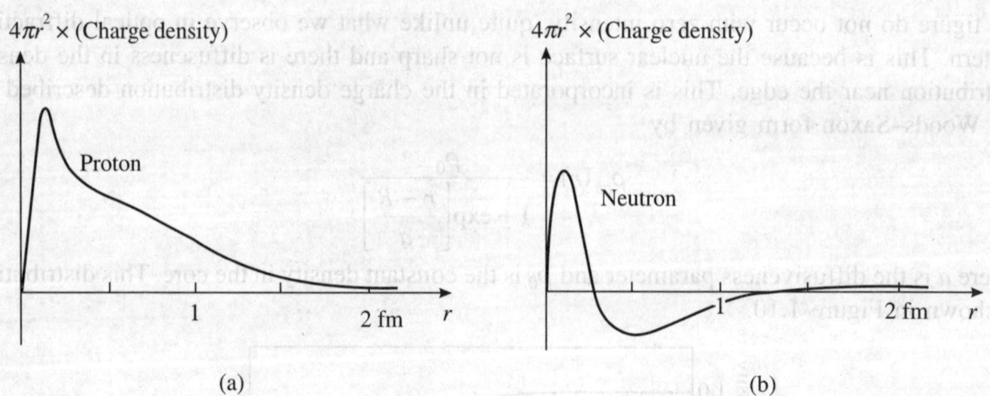

FIGURE 1.11 Charge density as a function of radius (a) in the proton, and (b) in the neutron.

We can take ρ_p as ρ_{ch}, then the total density of nucleons $\rho = \rho_n + \rho_p$ can be expressed as follows:

$$\rho = \left(\frac{A}{Z}\right)\rho_{ch} \tag{1.10}$$

where $A = Z + N$. The resulting nuclear matter densities for the nuclei also have the same shape as shown in Figure 1.10. On the basis of large data, the nucleon density ρ_0 is found to be ~0.17 nucleons/fm^3.

1.5 NOMENCLATURE

A species of atom characterized by the constituents of its nucleus is called a nuclide. A nuclide (nucleus) is a bound assembly of neutrons and protons. ${}^{A}_{Z}X_N$ denotes a nucleus of an atom of the element X containing A nucleons, of which Z are protons and N are neutrons. Nuclides or atoms having the same atomic number Z are called *isotopes*, e.g., ${}^{50}_{24}Cr$, ${}^{52}_{24}Cr$, ${}^{53}_{24}Cr$, ${}^{54}_{24}Cr$. They occupy the same position in the periodic table and are chemically similar. Nuclides having the same number of neutrons are called *isotones*, e.g., ${}^{13}_{6}C_7$ and ${}^{14}_{7}N_7$. Nuclides having the same mass number $A = Z + N$ are called *isobars*, e.g., ${}^{14}_{6}C_8$ and ${}^{14}_{7}N_7$.

Masses of the neutral atoms are experimentally measured by a mass spectrometer, and expressed in mass units (u), which are so defined that the mass of a ${}^{12}_{6}C_6$ nuclide, the most abundant isotope of carbon, is exactly 12u. The value of a mass unit is

$$1u = 1.66054 \times 10^{-27}\ \text{kg}$$

The energy equivalent of a mass unit is 931.49 MeV. The nuclear mass in terms of atomic mass M_A is given by[2]

$$m_{\text{nuclear}} = M_A - \{Zm_e - B_e(Z)\}$$

where m_e is thc mass of one electron and $B_e(Z)$ is the total binding energy. According to the Thomas–Fermi model of the atom, $B_e(Z) \cong 15.73Z^{7/3}$ eV. Atomic binding energy of an electron is negligible when compared with B/A (binding energy per nucleon) which is about 7–8 MeV.

Nuclear masses are generally not tabulated. Precise atomic masses are available online at <http://www.nndc.bnl.gov/wallet/>. Tabular mass excesses in micro u and energy units are also available. The *mass excess* is defined as the difference between the exact atomic mass M_A and the mass number A,

$$\text{Mass excess} = M_A - A \tag{1.11}$$

1.5.1 Binding Energy

Careful measurements have shown that, the mass of a nucleus is always smaller than the combined mass of its constituent nucleons. The difference between the two is called *mass defect* Δm or binding energy $B(= \Delta mc^2)$ of the nucleus[3].

If m, m_n and m_p are the masses of a nucleus ${}^{A}_{Z}X_N$, a neutron and a proton, respectively, then the mass defect (***the mass defect must not be confused with the mass excess***) is given by

$$\Delta m = Nm_n + Zm_p - m \tag{1.12}$$

And the binding energy of the nucleus is

2. By convention, masses of particles, nucleons and nuclei are represented by m, while atomic masses are represented by M.
3. In calculating the mass defect, it is important to use the full accuracy of mass measurements, because the difference in mass is small compared to the mass of the nucleus. *Note: Rounding off the masses of atoms and particles to three or four significant digits prior to the calculation will result in a calculated mass defect of zero.*

$$B = (Nm_n + Zm_p - m)c^2 \tag{1.13}$$

The binding energy of the state of infinite separation of nucleons at rest is taken to be zero.

It is practical to work with the mass of an atom rather than with that of the nucleus. Adding and subtracting Zm_ec^2 in the above equation, we get

$$B = [Nm_n + (Zm_p + Zm_e) - (m + Zm_e)]c^2$$

or

$$B = [Nm_n + ZM_H - M_A]c^2 \tag{1.14}$$

Thus, the mass changes are really changes of binding energy, i.e., there is no actual destruction of nucleons.

This energy (B) also represents the minimum energy required to separate a nucleus into protons and neutrons. It can also be understood as the amount of energy that would be released if the nucleus was formed from separate particles. Since, the total binding energy of the nucleus depends on the number of its constituent nucleons, a more useful measure of the cohesiveness is the binding energy per nucleon $\overline{B}(= B/A)$. The binding energy per nucleon varies with the mass number A as shown in Figure 1.12. It illustrates that as A increases, $\overline{B}$ decreases for $A > 60$. The general shape of the $\overline{B}$ vs. A curve can be explained using the general properties of nuclear forces. The nucleus is held together by extremely short range attractive forces that exist between nucleons, because if each nucleon is interacting with all the remaining nucleons of a nucleus, then the total number of interacting bonds will be:

$$\frac{A(A-1)}{2} \Rightarrow B \propto A^2$$

But $\overline{B}$ is almost constant in the range of $30 \le A \le 170$, which confirms the short range and saturation property of the nuclear forces. There are three characteristic regions in the $\overline{B}$ *vs.* A curve:

(i) *Region of greater stability*: This region exhibits a peak near $A = 60$. ^{56}Fe represents the most stable isotope.

(ii) *Region of fission reactions*: From the curve, it can be seen that the heaviest nuclei are less stable than the nuclei close to $A \equiv 60$, which suggests that the energy can be released if heavy nuclei split apart into more stable smaller nuclei. This process is called fission.

For example, in the following thermal neutron induced fission reaction, the energy released may be estimated from binding energy per nucleon curve.

$${}_0^1n + {}_{92}^{235}\text{U} \rightarrow {}_{92}^{236}\text{U}^* \rightarrow {}_{55}^{140}\text{Cs} + {}_{37}^{93}\text{Rb} + 3\,{}_0^1n$$

The total binding energy for a nucleus is the product of the binding energy per nucleon and the number of nucleons. The energy released will be equivalent to the difference in the binding energy ΔB between the reactants and the products. Neutrons are free, therefore, their binding energy has not been considered.

$$\Delta B = \Sigma(\overline{B}A)_{\text{products}} - \Sigma(\overline{B}A)_{\text{reactants}}$$

$$= (8.4 \times 140 + 8.7 \times 93) - (7.6 \times 236) \text{ MeV}$$
$$= (1176 + 809) - 1786 = 199 \text{ MeV}$$

Thus, each fission event gives rise to 199 MeV energy. Fission is described in detail in Chapter 5.

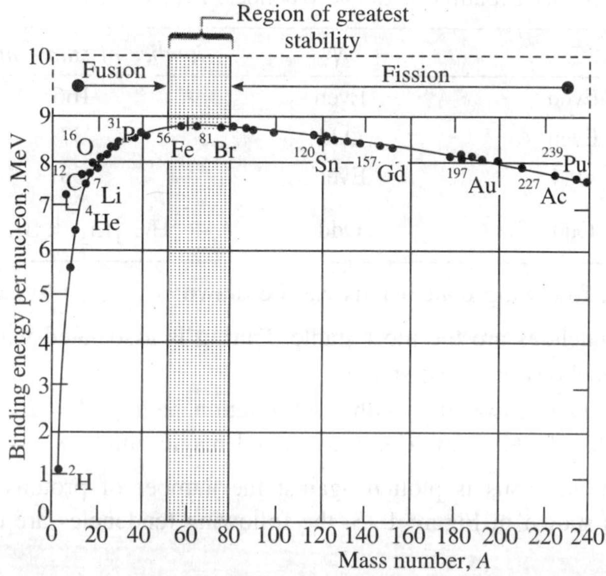

FIGURE 1.12 Binding energy per nucleon as a function of mass number.

(iii) *Region of fusion reactions*: The binding energy curve suggests a second way in which the energy could be released in a nuclear reaction. The lightest elements (like hydrogen and helium) have nuclei that are less stable than heavier one up to ^{56}Fe. If two light nuclei fuse together to form a heavier nucleus, a significant energy could be released. This promising process, which is called fusion (is exothermic only for nuclides of mass number below 60) could be used in future nuclear power reactors. Energy released in fusion may also be estimated from the binding energy curve. For example, for the following fusion reaction:

$${}^{2}_{1}\text{H} + {}^{2}_{1}\text{H} \rightarrow {}^{4}_{2}\text{He}$$

$$\Delta B = \Sigma(\bar{B}A)_{\text{products}} - \Sigma(\bar{B}A)_{\text{reactants}}$$

$$= (7 \times 4) - (1.1 \times 2 + 1.1 \times 2) = 23.6 \text{ MeV}$$

1.6 NUCLEAR STABILITY

The properties of a nucleus depend on evenness or oddness of Z and N, and consequently on $A(= Z + N)$. Since, free nucleons have (1/2) spin, they obey the Pauli exclusion principle, which allows two protons and two neutrons each with opposite spin to occupy a quantum state of a nucleus. There is a preference for having pairs of protons and neutrons, and is known as pairing of nucleons. The frequency distribution of stable nuclides is given in Table 1.3. The total number of nuclides is subject to change, as some of these might be determined to be very long-lived radioactive nuclides in the future.

Table 1.3 Distribution of Stable Nuclides: Effect of Pairing Nucleons

Z	*N*	*Number of stable nuclides*
Even	Even	166
Even	Odd	57
Odd	Even	53
Odd	Odd	4($^{2}_{1}H_1$, $^{6}_{3}Li_3$, $^{10}_{5}B_5$, $^{14}_{7}N_7$)

From these data, following conclusions can be drawn:

(i) Even–odd nuclides are the most stable. Pairing is associated with nuclear stability.
(ii) Odd–odd nuclides are least stable.
(iii) Extra stability is associated with configuration having Z = 2, 8, 20, 50, 82 and/or N = 2, 8, 20, 50, 82, 126. These are called magic numbers.

If the number of neutrons is plotted against the number of protons for all the nuclides existing in nature as shown in Figure 1.13, the following tendencies are observed.

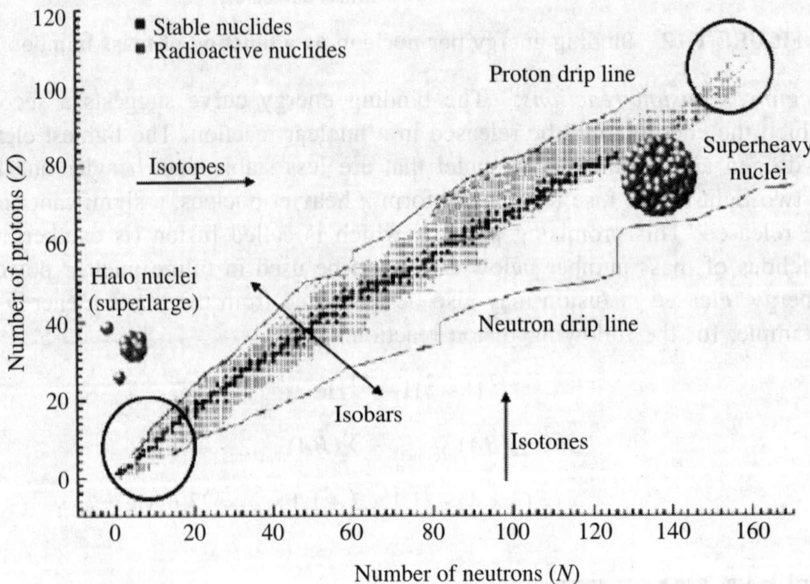

FIGURE 1.13 Stability curve showing proton drip line, the region where the decay of the unstable nuclide towards stability is by proton emission, and neutron drip line where the decay of the unstable nuclide is via neutron emission.

(i) For light nuclei ($A \leq 40$), Z and N are nearly equal. Up to ^{40}Ca, $Z = N$.

(ii) For heavier nuclei, more neutrons are needed to form a stable configuration and the ratio of N/Z approaches 1.5 for the heaviest nuclei. The nucleus is held together by attractive forces between the nucleons. These attractive nuclear forces are not completely understood, but it is known that they must be strong enough to overcome the electrostatic repulsion between the protons. As the Coulomb repulsion increases with the increasing number of protons, more number of neutrons is needed to provide the attractive force to maintain stability.

(iii) If a nucleus has too many or few neutrons, it is unstable and may spontaneously rearrange its constituents to make a stable formation via β-decay. In β^- decay, a neutron gets transformed into a proton and an electron and antineutrino are emitted. Symbolically, the process may be written as follows:

$$n \rightarrow p^+ + e^- + \overline{\nu}$$

In β^+ decay, a proton becomes a neutron and a positron along with neutrino is emitted. Symbolically, the process is represented as follows:

$$p^+ \rightarrow n + e^+ + \nu$$

A process that competes with β^+ decay is known as *electron capture*, in which the orbital electron is captured by a nuclear proton. The process is represented as follows:

$$p^+ + e^- \rightarrow n + \nu$$

(iv) All nuclei with $Z > 83$ and $A < 209$ spontaneously transform themselves into lighter ones through the emission of α-particles, which are doubly ionized helium ion ($^4_2He^{++}$). Beyond $A > 209$, some nuclides undergo spontaneous fission (SF).

(v) The neutron/proton ratio is not the only factor affecting nuclear stability. Adding neutrons to isotopes can also vary their nuclear spins and nuclear shapes.

(vi) By bombarding stable isotopes with neutrons or protons, N/Z ratio of the resulting nuclides is changed and often they are radioactive. This is the way, radioisotopes for various applications are produced.

(vii) Halo nuclei form at the extreme edge as shown in Figure 1.13. These nuclei have half-lives in milliseconds, and therefore, are studied shortly after their formation in radioactive ion beams based experimental facility.

An atomic nucleus is typically a tightly bound group of protons and neutrons. However, in some isotopes, there is an overabundance of the neutrons or the protons, in which case a nuclear core surrounded by a halo of orbiting protons or neutrons will then form. The radius of halo nucleus is appreciably larger than the value based on $R = r_0A^{1/3}$. Experimental confirmation of halo nuclei is quite recent and ongoing. Table 1.4 lists some of the examples of halo nuclides.

Table 1.4 List of some known Isotopes with Nuclear Halo

Z	*Name*	*Halo isotopes*	*Halo composition*	*Half-life* (ms)
2	Helium	$^{6}_{2}He_4$	2 neutrons	801
		$^{8}_{2}He_6$	4 neutrons	119.1
3	Lithium	$^{11}_{3}Li_8$	2 neutrons	8.75
4	Beryllium	$^{11}_{4}Be_7$	1 neutron	13818
		$^{14}_{4}Be_{10}$	4 neutrons	4.35
5	Boron	$^{8}_{5}B_3$	1 proton	770
		$^{17}_{5}B_{12}$	2 neutrons	5.08
		$^{19}_{5}B_{14}$	2 neutrons	2.92
6	Carbon	$^{19}_{6}C_{13}$	1 neutron	49
		$^{22}_{6}C_{16}$	2 neutrons	6.1
10	Neon	$^{17}_{10}Ne_7$	2 protons	109.2
15	Phosphorus	$^{26}_{15}P_{11}$	1 proton	43.7
16	Sulphur	$^{27}_{16}S_{11}$	2 protons	15.5

EXERCISE 1.5: The halo nucleus $^{11}_{3}Li$ has a halo composition of 2 neutrons. What is meant by this statement?

Solution: The above statement means that the halo nucleus $^{11}_{3}Li$ contains a core of 3 protons and 6 neutrons and a halo of two independent and loosely-bound neutrons.

1.6.1 Search for Superheavy Nuclides

The search for the superheavies started in the mid of sixties, when it was predicted by Glenn T. Seaborg that an 'Island of Stability' of superheavy elements existed. Normally, as the atomic number is increased for heavy nuclides, shorter half-lives are expected. This is because of the large electrostatic repulsion between protons which overcomes the attractive nuclear force and makes SF (spontaneous fission) of heavy nuclides more and more likely as their atomic numbers increase above 100. The probability of α-decay also increases as the Coulomb forces become stronger. The shell model of the nucleus, in which the protons and the neutrons are arranged in shells somewhat like atomic electrons, provides a way to circumvent this tendency towards shorter half-lives. When the neutron and the proton shells are filled, the extra stability due to the pairing leads to longer lifetimes. The magic numbers for which this closed shell condition would hold are thought to be 114, 120 and 126 protons and 184 neutrons. The longer predicted lifetimes for the superheavy elements (SHE) close to the 'Island of Stability' gave a big impetus to the research activities globally. The longest-lived isotopes that were observed of each of the heaviest elements are given in Table 1.5[4]

The existence of the superheavy elements has important implications for nuclear theory. If they are found, they would provide data to test theoretical models of the nucleus, such as the liquid drop and shell model used to predict lifetimes in this region. Their existence in nature would also indicate whether or not SHE were created by the process of nucleogenesis in stars. It is possible that these elements possess unusual chemical properties and the very small critical masses of SHE would become the source of energy.

Table 1.5 Longest-lived Isotopes for each of the Heaviest Elements

Z	*Name*	*Longest-lived isotope*	*Half-life*
100	Fermium	$^{257}_{100}Fm$	100.5 d
101	Mendelevium	$^{258}_{101}Md$	51.5 d
102	Nobelium	$^{259}_{102}No$	58 m
103	Lawrencium	$^{266}_{103}Lr$	~11 h
104	Rutherfordium	$^{267}_{104}Rf$	~1.3 h
105	Dubnium	$^{268}_{105}Db$	1.3 d
106	Seaborgium	$^{269}_{106}Sg$	~2.1 m
107	Bohrium	$^{270}_{107}Bh$	61 s
108	Hassium	$^{277m}_{108}Hs$	~12 m
109	Meitnerium	$^{278}_{109}Mt$	7.6 s
110	Darmstadtium	$^{281m}_{110}Ds$	~3.7 m
111	Roentgenium	$^{281}_{111}Rg$	26 s

(*Contd...*)

4. J. Emsley, *Nature's Building Blocks*, First Ed., Oxford University Press, pp. 143–144, 458 (2001); J. Khuyagbaatar, Phys. Rev. Lett. 112: 172501 (2014).

Table 1.5 Longest-lived Isotopes for each of the Heaviest Elements (*Contd.*)

Z	*Name*	*Longest-lived isotope*	*Half-life*
112	Copernicium	$^{285m}_{112}Cn$	~8.9 m
113	Ununtrium	$^{286}_{113}Uut$	19.6 s
114	Flerovium	$^{289m}_{114}Fl$	~1.1 m
115	Ununpentium	$^{289}_{115}Uup$	220 ms
116	Livermorium	$^{293}_{116}Lv$	61 ms
117	Ununseptium	$^{294}_{117}Uus$	78 ms
118	Ununoctium	$^{294}_{118}Uuo$	890 ms

1.6.2 Energy Levels of the Nucleus

The nucleons in the nucleus of an atom, like the electrons that surround the nucleus, exist in shells that correspond to the energy states. The energy shells of the nucleus are less defined and less understood in comparison to the electronic shells. There is a state of lowest energy (the ground state) and discrete possible excited states for a nucleus. The energy levels of the nucleus are typically measured in MeV.

A nucleus in an excited state will not remain at that energy level for an indefinite period. Like the electrons in an excited atom, the nucleons in an excited nucleus will transit towards their lowest energy configuration and in doing so emit discrete quanta of electromagnetic radiation, called gamma rays (γ-rays). The ground state and the excited state of a nucleus are depicted in a nuclear energy level diagram. Few examples are given in Figure 3.1 (Chapter 3) of this book.

1.7 MODERN VIEW OF NUCLEAR MAKEUP

Fundamental particles, called quarks, are held in association with the strong interaction in certain stable combinations of hadrons, called baryons, which manifest themselves as the neutrons and protons of the nucleus. The strong interaction has a short range but extends far enough from each baryon so as to bind the neutrons and the protons together against the Coulombian repulsive force among the protons. Protons and neutrons are fermions, with different strong isospin quantum numbers. Therefore, two protons and two neutrons can share the same space wave function, since they do not have identical quantum identities. The neutron has a positively charged core radius of ~0.3 fm, which is surrounded by a compensating negative charge between radius 0.3 fm and 2 fm. The proton has almost an exponentially decaying positive charge distribution with a mean square radius of ~0.8 fm (see Figure 1.11). Details on quarks and their combinations are given in Chapter 8 on 'Particle Physics.'

The nucleus of an atom is an extremely weak magnet that can point along two natural directions, either 'up' or 'down'. In the strange quantum world, the magnet can exist in both the states simultaneously—a feature known as the quantum superposition. The natural positions of the magnet are equivalent to the 'zero' and 'one' of the binary code, as used in the classical computers. Experiments are underway where researchers controlled the direction of the nucleus, in effect 'writing' a value on its spin, and then 'reading' the value out. Thus, the nucleus turns into a functioning quantum bit (qubit), which will be the building blocks of ultra powerful quantum computers of the future.[5]

1.8 OTHER PROPERTIES OF NUCLEI

1.8.1 Nuclear Angular Momentum

Like the electronic motion in the atom, the nuclear model used to explain the various nuclear properties has nucleons in the nucleus constantly orbiting around it and spinning on their axes. Since, the nucleus is a quantized system, the angular momentum that it has is also quantized. The angular momentum of individual nucleons due to these two motions, the associated quantum numbers and the laws governing them, their interaction leading to the resultant angular momentum of the nucleus as a whole and the electromagnetic effects arising from it, all of these will be discussed in the subsequent sections.

Orbital angular momentum of nucleons

The orbital motion of the nucleons, like those of the electrons, is quantized, i.e., the angular momentum of a nucleon due to its orbital motion can only have certain discrete values given by $\sqrt{l(l+1)}$ in units of $\hbar$. The integral l is referred to as the orbital angular momentum quantum number and can take values from 0, 1, 2, 3... The various nucleon states are designated as s, p, d, f,..., respectively, corresponding to the values of l. In a frame of spatial coordinate system, this angular momentum vector has components along x, y and z-directions. The component along the z-direction (direction of the applied magnetic field) l_z is quantized with eigen values given by the magnetic quantum number m_l, which can take $2l + 1$ values. Thus, in units of $\hbar$

$$m_l = l, l-1, l-2, \ldots, 0, -1, \ldots, -l$$

Spin angular momentum of nucleons

Nucleons (protons and neutrons) inside the nucleus are also spinning about their axes. This spin motion is also quantized and the magnitude of spin angular momentum is $\sqrt{s(s+1)}$ in units of $\hbar$, where s = 1/2 for nucleons.

5. R.P. Feynman, 'Simulating physics with computers', *International Journal of Theoretical Physics*, 21(6): 467–488 (1982).
D. Deutsch, *Quantum Theory, the Church-Turing Principle and the Universal Quantum Computer*, Proceedings of the Royal Society of London, A 400: 97–117 (1985).

The properties of this momentum along the z-axis has two components $m_s = +1/2$ and $m_s = -1/2$. These represent the direction of spinning motion of the nucleon.

Total angular momentum of nucleons

The total angular momentum of a nucleon is the vector sum of the orbital $(\vec{l})$ and spin $(\vec{s})$ angular momentum, i.e.,

$$\vec{j} = \vec{l} + \vec{s}$$

The magnitude of the total angular momentum is $\sqrt{j(j+1)}\,\hbar$. For particles with $s = 1/2$, only two values of j are permitted. Thus, $j = l + 1/2$ and $j = l - 1/2$. If $l = 0$, only one value of j is allowed, i.e., $j = 1/2$, since j value cannot be negative.

Total angular momentum of the nucleus

When two or more nucleons are able to come together to form a nucleus, the components of motion of each nucleon interact with one another, which leads to a resultant total angular momentum of the nucleus $\vec{J}$, often represented by $\vec{I}$ and loosely referred to as nuclear spin. Its magnitude is given by $\sqrt{I(I+1)}\,\hbar$.

The value of I can be calculated in two different ways depending upon the type of coupling between angular momenta by the nucleons. The two types of coupling are as follows:

(a) *L-S coupling*: Under this type of coupling, the interaction between the orbital motion of a nucleon with its own spin motion, is believed to be weak. On the other hand, the orbital motion of the different nucleons in the nucleus interacts strongly with one another such that the resultant orbital angular momentum $\vec{L} = \sum_i \vec{l}_i$.

In the same way, the individual spin motions of different nucleons interact strongly with one another such that $\vec{S} = \sum_i \vec{s}_i$.

Finally, the total angular momentum of the nucleus $\vec{I}$ is given by $\vec{I} = \vec{L} + \vec{S}$.

(b) *J-J coupling*: In this scheme, orbital $(\vec{l}_i)$ and spin $(\vec{s}_i)$ angular momenta of individual nucleons couple together to give the resultant angular momentum $(\vec{j}_i)$. Thus, $\vec{j}_i = \vec{l}_i + \vec{s}_i$. The different value of J_i couple together to give the nuclear spin I, i.e., $\vec{I} = \sum_i \vec{j}_i$.

The total angular momentum $\vec{I}$ of the nucleus can be measured directly. One should note that

$$I = \begin{cases} 0 \text{ for even–even nuclei} \\ \text{integral value for odd–odd nuclei} \\ (1/2) \text{ integral value for odd–even and even–odd nuclei} \end{cases}$$

1.8.2 Parity

In addition to the angular momentum I, another quantity that characterizes the different states of a nucleus is the parity, π. Parity is a physical property of a wave function which specifies the wave function behaviour under the space inversion, i.e., when x is replaced by $-x$, y by $-y$ and z by $-z$. Space inversion is thought to be the product of a reflection about the xy-plane followed by a rotation about the z-axis (Figure 1.14).

The mirror image of $\vec{r}$ is $\vec{r'}$ where

$$\vec{r} = x\hat{i} + y\hat{j} + z\hat{k}$$

$$\vec{r'} = x\hat{i} + y\hat{j} - z\hat{k}$$

Now, rotate $\vec{r'}$ about the z-axis through 180° to get $-\vec{r}$.

If the wave function ψ satisfies the following equation

$$\psi(x, y, z) = \psi(-x, -y, -z)$$

then it is said to have even parity. If on the other hand,

$$\psi(x, y, z) = -\psi(-x, -y, -z)$$

then the wave function is said to have odd parity.

Each nuclear state has a parity assigned to it. This indicates whether the wave function of the state is even or odd on space inversion. This is the composite of the parity of the states of all the composite nucleons, and if these are known, the parity could be determined. Except for fairly simple nuclei, this cannot be done, and the parity is determined via conservation laws applied to nuclear reactions.

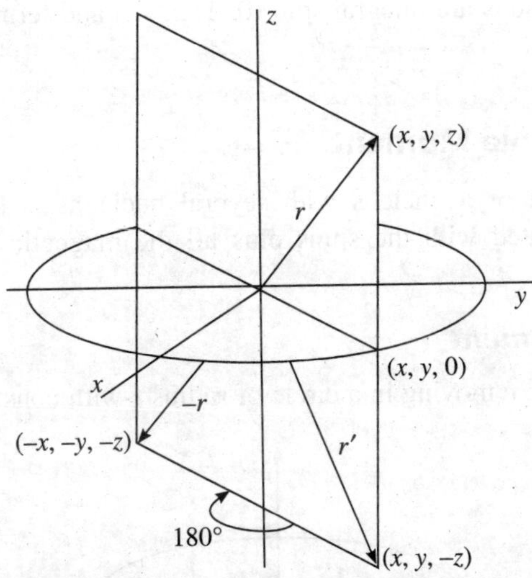

FIGURE 1.14 Mirror reflection about *xy* plane followed by rotation about *z*-axis by 180°.

1.8.3 Statistics

When we group together several particles to make a larger quantal system, like several nucleons inside nuclei, a new quantum effect arises if the particles are indistinguishable from one another.

Consider a system of two indistinguishable nucleons say 1 and 2. Suppose '1' nucleon is described by coordinate r_1 and is in the state $\psi_1(\vec{r_1})$ and '2' nucleon is described by coordinate r_2 and is in the state $\psi_2(\vec{r_2})$. The combined wave function of the two nucleon system is as follows:

$$\psi(1,2) = \psi_1(\vec{r_1})\,\psi_2(\vec{r_2})$$

If the particles are interchanged, then the new wave function is

$$\psi(2,1) = \psi_2(\vec{r_2})\,\psi_1(\vec{r_1})$$

There is no way to distinguish between '1' and '2' nucleon as they are indistinguishable. However, the probability density $|\psi(1, 2)|^2$ should be the same as $|\psi(2, 1)|^2$. The exchanged wave function $\psi(2, 1)$ can at most differ only in sign from the original wave function $\psi(1, 2)$, i.e.,

$$\psi(1,2) = +\psi(2,1)$$

or

$$\psi(1,2) = -\psi(2,1)$$

If the exchange does not change the sign, the wave function is symmetric. If the exchange changes the sign, the wave function is said to be antisymmetric. All particles in the universe (elementary or composite) can be grouped into the above two categories. The particles in the first category are called bosons and those in the second category are called fermions. Without any exception, bosons are integral spin (0, 1, 2, ...) and fermions are half-integral spin (1/2, 3/2, 5/2, ...) particles.

1.8.4 Magnetic Dipole Moment

The net magnetic moment of a nucleus with several nucleons is the vector sum of all the magnetic moments associated with the spins plus all the magnetic moments associated with the orbital motion.

Orbital magnetic moment

Consider a proton of charge e moving in a circle of radius r with constant speed v (Figure 1.15).

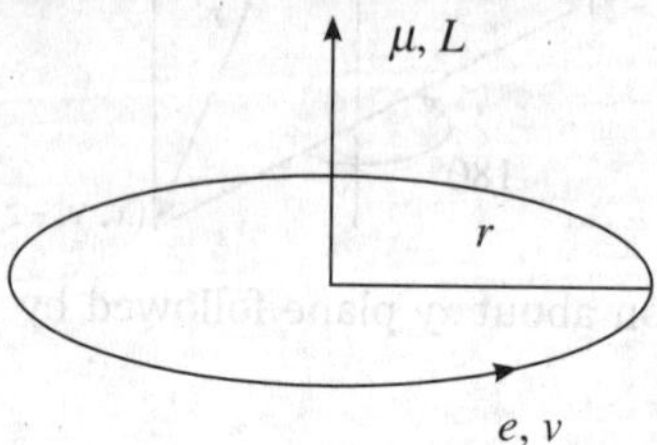

FIGURE 1.15 A proton of charge e moving in a circle of radius r with a veolcity v.

This moving proton is equivalent to a circular current loop with a current,

$$I = \frac{\text{charge}}{\text{time}} = \frac{e}{\left(\dfrac{2\pi r}{v}\right)}$$

$$= \frac{ev}{2\pi r}$$

The magnetic moment of such a current loop is

$$\mu = \text{Area} \times \text{Current} = (\pi r^2)\left(\frac{ev}{2\pi r}\right)$$

$$= \frac{e(m_p vr)}{2m_p}$$

But $m_p vr$ is the orbital angular momentum (L) of the proton. Both μ and L are in the same direction. Thus, in vector notation,

$$\vec{\mu} = \frac{e}{2m_p}\vec{L}$$

More generally,

$$\vec{\mu} = \frac{e}{2m_p} g\vec{L} \tag{1.15}$$

where g is a factor called the gyromatic ratio or Lande g-factor[6]. When charge and mass distribution of proton coincides in the orbital motion, $g = 1$.
In quantum mechanics, $L = l\hbar$,

$$\vec{\mu} = \left(\frac{e\hbar}{2m_p}\right) g\vec{l} \tag{1.16}$$

The quantity $e\hbar/2m_p$ is referred to as the nuclear magneton and it is the unit by which the nuclear magnetic moments are denoted.

$$\mu_N = \frac{e\hbar}{2m_p} = 3.1525 \times 10^{-8} \text{ eV/T}$$

The nuclear magneton μ_N is smaller than the Bohr magneton μ_B and their ratio is equal to the ratio of the mass of electron (m_e) to the mass of proton (m_p), i.e.,

$$\frac{\mu_N}{\mu_B} = \frac{m_e}{m_p} = \frac{1}{1826}$$

Thus, in terms of units of nuclear magnetons, the nuclear magnetic moment may be written as follows:

$$\vec{\mu} = g\vec{l}\,\mu_N \tag{1.17}$$

where $g = 1$ for protons and equal to zero for neutrons, since, they have no charge.

6. Section 6.6 – *Introduction to Nuclear Physics* by H. A. Enge, Addison-Wesley (1966).

If the nucleus contains Z protons, then

$$\vec{\mu}_{\text{orbital}} = \sum_{k=1}^{Z} \vec{l}_k \mu_N \tag{1.18}$$

Spin magnetic moment

The circulating protons are not the only source of magnetic fields in the nucleus. Neutrons have an intrinsic spin and the magnetic moment due to spin is a complex problem which can be understood only in terms of relativistic quantum theory.

From equation (1.17),

$$\vec{\mu}_{\text{spin}} = g_e \vec{s} \mu_N$$

For an electron, s =1/2, quantum electrodynamics gives a value of g_e = 2 and experiment confirms this. For nucleons, however, measurements give g_p = +5.5856912 ± 0.0000022 and g_n = –3.8260837 ± 0.0000018. Thus, μ_{spin} = 2.7928 μ_N for protons and μ_{spin} = –1.913 μ_N for neutrons.

These values of spin magnetic moment provide a reasonable evidence that the nucleons are not the fundamental particles (as is the electron), but are composite particles. In the standard model, they consist of three fractionally charged quarks. Figure 1.16 shows the spinning of a neutron having three quarks (*udd*). These moving quarks are equivalent to a current and produce a magnetic field. In a similar way, three quarks (*uud*) of a proton move during spin and produce the magnetic field. Theoretical treatment of this phenomenon is beyond the scope of this book.

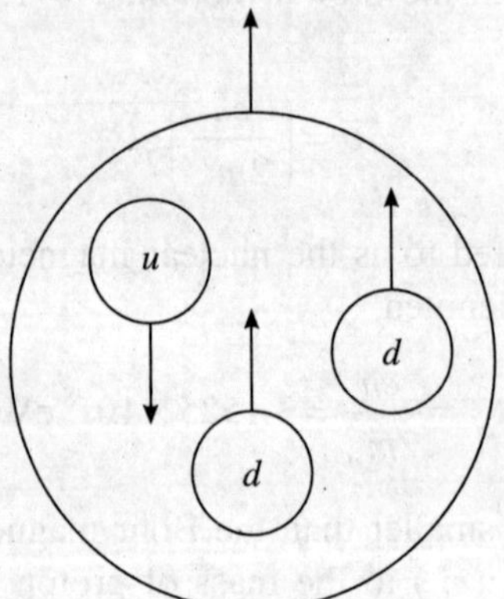

FIGURE 1.16 Spinning of a neutron, made up of three quarks *udd*.

The total spin magnetic moment of the nucleus due to all nucleons is:

$$\vec{\mu}_{\text{spin}} = \left[g_p \sum_{k=1}^{Z} \vec{s}_k + g_n \sum_{k=Z+1}^{A} \vec{s}_k \right] \mu_N \tag{1.19}$$

The net magnetic moment of the nucleus is:

$$\vec{\mu}_{\text{nucleus}} = \vec{\mu}_{\text{orbital}} + \vec{\mu}_{\text{spin}}$$

$$= \left[\sum_{k=1}^{Z} (\vec{l_k} + g_p \vec{s_k}) + g_n \sum_{k=Z+1}^{A} \vec{s_k} \right] \mu_N \tag{1.20}$$

An understanding of this vector sum requires an understanding of the detailed structure of nuclei as discussed in Chapter 2.

Methods of measuring magnetic moments of nuclei are based on atomic beam experiments, hyperfine structure of atomic spectral lines in the Zeeman effect, nuclear resonance in solids and liquids, etc.[7]

1.8.5 Electric Quadrupole Moment

It is important to know the way in which electric charge due to protons is distributed within the nucleus. Suppose the nucleus contains a proton of charge e as shown in Figure 1.17

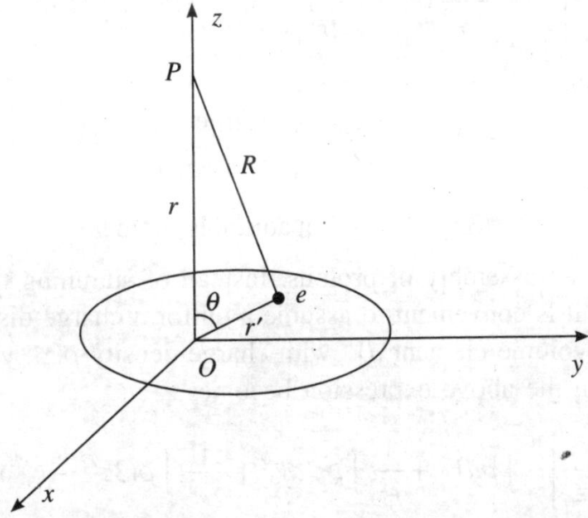

FIGURE 1.17 Non-spherical nucleus whose centre of mass is at origin O and does not coincide with the centre of charge.

The electric potential at point P situated outside the nucleus at a large distance from the origin in the z-direction is given by

$$\varphi(r) = \frac{1}{4\pi \epsilon_0} \frac{e}{R}$$

Using cosine law

$$R = \sqrt{r^2 + r'^2 - 2rr' \cos\theta}$$

Therefore,

$$\varphi(r) = \frac{1}{4\pi \epsilon_0} \frac{e}{\sqrt{r^2 + r'^2 - 2rr' \cos\theta}} \tag{1.21}$$

7. Brief description of these methods is given in the book – *Introduction to Nuclear Physics* by H. A. Enge, Addison-Wesley (1966).

For a distant point P, r' is much smaller than r. In that case,

$$(r^2 + r'^2 - 2rr'\cos\theta)^{-1/2} = \frac{1}{r}\left[1 + \left(\frac{r'^2}{r^2} - \frac{2r'}{r}\cos\theta\right)\right]^{-1/2}$$

Using binomial expansion $(1+\delta)^{-1/2} = 1 - \frac{1}{2}\delta + \frac{3}{8}\delta^2 + \ldots$, we get

$$\varphi(r)\Big|_{at\ P} = \frac{e}{4\pi \in_0 r}\left[1 + \frac{r'}{r}\cos\theta + \left(\frac{r'}{r}\right)^2\left(\frac{3\cos^2\theta - 1}{2}\right) + \text{terms of higher power}\right]$$

Since, $r'\cos\theta = z'$

$$\varphi(r)\Big|_{at\ P} = \frac{1}{4\pi \in_0}\left[\frac{1}{r}(e) + \frac{1}{r^2}(ez') + \frac{1}{2r^3}\{e(3z'^2 - r'^2)\} + \text{terms of higher power}\right] \quad (1.22)$$

In the above expression,

$$e = \text{net charge}$$

$$ez' = \text{dipole moment}$$

and
$$e(3z'^2 - r'^2) = \text{quadrupole moment}$$

A nucleus contains an assembly of protons. Instead of summing the potential at P due to each proton separately, it is convenient to assume a uniform charge distribution throughout the nucleus. Let us have a volume element dV' with charge density $\rho(x', y', z')$ instead of a single proton of charge e, then the above expression becomes

$$\varphi_P = \frac{1}{4\pi \in_0}\left[\frac{1}{r}\int \rho dV' + \frac{1}{r^2}\int \rho z' dV' + \frac{1}{2r^3}\int \rho(3z'^2 - r'^2)dV' + \ldots\right]$$

Because of the increasing power of r in the denominator, this is a rapidly converging series. The integral in the first term is the net charge and the potential is the monopole potential. The integral in the second term is the electric dipole term in the potential, which however due to symmetry is zero.

EXERCISE 1.6: Show that a nucleus must have zero electric dipole moment.

Solution The electric dipole moment D is

$$D = \int \rho z dV$$

Since, the charge distribution in a nucleus arises from the probability distribution of the protons, i.e.,

$$\rho \propto |\psi|^2$$

$|\psi|^2$ is an even function of z, i.e., it has the same value for $+z$ as for $-z$. The integrand in the above equation thus becomes an odd function of z. The integral from 0 to ∞ will therefore, be equal and opposite to the integral from $-\infty$ to 0. Hence, the integral in the above equation vanishes.

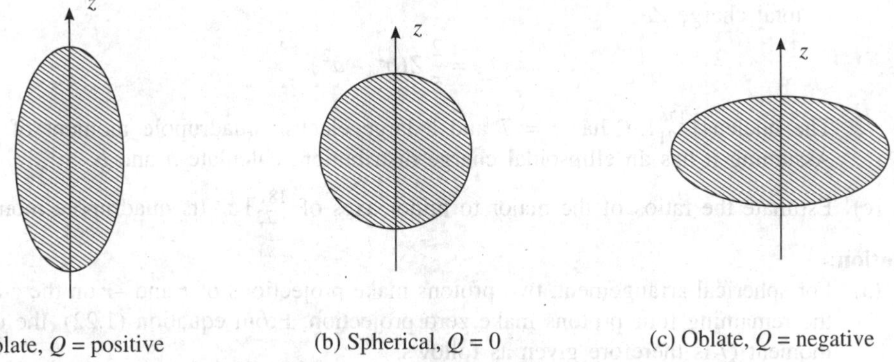

FIGURE 1.18 Quadrupole moment is (a) positive for prolate shaped nucleus, (b) zero for spherical nucleus, and (c) negative for oblate shaped nucleus.

The integral in the 3rd term is called the quadrupole moment Q, i.e.,

$$Q = \int \rho(3z'^2 - r'^2)dV' \tag{1.23}$$

The unit of quadrupole moment is thus Cm^2 though traditionally Q values are listed in units of area, i.e., in barns (= 10^{-28} m^2).

If the charge distribution is stretched along the z-axis (prolate shape), Q is positive. On the other hand, if it is stretched along the xy plane (oblate shape), then Q is negative. For spherical charge distribution, Q is zero. This is shown in Figure 1.18.

EXERCISE 1.7:

(a) Compare the quadrupole moment of spherically symmetric arrangement vis-à-vis prolate spheroidal arrangement of six protons. In spherical arrangement, each proton is at a distance r from the origin while in the prolate arrangement, two protons (1 and 3) are at a distance a from the centre and the rest four protons (2, 4, 5 and 6) are at a smaller distance b from the centre as shown in Figure 1.19.

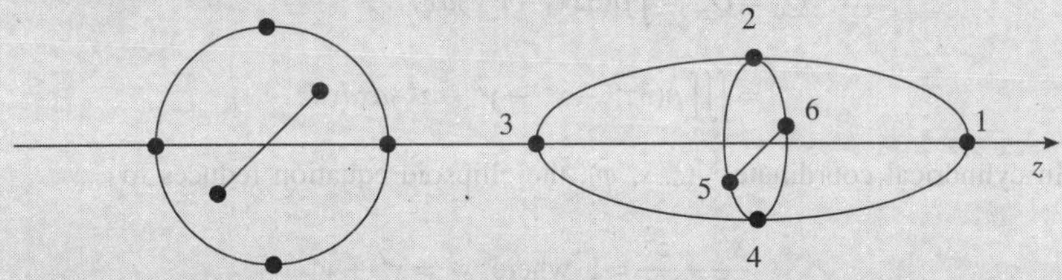

FIGURE 1.19 Spherical symmetric and prolate spheroidal arrangements of six protons.

(b) Show that for a uniform charged ellipsoid of revolution,

$$\frac{x^2 + y^2}{b^2} + \frac{z^2}{a^2} \le 1$$

of total charge Ze,

$$Q_{zz} = \frac{2}{5} Z(b^2 - a^2)$$

The nucleus $^{176}_{71}\text{Lu}$ has $j = 7$ and a large electric quadrupole moment of 4.92 barn. Assuming it has an ellipsoidal charge distribution, calculate a and b.

(c) Estimate the ratios of the major to minor axes of $^{181}_{73}\text{Ta}$. Its quadrupole moment is 6b.

Solution:

(a) For spherical arrangement, two protons make projections of r and $-r$ on the z-axis, while the remaining four protons make zero projection. From equation (1.22), the quadrupole moment Q is therefore given as follows:

$$Q = \sum_{i=1}^{6} e(3z_i^2 - r_i^2)$$

$$= e[2(3r^2 - r^2) + 4(-r^2)] = 0$$

For prolate arrangement, two protons that are a distance away from the centre of mass make z-projection of a and $-a$, while the remaining four protons make zero projection,

$$\therefore \qquad Q = e[2(3a^2 - a^2) + 4(-b^2)] = 4e(a^2 - b^2)$$

(b) The axis of the symmetry for the ellipsoid is z-axis. Assuming uniform charge distribution with density ρ equal to the total charge upon nuclear volume,

$$\rho = \frac{Ze}{\left(\frac{4}{3}\pi ab^2\right)} = \frac{3Ze}{4\pi ab^2}$$

From equation (1.23), the quadrupole moment is given as follows:

$$Q_0 = Q_{zz} = \int \rho(3z'^2 - r'^2)dV'$$

$$= \iiint \rho(3z^2 - x^2 - y^2 - z^2)dxdydz$$

In cylindrical coordinates[8] (z, s, φ), the ellipsoid equation reduces to

$$\frac{s^2}{b^2} + \frac{z^2}{a^2} = 1 \text{ where } s^2 = x^2 + y^2$$

or

$$s^2 = b^2\left(1 - \frac{z^2}{a^2}\right)$$

8. See Appendix E.

Therefore, the quadrupole moment is given as follows:

$$Q_{zz} = \frac{3Ze}{4\pi ab^2}\int_{-a}^{+a}\left(\int_0^{b\sqrt{1-\frac{z^2}{a^2}}}(2z^2 - s^2)sds\right)dz\int_0^{2\pi}d\varphi$$

$$= \frac{3Ze}{2ab^2}\int_{-a}^{+a}\left\{2z^2\frac{b^2}{2}\left(1-\frac{z^2}{a^2}\right)-\frac{b^4}{4}\left(1-\frac{z^2}{a^2}\right)^2\right\}dz$$

$$= \frac{3Ze}{2a}\left\{\frac{2a^3}{3}-\frac{2a^5}{5a^2}-\frac{2b^2a}{4}-\frac{b^2 2a^5}{20a^4}+\frac{b^2 2a^3}{6a^2}\right\}$$

$$\therefore \qquad \frac{Q_{zz}}{e} = \frac{2}{5}Z(a^2 - b^2) \qquad (1.24)$$

Using the given data,

$$(492 \text{ fm}^2) = \frac{2}{5}(71)(a^2 - b^2)$$

or

$$a^2 = b^2 + 17.32$$

Also, we know that the nuclear density ρ_0 ~ 0.17 nucleons/fm³. Therefore,

$$\rho_0 = \frac{A}{\left(\frac{4\pi a^2 b}{3}\right)}$$

or

$$a^2 b = \frac{176}{\left(\frac{4\pi \times 0.17}{3}\right)}$$

or

$$(b^2 + 17.32)b - 247.28 = 0$$

On solving the above equation using the following Matlab code, we get b = 5.36 fm.

```
X = fzero(@(x)(x^2+17.32)*x - 247.28,1)
```

Therefore, $a = \sqrt{5.36^2 + 17.32} = 6.79$ fm.

(c) We know that the average nuclear radius is equal to $1.2A^{1/3}$. Therefore,

$$R = \frac{a+b}{2} = 1.2 \times 181^{1/3} = 6.79 \text{ fm}$$

or

$$b = 13.58 - a$$

In equation (1.24), introducing the average radius term $R = (a + b)/2$, we get

$$\frac{Q_0}{e} = \frac{4}{5} ZR(a - b)$$

or
$$a - b = \frac{(500 \text{ fm}^2)}{\frac{4}{5} \times 73(6.79 \text{ fm})} = 1.26 \text{ fm}$$

or
$$2a - 13.58 = 1.26$$

$$a = 7.42 \text{ fm and } b = 6.16 \text{ fm}$$

$$\therefore \quad \frac{a}{b} = 1.205$$

Relation between nuclear spin and quadrupole moment

The measured value of the quadrupole moment depends on the orientation of the charge distribution relative to a definite direction of spin ($\vec{I}$). When the nuclear spin $I = 0$, there is no direction that can be specified; all directions are equivalent, therefore the quadrupole moment $Q = 0$. The simplest situation where Q is non-zero is for the ground state of a single-particle or single-hole nucleus. Since, closed shells have spherical symmetry, the entire value of Q is due to the orbit of the odd nucleon.

Assuming that a single nucleon is moving in a circular orbit of radius r in the plane $z = 0$, the quadrupole moment of a single particle can be written as:

$$Q_{sp} = e(-r^2)$$

The quantum-mechanical treatment makes the above expression as:

$$Q_{sp} = -\frac{2j - 1}{2(j + 1)} e \frac{3\overline{r^2}}{5}$$

where j is the total angular momentum of the unpaired nucleon, $\overline{r^2}$ is the average value of r^2 for the orbit, which is somewhat less than R^2.

$$\overline{r^2} = \frac{\int_0^\infty r^2 4\pi r^2 \rho(r) dr}{\int_0^\infty 4\pi r^2 \rho(r) dr} \quad \text{and} \quad R^2 = \frac{5}{3}\overline{r^2}$$

The above expression should be valid for a single proton nucleus. For a single neutron (odd neutron) nucleus, Q should be zero, since, the electric charge of the neutron is zero. It turns out, however, that a neutron passing through the nucleus has a small effect on the motions of all the protons in the nucleus and this induces a finite quadrupole moment.

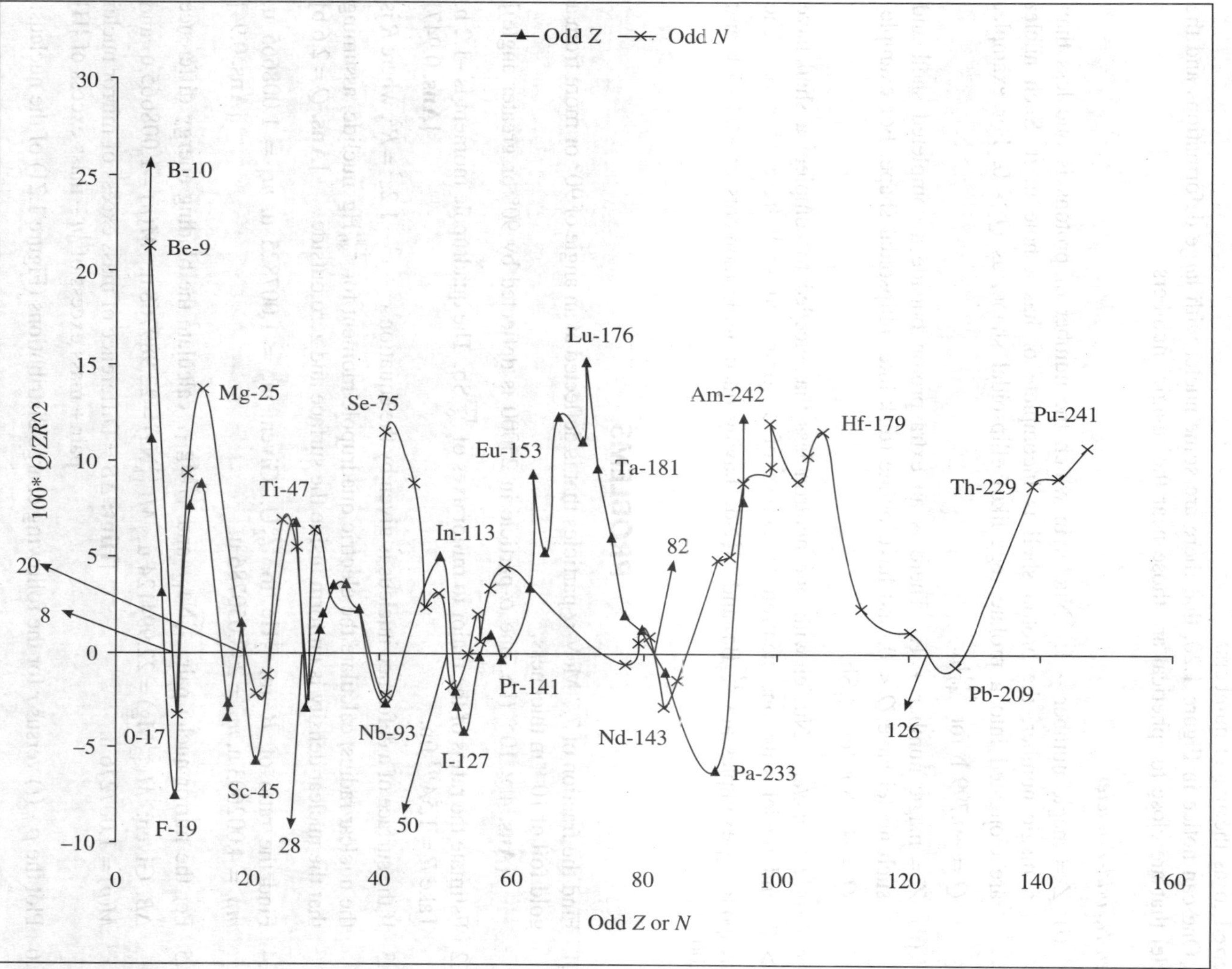

FIGURE 1.20 Deformation of nuclei vs. number of odd nucleons (*Z* or *N*). Near magic numbers, deformation is close to zero and the nuclei are nearly spherical. [Data from N. J. Stone, J. Phys. Chem., Ref. Data 44, 031215 (2015)]

There is additional stability and symmetry in shape if both Z and N happen to be even numbers and one of them (Z or N) happens to be equal to 2, 8, 20, 50, 82, 126 which are also referred to as the magic numbers.

One can notice in Figure 1.20, that there are some nuclei with large deformations and the nuclei that are close to spherical are those near the 'magic' numbers.

Even-N, odd-Z nuclei

(i) Z = magic number – 1: Nuclei in which the number of protons is one less than a magic number, the proton shell is incomplete or has a hole in it. Such nuclei are elongated into a prolate (egg-like) ellipsoidal shape, as $Q > 0$. For example, $Q = +0.799$ b for ${}^{113}_{49}\text{In}$.

(ii) Z = magic number + 1: There is an extra proton outside a completed shell and such nuclei have $Q < 0$, and have oblate (disk-like) ellipsoidal shape. For example, $Q = -0.36$ b for ${}^{121}_{51}\text{Sb}$.

Even-Z, odd-N nuclei: Nuclei with one neutron less than needed to complete a shell have $Q > 0$, and nuclei with one neutron in excess of a completed shell have $Q < 0$. The corresponding examples are ${}^{137}_{81}\text{Ba}$ and ${}^{148}_{83}\text{Nd}$, having quadrupole moments of +0.245 b and –0.063 b, respectively.

PROBLEMS

1.1 Find the fraction of 7.7 MeV α-particles that is deflected at an angle of 90° or more from a gold foil of 10^{-6} m thickness.
[**Ans.** 4×10^{-5} (i.e. one α-particle in 25000 is deflected by 90° or greater angle)]

1.2 Estimate the ratios of the major to minor axes of ${}^{123}_{51}\text{Sb}$. The quadrupole moment is –1.2 b. Take $R = 1.5A^{1/3}$ fm. [**Ans.** 0.947]

1.3 If the surface of a deformed nucleus is given by the equation $x^2 + y^2 + 1.2z^2 = R^2$, where R is the nuclear radius, calculate the electric quadrupole moment for ${}^{200}_{80}\text{Hg}$ nuclide, assuming that the nuclear density is uniform inside the surface and zero outside. [**Ans.** $Q = 2.6$ b]

1.4 Find the ratio of $\overline{B}$ for ${}^{4}_{2}\text{He}$ to ${}^{238}_{92}\text{U}$. Given: m_p = 1.007825 u, m_n = 1.008665 u, m_{He} = 4.002603 u, m_U = 238.050786 u. [**Ans.** 0.94]

1.5 For the mirror nuclei pair, ${}^{23}_{11}\text{Na}_{12}$ and ${}^{23}_{12}\text{Mg}_{11}$, calculate the binding energy difference ΔB. Given: $M({}^{23}_{12}\text{Mg}) = 22.994124$ u, $M({}^{23}_{11}\text{Na}) = 22.989768$ u, $M(n) = 1.008665$ u and $M(p) = 1.007276$ u. [**Hint:** ΔB = Difference of mass excess of mirror nuclei pair + mass excess of ${}^{1}_{0}n$ – mass excess of ${}^{1}_{1}\text{H}$]

1.6 Plot the $\rho_{ch}(r)$ versus r for the following charge distributions (Figure 1.21) of the nucleus:

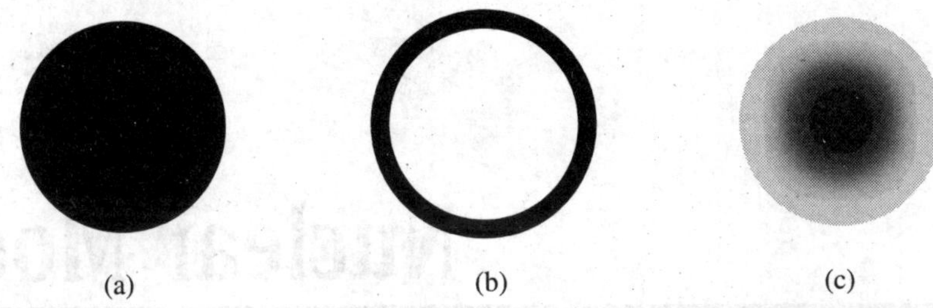

FIGURE 1.21 Models of the nucleus, showing different distributions of the electric charge within it.

BIBLIOGRAPHY

Beiser, A., *Concepts of Modern Physics*, McGraw-Hill, New York, 2003.

Cohen, B.L., *Concepts of Nuclear Physics*, McGraw-Hill, New York, 1971.

Devanathan, V., *Nuclear Physics*, Narosa, New Delhi, 2006.

Enge, H.A., *Introduction to Nuclear Physics*, Addison-Wesley, London, 1966.

Harvey, B.G., *Introduction to Nuclear Physics and Chemistry*, Prentice-Hall, New Jersey, 1970.

Sood, D.D. and Ramamoorthy, A. V. R. *Fundamentals of Radiochemistry*, IANCAS, Mumbai, 2000.

2

Nuclear Models

"Every sentence I utter must be understood not as an affirmation but as a question."

—**Niels Bohr**

Our understanding of the nuclear structure and the nuclear force is still primitive compared to the atomic structure and the electromagnetic interaction which binds the atom. To explain the various experimental observations related to the nuclides, over the years, a number of nuclear models have been proposed with different sets of simplifying assumptions by scientists. Liquid drop model (LDM), shell model or independent particle model, collective model, Fermi gas model, optical model, vibrational and rotational models are some of them. None of these theoretical models that are currently known, can explain all the experimental observations.

Some of the experimental observations related to the nuclides are listed below:

(i) Except near the surface, there is constant density throughout the nucleus.

(ii) In a family of isobars, the β-decay energy is related to the mass difference.

(iii) α-decay energies show systematic variation as a function of Z and N.

(iv) $^{235}_{92}$U undergoes nuclear fission with thermal neutrons.

(v) There is a finite upper boundary for Z and N of the nuclides produced in the nuclear reactions, and the nuclides heavier than $^{238}_{92}$U are non-existent in nature.

(vi) The number of stable nuclei with a given value of Z and N corresponding to the magic numbers is more than the number of stable nuclei with neighbouring values of Z and N.

(vii) Nuclear angular momenta (usually called spin I) of the ground state are zero for even–even nuclei, integral multiples of 1/2 for odd-A nuclei and non-zero integers for odd–odd nuclei.

(viii) Mirror nuclei have the same value of nuclear angular momentum.

(ix) Electric quadrupole moments vary systematically with Z and N.

(x) Parity change of the nuclei in β-decay is followed by γ de-excitation.

(xi) Presence of stable end products in the decay series of $4n$, $4n$+1, $4n$+2 and $4n$+3 corresponding to Pb(Z = 82) or Bi(N = 126).

(xii) There is a higher number of stable isotones with N = 82, 126 and stable isotopes with Z = 50 and 80.

(xiii) Occurrence of first excited state in large number of even–even nuclei at an energy much lower than the energy required for breaking the pair within the shell model domain.
(xiv) Observation of excited nuclear states at much lower energies in the nuclei with $150 < A < 190$ and $A > 230$.
(xv) Wide spacing of low lying excited levels in the nuclei and the expression for the nuclear level density in the higher energy domain.
(xvi) Anisotropic angular distribution of the emitted particles in a nuclear reaction.
(xvii) Prediction of the observed cross-section for the scattering and the reaction events for neutrons or protons as a projectile.

Observations (i) to (v) are well explained by Liquid drop model (LDM), (vi) to (xii) by shell model, (xiii) and (xiv) by collective model, (xv) by Fermi gas model and the last two by optical model.

In the following pages, some of the nuclear models mentioned above will be discussed in more detail.

2.1 LIQUID DROP MODEL (LDM)

Liquid drop model is based on the assumption that there are many similarities between a nucleus and a liquid drop, such as:

(i) Constant density which is independent of size.
(ii) Latent heat of vaporization similar to fairly constant B/A, after $Z = 10$.
(iii) Evaporation of liquid drop similar to the radioactive decay of nuclei.
(iv) Condensation of drops similar to the formation of a compound nucleus.

In a liquid drop, the molecules are only influenced by their immediate neighbours. The surface molecules in a spherical liquid drop are not so tightly bound as the inner molecules. Same is true for the nucleons in a nucleus.

2.1.1 Semi-empirical Mass Formula (SEMF)

Using the above ideas, it is possible to derive an expression for the binding energy (mass) of a nucleus in its ground state. The binding energy formula derived using LDM consists of 5 energy terms—volume energy (B_v), surface energy (B_s), Coulomb energy (B_c), asymmetry energy (B_a) and pairing energy (B_p), which are explained in the following sections. Asymmetry energy and pairing energy are the correction terms based on quantum effects.

Volume Energy

As already mentioned, B/A is found to be fairly constant for stable isotopes. The density is taken as constant in LDM, i.e., the volume of the nucleus is proportional to its mass, and hence, to its mass number (A). Therefore, the following formula can be written for volume energy:

$$B_v = +a_v A \tag{2.1}$$

The positive sign indicates the binding effect of the attractive forces. The value of a_v which is the constant of proportionality is evaluated using experimental binding energies of nuclei.

Surface Energy

The argument that all nucleons are equally attracted in all directions is not true for surface nucleons. Therefore, the volume energy term must be decreased by an amount proportional to the surface energy of the nucleus; which in turn depends upon the surface area of the spherical nucleus.

$$B_s \propto 4\pi R^2$$

As already mentioned in the first chapter, considering constant density of the nucleus,

$$R = r_0 A^{1/3}$$

$$\therefore \quad B_s = -a_s A^{2/3} \tag{2.2}$$

The negative sign indicates that the volume energy term overestimates the attractive forces of the surface nucleons. The value of a_s, which is the proportionality constant is evaluated using experimental binding energies of nuclei, just like it was done for a_v.

Coulomb Energy

The Z number of protons present in the nucleus repel each other by an electrostatic force of repulsion calculated using Coulomb's law, which opposes the binding nuclear force.

The potential energy of a pair of protons which are r distance apart is equal to

$$V = \frac{-e^2}{4\pi\varepsilon_0 r}$$

Since, there are $Z(Z-1)/2$ pairs of protons,

$$B_c = \frac{Z(Z-1)}{2} V$$

$$= -\frac{Z(Z-1)e^2}{8\pi\varepsilon_0}\left(\frac{1}{r}\right)_{\text{avg}}$$

Here $(1/r)_{\text{avg}}$ is the value of $(1/r)$ averaged over all the proton pairs. If the protons are uniformly distributed throughout a nucleus of radius R,

$$\left(\frac{1}{r}\right)_{\text{avg}} \propto \frac{1}{R} \propto \frac{1}{A^{1/3}}$$

$$\therefore \quad B_c = -\frac{a_c Z(Z-1)}{A^{1/3}} \tag{2.3}$$

As before, the value of a_c is evaluated using experimental binding energies of nuclei.

Thus, the binding energy per nucleon (without the correction terms) is

$$\bar{B} = \frac{B}{A} = a_v - \frac{a_s}{A^{1/3}} - \frac{a_c Z(Z-1)}{A^{4/3}} \tag{2.4}$$

EXERCISE 2.1:

(a) Determine the value of a_c from the data for mirror nuclei[1] presented in Table 2.1, using least square fitting. (Mass excess of ${}_{1}^{1}\text{H}$ = 7.2883 MeV, mass excess of ${}_{0}^{1}n$ = 8.0713 MeV)

Table 2.1 Mirror Nuclides for $A = 11$ to $A = 63$ (There is no mirror nuclide for $A > 63$)

A	*Nuclide*	Z	*Mass excess* (MeV)	*Nuclide*	Z	*Mass excess* (MeV)
11	B	5	8.6679	C	6	10.6503
13	C	6	3.125	N	7	5.3454
15	N	7	0.1014	O	8	2.856
17	O	8	−0.8087	F	9	1.9517
19	F	9	−1.4874	Ne	10	1.752
21	Ne	10	−5.7317	Na	11	−2.1846
23	Na	11	−9.5298	Mg	12	−5.4732
25	Mg	12	−13.1927	Al	13	−8.9161
27	Al	13	−17.1967	Si	14	−12.3843
29	Si	14	−21.895	P	15	−16.9526
31	P	15	−24.4405	S	16	−19.043
33	S	16	−26.5858	Cl	17	−21.0032
35	Cl	17	−29.0135	Ar	18	−23.0473
37	Ar	18	−30.9476	K	19	−24.8001
39	K	19	−33.8071	Ca	20	−27.2827
41	Ca	20	−35.1379	Sc	21	−28.6424
43	Sc	21	−36.1881	Ti	22	−29.3212
45	Ti	22	−39.0083	V	23	−31.8797
47	V	23	−42.0056	Cr	24	−34.5585
49	Cr	24	−45.3329	Mn	25	−37.6146
51	Mn	25	−48.2437	Fe	26	−40.2214
53	Fe	26	−50.9467	Co	27	−42.6585
55	Co	27	−54.0292	Ni	28	−45.3351
57	Ni	28	−56.0831	Cu	29	−47.3082
59	Cu	29	−56.3577	Zn	30	−47.2149
61	Zn	30	−56.3432	Ga	31	−47.0882
63	Ga	31	−56.547	Ge	32	−46.9212

1. Mirror nuclei are pairs of nuclei in which the proton number in one equals the neutron number in the other and vice versa. The mass number is same for the pair.

(b) Plot the binding energy per nucleon, as the sum of the volume, surface and Coulomb energy terms using the following nuclides. (Take a_v = 14.1 MeV and a_s = 13.0 MeV)

$^{1}_{1}H$ $^{4}_{2}He$ $^{9}_{4}Be$ $^{16}_{8}O$ $^{27}_{13}Al$ $^{32}_{16}S$ $^{56}_{26}Fe$ $^{63}_{29}Cu$ $^{98}_{42}Mo$ $^{127}_{53}I$ $^{195}_{78}Pt$ $^{238}_{92}U$ $^{245}_{97}Bk$

Solution:

(a) For the first nucleus of the mirror nuclei pair,

$$N_1 - Z_1 = 1$$

Also we already know that,

$$N_1 + Z_1 = A$$

$$\therefore \qquad Z_1 = \frac{A-1}{2}$$

Similarly, for the second nucleus of the mirror nuclei pair,

$$N_1 - Z_1 = -1$$

$$\therefore \qquad Z_2 = \frac{A+1}{2}$$

Comparing the binding energies of the mirror nuclides using the semi-empirical mass formula [equation (2.4)], we thus get,

$$\Delta B = -\frac{a_c}{A^{1/3}}\left\{\left(\frac{A-1}{2}\right)\left(\frac{A-1}{2}-1\right)-\left(\frac{A+1}{2}\right)\left(\frac{A+1}{2}-1\right)\right\}$$

$$\therefore \qquad \Delta B = \frac{a_c}{A^{1/3}}(A-1)$$

Now, we plot ΔB vs. $(A - 1)/A^{1/3}$ in MS Excel using the values given in Table 2.1. A straight line (Figure 2.1) passing through the origin is expected for ΔB vs. $(A - 1)/A^{1/3}$ plot. The slope of this line which is constrained to pass through the origin should be the value of the Coulomb coefficient, a_c. Thus, we get, a_c = 0.646 MeV.

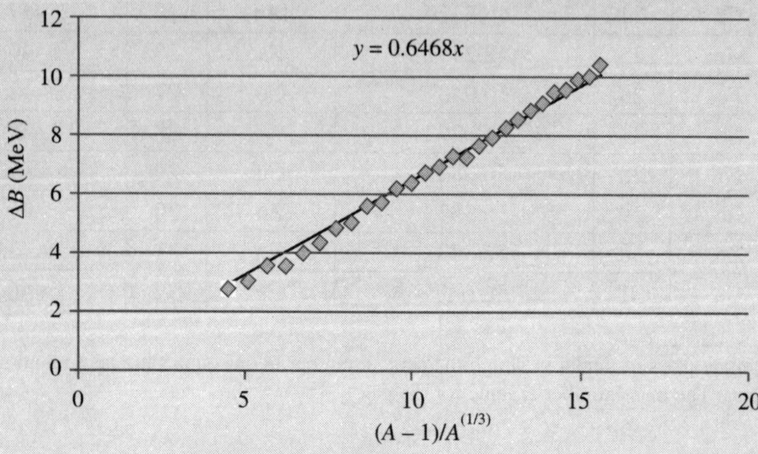

FIGURE 2.1 ΔB vs. $(A - 1)/A^{1/3}$ curve.

(b) Using Excel spreadsheet, the volume, surface and Coulomb energy terms are calculated, and Figure 2.2 is obtained for the total binding energy per nucleon.

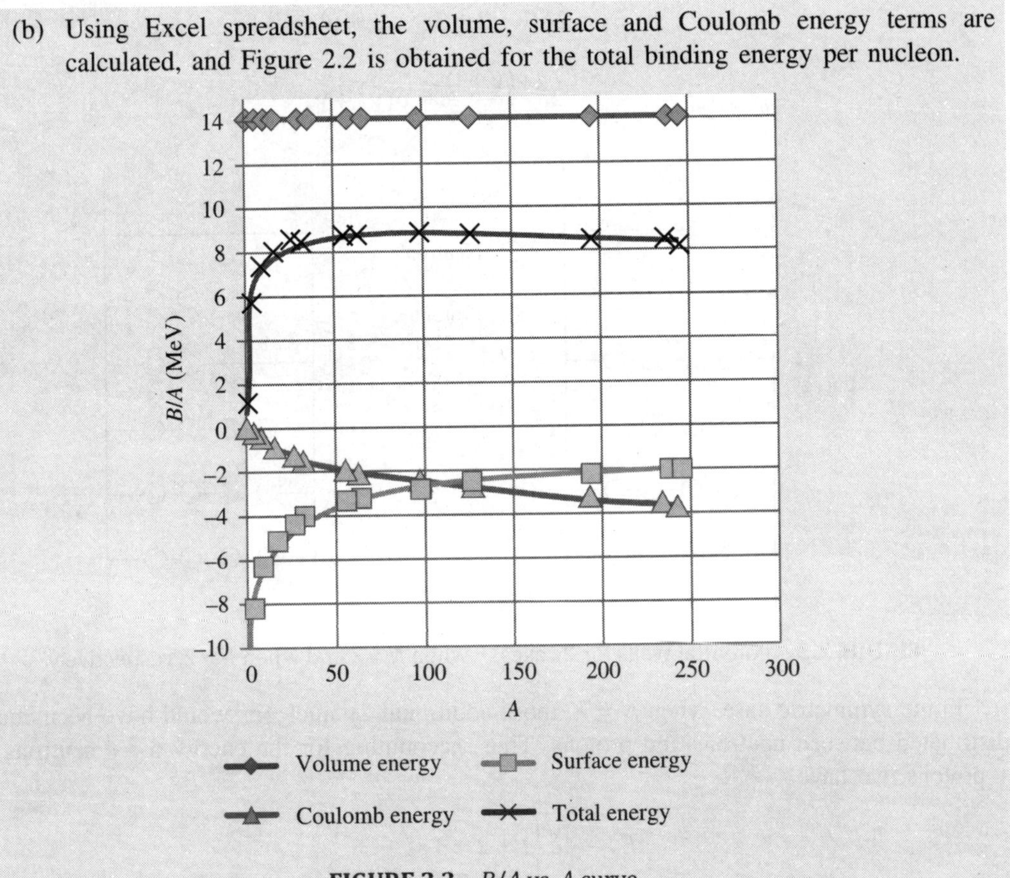

FIGURE 2.2 B/A vs. A curve.

Quantum Mechanical Corrections to the Formula

The binding energy formula, equation (2.4) can be improved by taking into account two effects based on quantum mechanics that are not explained by simple liquid drop model.

Asymmetry Energy: This correction term is needed due to the occupancy of neutrons in the higher energy levels when $N > Z$ as compared to the lighter nuclei when N and Z are equal. For heavier nuclei, the Coulomb repulsion effect starts showing up and to compensate for that N becomes greater than Z and the nuclear force becomes less effective.

This term is based on the shell model description of fermions (Fermi gas model). Let us assume that both neutron and proton independently constitute Fermi gas which is confined in potential wells of equal depth having energy level spacing equal to δ (Figure 2.3). The levels (orbitals) are filled in tune with Pauli exclusion principle. Therefore, the maximum number of nucleons that each level can occupy is 4 (2 proton $\uparrow\downarrow$ + 2 neutrons $\uparrow\downarrow$).

In the asymmetry case, when $N > Z$, if there are 2ν additional neutrons left after $(A/2 - \nu)$ neutrons and the same number of protons has been filled, then the energy for additional 2ν neutrons can be written as follows:

$$E_1 = 2\delta\{1 + 2 + 3 + \ldots + \nu\}$$

$$= 2\delta \frac{\nu(\nu+1)}{2} = \delta(\nu^2 + \nu)$$

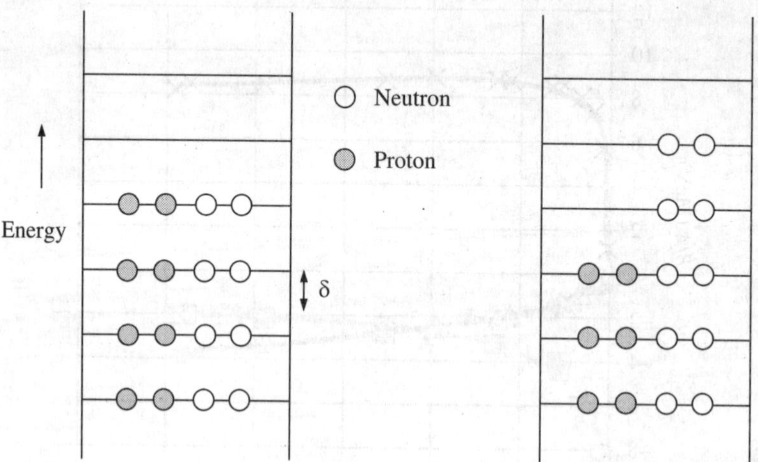

FIGURE 2.3 Potential wells for 2 cases: when $N = Z$ and when $N > Z$, respectively.

In the symmetric case, when $N = Z$, those additional 2ν nucleons would have been equally distributed between neutrons and protons. Thus, accounting for the energy for ν neutrons and ν protons, we have

$$E_2 = 4\delta\left\{1 + 2 + 3 + \cdots + \frac{\nu}{2}\right\}$$

$$= 4\delta \frac{\frac{\nu}{2}\left(\frac{\nu}{2}+1\right)}{2} = \delta\left(\frac{\nu^2}{2} + \nu\right)$$

Therefore, the difference in the energies in the two cases should be the asymmetry energy, i.e.,

$$E_1 - E_2 = \frac{\delta\nu^2}{2} \tag{2.5}$$

The energy of the nucleus will increase in the asymmetry case, as can be seen from the above equation.

As already mentioned in the asymmetry case

$$N = \frac{A}{2} + \nu \quad \text{and} \quad Z = \frac{A}{2} - \nu$$

$$\therefore \qquad \nu = \left(\frac{A - 2Z}{2}\right) \tag{2.6}$$

Now, if for some energy, it is assumed that the A nucleons are equally divided, then it can be seen that greater the number of nucleons in a nucleus, the smaller is the energy level spacing δ, i.e.,

$$\delta \propto \frac{1}{A} \tag{2.7}$$

Also, we know that as the nuclear mass energy goes up, the binding energy decreases. Hence, the asymmetry correction term can be written as follows:

$$B_{\text{asym}} = -\frac{a_{\text{asym}}(A-2Z)^2}{A} \tag{2.8}$$

where a_{asym} is the proportionality constant.

Pairing Energy: From the pattern of abundance of stable nuclides, it is seen that even–even nuclides are more stable than odd–even, even–odd or odd–odd nuclides. Odd–odd nuclei have low binding energies due to the last unpaired proton and neutron. Since, pairing lowers the energy of the nucleus, a pairing energy correction term is added to the binding energy formula as follows:

$$B_p = (\pm, 0)\delta \qquad \left.\begin{array}{l}(+) \text{ for e-e nuclides}\\(-) \text{ for o-o nuclides}\\(0) \text{ for o-e and e-o nuclides}\end{array}\right\} \tag{2.9}$$

The pairing energy seems to vary with A as $A^{-3/4}$. Therefore, in the above expression $\delta = a_p A^{-3/4}$, where a_p is the proportionality constant.

Thus, the final expression for SEMF (Weizsacker expression) is as follows:

$$B(A, Z) = a_v A - a_s A^{2/3} - \frac{a_c Z(Z-1)}{A^{1/3}} - \frac{a_{\text{asym}}(A-2Z)^2}{A} (\pm, 0)\, a_p A^{-3/4} \tag{2.10}$$

EXERCISE 2.2:

(a) Identify the stable radionuclide(s) and determine the asymmetry coefficient (a_{asym}) from the position of maximum β-stability by fitting the masses of isobars (nuclides having the same mass number A) with parabola for $A = 65$.

Table 2.2 Mass Excess (MeV) for (A = 65) Isobar Family

Nuclides	$^{65}_{28}$Ni	$^{65}_{29}$Cu	$^{65}_{30}$Zn	$^{65}_{31}$Ga	$^{65}_{32}$Ge
Mass excess (MeV)	−65.137	−67.266	−65.917	−62.658	−56.260

(b) Using the mass parabola for A = 90, determine the pairing coefficient (a_p).

Table 2.3 Mass Excess (MeV) for (A = 90) Isobar Family

Z	36	37	38	39	40	41	42
Mass excess (MeV)	−74.780	−79.340	−85.932	−86.476	−88.770	−82.660	−80.160

(c) Finally, estimate the values of a_v and a_s using the following nuclides whose binding energies are given in Table 2.4.

Table 2.4 Binding Energies (MeV) of Nuclides

Nuclides	$^{9}_{4}Be$	$^{16}_{8}O$	$^{27}_{13}Al$	$^{32}_{16}S$	$^{56}_{26}Fe$	$^{63}_{29}Cu$
B.E. (MeV)	58.162	127.620	224.953	271.779	492.262	551.393
Nuclides	$^{98}_{42}Mo$	$^{127}_{53}I$	$^{195}_{78}Pt$	$^{238}_{92}U$	$^{245}_{97}Bk$	
B.E. (MeV)	846.248	1072.587	1545.675	1801.726	1839.885	

(d) Using the values of the coefficients that have been calculated, compare the binding energy of $^{56}_{26}Fe$ with the prediction made from SEMF.

Solution: (a) The mass of a nuclide is given by

$$M(A, Z) = ZM_H + (A - Z)M_n - B(A, Z)$$

Neglecting 1 in comparison to Z in the Coulomb energy term and rearranging the semi-empirical mass formula [equation (2.10)], we get

$$M(A, Z) = aA + bZ + cZ^2 + \delta \tag{2.11}$$

where

$$a = M_n - \left\{a_v - a_{\text{asym}} - \frac{a_s}{A^{1/3}}\right\}$$

$$b = -4a_{\text{asym}} - (M_n - M_H)$$

$$c = \left\{\frac{4a_{\text{asym}}}{A} + \frac{a_c}{A^{1/3}}\right\}$$

Using Excel spreadsheet, $M(A, Z)$ values are calculated using the given values for mass excess, and the following plot (Figure 2.4) is obtained. For fitting a parabola, a polynomial trend line (2nd order) is chosen to get the values of a, b and c.

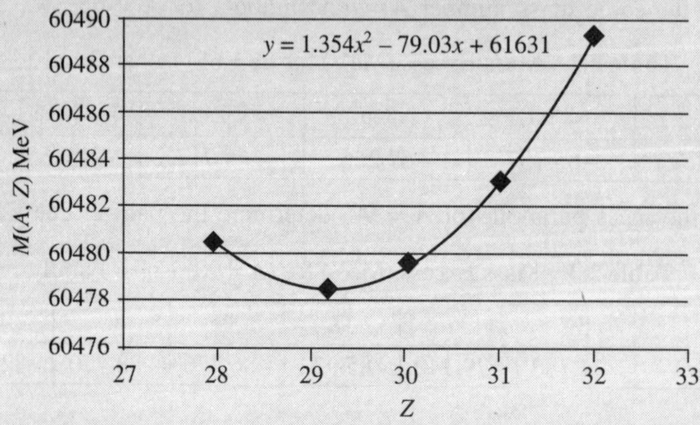

FIGURE 2.4 Mass parabola (fitted) for $A = 65$ isobar family.

Now, to find the position of maximum β-stability

$$\left.\frac{\partial M(A,Z)}{\partial Z}\right|_{A=\text{constant}} = 0$$

or
$$Z_0 = -\frac{b}{2c} = 29.18$$

i.e., $^{65}_{29}$Cu is the most stable nuclide in this isobar family.

The asymmetry coefficient can be calculated based on the value of b.

$$a_{\text{asym}} = \frac{(M_n - M_H) - b}{4} = \frac{0.783 + 79.03}{4} = 19.95 \text{ MeV}$$

Similar procedure can be repeated for different isobar families to get an average value for a_{asym}.

(b) Two separate parabolas for e–e nuclides and o–o nuclides are fitted to obtain the value of the pairing term δ.

$$Z_{\text{even}} = (36, 38, 40, 42) \text{ and } Z_{\text{odd}} = (37, 39, 41)$$

The stability rules suggest that the mass parabolas of o–o nuclides are shifted upward from the mass parabolas of the e–e nuclides. The difference between the two parabolas at the most stable Z value is taken to be 2δ.

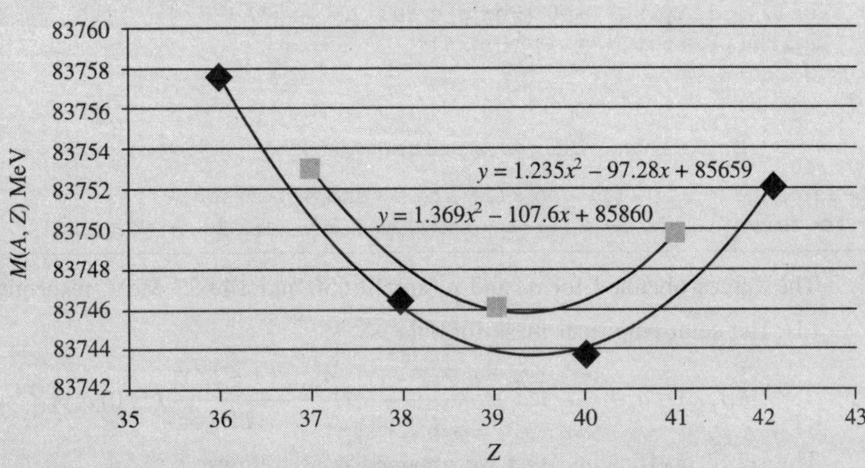

FIGURE 2.5 Mass parabolas (fitted) for even-even or odd-odd nuclides.

$$Z_0 = -b/2c \sim 39$$

Hence,

$$2\delta = 39^2 \times (1.369 - 1.235) - 39 \times (107.6 - 97.28) + (85860 - 85659)$$

$$\therefore \qquad \delta = 1.167 \text{ MeV and } a_p = \delta \times A^{3/4} = 34.1 \text{ MeV}$$

(c) Using the final expression of SEMF,

$$B(A,\ Z) + \frac{a_c Z(Z-1)}{A^{1/3}} + \frac{a_{\text{asym}}(A-2Z)^2}{A}(\mp,\ 0)\delta = a_v A - a_s A^{2/3}$$

The LHS can be calculated for different A and Z. Using the least square curve fitting function in Matlab, the values of a_v and a_s can be estimated.

```
clear all
% Coulumb coef, asymmetry coef, pairing coef (MeV)
ac = 0.646; aasym = 19.95; ap = 34.1;

A = [9;16;27;32;56;63;98;127;195;238;245]; % Mass numbers (A)
Z = [4;8;13;16;26;29;42;53;78;92;97]; % Atomic numbers (Z)

B = [58.162;127.620;224.953;271.779;492.262;551.393; ...
     846.248;1072.587;1545.675;1801.726;1839.885]; % B.E. (MeV)

Bc = -ac*Z.*(Z-1).*A.^(-1/3); % Coulomb energy (MeV)
Basym = -aasym*(A-2*Z).^2.*A.^(-1); % Asymmetry energy (MeV)

% Pairing energy term (MeV)
delta = zeros(11,1); %For e-o and o-e delta = 0
for i=1:11
    if mod(Z(i),2)==0 && mod(A(i),2)==0 %For e-e delta = ap*A(-3/4)
      delta(i) = ap*(A(i))^(-3/4);
    elseif mod(A(i),2)==0 %For o-o delta = -ap*A(-3/4)
      delta(i) = -ap*(A(i))^(-3/4);
    end
end

ydata = B - Bc - Basym - delta; % DeltaB (MeV)

% Least square fit: x(1) and x(2) are av and as, respectively
x = lsqcurvefit(@(x,A) x(1)*A - x(2)*A.^(2/3),[1 1],A,ydata)
```

The values obtained for a_v and a_s are 14.656 and 14.687 MeV, respectively.

(d) The semi-empirical mass formula is

$$B(A, Z) = a_v A - a_s A^{2/3} - \frac{a_c Z(Z-1)}{A^{1/3}} - \frac{a_{\text{asym}}(A-2Z)^2}{A}(\pm,\ 0)\ a_p A^{-3/4}$$

The set of coefficients[2] that we obtained is as follows:

a_v = 14.656 MeV a_s = 14.687 MeV a_c = 0.646 MeV
a_{asym} = 19.95 MeV a_p = 34.1 MeV

Plugging in the values, we get

$$B(A, Z) = 14.656\times 56 - 14.687\times 56^{2/3} - \frac{0.646\times 26\times(26-1)}{56^{1/3}} - \frac{19.95\times(56-2\times 26)^2}{56}$$
$$+\ 34.1\times 56^{-3/4} = 491.9718 \text{ MeV}$$

2. This set of coefficients is not unique. There can be other sets too that can give good fit with the data.

$$\text{Percentage error} = \frac{492.262 - 491.9718}{492.262} \times 100 = 0.059$$

Thus, the difference between the observed and the calculated values of BE is less than 0.06 per cent in this case.

2.1.2 Applications of LDM

The semi-empirical mass formula, based on liquid drop model, is very useful to calculate the masses of unknown nuclei. Using LDM, predictions can be made regarding nuclear stability with respect to α-decay, β-decay and spontaneous fission. LDM explains the basic features of the fission process. One can also draw the neutron and proton drip lines based on SEMF.

EXERCISE 2.3:

(a) Using SEMF, explain why ${}^{238}_{92}\text{U}$ nuclide is an α-emitter and not a β^- emitter?

(b) Show that nuclei with $A > 150$ are energetically unstable against α-decay.

(c) From SEMF, show that only for heavier nuclei, the Q-value or kinetic energy released in the fission > 0.

(d) Calculate the separation energy of neutron from SEMF and draw neutron drip line using any four Z-values.

Solution:

(a) If α emission is possible, then the following reaction is possible:

$${}^{238}_{92}\text{U} \rightarrow {}^{234}_{90}\text{X} + {}^{4}_{2}\text{He}$$

The Q_α value or the decay energy associated with α-decay of a nucleus ${}^{A}_{Z}\text{X}$ can be calculated from the difference of the mass of ${}^{A}_{Z}\text{X}$ nucleus and the combined mass of product nucleus ${}^{A-4}_{Z-2}\text{Y}$ and the α-particle. Thus, the condition for α-decay is:

$$Q = M(A, Z) - M(A-4, Z-2) - M({}^{4}_{2}\text{He}^{++}) > 0$$

The mass of a nuclide is given by

$$M(A, Z) = ZM_H + (A - Z)M_n - B(A, Z)$$

Therefore, we get

$$Q = -B(A, Z) + B(A-4, Z-2) + B({}^{4}_{2}\text{He}^{++})$$

From the above relation along with equation (2.10) and substituting A, Z and coefficient values that we obtained before in semi-empirical mass formula, we get

$$Q = -1807.5 + (1784 + 32.85) = 9.35 > 0$$

Hence, the condition for α-emission is satisfied.

Similarly, if β^- emission were to take place, the following reaction should be possible:

$${}^{238}_{92}\text{U} \rightarrow {}^{238}_{93}\text{X} + \beta^-$$

The condition for β^- emission is

$$Q = M(A, Z) - M(A, Z+1) > 0$$

Therefore, we get

$$Q = -B(A, Z) + B\left(A, Z+1\right) + (M_H - M_n)$$

Substituting the A, Z and coefficient values in semi-empirical mass formula, we get

$$Q = -1807.5 + 1805 + (-0.78) = -3.28 < 0$$

Hence, β^- emission is not possible.

(b) The condition for α-decay is:

$$Q = -B(A, Z) + B(A-4, Z-2) + B({}_2^4\text{He}^{++}) > 0$$

If the naturally occurring nuclides are stable against β-decay, these may decay by emitting α-particles. As already mentioned, for β-stable nuclides:

$$Z = Z_0 = -\frac{b}{2c} = \frac{4a_{\text{asym}} + (M_n - M_H)}{2\left\{\dfrac{4a_{\text{asym}}}{A} + \dfrac{a_c}{A^{1/3}}\right\}}$$

Using the above expression for Z, we get an expression for Q as an exclusive function of A, which can be used to see how Q varies with A.

In the following Matlab code, the value of A is continuously incremented and it is found that for A = 152 onward, $Q > 0$, i.e., such nuclides are energetically unstable against α-decay.

```
clear all
% Volume, surface, Coulomb, asym. and pairing coef (MeV)
av=14.656; as=14.687; ac=0.646; aasym=19.95; ap=34.1;

% Mass of proton and neutron; B.E. of alpha (MeV)
MH = 938.7583; Mn = 939.5413; Balpha=28.296;

for A=5:300
   Z =(4*aasym + (Mn-MH))/(2*(4*aasym/A+ac*A^(-1/3)));

   B = av*A-as*A^(2/3)-(ac*Z*(Z-1))/A^(1/3)-(aasym*(A-2*Z)^2)/A;
   B2 = av*(A-4)-as*(A-4)^(2/3)-(ac*(Z-2)*(Z-3))/(A-4)^(1/3)...
      -(aasym*((A-4)-2*(Z-2))^2)/(A-4);

   Q(A) = -B + B2 + Balpha; %Q alpha (MeV)
end

plot([0 A],[0 0],'k-') %x-axis
hold on
grid minor
plot(Q,'r-')
xlabel('A')
ylabel('Q alpha (MeV)')
```

The following plot (Figure 2.6) is obtained when the Matlab code is run.

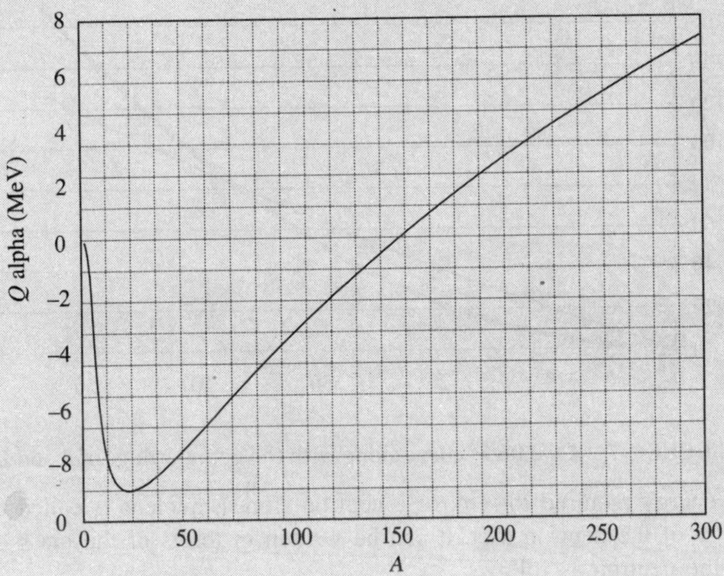

FIGURE 2.6 Q-values for α decay vs. A.

(c) Assuming symmetric fission for simplicity, the expression for Q becomes

$$Q = -B(A, Z) + 2B(A/2, Z/2)$$

Substituting SEMF and ignoring the pairing energy term, we get

$$Q = a_s\left\{A^{2/3} - 2(A/2)^{2/3}\right\} + a_c\left\{\frac{Z(Z-1)}{A^{1/3}} - 2\frac{(Z/2)((Z/2)-1)}{(A/2)^{1/3}}\right\}$$

Using the approximation, $Z(Z-1) \to Z^2$, we get

$$Q = a_s A^{2/3}\left\{1 - 2 \times (1/2)^{2/3}\right\} + a_c \frac{Z^2}{A^{1/3}}\left\{1 - 2 \times \frac{1/4}{(1/2)^{1/3}}\right\}$$

or

$$Q = A^{-1/3}\{0.37\, a_c Z^2 - 0.26\, a_s A\}$$

For fission to be possible, $Q > 0$, which means

$$0.37\, a_c Z^2 > 0.26\, a_s A$$

or

$$\frac{A}{Z^2} < \frac{0.37 \times 0.646}{0.26 \times 14.687} \approx 0.06$$

Now, $A = 0.06Z^2$ curve is plotted along with the actual values of mass number and atomic mass from the Periodic table. It can be seen that for nuclides with mass number $A > 85$, condition $A/Z^2 < 0.06$ is satisfied, which implies that for heavier nuclei, Q-value for the fission > 0. For example, $A/Z^2 = 0.028$ for $^{235}_{92}U$ and therefore, fission can take place.

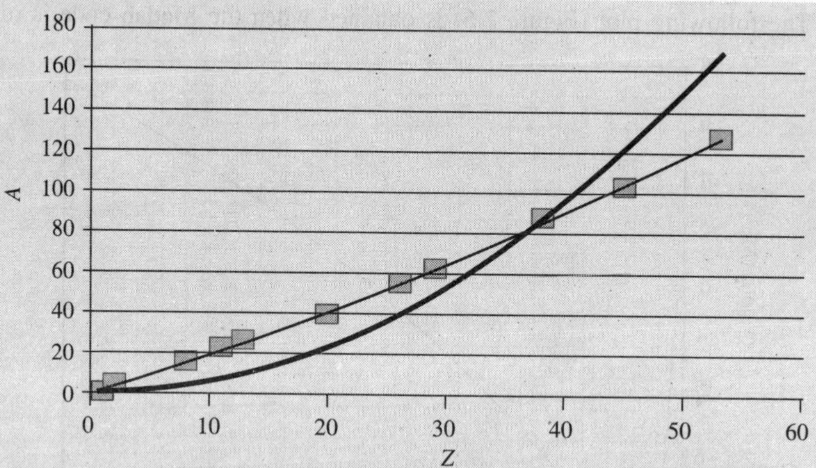

FIGURE 2.7 $A = 0.06\,Z^2$ curve along with the actual values of *A* and *Z*.

(d) The energy required to remove a neutron from a nucleus is called the separation energy of the neutron (S_n). It can be written in terms of the mass of the nuclide and the neutron as follows:

$$S_n(Z, A) = M(A-1, Z) + M_n - M(A, Z)$$

In terms of the binding energies, the expression for the separation energy for neutron becomes

$$S_n(Z, A) = B(A, Z) - B(A-1, Z) \tag{2.12}$$

For a particular *Z*-value, there is a limit to how many neutrons can go in the nuclei and at a certain point, the neutron emission becomes energetically favourable. The physics close to the drip line is an active research area, and still in its nascent stage. For the purpose of showing the general trend of the drip line using SEMF, when the neutron separation energy stays negative for a series of nuclides that value is taken as the coordinate for the neutron drip line.

The Matlab code given below is used to calculate the neutron drip line coordinate for each *Z*-value.

```
clear all

% Volume, surface, Coulomb, asym. and pairing coef (MeV)
av=14.656; as=14.687; ac=0.646; aasym=19.95; ap=34.1;

Z = 10;

for A=2*Z:5*Z
  if mod(Z,2)==0 && mod(A,2)==0 %For e-e delta = +ap*A(-3/4)
    delta = 34.1*A^(-3/4);
  elseif mod(A,2)==0 %For o-o delta = -ap*A(-3/4)
    delta = -34.1*A^(-3/4);
  else delta = 0;
  end
```

```
B = av*A-as*A^(2/3)-(ac*Z*(Z-1))/A^(1/3)...
    -(aasym*(A-2*Z)^2)/A + delta; % B(A,Z)

  if mod(Z,2)==0 && mod(A-1,2)==0 %For e-e delta = +ap*A(-3/4)
    delta = 34.1*(A-1)^(-3/4);
  elseif mod(A-1,2)==0 %For o-o delta = -ap*A(-3/4)
    delta = -34.1*(A-1)^(-3/4);
  else delta = 0;
  end
   B2 = av*(A-1)-as*(A-1)^(2/3)-(ac*Z*(Z-1))/(A-1)^(1/3)...
      -(aasym*((A-1)-2*Z)^2)/(A-1) + delta; % B(A-1,Z)

   Sn(A) = B - B2; %Separation energy for neutron (MeV)
end

plot([0 A],[0 0],'k-') %x-axis
hold on
grid minor
plot(Sn,'r-')
xlabel('A')
ylabel('Separation energy of neutron (MeV)')
```

The following plot (Figure 2.8) is obtained when this code is run. The saw-tooth plot that we see below is because of the nucleon pairing effect. For the $Z = 10$ case, the drip line coordinate for our purpose is taken as $A = 35$ or $N = 25$.

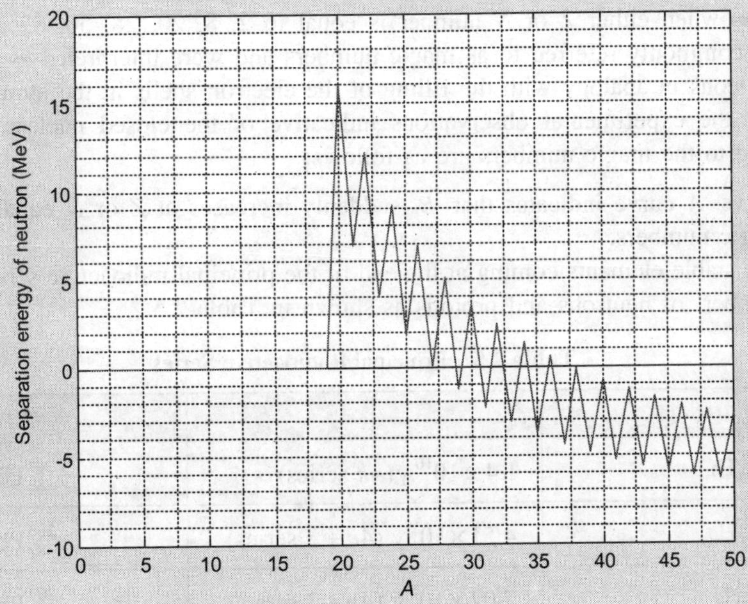

FIGURE 2.8 Separation energy of neutron vs. *A*.

Similarly, we get the points for $Z = 20$, 30 and 40 and the following plot (Figure 2.9) for the neutron drip line is drawn.

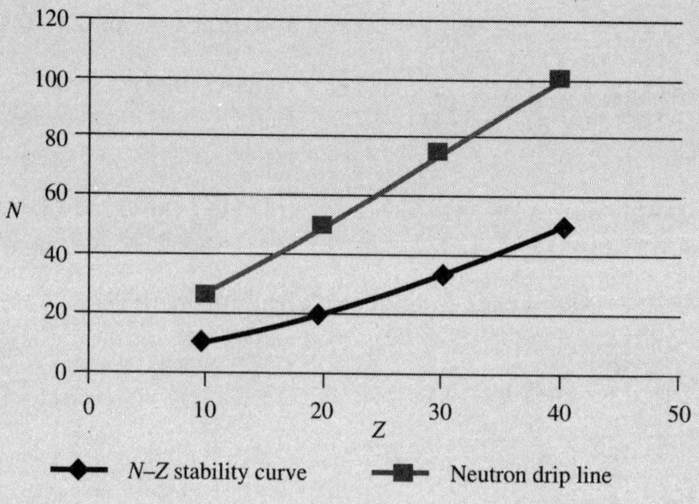

FIGURE 2.9 Neutron drip line along with the stability curve.

2.2 SHELL MODEL

The nuclear shell model is one of the most important and useful models of the nuclear structure. It evolved during 1945–50 to explain the fact that many nuclear properties show marked discontinuities when either Z or N number is equal to 2, 8, 20, 28, 50, 82 and 126. These numbers are commonly referred to as magic numbers and were interpreted as forming closed shells of nucleons in analogy with the filling of the electron shells in the atoms.

Some of the experimental observations indicative of the closed nuclear shell structure corresponding to the magic numbers are as follows:

(a) $\overline{B}$ vs. A curve indicates that $\overline{B}$ suddenly increases at Z or N equal to one of the magic numbers.
(b) The stable elements coming at the end of the principal radioactive series have 'magic number' of neutrons and protons as shown in Table 2.5.

Table 2.5 Principal Radioactive Series

Parent		*Decay product*
Thorium $^{232}_{90}\text{Th}$	1.4×10^{10} y ($4n$ series)	$^{208}_{82}\text{Pb}_{126}$
Uranium $^{238}_{92}\text{U}$	4.51×10^{9} y ($4n + 2$ series)	$^{206}_{82}\text{Pb}_{124}$
Actinium $^{235}_{92}\text{U}$	7.07×10^{8} y ($4n + 3$ series)	$^{207}_{82}\text{Pb}_{125}$
Neptunium $^{237}_{93}\text{Np}$	2.25×10^{6} y ($4n + 1$ series)	$^{209}_{83}\text{Bi}_{126}$

(c) Number of stable isotones is maximum for $N = 50(6)$ and $N = 82(7)$.

(d) Number of stable isotopes for $Z = 20$, 50 and 82 are much larger than their neighbours with $(Z \pm 1)$.

(e) The most abundant nuclides in the universe are those with Z or N or both equal to the magic numbers, e.g., ${}^{16}_{8}O_8$, ${}^{40}_{20}Ca_{20}$, ${}^{88}_{38}Sr_{50}$, ${}^{89}_{39}Y_{50}$, ${}^{90}_{40}Zr_{50}$, ${}^{118}_{50}Sn_{68}$, ${}^{138}_{56}Ba_{32}$, ${}^{139}_{57}La_{82}$, ${}^{140}_{58}Ce_{82}$, ${}^{208}_{82}Pb_{126}$, etc.

(f) Neutron absorption cross-section σ is lower by two orders of magnitude for nuclides having magic number of neutrons. This indicates that these nuclei are less likely to absorb an additional neutron.

(g) α-decay energies are rather smooth functions of A for a given Z, but it shows striking discontinuities at $N = 126$ or $Z = 82$, when the energy of the α-particles (E_α) increases, e.g.

$${}^{212m}_{84}Po_{128} \rightarrow {}^{208}_{82}Pb_{126} + \alpha(E_\alpha = 8.8 \text{ MeV}, T_{1/2} = 42.1 \text{ s})$$

$${}^{210}_{84}Po_{126} \rightarrow {}^{206}_{82}Pb_{124} + \alpha(E_\alpha = 5.31 \text{ MeV}, T_{1/2} = 138.376 \text{d})$$

(h) Separation energy of the last neutron or proton S_n/S_p having magic number is very high analogous to the high ionization potential of noble gases.

Once the evidence of shell structure was found based on the above experimental observations, the challenge before the nuclear scientists was to explain the existence of magic numbers. Besides magic numbers, the nuclear properties like I^π (spin and parity of ground and excited nuclear states), μ (magnetic moment) also motivated them to invoke a model based on a single particle behaviour of nucleons.

The shell model is based on the assumption that each nucleon moves in its orbit within the nucleus independent of all other nucleons. The orbits corresponding to the definite states of energy and angular momentum are determined by a potential energy function $V(r)$, representing the average effect of all the interactions with other nucleons. Each nucleon is regarded as an independent particle and the residual interaction between nucleons, if any, is considered to be small perturbation on the interaction between a nucleon and the potential field. This model is also known as *independent particle model*.

This is in sharp contrast to the assumption of LDM, where it is assumed that the interaction among the nucleons is strong. However, it should be realized that in an energy level model, the nucleons will fill the levels up to Fermi level and although there are levels vacant at higher energy, no lower level remains into which a nucleon may be scattered. Thus, one can say that the Pauli's exclusion principle effectively suppresses any scattering (collision) and ensures reasonably well-defined orbits. This restriction does not apply to a higher energy nucleon coming from outside, which can readily collide and raise the energy of the constituents of the nucleus.

2.2.1 Shell Model Potential

As already mentioned, the shell model is based on the single particle assumption in which the effect of all other nucleons is smoothened and the force on any nucleon inside the nucleus is represented by a potential well due to all other nucleons. This central potential $V(\vec{r}) = V(r)$ is then used in Schrödinger equation to determine the eigen values (energies) of the nuclear levels.

Square Well Potential

A three-dimensional infinite square well central potential is shown in Figure 2.10, where

$$V(\vec{r}) = 0, \text{ for } r < R$$
$$= \infty, \text{ for } r > R$$

where R is the nuclear radius.

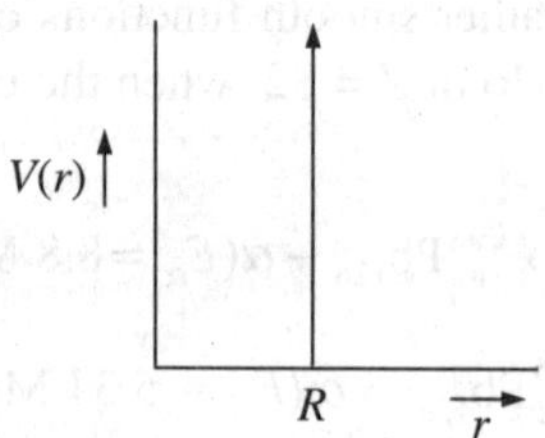

FIGURE 2.10 Infinite square well central potential.

The general solution of the one-particle Schrödinger equation in a spherically symmetric potential can be written as follows:

$$\psi(r, \theta, \varphi) = \frac{u(r)}{r} Y_l^m(\theta, \varphi)$$

where $u(r)$ is the radial function, and $Y_l^m(\theta, \varphi)$ are the spherical harmonics. Since, the assumed potential is central, no directional part, i.e., (θ, φ), is needed. The corresponding radial equation becomes:

$$-\frac{\hbar^2}{2m}\frac{d^2u}{dr^2} + (V(r) + \text{centrifugal barrier due to angular momentum } l)u = Eu$$

or
$$-\frac{\hbar^2}{2m}\frac{d^2u}{dr^2} + \left(0 + \frac{l(l+1)\hbar^2}{2mr^2}\right)u = Eu \quad \text{for } r < R$$

The solution to the above equation is:

$$u = rj_l(kr)$$

where $k = \sqrt{\frac{2mE}{\hbar^2}}$, j_l are the spherical Bessel functions.

Energy eigen values are obtained from the boundary conditions, $u(0) = 0$ and $u(R) = 0$; i.e., $j_l(kR) = 0$. For each l-value, the spherical Bessel function $j_l(kR)$ is equal to zero for various values of k. Each of these k values is related to the energy of the levels. Thus, for $l = 0$, $n = 1, 2, 3, \ldots$; for $l = 1$, $n = 1, 2, 3, \ldots$, and so on.

Note: This (n, l) relationship is different from what we use in the atomic physics.

As expected only certain discrete values of E are allowed. The ordering of the levels comes out to be:

$$1s \quad 1p \quad 1d \quad 2s \quad 1f \quad 2p \quad 1g \quad 2d \quad 1h \quad 3s$$

where s, p, d, f, g, h, …, etc. stand for usual spectroscopic notation for l = 0, 1, 2, 3, 4, 5, …, respectively.

The total number of protons or neutrons in a given l state for two possible orientations of the nucleon spin and the magnetic quantum number m_l ranging from $-l$ to $+l$ with integer steps between them, is given by $2(2l + 1)$.

EXERCISE 2.4:

What are the shell closures and the magic numbers predicted by square well potential?

Solution: The underlined numbers in the shell closures column are the magic numbers. It can be seen (Table 2.6), the square well potential predicts only 3 magic numbers.

Table 2.6 Prediction of Magic Numbers by Square Well Potential

Levels	*Magnetic quantum number m_l*	*Number of nucleons $2(2l + 1)$*	*Shell closures*
1*s*	0	$2(2 \times 0 + 1) = 2$	<u>2</u>
1*p*	–1, 0, +1	$2(2 \times 1 + 1) = 6$	<u>8</u>
1*d*	–2, –1, 0, +1, +2	$2(2 \times 2 + 1) = 10$	18
2*s*	0	$2(2 \times 0 + 1) = 2$	<u>20</u>
1*f*	–3, –2, –1, 0, +1, +2, +3	$2(2 \times 3 + 1) = 14$	34
2*p*	–1, 0, +1	$2(2 \times 1 + 1) = 6$	40
1*g*	–4, –3, –2, –1, 0, +1, +2, +3, +4	$2(2 \times 4 + 1) = 18$	58
2*d*	–2, –1, 0, +1, +2	$2(2 \times 2 + 1) = 10$	68
1*h*	–5, –4, –3, –2, –1, 0, +1, +2, +3, +4, +5	$2(2 \times 5 + 1) = 22$	90
3*s*	0	$2(2 \times 0 + 1) = 2$	92

Harmonic Oscillator Potential

A 3-dimensional harmonic oscillator potential is shown in Figure 2.11.

$$V(\vec{r}) = -V_0 + \frac{1}{2} m\omega^2 r^2$$

where $\omega^2 = \frac{k}{m}$ is the harmonic oscillator frequency.

The harmonic oscillator potential diminishes steadily at the edges in contrast to the infinite sharp edges in the square well type potential.

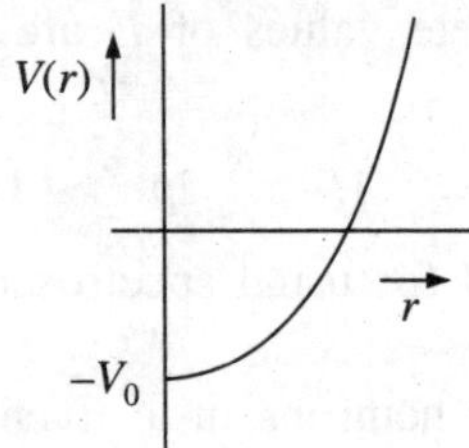

FIGURE 2.11 Harmonic oscillator potential.

Again solving the Schrödinger equation for the eigen values, the energy is written as:

$$E = \left(N + \frac{3}{2}\right)\hbar\omega$$

where $N = 2(n-1) + l$, $n = 1, 2, 3, \ldots$ and $l = 0, 1, 2, \ldots$

For a specific oscillator level (n, l), there is a large degeneracy related to the energy characterized by the quantum number N, i.e., one has to consider all possible (n, l) values such that $N = 2(n - 1) + l$.

EXERCISE 2.5:

Show that for a specific level (n, l), there exists a large degeneracy relative to the energy characterized by the quantum number N. Find the shell closures and the magic numbers predicted by harmonic oscillator potential.

Solution:

Table 2.7 Prediction of Magic Numbers by Harmonic Oscillator Potential

N	*Levels* (n, l)	*Number of nucleons* $\sum_{n,l} 2(2l+1)$	*Shell closures*
0	1*s*	2	**2**
1	1*p*	6	**8**
2	2*s*, 1*d*	12	**20**
3	2*p*, 1*f*	20	40
4	3*s*, 2*d*, 1*g*	30	70
5	3*p*, 2*f*, 1*h*	42	112
6	4*s*, 3*d*, 2*g*, 1*i*	56	168

It can be seen from Table 2.7, for a particular N, there can be more than one levels (n and l values) having the same energy. The underlined numbers in the shell closures column are the magic numbers. Like square well potential before, harmonic oscillator potential also does not reproduce the experimentally observed magic numbers and predicts only 3 magic numbers.

Woods and Saxon Potential

The other potential which is a compromise between square well and harmonic oscillator potential is

$$V(\vec{r}) = -V_0 \frac{1}{1 + \exp\left(\dfrac{r-R}{a}\right)}$$

where the nuclear radius $R = r_0 A^{1/3}$, surface thickness $a = 0.52$ fm and the potential depth $V_0 = 50$ MeV. These values are obtained from scattering experiments and reproduce the right kind of energy spacing.

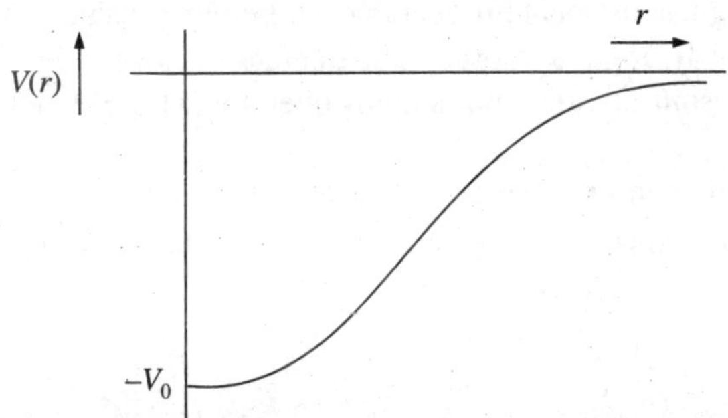

FIGURE 2.12 Woods-Saxon potential.

Unlike square well potential, the Woods–Saxon potential does not have any sharp edges. The shape of this potential is given in Figure 2.12. This potential closely represents the nuclear charge and matter distributions, falling gradually to zero beyond the mean radius R. When the Schrödinger equation is solved for this potential, it predicts 2, 8, 20, 40, 58, 92 and 112 as shell closures. Again 2, 8 and 20 are the common shell closures and the numbers correspond to the particularly stable nuclei ${}^{4}_{2}He_2$, ${}^{16}_{8}O_8$ and ${}^{40}_{20}Ca_{20}$.

It is clear that something fundamental is missing.

2.2.2 Spin–Orbit Coupling

It was pointed out [independently by Mayer (1949–50) and Haxel, Jensen and Suess (1949–50)] that a contribution to the average field, experienced by each individual nucleon, should contain a spin–orbit term. The corrected potential then becomes

$$V(\vec{r}) = -V_0 + \frac{1}{2} m\omega^2 r^2 + V_{l,s}(r) \tag{2.13}$$

where $V_{l,s}(r) = -f(r)(\vec{l} \cdot \vec{s})$ and $f(r)$ has Woods and Saxon form.

The spin–orbit interaction causes a splitting of the energy levels and consequently changes the order of the energy levels which were obtained without the spin–orbit interaction term in the assumed central potential.

EXERCISE 2.6: What is the splitting in the energy levels due to spin–orbit coupling?

Solution: The intrinsic spin (s) and the orbital angular momentum (l) can couple to give the total nucleon angular momentum, $\vec{j} = (\vec{l} + \vec{s})$. From the cosine law, the spin–orbit term can be rewritten in terms of the total angular momentum as follows:

$$\vec{l}.\vec{s} = \frac{1}{2}(j^2 - l^2 - s^2) \tag{2.14}$$

where j^2 is the total angular momentum operator whose eigen values are given by $j(j + 1)\hbar^2$. Similarly, l^2 is the orbital angular momentum operator whose eigen values are given by $l(l + 1)\hbar^2$ and s^2 is the spin angular momentum operator whose eigen values are given by $s(s + 1)\hbar^2$.

The total angular momentum j depends on the relative orientation (parallel and anti-parallel) of l and s. The expectation value of $\vec{l} \cdot \vec{s}$ in a single particle state with the total angular momentum $\vec{j}$ is as follows:

Stretch Case: $s = 1/2,\ j = l + 1/2$

$$\langle \vec{l} \cdot \vec{s} \rangle = \frac{j(j+1) - l(l+1) - \frac{1}{2}\left(\frac{1}{2}+1\right)}{2}\hbar^2$$

$$= \frac{\left(l+\frac{1}{2}\right)\left(l+\frac{3}{4}\right) - l(l+1) - \frac{3}{4}}{2}\hbar^2$$

or

$$\langle \vec{l}.\vec{s} \rangle = +\frac{\hbar^2}{2}(l) \tag{2.15}$$

Jackknife Case: $s = 1/2,\ j = l - 1/2$

$$\langle \vec{l} \cdot \vec{s} \rangle = \frac{\left(l-\frac{1}{2}\right)\left(l+\frac{1}{2}\right) - l(l+1) - \frac{1}{2}\left(\frac{1}{2}+1\right)}{2}\hbar^2$$

or

$$\langle \vec{l} \cdot \vec{s} \rangle = -\frac{\hbar^2}{2}(l+1) \tag{2.16}$$

It can be seen, each level with l splits up into two levels having different energies. From equations (2.15) and (2.16), the splitting in the energy levels due to spin–orbit coupling is given by:

$$\Delta E_{\vec{l}\cdot\vec{s}} \frac{\hbar^2}{2}(2l+1)\cdot f(r)$$

Larger the l more is the splitting. $f(r)$ is determined by comparison with the experimental data.

Note:

(i) For $l = 0$ or s-state, $j = 1/2$ and there is no splitting of the energy level ($s_{1/2}$). All other states are split, e.g., p-state splits into $j = 3/2$ and $j = 1/2$, i.e., $p_{3/2}$ and $p_{1/2}$.

(ii) Originally the introduction of the spin–orbit interaction term was phenomenological in order to predict the magic numbers but now it is believed to arise from the tensor force of the nucleon-nucleon interaction.

The inclusion of $(\vec{l}\cdot\vec{s})$ potential by Mayer and Jensen modifies the energy levels as shown in Figure 2.13, and reproduced all the magic numbers. This success was recognized with the award of Nobel Prize[3] (Physics) in 1963 to Mayer and Jensen together with Eugene Wigner.

2.2.3 Nuclear Ground State Configuration and Spin-parity

On the basis of the nuclear shell model, it is possible to write down the nuclear ground state configuration of a given nuclide by filling the neutrons and protons separately in accordance with the Pauli's exclusion principle in the shells as shown in Figure 2.13, beginning from the bottom of the well till all the protons and neutrons are accommodated, and determine its spin-parity. In view of the charge independence of the nuclear force, levels for neutrons are similar to that of protons except for the effect of Coulomb interaction, which makes the depth of the potential well less for protons than neutrons.

An energy level can accommodate $(2j + 1)$ nucleons. Since, the pairing interaction is very strong between the nucleons and they cancel each other's spin, the nuclear spin-parity is represented as J^π or I^π, corresponds to the total angular momentum of the valence nucleon, which is the last unpaired nucleon left outside the closed shell. The parity of a nucleon in the given orbital is given by $(-1)^l$, where l is the orbital quantum number.

3. "Nobel Prize in Physics 1963—Presentation Speech" (<http://www.nobelprize.org/nobel_prizes/physics/laureates/1963/press.html>)

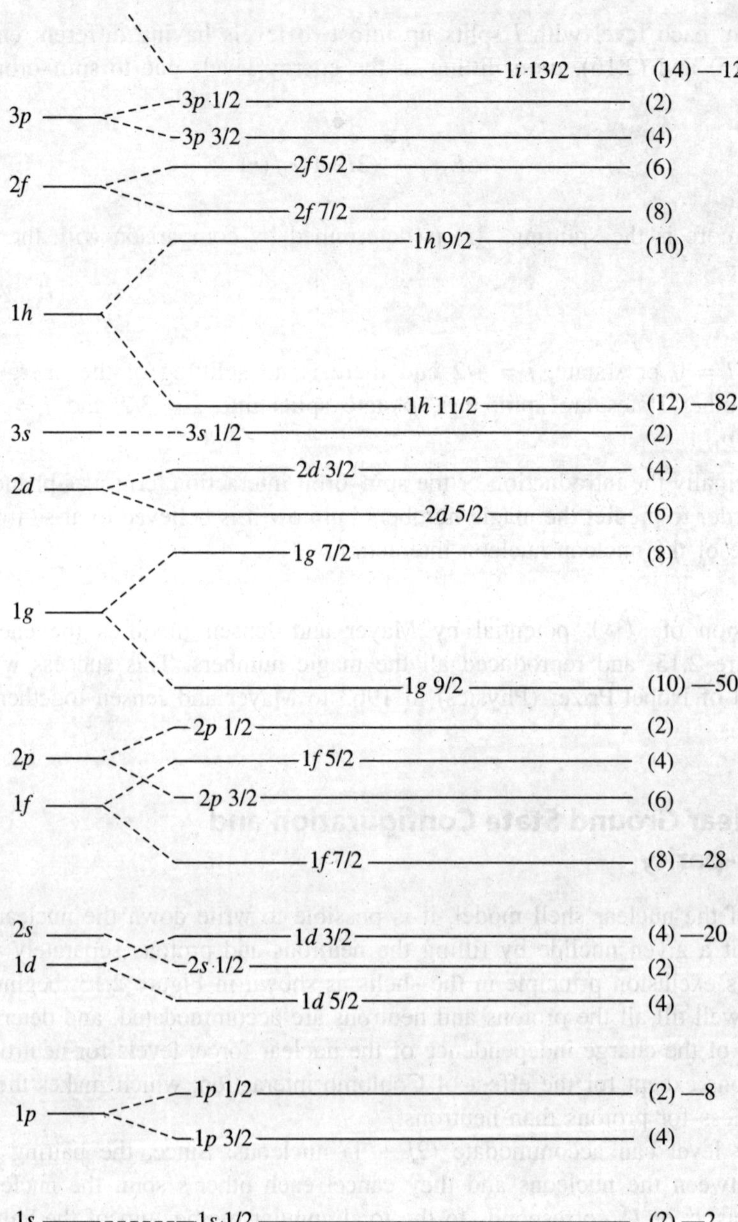

FIGURE 2.13 The splitting of the energy levels of 3-dimensional harmonic oscillator due to spin–orbit coupling. The numbers in the brackets are number of nucleons that each shell can accommodate; the numbers on the right are the major shell closures.

EXERCISE 2.7:

(a) What is the spin-parity for an even–even nuclide?

(b) Determine the spin-parity for the following odd nuclei (with odd Z and even N or vice versa)

$$^{17}_{8}\text{O}_9,\ ^{63}_{29}\text{Cu}_{34} \text{ and } ^{75}_{33}\text{As}_{42}.$$

Solution:

(a) All nuclides with filled j-shells have spin zero. In absence of any unpaired nucleons, all even–even nuclei have ground state $I^\pi = 0^+$, with no exceptions to this rule.

(b) **Case 1:** $^{17}_{8}\text{O}_9$ (Here the neutron number is odd)
Eight protons and 8 neutrons are paired up, so the configuration of the 9th neutron can be written as follows using Figure 2.13.

$$(1s_{1/2})^2\Big|(1p_{3/2})^4(1p_{1/2})^2\Big|(1d_{5/2})^1$$

Here, the vertical line (|) shows that the shell is completely filled. The number of nucleons in each state is shown by the superscript number above that state.

The last neutron has $d_{5/2}$ configuration, therefore, $I = 5/2$ and parity $(-1)^2$ = positive (even parity). So, the ground state of $^{17}_{8}\text{O}_9$ nucleus is $I^\pi = 5/2^+$.

Case 2: $^{63}_{29}\text{Cu}_{34}$ (Here the proton number is odd)
The configuration of the 29th proton can be written as follows:

$$(1s_{1/2})^2\Big|(1p_{3/2})^4(1p_{1/2})^2\Big|(1d_{5/2})^6(2s_{1/2})^2(1d_{3/2})^4\,|\,(1f_{7/2})^8(2p_{3/2})^1$$

The last proton has $p_{3/2}$ configuration, therefore, $I = 3/2$ and parity $(-1)^1$ = negative (odd parity). So, the ground state of $^{63}_{29}\text{Cu}_{34}$ nucleus is $I^\pi = 3/2^-$.

Case 3: $^{75}_{33}\text{As}_{42}$ (Here the proton number is odd)
The configuration of the 33rd proton can be written as follows:

$$(1s_{1/2})^2\Big|(1p_{3/2})^4(1p_{1/2})^2\Big|(1d_{5/2})^6(2s_{1/2})^2(1d_{3/2})^4\,|\,(1f_{7/2})^8(2p_{3/2})^4(1f_{5/2})^1$$

Therefore, $I = 5/2$ and parity = $(-1)^3$ = negative (odd parity). So, the ground state of $^{75}_{33}\text{As}_{42}$ nucleus is $I^\pi = 5/2^-$.

But experimentally, it is found to be $3/2^-$. Hence, the following modified rule has to be applied.

If high spin shell comes after low spin shell, the high spin shell fills faster, pairing its nucleons, before the low spin shell gets filled completely.

Applying the aforementioned rule, the final configuration is:

$$(1s_{1/2})^2\Big|(1p_{3/2})^4(1p_{1/2})^2\Big|(1d_{5/2})^6(2s_{1/2})^2(1d_{3/2})^4\,|\,(1f_{7/2})^8(2p_{3/2})^3(1f_{5/2})^2$$

Therefore, $I = 3/2$ and parity = $(-1)^1$ = negative (odd parity). So, the ground state of $^{75}_{33}\text{As}_{42}$ nucleus is $I^\pi = 3/2^-$.

2.2.4 Nuclear Moment (μ)

The nuclear magnetic dipole moment is also due to the valence nucleon. For even–even nuclide, the magnetic moment, just like spin, is zero. For odd-proton and odd-neutron nuclides, Schmidt derived the limiting values of magnetic moment μ in nuclear magnetons (μ_N), which are given in Table 2.8.

Table 2.8 Schmidt's Predictions for Magnetic Moment

Nuclide	$I = l + s$	$I = l - s$
Odd proton	$\mu = I + 2.293$	$\mu = I - 2.293\left(\frac{I}{I+1}\right)$
Odd neutron	$\mu = -1.913$	$\mu = 1.913\left(\frac{I}{I+1}\right)$

EXERCISE 2.8:

(a) What are the predictions for magnetic moment values for the nuclides given in the previous exercise?

(b) Determine the spin-parity and magnetic dipole moment for ${}^{16}_{7}N_9$.

(c) Predict the spin-parity and magnetic dipole moment of ${}^{6}_{3}Li_3$, given that $g\left({}^{4}_{2}He\right) = 0$ and $g\left({}^{2}_{1}H\right) = 0.86$.

(d) The electric quadrupole moment (Q) of a nuclide arises because of the non-spherical charge distribution. The single-particle quadrupole moment of an odd proton nuclide in a shell state with spin I, using quantum mechanical calculation is given as follows:

$$Q_{sp} = -\frac{2I-1}{2(I+1)}\langle r^2 \rangle$$

For uniform density spherical nucleus, the mean square radius $\langle r^2 \rangle = \frac{3}{5}(1.2A^{1/3})^2 \text{ fm}^2$.

In case of an odd uncharged neutron, there should not be any quadrupole moment. However, relatively smaller values of the quadrupole moment are found experimentally. It is believed to arise due to the polarization of the neutron relative to the nucleus. The value of quadrupole moment of an odd neutron nuclide is given as follows:

$$Q_n \approx \frac{Z}{(A-1)^2} Q_{sp}$$

Now, estimate the quadrupole moment for ${}^{63}_{29}Cu_{34}$ nuclide with an odd proton.

(e) When a subshell contains more than a single nucleon, all of the nucleons in the subshell contribute to the quadrupole moment. For (n) nucleons in an unfilled subshell, the corresponding quadrupole moment is given by:

$$Q_h = \left(1 - 2\frac{n-1}{2j-1}\right)Q_{sp}$$

Now, derive a relation to estimate the quadrupole moment for nuclides with a single hole. (A hole is defined as the number of missing nucleons corresponding to the filled subshell.)

(f) For certain nuclei in the rare earth regions, the quadrupole moments are very large. Similarly, certain medium and light-weight nuclei have large quadrupole moments. Quadrupole moment for ${}^{167}_{68}Er_{99}$ is found to be about 3.56 barns.
Compare the calculated value of the quadrupole moment for ${}^{167}_{68}Er_{99}$ nuclide with its measured value.

Solution:

(a) **Case 1:** ${}^{17}_{8}O_9$ (odd neutron)
The last neutron has $d_{5/2}$ configuration, therefore, I = 5/2 and l = 2. Using Table 2.8, the nuclear moment $\mu = -1.913\ \mu_N$.

Case 2: ${}^{63}_{29}Cu_{34}$ (odd proton)
The last proton has $p_{3/2}$ configuration, therefore, I = 3/2 and l = 1. Using Table 2.8, the nuclear moment μ = 3/2 + 2.293 = 3.793 μ_N.

Case 3: ${}^{75}_{33}As_{42}$ (odd proton)
The last proton has $p_{3/2}$ configuration, therefore, I = 3/2 and l = 1. Using Table 2.8, the nuclear moment μ = 3/2 + 2.293 = 3.793 μ_N.

(b) ${}^{16}_{7}N_9$ is a case of odd–odd nuclide. In such cases, we first find the orbitals of unpaired proton and neutron.
The last proton has $p_{1/2}$ configuration.

$$\therefore \qquad j_p = 1/2 \text{ and } l_p = 1$$

The last neutron has $d_{5/2}$ configuration

$$\therefore \qquad j_n = 5/2 \text{ and } l_n = 2$$

Now, the following *Nordheim's rule* is used to determine the value of nuclear spin.

Strong rule: $I = |j_p - j_n| \quad \text{if}\{(j_p - l_p) + (j_n - l_n)\} = 0$

Weak rule: $I = (j_p + j_n) \quad \text{if}\{(j_p - l_p) + (j_n - l_n)\} = \pm 1$

Here, $\{(j_p - l_p) + (j_n - l_n)\} = \{(1/2 - 1) + (5/2 - 2)\} = 0$. Therefore, the strong rule will be applicable and $I = |j_p - j_n| = |1/2 - 5/2| = 2$.

The parity for the odd–odd nuclide case is given by $(-1)^{l_p + l_n}$

$$= (-1)^{1+2} = \text{negative (odd parity)}$$

Therefore, the ground state of ${}^{16}_{7}N_9$ nucleus is $I^\pi = 2^-$.
The nuclear moment for the odd–odd nuclide is given by $\mu = \mu_n + \mu_p$.

For proton,

$$\mu_p = I - 2.293\left(\frac{j_p}{j_p + 1}\right) = 1/2 - 2.293\left(\frac{1/2}{1/2 + 1}\right) = -\,0.263\,\mu_N$$

For neutron,

$$\mu_n = -1.913\ \mu_N$$

$$\therefore \qquad \mu \text{ for } {}^{16}_{7}\text{N}_9 = -\,0.263 - 1.913 = -\,2.176\,\mu_N$$

(c) ${}^{6}_{3}\text{Li}_3$ nuclide can be treated as the combination of ${}^{4}_{2}\text{He}_2$ and ${}^{2}_{1}\text{H}_1$

$$I_1^{\pi}({}^{4}_{2}\text{He}_2) = 0^{+} \text{ and } I_2^{\pi}({}^{2}_{1}\text{H}_1) = 1^{+}$$

Parity is multiplicative while angular momentum is additive,

$$\therefore \qquad I^{\pi}({}^{6}_{3}\text{Li}_3) = 1^{+}$$

The nuclear magnetic moment is given by $\vec{\mu} = \mu_N g \vec{I}$, where g is the Lande factor. When two particles of Lande factors g_1 and g_2 combine into a new particle of Lande factor g (assuming the orbital angular momentum of relative motion is zero), then

$$g = g_1 \frac{I(I+1) + I_1(I_1+1) - I_2(I_2+1)}{2I(I+1)} + g_2 \frac{I(I+1) + I_2(I_2+1) - I_1(I_1+1)}{2I(I+1)}$$

where I is the spin of the new particle, I_1 and I_2 are the spins of the constituent particles.

or
$$g = g_1 \frac{1(1+1) + 0 - 1(1+1)}{2(1+1)} + g_2 \frac{1(1+1) + 1(1+1) - 0}{2(1+1)}$$

or
$$g = g_2 = 0.86$$

$$\therefore \qquad \mu({}^{6}_{3}\text{Li}_3) = \mu_N g({}^{6}_{3}\text{Li}_3) \times 1 = 0.86\,\mu_N$$

(d) The ground state of ${}^{63}_{29}\text{Cu}_{34}$ nucleus is $I^{\pi} = 3/2^{-}$

$$Q = -\frac{2(3/2) - 1}{2(3/2 + 1)} \times \frac{3}{5}(1.2 \times 63^{1/3})^2 = -5.47 \text{ fm}^2 = -\,0.0547 \text{ b}$$

(e) Since, the capacity of a subshell is $(2j + 1)$, the number of nucleons in an unfilled subshell will range from 1 to $2j$. For a single hole, $n = 2j$, as there is only one missing nucleon corresponding to the filled subshell.

$$\therefore \qquad Q_h = \left(1 - 2\frac{2j-1}{2j-1}\right) Q_{sp} = -\,Q_{sp}$$

Therefore, for a single hole state, only the sign becomes opposite vis-a-vis single particle quadrupole moment.

(f) In $^{167}_{68}\text{Er}_{99}$ nuclide, using Figure 2.13, the single neutron hole corresponds to $2f_{7/2}$ level.

$$Q_h \approx -\frac{Z}{(A-1)^2} Q_{sp}$$

or
$$Q_h \approx -\frac{68}{(167-1)^2}\left(-\frac{2\left(\frac{7}{2}\right)-1}{2\left(\frac{7}{2}+1\right)} \times \frac{3}{5}(1.2 \times 167^{1/3})^2\right)$$

$$= 0.0431 \text{ fm}^2 = 4.31 \times 10^{-4} \text{ b}$$

As can be seen clearly in this example, the experimental value is several orders of magnitude larger than predicted by the shell model. This is just one of the numerous cases, where the shell model fails miserably to correctly predict the quadrupole moments. Such failures of the shell model gave insight into other aspects of the nuclear structure, which will be discussed later in this chapter.

2.2.5 Nuclear Isomerism (Islands of Isomeric State)

The shell closures at 28, 50, 82 and 126 correspond to $1f_{7/2}$, $1g_{9/2}$, $1h_{11/2}$, $1i_{13/2}$. Long lived nuclear states, known as isomers, occur most frequently in those regions of Z or N which are just below the magic numbers.

$19 \le N$ or $Z \le 27$	$1f_{7/2}$ and $1d_{3/2}$ levels
$39 \le N$ or $Z \le 49$	$1g_{9/2}$ and $2p_{1/2}$ levels
$69 \le N$ or $Z \le 81$	$1h_{11/2}$ and $3s_{1/2}$ levels
$111 \le N$ or $Z \le 125$	$1i_{13/2}$ and $3p_{1/2}$ levels

The energy gap between $j = (l \pm 1/2)$ states increases with l values. A direct consequence of this is close spaced energy levels of significantly different spin values just before the shell closure configuration. In such cases, for nuclei having odd number of nucleons, the last nucleon can occupy either the expected low spin state or high spin state of slightly higher energy. De-excitation from the isomeric state to the ground state, which results in the emission of γ-rays, is hindered because of large angular momentum carried by γ-ray. For example, $^{95m}_{41}\text{Nb}$ nuclide has I^π (ground state) = $9/2^+$ and I^π (isomeric state) = $1/2^-$; therefore, γ-rays will carry large angular momentum ($L = 4$) and the probability of this transition will be small. Detail of transition rates for γ-decay is discussed in Chapter 3.

2.2.6 Predictions and Failures of the Shell Model

Shell model explains the extra stability of magic nuclei. It predicts the ground state I^π all even–even nuclei and most of the odd-A nuclei. It also successfully predicts I^π of odd–odd nuclei. The qualitative features of magnetic dipole and electric quadrupole moments of various nuclei can also be explained using shell model.

Shell model fails to explain ground state I^π for certain nuclei (e.g., for ${}^{6}_{3}Li_3$ nuclide, the predicted $I^\pi = 3^+$, while experimental $I^\pi = 1^+$) and is unable to explain I^π and energy of the first excited states in even–even nuclei. The model also fails to predict quadrupole moments of many nuclei and is unable to predict magnetic moments of some nuclei.

EXERCISE 2.9:

(a) The level scheme of ${}^{17}_{8}O_9$ is shown in Figure 2.14(a). Predict I^π for all the levels based on the shell model.

(b) The experimental energy levels of ${}^{38}_{18}Ar_{20}$ are shown in Figure 2.14(b). Can you predict I^π for all the levels based on the shell model?

(c) The total binding energy of ${}^{15}_{8}O$, ${}^{16}_{8}O$ and ${}^{17}_{8}O$ are 111.96 MeV, 127.62 MeV and 131.76 MeV, respectively. Determine the energy gap between $1p_{1/2}$ and $1d_{5/2}$ neutron shells for the nuclide whose mass number is close to 16.

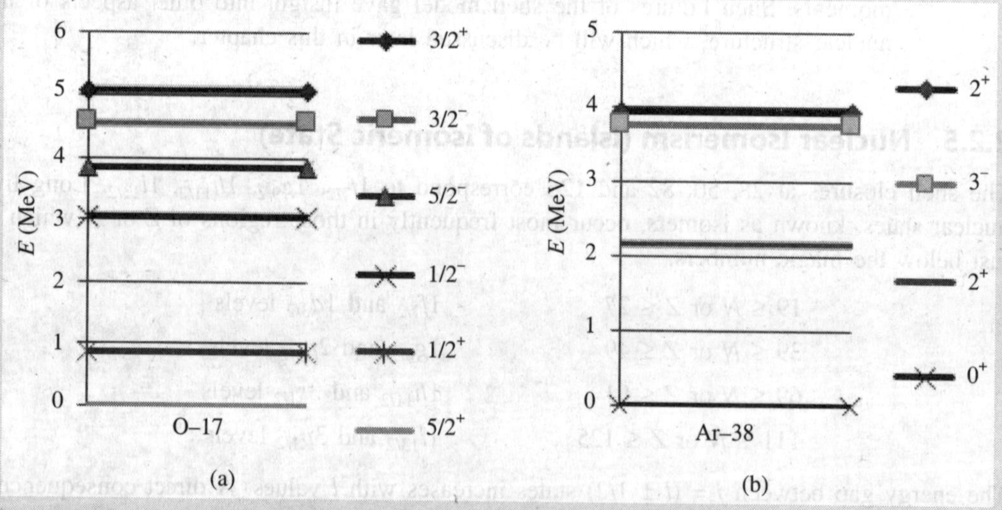

FIGURE 2.14 Energy levels of ${}^{17}_{8}O_9$ and ${}^{38}_{18}Ar_{20}$.

Solution:

(a) The 9th neutron of ${}^{17}_{8}O_9$ nuclide is in $1d_{5/2}$, therefore, the ground state $I^\pi = 5/2^+$. From Figure 2.13, in excited state, the last neutron jumps to a higher state, i.e., $2s_{1/2}$, so the excited state $I^\pi = 1/2^+$. If the last neutron jumps to $1d_{3/2}$, the excited state $I^\pi = 3/2^-$. Deviating from the extreme single particle model which says that everything is decided by the last unpaired nucleon, if the pair in $1p_{1/2}$ state is broken and a neutron jumps to $1d_{5/2}$ state, the excited state $I^\pi = 1/2^-$. Also, considering that breaking a pair needs ~2 MeV energy, the excited state $I^\pi = 1/2^-$ can be explained.

However, there is no simple way to explain the $I^\pi = 5/2^-$ state. To get $I^\pi = 5/2^-$ state, the neutron will have to skip many levels and jump to $1f_{5/2}$, a much higher energy level, which is not reasonable as the energy of $I^\pi = 5/2^-$ state is close to that of $I^\pi = 1/2^-$. The $I^\pi = 3/2^+$ state can be explained if the neutron jumps to $1d_{3/2}$ state, and considering that $I^\pi = 5/2^-$ state is much below that, the neutron jumping to $1f_{5/2}$ is not feasible. If there are 3 unpaired neutrons in $1p_{1/2}$, $1d_{5/2}$ and $2s_{1/2}$ and if their $j(s)$ are combined, then one can see that $I^\pi = 5/2^-$ state can be got.

(b) ${}^{38}_{18}Ar_{20}$ is a case of even–even nuclei. The ground state $I^\pi = 0^+$. The neutrons form a closed shell, while pair of protons lie in the $1d_{3/2}$ state. If a proton jumps to $1f_{7/2}$, then two unpaired protons can combine to give $I^\pi = 3^-$. Taking into consideration the energies, the higher 2^+ state can be explained if the proton pair in $2s_{1/2}$ is broken and a proton jumps to the $1d_{3/2}$ state. But, using the shell model, there is no way, we can explain the first excited 2^+ state.

(c) The ordering to the nuclear energy levels according to the shell model is:

$$(1s_{1/2})^2\left|(1p_{3/2})^4(1p_{1/2})^2\right|(1d_{5/2})^6$$

As we can see, the number of neutrons rises from 7 to 9 in the oxygen isotopes given in the question. Therefore, the B.E. of a neutron in the $1p_{1/2}$ level

$$= 127.62 - 111.96 = 15.66 \text{ MeV}$$

and the B.E. of a neutron in the $1d_{5/2}$ level

$$= 131.76 - 127.62 = 4.14 \text{ MeV}$$

Thus, the $(1p_{1/2} - 1d_{5/2})$ energy difference

$$= 15.66 - 4.14 = 11.52 \text{ MeV}$$

2.3 COLLECTIVE MODEL

The shell model is based on the idea of constituent nucleons inside the nucleus behaving independent of each other. LDM, on the other hand assumes the opposite, that the constituent particles of the nucleus are indeed influenced by their neighbours. Both these models have limited domain of applicability and are not able to explain all the observations. The single particle shell model explains many general features of the nuclear structure, hence, needs to be modified and reconciled with LDM, thus returning to the picture of nucleus seen as a collective body, in order to explain the large quadrupole moments. This is the basis for the collective model, where it is assumed that the motion of the nucleons on the surface of the nucleus leads to deformation of the potential from strictly spherical form used in the shell model, which in turn alters the spacing of the energy levels depending on the magnitude of the distortion. Like the shell model, collective model too is a fully quantum mechanical model.

Two kinds of collective motion may be visualized—quantized rotations and vibrations. The even–even nuclei with $A < 150$ usually exhibit the vibrational spectra, while the rotational spectra, which occurs at relatively lower energies, is more common in nuclei with $150 < A < 190$ and $A > 230$.

Figure 2.15 shows the excitation energies of the first excited state in even–even nuclei as a function of A. One can see that there are low lying 2^+ levels in all even-even nuclei. The level energies generally decrease slowly with Z except for sharp irregularities at closed and nearly closed shell nuclei. For nuclei with A ranging from ~30 to 150, $E(2^+)$ is of the order of 1 MeV. In the mass regions 150–190 and above 230, the value of $E(2^+)$ is much lower and changes little with A. The shell model fails to explain the 2^+ levels. These states are due to collective motion of many nucleons. Below $A \sim 150$, the motion is mainly vibrational in character. In mass regions, $150 \leq A \leq 190$ and $A \geq 230$, it is rotational.

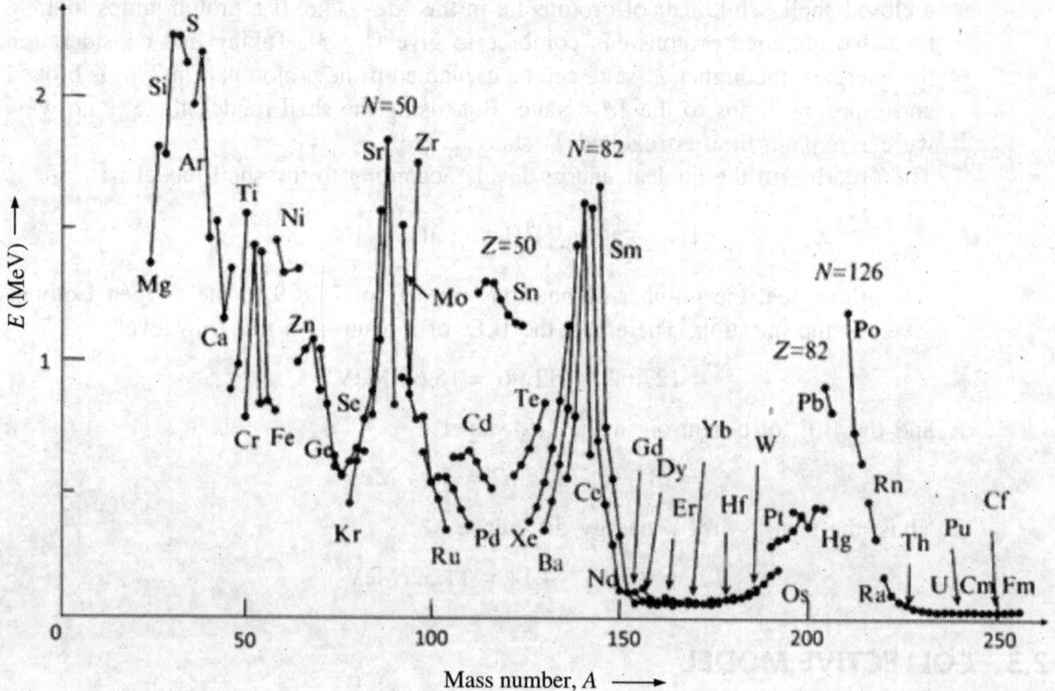

FIGURE 2.15 Excitation energies of lowest 2^+ states of even–even nuclei. Isotopic sequences have been shown by solid lines.

2.3.1 Vibrational Model

In this model, the nucleus is considered to be a liquid drop with surface oscillations. The basic assumptions of the model are as follows:

The liquid drop has a well-defined surface with radius

$$R(\theta, \varphi, t) = \bar{R}\left\{1 + \sum_{\lambda=0}^{\infty} \sum_{\mu=-\lambda}^{\lambda} \alpha_{\lambda,\mu}(t) Y_{\lambda}^{\mu}(\theta, \varphi)\right\} = \bar{R} + \Delta R \tag{2.17}$$

where $\bar{R}$ is the radius of a sphere which has the same volume as the deformed nucleus, $Y_{\lambda}^{\mu}(\theta, \varphi)$ are the spherical harmonics and form a basis for (θ, φ) space, $\alpha_{\lambda,\mu}$ are time dependent deformation parameters which determine the nuclear shape. (l and m subscripts are avoided here because these are used in the shell model orbitals).

(i) The nuclear deformation is small, i.e., $\overline{R} >> \Delta R$.
(ii) The liquid drop is incompressible, i.e., the volume of the droplet does not change during oscillations.
(iii) The oscillations are slow, i.e., $\omega_{\text{osc}} << \omega_{\text{particle}}$.

In equation (2.17), for $\lambda = 0$, we get the spherical shape. The $\lambda = 1$ mode corresponds to dipole vibrations, resulting in the shifting of the C.M. of the nucleus, which is not possible due to internal nuclear forces. Thus, the first non-vanishing $\lambda = 2$, corresponds to quadrupole mode of deformation with $\mu = -2, -1, 0, +1, +2$, representing ellipsoidal shapes (Figure 2.16). The $\lambda = 3$ mode corresponds to octupole vibrations.

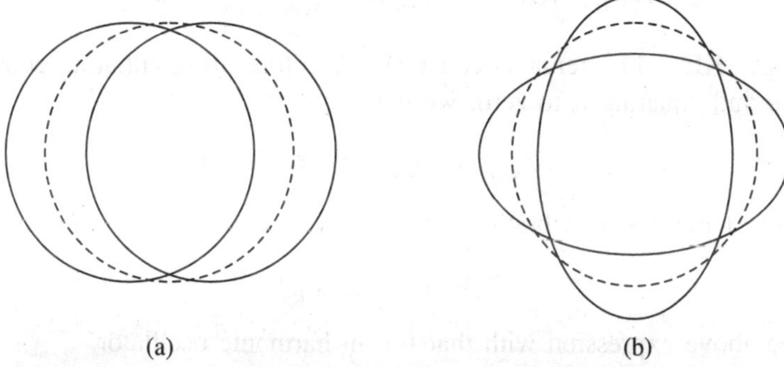

FIGURE 2.16 (a) Dipole vibration; (b) quadrupole vibration.

$\alpha_{\lambda,\mu}$ represents the state of deformation of the nuclear surface and gives rise to the total energy of the deformed nucleus as a function of time. The variation of $\alpha_{\lambda,\mu}$ with time gives the KE of the vibration.

$$\text{KE} = \frac{1}{2}\sum_{\lambda,\mu} B_\lambda \left|\dot{\alpha}_{\lambda,\mu}\right|^2$$

where B_λ corresponds to the moment of inertia of the nucleus with respect to the changes in deformation and can be calculated as follows:

$$B_\lambda = \frac{\rho \overline{R}^5}{\lambda}$$

(ρ is the density of the nuclear matter).

The displacement from the equilibrium position gives the PE of the vibration

$$\text{PE} = \frac{1}{2}\sum_{\lambda,\mu} C_\lambda \left|\alpha_{\lambda,\mu}\right|^2$$

where C_λ are the deformation coefficients which measure the resistance of the nucleus against deformation. The value of C_λ coefficient depends on the surface tension and Coulomb energy, and is given by Bohr and Wheeler as follows (S is the surface tension):

$$C_\lambda = S\bar{R}^2(\lambda-1)(\lambda+2) - \frac{3}{2\pi}\frac{(Ze)^2}{\bar{R}}\left(\frac{\lambda-1}{2\lambda+1}\right)$$

The total Hamiltonian H is given by

$$H = E_0 + \sum_{\lambda,\mu} H_{\lambda,\mu}$$

where E_0 is the energy of the nucleus for a spherically symmetric shape and

$$H_{\lambda,\mu} = \text{KE} + \text{PE} = \frac{1}{2}B_\lambda\left|\dot{\alpha}_{\lambda,\mu}\right|^2 + \frac{1}{2}C_\lambda\left|\alpha_{\lambda,\mu}\right|^2 \tag{2.18}$$

The total energy (KE + PE) remains constant. Therefore, differentiating equation (2.18) with respect to time and equating it to zero, we get

$$\dot{\alpha}_{\lambda,\mu}(B_\lambda\ddot{\alpha}_{\lambda,\mu} + C_\lambda\alpha_{\lambda,\mu}) = 0$$

If $\dot{\alpha}_{\lambda,\mu}$ is zero, it implies no vibrations and therefore this solution is not of our interest, i.e.,

$$(B_\lambda\ddot{\alpha}_{\lambda,\mu} + C_\lambda\alpha_{\lambda,\mu}) = 0$$

Comparing the above expression with that for an harmonic oscillator,

$$\ddot{x} + \omega^2 x = 0$$

for small oscillations, we get the frequency of harmonic oscillation for amplitude $\alpha_{\lambda,\mu}$, as:

$$\omega_\lambda = \left(\frac{C_\lambda}{B_\lambda}\right)^{1/2}$$

Substituting the values of C_λ and B_λ, we get

$$\omega_\lambda = \sqrt{\frac{\lambda(\lambda-1)}{\rho\bar{R}^3}\left[S(\lambda+2) - \frac{3}{2\pi}\frac{Z^2e^2}{\bar{R}^3}\frac{}{(2\lambda+1)}\right]}$$

For all nuclei found in nature, the first term (surface energy) is always greater than the second term (Coulomb energy), otherwise, spontaneous fission will take place. Thus, neglecting the Coulomb energy term, we get

$$\omega_\lambda = \sqrt{\frac{S(\lambda-1)(\lambda+2)\lambda}{\rho\bar{R}^3}} \tag{2.19}$$

The energy eigen values of the harmonic oscillations are given by:

$$E = E_0 + \sum_{\lambda,\mu}\left(n_{\lambda,\mu} + \frac{1}{2}\right)\hbar\omega_\lambda \tag{2.20}$$

where $n_{\lambda,\mu}$ is the number of oscillations or phonons[4] in λ, μ mode of oscillation and $\hbar\omega_\lambda$ is a quantum of vibrational energy for multipole λ.

EXERCISE 2.10:

(a) $\lambda = 2$ is the quadrupole mode of vibration, which is the dominant mode to be considered for low energy excitations of even–even nuclei. Hence, the one-phonon state for even–even nuclei is 2^+. Now, show that only states with even *J*-values are allowed, when two identical quadrupole phonons are coupled together.

(b) What are the allowed states when three identical quadrupole phonons are coupled together?

(c) Explain the 3^- state using vibrational model.

(d) Sketch the schematic vibrational energy-level diagram for even-even nuclei.

(e) What is roughly the energy of 3^- state (one-octupole phonon state) with respect to one-quadrupole phonon state?

Solution:

(a) Two quadrupole phonons, i.e., $\lambda_1 = 2$ and $\lambda_2 = 2$. The allowed values of μ_1 and μ_2 are –2, –1, 0, 1, 2. Since, both the phonons are identical, the angular momentum coupling method should be based on $\mu = \mu_1 + \mu_2$ (magnetic projection quantum number).

In Table 2.9, the allowed values of μ are populated. '×' sign indicates that a particular set of values (μ_1, μ_2) has already been counted, for example, (2, 1) set is same as (1, 2) set because the phonons are identical, and hence the sets should be counted only once. Using the table, $J^\pi = 4^+$, 2^+ and 0^+ states corresponding to [4 3 2 1 0 –1 –2 –3 –4], [2 1 0 –1 –2] and [0] sets of magnetic projection quantum numbers are allowed.

Table 2.9 Allowed states for two quadrupole phonons.

μ_2 \ μ_1	2	1	0	–1	–2
2	4	×	×	×	×
1	3	2	×	×	×
0	2	1	0	×	×
–1	1	0	–1	–2	×
–2	0	–1	–2	–3	–4

4. A phonon is the vibrational quantum of energy. Each phonon is a boson (any of a class of elementary particles not subject to the exclusion principle that have spins of zero or an integral number, such as photons) carrying $\lambda\hbar$ units of angular momentum and having a parity of $(-1)^\lambda$. For even–even nuclei, the ground state (0^+) may be regarded as the zero-phonon state. The lowest vibrational state built upon the ground state has angular momentum quantum number of λ and parity of $(-1)^\lambda$.

(b) For coupling of three quadrupole phonons, there will be many more cases. Hence, rather than counting manually, the following Matlab code is used which employs the same method as in the previous part, to get the allowed states.

```
clear all

 j = 2; %Angular momentum of a single particle
 m1 = j; m2 = j; m3 = j; %Magnetic quantum numbers

 i=1;
 while m1>=-j %Loop to get the allowed values of Mj
   Mu(i) = m1+m2+m3; %Mu = magnetic projection quantum number
   i = i+1;
   if m3>-j
     m3 = m3 - 1;
   elseif m2>-j
     m2 = m2 -1;
     m3 = m2;
   else
     m1 = m1 - 1;
     m2 = m1; m3 = m1;
   end
 end
 Mu

 i=1;
 while max(Mu)~=-100 %Loop to get allowed J+ states from Mu array
     [J(i) idx] = max(Mu);
     Mu(idx)=-100;

     x = J(i);
     while x>-J(i)
         x = x-1;
         idx = find(Mu == x);
         Mu(idx(1)) = -100;
     end

     i = i+1;
 end
 J
```

Thus, $J^\pi = 6^+, 4^+, 3^+, 2^+$ and 0^+ states are allowed.

(c) $\lambda = 3$ is the octupole mode of vibration. Hence, the first excited state with one octupole phonon is 3^-. 3^- states are commonly found in vibrational nuclei at an energy above the two-quadrupole phonon states.

(d) Figure 2.17 is the schematic diagram of low-lying vibrational states for even–even nuclei.

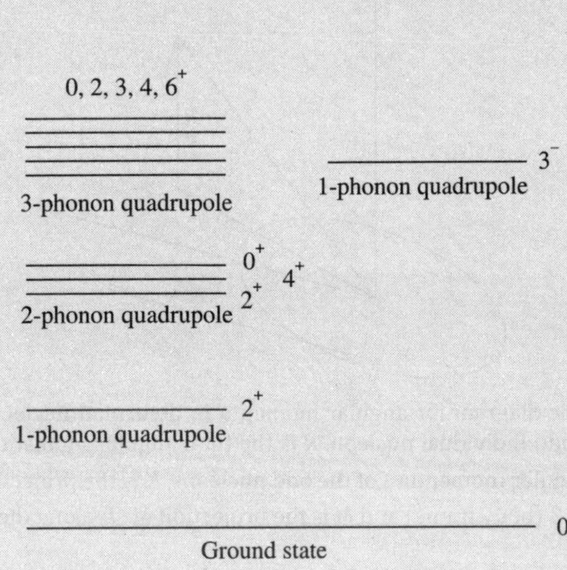

0, 2, 3, 4, 6^+

3-phonon quadrupole

3^-

1-phonon quadrupole

0^+ 4^+ 2^+

2-phonon quadrupole

2^+

1-phonon quadrupole

0^+

Ground state

FIGURE 2.17 Low-lying vibrational states of even–even nuclei.

(e) Energy of one-quadrupole phonon state is $\hbar\omega_2$ and the energy of one-octupole phonon state is $\hbar\omega_3$. Using equation (2.19), we get

$$\hbar\omega_3 = \left(\sqrt{\frac{(3-1)(3+2)3}{(2-1)(2+2)2}}\right)\hbar\omega_2$$

$$= \left(\frac{5.47}{2.82}\right)\hbar\omega_2 \approx 2\hbar\omega_2$$

2.3.2 Rotational Model

So far we have assumed that the basic shape of a nucleus is spherical and small surface oscillations lead to the vibrational spectra over the single particle states. The interplay among short range nuclear interaction, long range Coulomb interaction and the centrifugal force acting on a rotating nucleus may also result in a deformed equilibrium shape of the nucleus.Regions of relatively large deformation exist, e.g., in medium heavy nuclei with $150 < A < 190$ and heavy nuclei with $A > 230$, where rotational bands have been observed experimentally.

Let us consider the collective rotational motion of deformed nuclei which have axial symmetry. Figure 2.18 shows two sets of orthogonal systems of axes: (i) body-fixed reference frame: 1, 2, 3 axes; and (ii) laboratory-fixed reference frame: x, y, z axes. In the first system, the body-fixed reference frame is attached to the rotating body and 3-axis is used as the axis of symmetry.

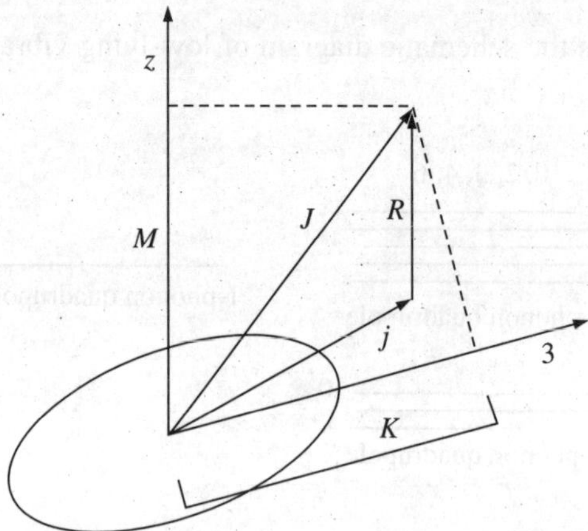

FIGURE 2.18 Schematic diagram for angular momenta in deformed nuclei. Here j is the total angular momentum of an odd individual nucleon, R is the total angular momentum of even–even core, $\vec{J} = \vec{j} + \vec{R}$ is the total angular momentum of the odd nucleons, K is the projection (component) of $\vec{J}$ along the symmetry axis 3 (body frame) and M is the projection of $\vec{J}$ along the z-axis (lab frame).

Classically, the rotational energy E_J is given as (I is the moment of inertia and J is the angular momentum)

$$E_J = \frac{1}{2} I\omega^2 = \frac{1}{2} I\left(\frac{J}{I}\right)^2 = \frac{J^2}{2I}$$

Quantum mechanically, the rotational Hamiltonian may be written by analogy in the form:

$$H = \sum_{i=1}^{3} \frac{\hbar^2}{2I_i} J_i^2$$

where I_i is the moment of inertia along the ith axis. For an axially symmetric object, $I_1 = I_2 = I$ and $I_3 \neq I$ (for non-spherical nucleus), where I is the moment of inertia for the rotating nuclei on axis perpendicular to 3-axis. Therefore,

$$H = \frac{\hbar^2}{2I}(J^2 - J_3^2) + \frac{\hbar^2}{2I_3} J_3^2$$

In classical mechanics, a rotating body requires three Euler angles (α, β, γ) to specify its orientation in space. By analogy, quantum mechanics requires three independent quantum numbers to describe the rotational state. In this case,

$J(J + 1)$ the eigen value of J^2
M is the eigen value of J_z (z is the quantization axis)
K is the eigen value of J_3

Therefore, the energy of a rotational state corresponding to a simple rotational Hamiltonian is given by:

$$E_J = \frac{\hbar^2}{2I} J(J+1) + E_K \tag{2.21}$$

where E_K represents contribution from the intrinsic part of the wave function a denotes the position where the band starts. For even–even nuclei, the value of E_K can be neglected.

EXERCISE 2.11:

The experimental excitation energies of ${}^{170}_{72}Hf_{98}$ for 2^+, 4^+, 6^+, 8^+, 10^+, 12^+ and 14^+ states are 0.1, 0.32, 0.64, 1.04, 1.50, 2.01 and 2.56 MeV, respectively. Calculate the moment of inertia from the given rotational spectra.

Solution: According to equation (2.21), the excitation energies should be proportional to $J(J + 1)$. However, when the values are plotted in Excel, we see that the data points do not perfectly fall on the linear trend line, as shown below. The slight curvature (Figure 2.19) in the plot indicates that the moment of inertia does not remain constant and increases as the angular momentum of rotation increases, which may be because of the centrifugal stretching of the nucleus.

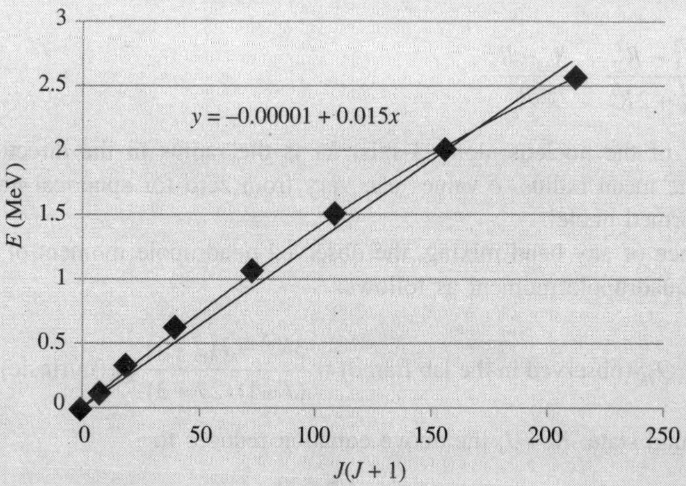

FIGURE 2.19 Excitation energies of rotational states for Hf-170 nuclide vs. $J(J + 1)$.

As can be seen in the above plot, a second order polynomial trend line fits well with the data, which suggests that a second order correction term needs to be included as compensation, in order to reproduce the experimental results (assuming moment of inertia to be constant). Therefore, we get

$$E_J = \frac{\hbar^2}{2I} J(J+1) - BJ^2(J+1)^2 \tag{2.22}$$

The value of B is adjusted to obtain the best fit curve. Comparing equation (2.22) and the equation of the second order polynomial trend line, we get

$$\frac{\hbar^2}{2I} = 0.015$$

$$\therefore \qquad I = 33.33\ \hbar^2$$

2.3.3 Quadrupole Moment (Q)

Electric quadrupole moment of nuclei arises due to the deviation of charge distribution from the spherical symmetry. It is given by

$$Q_0 = \frac{1}{e}\int (3z^2 - r^2)\,\rho(\vec{r})\,dv \qquad \text{(see section 1.8.5)}$$

For axially symmetric shape of the nucleus, it is related to the deformation parameter δ, as follows:

$$Q_0 = \frac{4}{3}\left\langle \sum_{i=1}^{A} r_i^2 \right\rangle \delta \approx \frac{4}{3} A\langle r^2\rangle\,\delta \tag{2.23}$$

where $\delta = \dfrac{3}{2}\dfrac{R_3^2 - R_\perp^2}{R_3^2 + 2R_\perp^2} \approx \dfrac{R_3 - R_\perp}{R}$

R_3 is the radius of the nucleus along 3-axis, $R_\perp$ is the radius in the direction perpendicular to it and R is the mean radius. δ values can vary from zero for spherical nuclei to values of $\delta \sim 0.5$ for deformed nuclei.

In the absence of any band mixing, the observed quadrupole moment of a state is related to the intrinsic quadrupole moment as follows:

$$Q_{JK}(\text{observed in the lab frame}) = \frac{3K^2 - J(J+1)}{(J+1)(2J+3)} Q_0(\text{intrinsic})$$

For the ground state, $K = J$; the above equation reduces to

$$Q_{JK} = \frac{J(2J-1)}{(J+1)(2J+3)} Q_0 \tag{2.24}$$

EXERCISE 2.12:

(a) What is the mean square radius of a uniformly charged sphere? By extension, what should be the mean square radius of a uniformly charged spherical nucleus?

(b) The ground state of ${}^{152}_{63}\text{Eu}_{89}$ is known to be 3^- with quadrupole moment of $+316$ efm^2. Predict the shape of the nucleus. Also calculate the values of R_3 and $R_\perp$.

(c) Assuming no configuration stretching, predict the quadrupole moment of the 5^- state for the above nuclide.

Solution:

(a) Since, the charge density is constant, we have spherical symmetry and the mean square radius will be given by the volume average of r^2 for the sphere with radius R. Dividing the sphere into spherical shells, we get

$$\langle r^2 \rangle = \frac{\int_0^R r^2 \rho(\vec{r})\, dv}{\int_0^R \rho(\vec{r})\, dv} = \frac{\int_0^R r^2\, 4\pi r^2\, dr}{\frac{4\pi}{3} R^3}$$

$$\therefore \qquad \langle r^2 \rangle = \frac{3}{5} R^2$$

The empirical relation for the nuclear radius is given by $R \approx 1.2\, A^{1/3}$ fm, where A is the mass number. Therefore, for a spherical nucleus:

$$\langle r^2 \rangle = \frac{3}{5}(1.2 A^{1/3})^2$$

(b) From equation (2.24), we have

$$Q_0 = \frac{(J+1)(2J+3)}{J(2J-1)} Q_{JK} = \frac{4 \times 9}{3 \times 5} \times 316 = 758 \text{ efm}^2$$

From equation (2.23), for small deformation

$$\delta \approx \frac{Q_0 \text{ efm}^2}{\frac{4}{3} A \left\{ \frac{3}{5} \left(1.2\, A^{\frac{1}{3}} \right)^2 \text{fm}^2 \right\}} = \frac{758}{\frac{4}{3} \times 152 \left\{ \frac{3}{5} \times (1.2 \times (152)^{1/3})^2 \right\}} = 0.152$$

Since, $\delta > 0$ (which is always the case, for positive quadrupole moment), the shape of the nucleus is prolate ($R_3 > R_\perp$).
Now,

$$R = \frac{1}{3}(R_3 + 2R_\perp) = 1.2 A^{1/3} = 6.40 \text{ fm} \tag{2.25}$$

Also,

$$R_3 - R_\perp = \delta R = 0.97 \text{ fm} \tag{2.26}$$

From equations (2.25) and (2.26), we get $R_\perp = 6.08$ fm and $R_3 = 7.05$ fm.

(c) Since, 3^- state is the lowest member of a rotational band, the band is designated as $K = 3$. If Q_0 is assumed to be constant, the observed quadrupole moment for 5^- state is given by:

$$Q_{53} = \frac{3K^2 - J(J+1)}{(J+1)(2J+3)} Q_0 = \frac{3(3)^2 - 5 \times 6}{6 \times 13} (758) = -29 \text{ efm}^2$$

2.3.4 Rotational Bands in Deformed Odd–A Nuclei

Almost all the low lying states in deformed odd–*A* nuclei can be understood on the basis of single particle states, on which the rotational bands are built. Odd–*A* nuclei consist of an even–even core and a loosely bound (unpaired) nucleon. If $\vec{j}$ represents the intrinsic angular momentum of the nucleon, $\vec{R}$ is the total angular momentum of even–even core, then coupling of $\vec{j}$ and $\vec{R}$ gives $\vec{J}$ (total angular momentum of the odd nucleus). The projection of $\vec{J}$ on the symmetry axis is *K*, which is constant of motion. Rotational energy bands are designated with *J* with values equal to *K*, *K*+1, *K*+2, ...

The total Hamiltonian for such a deformed nucleus is given by

$$H = H_{\text{rot}} + H_p + V(\vec{r})$$

where H_{rot} is the Hamiltonian for the rotator, H_p is the Hamiltonian for the odd nucleon, $V(\vec{r})$ is the interaction potential between the unpaired nucleon and the rotator and $\vec{r}$ is the radius vector in the body fixed reference frame.

The energy levels in any single band has the same parity and is given as follows:

$$E_J = \frac{\hbar^2}{2I}\left\{J(J+1) + \delta_{K,\,1/2}\, a(-1)^{J+1/2}\left(J + \frac{1}{2}\right)\right\} + E_K \tag{2.27}$$

where E_K is the starting energy level of a given band and *a* is the decoupling parameter, which is operative for the band with *K* = 1/2 only. The rotational motion of even–even core is strongly coupled to the angular momentum states of particles for which $l \neq 0$. If *K* = 1/2 (i.e., *J* = 1/2) which implies *l* = 0 and *s* = 1/2, leads to partial decoupling of particle motion from the rotator.

The rotational model successfully explains the energy spectrum of odd–*A* nuclei in all mass regions.

EXERCISE 2.13:

(a) The experimental excitation energies of ${}^{19}_{9}F_{10}$ for $1/2^+$, $1/2^-$, $5/2^+$, $5/2^-$, $3/2^-$, $3/2^+$, $9/2^+$, $7/2^-$, $9/2^-$, $13/2^+$ and $7/2^+$ rotational levels are 0.0, 0.11, 0.197, 1.346, 1.459, 1.554, 2.78, 3.999, 4.033, 4.648 and 5.465 MeV, respectively. Plot the energy level diagram of the rotational bands and discuss any peculiar behaviour that you see.

(b) Calculate the moment of inertia and the decoupling parameter for each band.

(c) Comment on the likelihood of a $11/2^+$ level at 6.0 MeV to be a member of the $K^\pi = 1/2^+$ band.

Solution:

(a) A rotational band for a particular K^π is identified by the lowest *J*-value whose value is equal to *K*. Therefore in this exercise, we have two rotational bands given with $K^\pi = 1/2^+$ and $K^\pi = 1/2^-$.

The energy level diagram (Figure 2.20) for both the bands is plotted using Excel. It is clear from the energy level diagram that the decoupling parameter *a* is large and therefore, higher *J*-levels appear below lower *J*-levels.

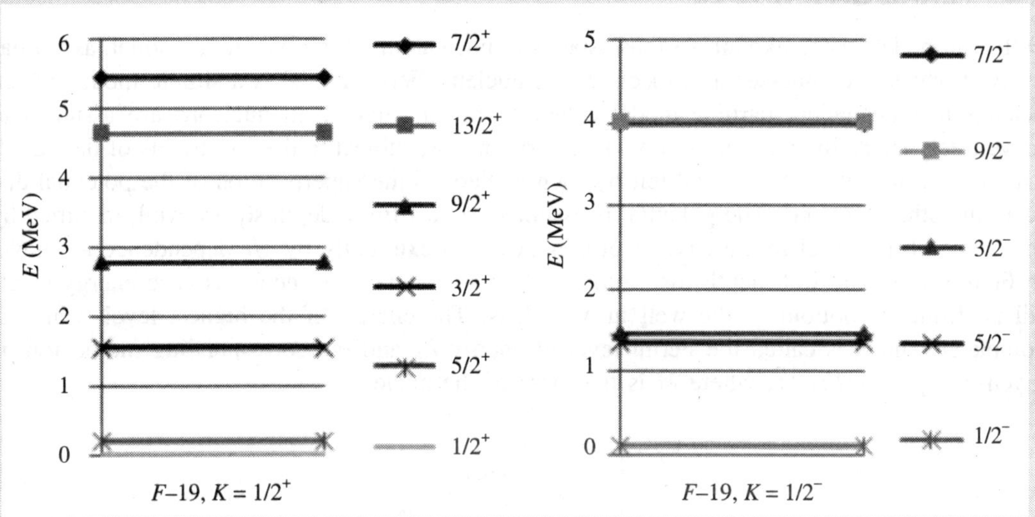

FIGURE 2.20 Energy level diagram of the rotational bands for F-19.

(b) Equation (2.23) can be written in the following form:

$$E_J = AJ(J+1) + B(-1)^{J+\frac{1}{2}}\left(J+\frac{1}{2}\right) + C$$

Now, using least square curve fitting function in Matlab, the values of A, B and C can be found out for both the bands. The Matlab code given below is used for the $K^\pi = 1/2^+$ case. Thus, for $K^\pi = 1/2^+$ band, the decoupling parameter,

$$a = B/A = 0.514/0.1697 = 3.029$$

and moment of inertia $= \hbar^2/(2A) = 2.946\hbar^2$

Similarly, for $K^\pi = 1/2^-$ band, $a = 1.016$ and $I = 2.504\ \hbar^2$

```
clear all

xdata = [0.5;1.5;2.5;3.5;4.5;6.5]; % J
ydata = [0;1.554;0.197;5.465;2.78;4.648]; % Ej
sdata = [-1;1;-1;1;-1;-1]; %Signs for different values of J

x = lsqcurvefit(@(x,xdata) x(1)*(xdata.*(xdata+1)) + ...
 x(2)*(sdata.*(xdata+0.5)) + x(3),[1 1 1],xdata,ydata)
```

(c) Substituting the values of A, B and C for $K^\pi = 1/2^+$ band, we get

$$E_{11/2} = 0.1697 \times \frac{11}{2}\left(\frac{11}{2}+1\right) + 0.514\left(\frac{11}{2}+\frac{1}{2}\right) + 0.3981 = 9.55 \text{ MeV}$$

Hence, we are unlikely to find $11/2^+$ level at 6.0 MeV, for $K^\pi = 1/2^+$ band.

2.4 FERMI GAS MODEL

Different models try to explain various aspects of nuclear structure. No single model, as yet has evolved which encompasses all aspects of the nucleus. Fermi model is a simple model, which belongs to independent particle model category. In this model, all nucleons are assumed to move freely inside the nuclear volume, like Fermion gas, subject to the constraints of the Pauli's exclusion principle. The potential felt by each nucleon is the superposition of the potential due to all the other nucleons. The potential is assumed to be a finite depth square well, modified by the Coulomb potential in the case of proton extended externally by $1/r$ dependence as shown in Figure 2.21. The well depth for protons is less as compared to neutrons. The energy levels fill up from the bottom of the well in a nucleus. The energy of the highest level, which is completely filled, is called the Fermi level of energy E_F and the corresponding momentum is given by $p_F = (2ME_F)^{1/2}$, where M is the mass of the nucleon.

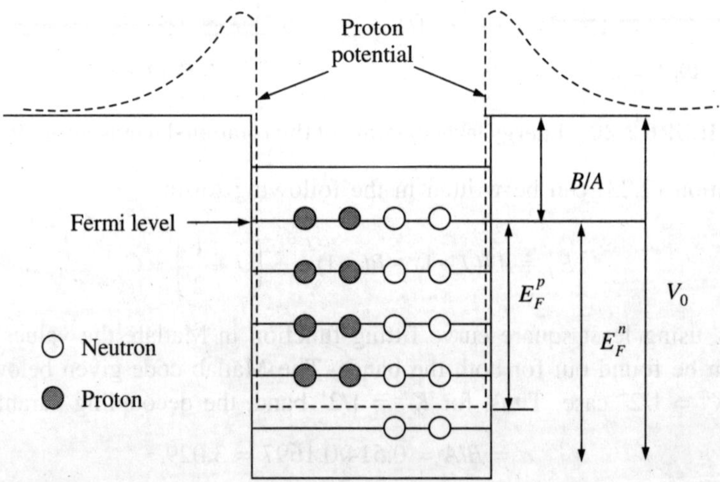

FIGURE 2.21 Proton and neutron potential wells in Fermi gas model.

EXERCISE 2.14:

Estimate the depth of the nuclear potential on the basis of Fermi gas model.

Solution:

Number of states within a nuclear volume $V = \frac{4}{3}\pi R^3$, where $R = r_0 A^{1/3}$, with momentum between p and dp is given by[5]

$$n(p)dp = \frac{4\pi V}{(2\pi\hbar)^3} p^2 dp$$

5. See Appendix D.

Since, every state can contain two similar nucleons (↑↓), we get

$$n = 2\int_0^{p_F} n(p)\,dp = \frac{V}{\pi^2\hbar^3}\left[\frac{p_F^3}{3}\right]$$

Therefore, the number of neutrons and protons is given by

$$N = \frac{V(p_F^n)^3}{3\pi^2\hbar^3} \text{ and } Z = \frac{V(p_F^p)^3}{3\pi^2\hbar^3}$$

Assuming, the depth of the neutron and proton wells to be roughly equal, the Fermi momentum for a nucleus with $Z = N = A/2$ is given by

$$p_F = p_F^n = p_F^p = \frac{\hbar}{r_0}\left(\frac{9\pi}{8}\right)^{1/3}$$

Using $r_0 = 1.21$ fm determined from fast electron scattering experiment and $\hbar c = 197$ MeV-fm, we get $p_F \approx 250$ MeV/c. The corresponding Fermi kinetic energy is given as follows:

$$E_F = \frac{p_F^2}{2M} = \frac{\left(250\,\dfrac{\text{MeV}}{c}\right)^2}{2\times\left(940\,\dfrac{\text{MeV}}{c^2}\right)} \approx 33 \text{ MeV}$$

The difference between the top of the well and the Fermi level is constant for most heavy nuclei and can be taken as the average binding energy per nucleon ≈ 7 MeV. Therefore, the total well depth is given by:

$$V_0 = E_F + \bar{B} = 33 + 7 \approx 40 \text{ MeV}$$

PROBLEMS

2.1 Calculate the separation energy of the proton from SEMF and draw the proton drip line using any four Z-values.

2.2 The mass excess of $^{65}_{29}$Cu nuclide is –67.266 MeV. Using the values of the coefficients that have been calculated in the chapter, compare the mass excess with the prediction made from SEMF.

2.3 A certain odd-parity shell model state can accommodate up to a maximum of 12 nucleons. What are its j and l values? [**Ans.** $j = l \pm 1/2$ and $l = 3$]

2.4 A nuclide with $Z = 72$, $A = 170$ is radioactive. Determine whether it is an α-emitter or a β-emitter.

2.5 Which of the following experimental characteristics of nuclei can be explained by (i) LDM, (ii) single particle model of the nucleus?

(a) Approximately constant density of the nuclei

(b) Discontinuities in the nuclear binding energy curves

(c) Distribution characteristics of the stable isotopes

(d) As A increases, approximate constancy of the binding energy per nucleon

2.6 State the various assumptions involved in (a) LDM, (b) shell model, (c) collective model, (d) Fermi gas model.

2.7 Consider nuclei with small nucleon number A, such that $Z = N = A/2$. Neglecting the pairing term, derive an expression for the binding energy per nucleon. Treating A as a continuous variable, show that it reaches a maximum for $Z = A/2 = 26$ (iron).

2.8 What are the expected ground state spins and parities for the following nuclides based on the single particle model? Also calculate their dipole and quadrupole moments.

(a) ${}^{45}_{21}\text{Sc}_{24}$ (b) ${}^{39}_{19}\text{K}_{20}$ (c) ${}^{7}_{3}\text{Li}_{4}$

2.9 How does the shell model predict $7/2^-$ for the ground state spin parity of ${}^{41}_{20}\text{Ca}$. What does the model predict about the spin and parities of the ground states of ${}^{30}_{14}\text{Si}$ and ${}^{14}_{7}\text{N}$?

2.10 Write a computer program to find the allowed states when two identical quadrupole phonons are coupled together.

2.11 The rotational levels for an even–even nuclide are $44(2^+)$ and 146, 304 and 514 keV, respectively, for the next 3 excited states. Assign J^πs to the aforementioned states. [*Hint:* $E_J \propto J(J+1)$]

2.12 The experimental excitation energies of ${}^{169}_{70}\text{Yb}_{99}$ for $7/2^+$, $1/2^-$, $9/2^+$, $3/2^-$ and $5/2^-$ rotational levels are 0.0, 24.3, 70.9, 82.0 and 99.3 keV, respectively. Calculate I and E_K for all the bands. At what energy, would you look for $11/2^+$ level? **[Ans.** 161.7 keV]

2.13 Estimate the depth of nuclear potential for ${}^{238}_{92}\text{U}_{146}$ nuclide on the basis of Fermi gas model. [*Hint:* $V_0 = E_F^n + \overline{B}$]

2.14 In the Schmidt model of nuclear magnetic moments, we have $\vec{\mu} = \mu_N(g_l\vec{l} + g_s\vec{s})$. For the case of $j = l + 1/2$, where j is the total angular momentum of odd nucleon, find the expectation value of $\vec{s}.\vec{j}$. **[Ans.** $\langle \vec{s}.\vec{j} \rangle = 1/2\{j+1\}$]

2.15 Predict the spins, parities and magnetic moments of the following nuclei based on the shell model.

$${}^{43}_{20}\text{Ca},\ {}^{93}_{41}\text{Nb},\ {}^{137}_{56}\text{Ba},\ {}^{197}_{79}\text{Au and } {}^{26}_{13}\text{Al}$$

BIBLIOGRAPHY

Beiser, A., *Concepts of Modern Physics*, McGraw-Hill, New York, 2003.

Devanathan, V., *Nuclear Physics*, Narosa, New Delhi, 2006.

Garg, J.B., *Nuclear Physics: Basic Concepts*, Macmillan, New Delhi, 2011.

Harvey, B.G., *Introduction to Nuclear Physics and Chemistry*, Prentice-Hall, New Jersey, 1970.

Heyde, K., *Basic Ideas and Concepts in Nuclear Physics*, Overseas Press India, New Delhi, 2005.

Krane, K.S., *Introductory Nuclear Physics*, Wiley, New York, 2008.

Lilley, J., *Nuclear Physics: Principles and Applications*, Wiley, New York, 2002.

Martin, B.R., *Nuclear and Particle Physics*, Wiley, New York, 2006.

Sood, D.D. and Ramamoorthy, A.V.R., *Fundamentals of Radiochemistry*, IANCAS, Mumbai, 2000.

Wong, S.S.M., *Introductory Nuclear Physics*, Prentice-Hall of India, New Delhi, 2005.

3

Radioactivity

"We must not forget that when radium was discovered no one knew that it would prove useful in hospitals. The work was one of pure science. And this is a proof that scientific work must not be considered from the point of view of the direct usefulness of it. It must be done for itself, for the beauty of science, and then there is always the chance that a scientific discovery may become like the radium a benefit for humanity."

—Marie Curie

The spontaneous nuclear transformation of an unstable nuclide by emission of an alpha(α) particle, a beta(β^-) particle or a positron(β^+) and by orbital electron capture, neutron emission or proton emission to a more stable nuclide, is called radioactivity. Radioactivity is a nuclear process that originates in the nucleus and is therefore not influenced by the chemical or the physical state of the atom. It may be either natural or artificial.

The phenomenon of radioactivity was discovered by Henri Becquerel[1] in 1896 accidently when he noticed fogging of the photographic plates by an unknown radiation emanating from uranium salts. Becquerel called this new radiation as 'U rays,' later renamed radioactivity. An era of radioactivity thus began. Pierre and Marie Curie[2] working in Becquerel laboratory succeeded in isolating a new material from uranium minerals such as pitchblende which was million times more radioactive than uranium, that Marie Curie named 'polonium.' They further isolated another radioactive substance for which they suggested the name 'radium.' The discovery of artificial radioactivity in 1934, by Irene and Frederic Joliot-Curie was a milestone in the use and control of radioactivity. For this discovery, they were awarded the Nobel Prize (Chemistry) in 1935.

The most far reaching advances in the field of radioactivity were made by Rutherford and co-workers. The scattering of the α-particles by atoms led to the idea of nucleus inside the

1. Henri Becquerel (1852–1908) was a French physicist who discovered radioactivity. For his work he, along with Marie Curie and Pierre Curie, received the Nobel Prize (Physics) in 1903. The SI unit for radioactivity, the becquerel (Bq), is named after him.
2. The Nobel Prize in Chemistry (1911) was awarded to Marie Curie in recognition of her services to the advancement of chemistry by the discovery of radium and polonium, and the study of their nature and compounds. Marie Curie was the first person to win two Nobel Prizes.

atom (see Chapter 1). All the laws governing radioactive decay were established. It was shown that α and β-decays change the nature of the element; while γ-rays are high energy photons.

Convincing proof of the isotopes came from the analyses of the chemical properties among the naturally occurring radioactive elements. James Chadwick[3] proposed the existence of the neutron after studying the highly penetrating uncharged radiation from the α-particle bombardment of beryllium. This discovery further established the concept of the nucleus.

Production of radioisotopes by bombarding various atoms by different projectiles further added a dimension of practical importance. The detection of radiations (α, β and γ) paved the way of visualizing the nuclear energy levels similar to those in the atoms. Thus, the measurement of radioactivity is closely related to the understanding of the nuclear structure and the systematic of nuclear processes. Therefore, it is important to understand the type of decays, their kinematics and the underlying theories. Few of these aspects will be discussed in this chapter.

3.1 MECHANISM OF RADIOACTIVE DECAY

The radioactive decay mode depends on the following two factors:

(i) The particular type of nuclear instability as a result of the high or too low neutron-to-proton ratio.

(ii) The mass–energy relationship among the parent nucleus and the decay products (daughter nucleus and the ejectile).

Radioactive decay of a nucleus changes the arrangement of its nucleons. The types of decay processes are as follows:

Alpha Decay

In this process, the parent nucleus emits an α-particle (${}^{4}_{2}\text{He}^{++}$) and thus, the daughter nucleus mass reduces to (*A*–4) and its proton number *Z* reduces by 2. It can be written symbolically as follows:

$$ {}^{A}_{Z}\text{X} \rightarrow {}^{A-4}_{Z-2}\text{Y} + {}^{4}_{2}\text{He}^{++} $$

${}^{4}_{2}\text{He}^{++}$ nuclide, which represents the α-particle, is usually taken as ${}^{4}_{2}\text{He}$ for most calculations. A typical example of this decay mode is

$$ {}^{238}_{92}\text{U} \rightarrow {}^{234}_{90}\text{Th} + {}^{4}_{2}\text{He} $$

Beta Decay

The radioactive decay processes which are designated by the general term β-decay include negatron emission (β^- or electron), positron emission (β^+ or positron) and electron capture (EC). Typical examples include the following:

3. For his discovery, he was awarded the Nobel Prize for Physics in 1935. “James Chadwick - Nobel Lecture: The Neutron and Its Properties” <http://www.nobelprize.org/nobel_prizes/physics/laureates/1935/chadwick-lecture.html>

$$^{14}_{6}C \rightarrow {}^{14}_{7}N + \beta^- + \bar{\nu}\text{(antineutrino)}$$

$$^{22}_{11}Na \rightarrow {}^{22}_{10}Ne + \beta^+ + \nu\text{(neutrino)}$$

$$^{54}_{25}Mn + e^-\text{(orbital electron)} \rightarrow {}^{54}_{24}Cr + \nu\text{(neutrino)}$$

In all these processes, the charge of the resulting product nucleus differs from the parent nucleus by one unit and there is no change in the mass number (A).

Gamma Decay

Alpha and β-decay usually leave the daughter (product) nucleus in excited states. If the excitation energy available with the daughter nucleus is insufficient for particle emission, it loses its energy by emitting electromagnetic radiations, known as γ-rays. One such example is:

$$^{60}_{29}Co^* \rightarrow {}^{60}_{29}Co + \gamma$$

The star (*) on $^{60}_{29}Co$ indicates that it is in excited state.

Spontaneous Fission

As the nuclear charge increases to large values beyond uranium, nuclei become unstable and undergo spontaneous division into two fragments along with the emission of 2–3 neutrons. The process may be written as follows:

$$^{A}_{Z}X \rightarrow {}^{A_1}_{Z_1}X_1 + {}^{A_2}_{Z_2}X_2 + 2 \text{ or } 3\ {}^{1}_{0}n$$

In fact, spontaneous fission becomes the dominating decay mode for the heavier nuclei (see Figure 1.13 from Chapter 1).

All decay modes are subjected to satisfy following conservation laws which are a direct consequence of the symmetries in nature that require certain variables to stay unchanged.

Conservation of mass-energy: In the process of radioactive decay, the difference in masses before and after the decay is emitted as the energy of the emitted particles. It may be shown that the conservation of mass and energy holds for the radioactive decay as a result of the symmetry in time. Every experiment performed at different time will give the same decay results for the same nucleus.

Conservation of momentum: The sum of the linear momentum before and after the decay must be the same. This law is the outcome of the symmetry in space.

Conservation of angular momentum: The sum of the orbital angular momenta and spins must be conserved in the decay. This is due to the isotropy of space.

Conservation of charge: The sum of the charges before and after the decay remains the same. According to this law, an electron cannot appear or disappear on its own and assures the stability of the electron and thus the stability of matter.

Conservation of nucleons: In any decay mode, the total number of nucleons ($Z + N = A$) must be conserved. This law inhibits the neutrons and protons within a nucleus to decay into other particles and thus, ensuring the stability of matter.

3.2 KINEMATICS OF RADIOACTIVE DECAY

Based on their experiments involving various radioactive isotopes, Rutherford and Soddy formulated for the first time, the radioactive decay law.

Decay Constant, Mean-Life and Half-Life

A radioactive nucleus is characterized by the rate at which it disintegrates. The decay constant, mean-life and half-life are equivalent ways of characterizing this decay. The decay constant $\lambda(T^{-1}$ is its dimension) represents the probability of a radionuclide decaying in unit time. Therefore, the probability of a radionuclide decaying in time dt is λdt. The characteristics of λ that have been confirmed experimentally are as follows:

(i) For all nuclei of a given atom, the decay constant is the same. It cannot be changed by the surrounding pressure or temperature.

(ii) The decay constant does not depend on the age of the nuclides, i.e., it does not change with time.

The mean-life τ is the average lifetime of all the nuclei of a particular radioactive substance. It is the sum of the lifetimes of all the individual unstable nuclei in a sample, divided by the total number of unstable nuclei present. The half-life $T_{1/2}$ is defined as the time taken for half the nuclei in a radioactive substance to disintegrate.

Radioactive Decay

Radioactive decay is random in its nature and therefore, demands statistical treatment. We can not say anything particular for a given nucleus; however, there is an equal probability for all nuclei of a radioactive element to decay.

The rate of decay of a radionuclide is proportional to the number of nuclides present at that instant, i.e.,

$$\frac{dN}{dt} \propto N$$

where N is the number of nuclides at any time t

or
$$-\frac{dN}{dt} = \lambda N \tag{3.1}$$

The negative sign indicates that as t increases N decreases.

Rewriting equation (3.1) and integrating both sides, we get

$$\int \frac{dN}{N} = -\lambda \int dt$$

or
$$\ln(N) = -\lambda t + C$$

where C is the integration constant. If we suppose that at $t = 0$, the number of radioactive nuclides is N_0, then $C = \ln(N_0)$. Substituting the value of C in the above equation, we get

$$\ln\left(\frac{N}{N_0}\right) = -\lambda t$$

or

$$N = N_0 e^{-\lambda t} \tag{3.2}$$

EXERCISE 3.1:

(a) Prove that the average life (or the mean-life τ) of a radionuclide is given by $\tau = 1/\lambda$.

(b) Derive a time [$t = T_{1/2}$ (half-life)] at which the number of radioactive nuclei has decreased to half of its original value, i.e., $N_0/2$.

(c) Suppose we initially have 1 g of radioactive strontium, which decays by β-decay,

$$^{90}_{38}\text{Sr} \rightarrow {}^{90}_{39}\text{Y} + \beta^- + \bar{\nu}$$

Measurements show that it takes 29 years for one half of the initial amount of parent to decay, i.e., $T_{1/2} = 29$ y. It then follows from equation (3.2), that during the next 29 y, one half of the remainder will decay. Hence, the amounts of the parent left at $t = 0, 29, 58, 87$ y, etc. will be 1, 1/2, 1/4, 1/8 g, etc. respectively.

Plot $N(t)$ (number of nuclides left at a particular time t) vs. time.

Solution:

(a) Suppose the number of nuclei which have survived time t is N, and the number decaying in the next small time interval Δt is ΔN, then the lifetime of ΔN decaying atoms is t. Therefore, we get

$$\tau = \frac{\sum \Delta N t}{N_0} = \frac{\int_{N_0}^{0} t dN}{N_0}$$

$$= \frac{1}{N_0} \int_0^\infty t(-\lambda N_0 e^{-\lambda t}\, dt)$$

$$= -\lambda \left[\frac{t(-e^{-\lambda t})}{\lambda} - \int \frac{(1)(-e^{-\lambda t})}{\lambda} dt \right]_0^\infty$$

$$= -\lambda \left[\frac{-te^{-\lambda t}}{\lambda} - \frac{e^{-\lambda t}}{\lambda^2} \right]_0^\infty$$

$$\therefore \quad \tau = \frac{1}{\lambda}$$

(b) Substituting $N = N_0/2$ in equation (3.2), we get

$$T_{1/2} = \frac{\ln 2}{\lambda} = \frac{0.693}{\lambda}$$

or

$$T_{1/2} = 0.693\,\tau$$

(c) Mathematically, one can write

$$N(t) = N_0 \left(\frac{1}{2}\right)^{t/(29)}$$

In terms of the half-life, we can write the law of radioactive decay as follows:

$$N(t) = N_0 \left(\frac{1}{2}\right)^{t/T_{1/2}} \tag{3.3}$$

Thus, Figure 3.1 for the amount of radioactive strontium as a function of time is obtained.

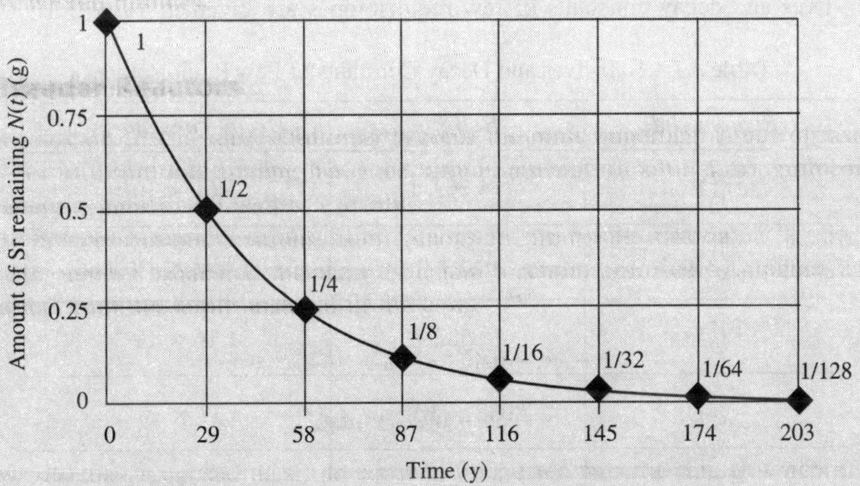

FIGURE 3.1 Amount of radioactive strontium as a function of time.

Experimentally, the half-life of the radioisotope is determined from the activity profile. The activity of a radionuclide is defined as follows:

$$A = -\,(\text{rate of decay}) = -\frac{dN}{dt}$$

$$\therefore \qquad A = \lambda N \tag{3.4}$$

From the above equation, the activity at time $t = 0$ is $A_0 = \lambda N_0$, therefore, from equations (3.2) and (3.4), we get

$$A = A_0 e^{-\lambda t} \tag{3.5}$$

The traditional units for the activity:

$$1 \text{ curie (Ci)} = 3.7 \times 10^{10} \text{ disintegration per second (dps)}$$

$$1 \text{ rutherford (R)} = 10^6 \text{ dps}$$

The SI unit for the activity:

$$1 \text{ becquerel (Bq)} = 1 \text{ dps}$$

Taking logarithm of equation (3.5), we get

$$\ln A = \ln A_0 - \lambda t \quad \text{or} \quad \log A = \log A_0 - \lambda t/2.303$$

Thus, we should get a straight line when log A is plotted as a function of time on a linear scale with slope as $(-\lambda/2.303)$. On a semi-log scale, a linear plot of A as a function of time should be obtained.

EXERCISE 3.2:

(a) Half-lives can range from less than a millionth of a second to million years. Half-lives and decay constants of few radioisotopes are given in Table 3.1.

Table 3.1 Half-lives and Decay Constants of Few Radioisotopes

Radionuclide	*Half-life*	*Decay constant* (s^{-1})
$^{32}_{15}P$	14.262 d	5.61×10^{-7}
$^{99}_{42}Mo$	65.94 h	2.92×10^{-6}
$^{113}_{49}In$	1.6582 h	1.16×10^{-4}
$^{235}_{92}U$	7.038×10^{8} y	3.12×10^{-17}
$^{259}_{100}Fm$	1.5 s	0.462

Using the aforementioned table, plot A as a function of time on a semi-log paper for $^{235}_{92}U$ nuclide, assuming the initial activity to be 1000 counts/s.

(b) The ratio of the nuclide activity to the total mass of the element present is called the specific activity of the sample.

Now, if a sample of $^{113}_{49}In$ radionuclide weighs 2 μg, calculate the number of atoms remaining and the specific activity of the sample after 4 h. [The product nuclide is stable in this case.]

Solution:

(a) Similar to equation (3.3), we can also write

$$A(t) = A_0\left(\frac{1}{2}\right)^{t/T_{1/2}} \tag{3.6}$$

The following plot (Figure 3.2) of the activity (on log scale) as a function of time is obtained for $^{235}_{92}U$.

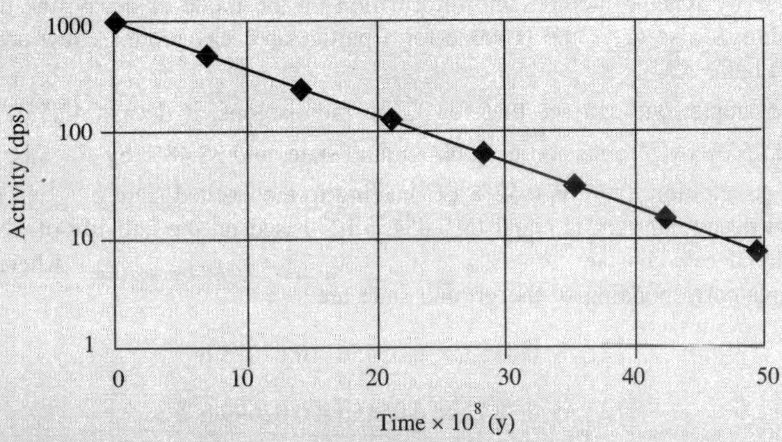

FIGURE 3.2 Activity of U-235 vs. time.

In radioactive research, ten half-lives is usually considered as the useful lifetime for a radioactive species, since $N = N_0/(2^{10}) \approx 10^{-3}N_0$, i.e., the remaining activity is (1/1000) times the original activity.

(b) Number of atoms present in the sample

$$N_0 = \frac{\text{Mass of sample } (g)}{\text{Mass of atom}\left(\dfrac{g}{\text{mole}}\right)}\left[\text{Avogadro number}\left(\frac{\text{atoms}}{\text{mole}}\right)\right]$$

$$= \frac{(2\times10^{-6})}{113}(6.023 \times 10^{23}) = 1.066 \times 10^{16} \text{ atoms}$$

Therefore, the number of atoms remaining after 4 h,

$$N = N_0e^{-\lambda t} = 1.066\times10^{16}e^{-(1.16 \times 10^{-4})(4 \times 3600)} = 2.006\times10^{15} \text{ atoms}$$

Activity of the sample after 4 h

$$A = N\lambda = (2.006 \times 10^{15})(1.16 \times 10^{-4}) = 2.32 \times 10^{11} \text{ dps or Bq}$$

Hence, the specific activity of the sample after 4 h is

$$SA = \frac{\text{Activity of sample}}{\text{Mass of sample}} = \frac{2.32 \times 10^{11} \text{ dps}}{2\ \mu g} = 1.16 \times 10^{11} \text{ dps}/\mu g$$

Radioactive Series

There are about 65 naturally occurring radioisotopes. These radioactive nuclides decay by α and β emission followed by γ emission, until a stable nucleus is reached. α-decay changes the mass number by $4u$ while β-decay does not change A, therefore, we have four independent decay chains with mass numbers $4n$, $4n+1$, $4n+2$ and $4n+3$, where n is an integer. These four series are listed in Table 2.5, in previous chapter.

Decay Schemes and Isotope Charts

The nuclear decay scheme includes the information on the mode of decay, the decay energy and the half-life. It also gives the Q-value for a particular decay route. A few decay schemes are given in Figure 3.3.

As an example, one can see that for ${}^{64}_{29}\text{Cu}$ radioisotope, it decays 43.53% by electron capture and 17.52% by β^+ emission to ${}^{64}_{28}\text{Ni}$ ground state, and 38.48% by β^- emission to ${}^{64}_{30}\text{Zn}$ ground state. In addition, there is 0.47% EC leading to the excited state of ${}^{64}_{28}\text{Ni}$ nuclide. The observed total decay constant is equal to 0.05456 h^{-1} based on the half-life of 12.7004 h.

The total decay constant $\lambda = [\lambda_{EC} + \lambda_{\beta^-} + \lambda_{\beta^+}]_{gs} + [\lambda_{EC}]_{\text{excited state}}$, where the partial decay constants corresponding to the ground state are

$$\lambda_{EC} = 0.4353 \times 0.05456 = 0.02375 \text{ h}^{-1}$$

$$\lambda_{\beta^-} = 0.3848 \times 0.05456 = 0.02099 \text{ h}^{-1}$$

$$\lambda_{\beta^+} = 0.1752 \times 0.05456 = 0.00956 \text{ h}^{-1}$$

The partial decay constants correspond to the following partial half-lives:

$$T_{1/2,\,EC} = \frac{0.693}{0.02375} = 29.1789 \text{ h}$$

$$T_{1/2,\,\beta^-} = \frac{0.693}{0.02099} = 33.0157 \text{ h}$$

$$T_{1/2,\,\beta^+} = \frac{0.693}{0.00956} = 72.4895 \text{ h}$$

Isotope charts can be considered as condensed isotope tables showing stable and radioactive nuclei[4]. Normally, the nuclear charge Z is on the vertical axis and the neutron number N is on the horizontal axis. Such isotope charts are quite useful for rapid scanning of ways to produce a certain nuclei and to follow its decay modes. Figure 3.4 shows a small section of the isotope chart.

---

4. Interactive chart of nuclides can be found at National Nuclear Data Center, Brookhaven National Laboratory (Website – <http://www.nndc.bnl.gov/chart/>)

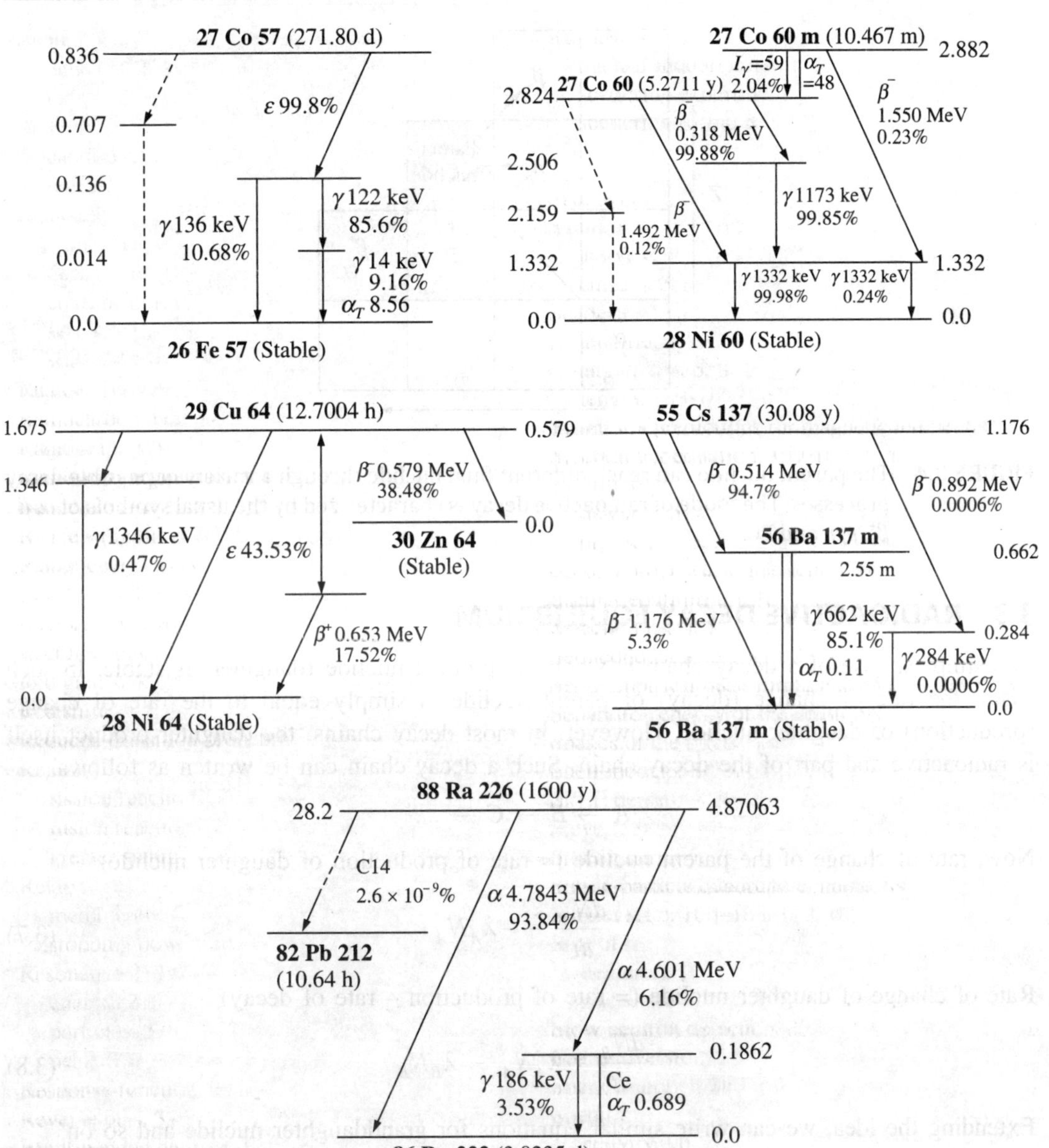

FIGURE 3.3 Few examples of decay schemes[5]. The mode of radioactive decay is characterized by the usual symbols of α, β, γ and ε for electron capture (EC). Next to this symbol is given decay energy for that particular transition along with its intensity if it is < 100%.

5. From <http://www.nucleonica.net/wiki/index.php?title=Decay_Schemes/>

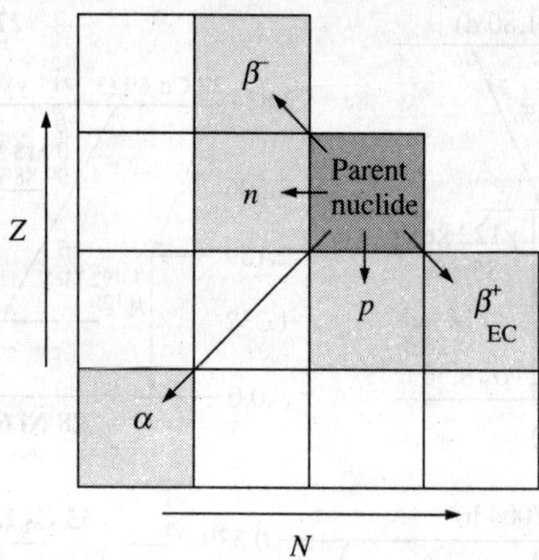

FIGURE 3.4 The parent nuclide can reach different final nuclide through a variety of possible decay processes. The mode of radioactive decay is characterized by the usual symbols of α, β^-, β^+, p, n and EC.

3.3 RADIOACTIVE DECAY EQUILIBRIUM

The simplest case of a decay chain is when the product nuclide (daughter) is stable. In such cases, the rate of change (decay) of parent nuclide is simply equal to the rate of change (production) of daughter nuclide. However, in most decay chains, the daughter product itself is radioactive and part of the decay chain. Such a decay chain can be written as follows:

$$A \rightarrow B \rightarrow C \rightarrow ....$$

Now, rate of change of the parent nuclide (= rate of production of daughter nuclide)

$$\frac{dN_A}{dt} = -\lambda_A N_A \tag{3.7}$$

Rate of change of daughter nuclide (= rate of production – rate of decay)

$$\frac{dN_B}{dt} = \lambda_A N_A - \lambda_B N_B \tag{3.8}$$

Extending the idea, we can write similar equations for granddaughter nuclide and so on.

$$\frac{dN_C}{dt} = \lambda_B N_B - \lambda_C N_C$$

If the granddaughter nuclide happens to be stable, then λ_C is zero, and the above equation reduces to

$$\frac{dN_C}{dt} = \lambda_B N_B$$

EXERCISE 3.3:

(a) Derive the expression for the accumulation of the daughter nuclide.

(b) Estimate the time at which the daughter nuclide reaches its maximum activity. Assume the initial activity of B to be zero.

(c) Now, assuming that initially we have a pure sample of A (i.e., $N_{B0} = N_{C0} = 0$), derive the expression for the accumulation of the granddaughter nuclide. If the granddaughter nuclide is stable, what does the expression reduce to?

(d) ${}^{20}_{8}O$ radioisotope undergoes β^- decay according to the following decay chain:

$$ {}^{20}_{8}O \xrightarrow{T_{1/2}=13.51\,s} {}^{20}_{9}F \xrightarrow{T_{1/2}=11.163\,s} {}^{20}_{10}Ne\,(\text{stable}) $$

Assuming that initially, the sample consists of ${}^{20}_{8}O$ only; calculate the specific activity of the sample after one minute. The sample weighs 2 μg.

Solution:

(a) Using equation (3.8), we have

$$ \frac{dN_B}{dt} + \lambda_B N_B = \lambda_A N_{A0} e^{-\lambda_A t} $$

Multiplying both sides by $e^{\lambda_B t}$, we get

$$ e^{\lambda_B t}\frac{dN_B}{dt} + e^{\lambda_B t}\lambda_B N_B = \lambda_A N_{A0} e^{(\lambda_B-\lambda_A)t} $$

or
$$ d(N_B e^{\lambda_B t}) = \lambda_A N_{A0} e^{(\lambda_B-\lambda_A)t}\,dt $$

Integrating both the sides, we get

$$ N_B e^{\lambda_B t} = \frac{\lambda_A N_{A0} e^{(\lambda_B-\lambda_A)t}}{\lambda_B-\lambda_A} + C $$

Applying the initial condition, at $t = 0$, $N_B = N_{B0}$ and $N_A = N_{A0}$, we get

$$ C = N_{B0} - \frac{\lambda_A N_{A0}}{\lambda_B-\lambda_A} $$

Substituting its value, we get

$$ N_B(t) = N_{A0}\frac{\lambda_A}{\lambda_B-\lambda_A}\left[e^{-\lambda_A t} - e^{-\lambda_B t}\right] + N_{B0}e^{\lambda_B t} \qquad (3.9) $$

(b) In terms of activity, equation (3.8) can be written as:

$$ A_B = A_{A0}\frac{\lambda_B}{\lambda_B-\lambda_A}\left[e^{-\lambda_A t} - e^{-\lambda_B t}\right] + A_{B0}e^{\lambda_B t} \qquad (3.10) $$

The above equation is known as ***Bateman equation***.

Now, at point of maximum activity of B,

$$\frac{dA_B}{dt} = 0$$

We are given $A_{B0} = 0$, therefore we get

$$[-\lambda_A e^{-\lambda_A t} + \lambda_B e^{-\lambda_B t}] = 0$$

$$\therefore \qquad t_{\max} = \frac{1}{\lambda_B - \lambda_A} \ln\left(\frac{\lambda_B}{\lambda_A}\right) = \frac{2.303}{\lambda_B - \lambda_A} \log\left(\frac{\lambda_B}{\lambda_A}\right) \tag{3.11}$$

(c) The granddaughter nuclide is unstable, therefore

$$\frac{dN_C}{dt} = \lambda_B N_B - \lambda_C N_C$$

$$\frac{dN_C}{dt} + \lambda_C N_C = N_{A0} \frac{\lambda_A \lambda_B}{\lambda_B - \lambda_A} [e^{-\lambda_A t} - e^{-\lambda_B t}]$$

Multiplying both sides by $e^{\lambda_C t}$, we get

$$d(N_C e^{\lambda_C t}) = N_{A0} \frac{\lambda_A \lambda_B}{\lambda_B - \lambda_A} \left[e^{(\lambda_C - \lambda_A)t} - e^{(\lambda_C - \lambda_B)t}\right] dt$$

On integrating and applying the initial condition, we finally get

$$N_C(t) = N_{A0} \lambda_A \lambda_B \left[\frac{e^{-\lambda_A t}}{(\lambda_A - \lambda_B)(\lambda_A - \lambda_C)} + \frac{e^{-\lambda_B t}}{(\lambda_B - \lambda_C)(\lambda_B - \lambda_A)} + \frac{e^{-\lambda_C t}}{(\lambda_C - \lambda_A)(\lambda_C - \lambda_B)}\right]$$

If the granddaughter nuclide is stable, $\lambda_C = 0$. Thus, the above equation reduces to

$$N_C = N_{A0} \frac{\lambda_A \lambda_B}{\lambda_B - \lambda_A} \left[\frac{1 - e^{-\lambda_A t}}{\lambda_A} + \frac{1 - e^{-\lambda_B t}}{\lambda_B}\right]$$

(d) The activity of a sample (mixture) is the sum of the activities of its constituents, assuming the activities to be mutually independent,

$$A_{\text{sample}} = A_A + A_B = A_{A0} e^{-\lambda_A t} + A_{A0} \frac{\lambda_B}{\lambda_B - \lambda_A} \left[e^{-\lambda_A t} - e^{-\lambda_B t}\right]$$

$$A_{A0} = \lambda_A \frac{\text{Mass of sample } (g)}{\text{Mass of atom}\left(\dfrac{g}{\text{mole}}\right)} \left[\text{Avogadro number}\left(\frac{\text{atoms}}{\text{mole}}\right)\right]$$

$$= \left(\frac{0.693}{13.51}\right) \frac{(2 \times 10^{-6})}{(20)} [6.023 \times 10^{23}] = 3.09 \times 10^{15}\ \text{Bq} = 8.35 \times 10^{4}\ \text{Ci}$$

Thus, at $t = 60$ s

$$A_{sample} = 1.44 \times 10^4 \text{ Ci}$$

The specific activity of the sample is,

$$SA = \frac{1.44 \times 10^4 \text{ Ci}}{2\,\mu g} = 7.2 \times 10^3 \text{ Ci}/\mu g$$

The radioactive decay equilibrium is attained when the ratio between the successive members (for our purpose, the parent and the daughter nuclide) in a decay chain remains constant, which only happens after t_{max}. Depending on the half-lives of the parent and the daughter nuclide, the Bateman equation is analyzed for the following three cases:

Transient Equilibrium: The half-life of the parent nuclide is of the same order, but longer than that of the daughter. The total activity rises initially, peaks and then decays at about the same rate as the parent nuclide.

No Equilibrium: The half-life of the parent is shorter than that of the daughter nuclide. Essentially, all parent nuclei transform into daughter nuclei and the activity of the sample comes mainly from the daughter nuclide.

Secular Equilibrium: The half-life of the parent nuclide is much longer than that of the daughter. The activity of the parent and the daughter nuclide reaches the maximum and thereafter, remains almost constant for several half-lives of the daughter nuclide.

EXERCISE 3.4: Calculate the activities of the parent and the daughter nuclides as a function of time for the following decay chains using numerical solution of the Bateman equations:

(a) Transient equilibrium example: 87% Mo-99 decays into Tc-99 m.

$$^{99}_{42}\text{Mo} \xrightarrow{T_{1/2} = 65.94 \text{ h}} {}^{99m}_{43}\text{Tc} \xrightarrow{T_{1/2} = 6.01 \text{ h}} {}^{99}_{43}\text{Tc}$$

(b) No equilibrium example:

$$^{138}_{54}\text{Xe} \xrightarrow{T_{1/2} = 14.08 \text{ m}} {}^{138}_{55}\text{Cs} \xrightarrow{T_{1/2} = 33.41 \text{ m}} {}^{138}_{56}\text{Ba}$$

(c) Secular equilibrium example:

$$^{144}_{58}\text{Ce} \xrightarrow{T_{1/2} = 284.893 \text{ d}} {}^{144}_{59}\text{Pr} \xrightarrow{T_{1/2} = 17.28 \text{ m}} {}^{144}_{60}\text{Nd}$$

Solution:

(a) Ordinary differential equation solvers (ode45, ode23, ode113, etc.) in Matlab can be used to arrive at the solution. The following Matlab code (written in 2 separate files) is used to plot the activities of the parent and the daughter nuclides.

Main file

```
clear all

global lambda_p lambda_d
Thalf_p = 65.94; %half-life of parent in h
Thalf_d = 6.01; %half-life of daughter in h
lambda_p = log(2)/Thalf_p; %h^-1
lambda_d = log(2)/Thalf_d; %h^-1

%A_p(0) = 1 Ci, thus initial parent nuclei = 1/lamda_p
%Initial daughter nuclei = zero
%For time from 0 to 100 h
[t,N] = ode45(@Bateman1,[0 100],[1/lambda_p 0]);%Numerical Sol

A_p = lambda_p*N(:,1); %Activity of the parent nuclide
A_d = 0.87*lambda_d*N(:,2); %Activity of the daughter nuclide

plot(t,A_p,'b-.')
hold on
plot(t,A_d,'r-')
xlabel('Time (s)')
ylabel('Activity (Ci)')
legend('Parent: Mo-99','Daughter: Tc-99m')
title('Transient Equilibrium Example')
```

Dependent file (Bateman1.m)

```
function dN = Bateman1(t,N)
global lambda_p lambda_d

%Initialization
dN = zeros(2,1);

%Diff. Eqns.
dN(1) = -lambda_p*N(1);
dN(2) = lambda_p*N(1) - lambda_d*N(2);
```

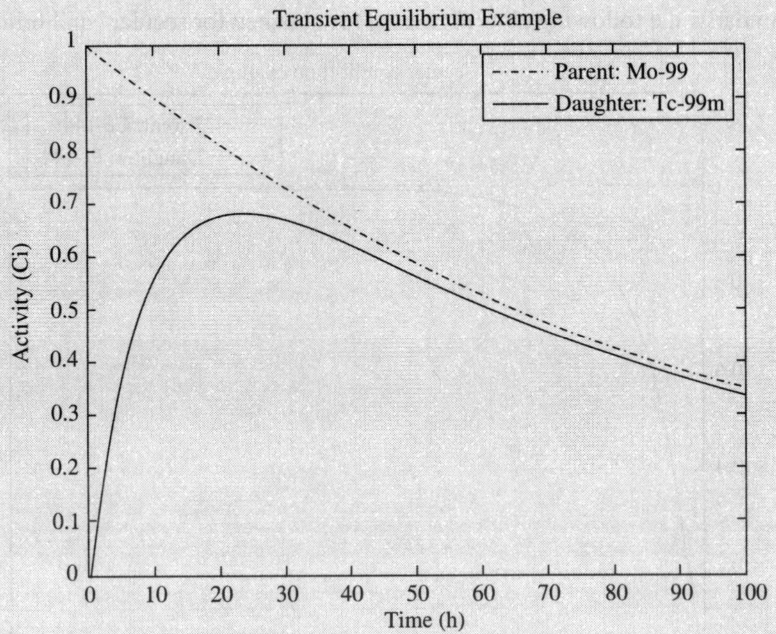

FIGURE 3.5 Activity profile in a transient equilibrium case.

The above activity plot (Figure 3.5) is obtained when the code is run.

(b) Similar code is written for no equilibrium example and the following plot (Figure 3.6) is obtained.

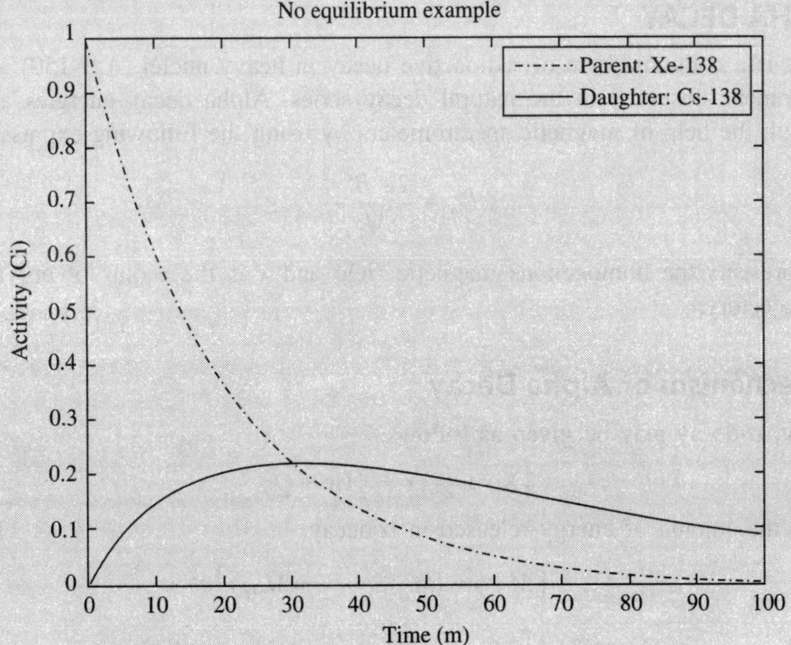

FIGURE 3.6 Activity profile in a no equilibrium case.

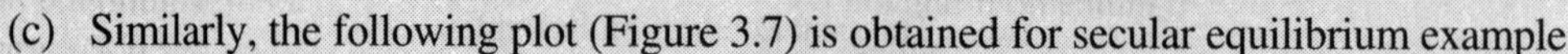

(c) Similarly, the following plot (Figure 3.7) is obtained for secular equilibrium example.

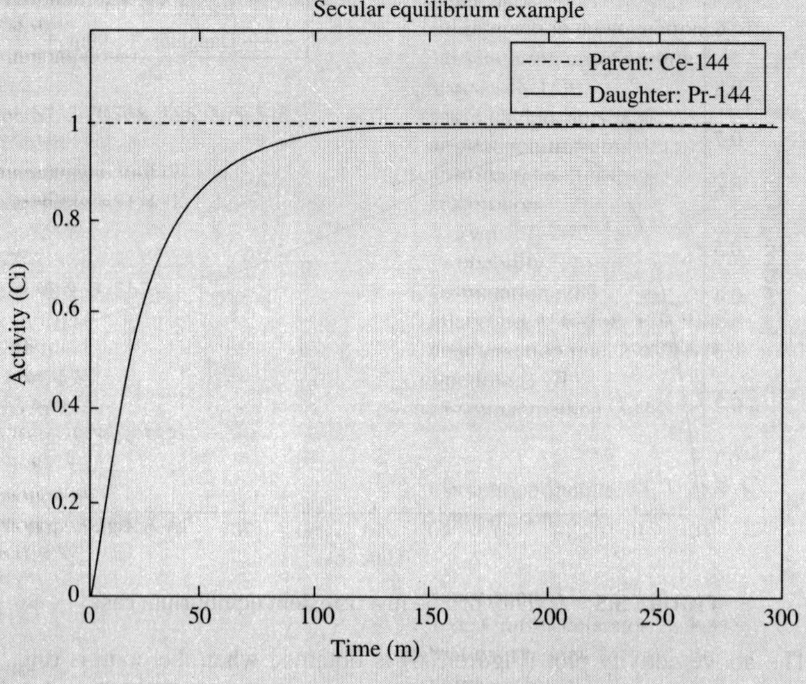

FIGURE 3.7 Activity profile in a secular equilibrium case.

3.4 ALPHA DECAY

Alpha decay is a common form of radioactive decay in heavy nuclei ($A > 150$) and was one of the phenomena observed in the natural decay series. Alpha decay energies are precisely measured with the help of magnetic spectrometers by using the following expression:

$$E_\alpha = \frac{2e^2B^2r^2}{M_{\text{He}}}$$

where B represents the homogenous magnetic field and r is the radius of curvature of the α-particle trajectory.

3.4.1 Mechanism of Alpha Decay

Symbolically, α-decay may be given as follows:

$$^{A}_{Z}\text{X} \rightarrow {}^{A-4}_{Z-2}\text{Y} + {}^{4}_{2}\text{He} + Q_\alpha$$

where Q_α is the amount of energy released in α-decay.

$$Q_\alpha = \left[M_{A,Z} - (M_{A-4,Z-2} + M_{\text{He}})\right]c^2$$

or

$$Q_\alpha = 931.5\left[M_{A,Z} - (M_{A-4,Z-2} + M_{\text{He}})\right] \text{MeV} \tag{3.12}$$

For example, for the following reaction:

$$^{238}_{92}\text{U}(238.050770\,\text{u}) \rightarrow {}^{234}_{90}\text{Th}(234.043583\,\text{u}) + {}^{4}_{2}\text{He}(4.002604\,\text{u})$$

$$Q_\alpha = 931.5\,[238.050770 - (234.043583 + 4.002604)] = 4.269\text{ MeV}$$

The equivalent form of equation (3.12) in terms of the binding energies of the parent and the daughter nuclei is as follows:

$$Q_\alpha = -B(A, Z) + B(A - 4, Z - 2) + 28.296\text{ MeV}$$

where, 28.296 MeV is the experimental binding energy of α-particle.

Note: One of the main reasons for this decay mode to appear as spontaneous process is the large binding energy of the α-particle. B/A ratio in heavy nuclei is about 7 MeV, which means that in order to remove four nucleons, the energy required will be $7 \times 4 = 28$ MeV; whereas when an α-particle is formed, it releases 28.3 MeV energy, thereby making the emission of an α-particle energetically possible for nuclei with $A > 150$.

Based on SEMF (as discussed in Exercise 2.3(b) in the previous chapter), Q_α may be expressed in terms of Z, after neglecting the pairing energy term whose effects are small. The resulting smooth $Q_\alpha(Z)$ is plotted and compared with the experimental values of Q_α in Figure 3.8. It is seen that the trend of the experimental points agrees with the SEMF predictions. Negative values of Q_α imply absolute stability against α-decay.

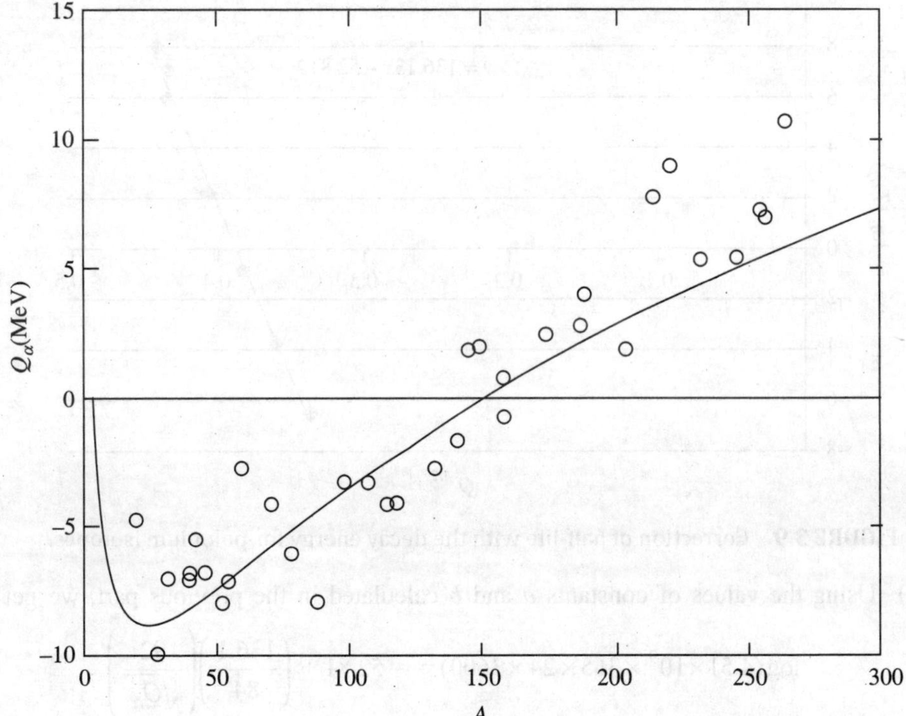

FIGURE 3.8 Comparison of experimental and SEMF predicted alpha energies as a function of A.

Geiger and Nuttall in 1911 noticed another important aspect of α-decay. They found a very strong correlation of half-life with the decay energy Q_α values. The empirical relation for even–even nuclei is as follows:

$$\log T_{1/2} = a + b\left(\frac{Z}{\sqrt{Q_\alpha}}\right) \tag{3.13}$$

The above relation suggests that a plot of log $T_{1/2}$ vs. $Q_\alpha^{-1/2}$ for nuclei with the same isotope series should be roughly a straight line.

EXERCISE 3.5:

(a) The half-lives and Q_α values of ${}^{208}_{84}\text{Po}$, ${}^{206}_{84}\text{Po}$, ${}^{210}_{84}\text{Po}$, ${}^{218}_{84}\text{Po}$, ${}^{216}_{84}\text{Po}$, ${}^{214}_{84}\text{Po}$ and ${}^{212}_{84}\text{Po}$ isotopes are 2.898 y, 8.8 d, 138.4 d, 3.05 m, 0.15 s, 164 μs, 0.3 μs and 5.21, 5.32, 5.41, 6.12, 6.91, 7.83, 8.95 MeV, respectively. Using the Geiger-Nuttall law, estimate the values of constants a and b.

(b) Predict the Q_α value for ${}^{238}_{92}\text{U}$ α-emitter nuclide. Its half-life is 4.51×10^9 y. See how it compares with the observed value of 4.27 MeV.

Solution:

(a) Fitting a straight line in a plot (Figure 3.9) of log $T_{1/2}$ vs. $Q_\alpha^{-1/2}$, we get the constants $a = -52.81$ and $b = 136.1/84$.

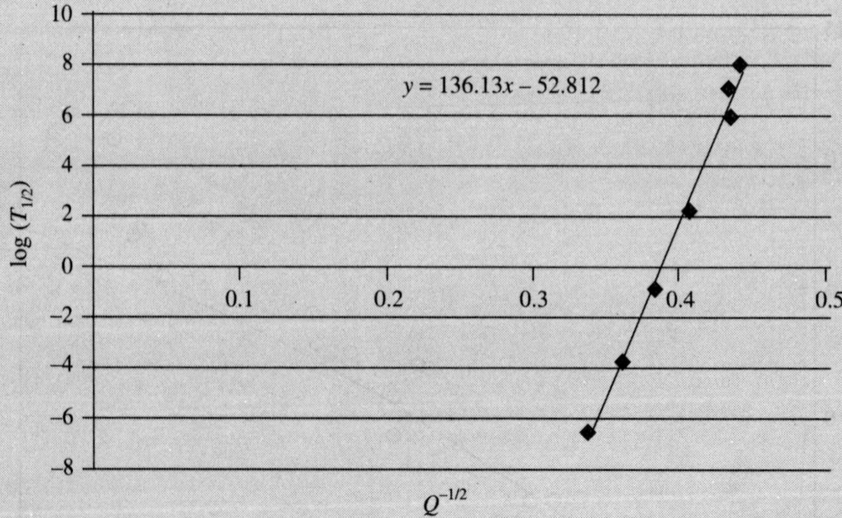

FIGURE 3.9 Correction of half-life with the decay energy for polonium isotopes.

(b) Using the values of constants a and b calculated in the previous part, we get

$$\log(4.51\times10^9\times365\times24\times3600) = -52.81 + \left(\frac{136.1}{84}\right)\left(\frac{92}{\sqrt{Q_\alpha}}\right)$$

$$Q_\alpha = 4.54 \text{ MeV}$$

The observed Q_α value for $^{238}_{92}U$ nuclide is 4.27 MeV and the predicted value is close to the observed value, though it must be noted that the most striking feature of this plot is the degree of sensitivity; a small change in Q gives rise to a change of much larger magnitude in the life-time.

It was puzzling though, that $^{238}_{92}U$ emits spontaneously an α-particle of ~4.2 MeV, while an α-particle incident on a heavy nucleus like uranium, experiences a Coulomb potential barrier of ~28 MeV. How could an α-particle escape through the high Coulomb barrier? Classically, it is not possible to explain this. A theoretical basis of understanding α-decay was not available until the advent of quantum mechanics. In 1928, Gamow and Condon independently proposed the mechanism of α-decay and explained the empirical relationship between the kinetic energy of the emitted alpha particle (E_α) and $T_{1/2}$ on the basis of quantum mechanics.

3.4.2 Gamow's Theory of Alpha Decay

Gamow's theory of α-decay is based on the following postulates:

(i) α-particles do not exist as such inside the heavy nucleus. Instead, they are formed, exist for some time, and disintegrate, over and over again.

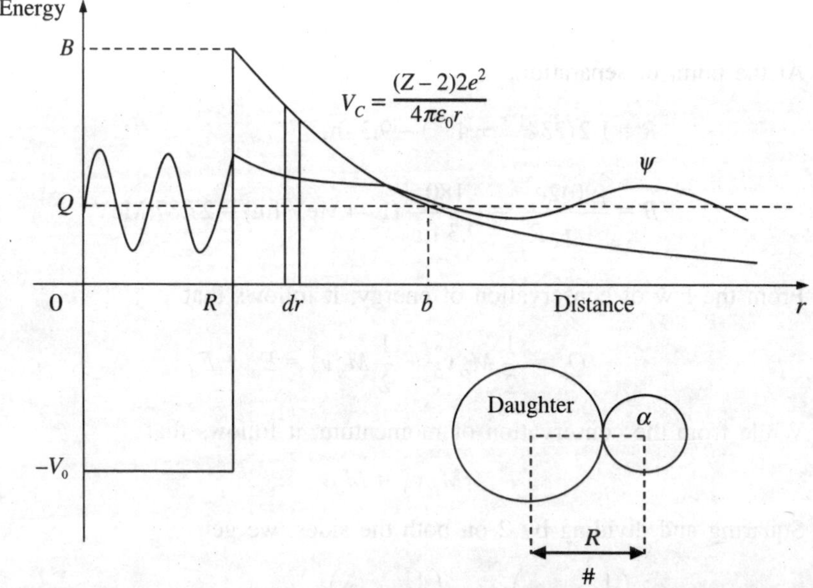

FIGURE 3.10 Schematic diagram of barrier penetration in alpha decay. The alpha particle wave function is oscillating inside the nucleus ($r < R$) and at large distance ($r > b$). Within the barrier region ($R < r < b$), the alpha particle cannot escape classically. However, it can tunnel quantum mechanically through the barrier with small but finite probability. Therefore, within the barrier region, the wave function is a decreasing function of r. The barrier can be considered to be series of narrow rectangular barriers of width dr, which the alpha particle has to tunnel in order to escape from the nucleus. For ($r > R$), $V_C(r)$ varies as $1/r$. '0' represents the centre of the nucleus. '#' represents the alpha particle and the daughter nucleus (A–4, Z–2) at the point of separation.

(ii) α-particles, in a negative energy level (bound) cannot penetrate the Coulomb barrier, but those in a positive energy level may tunnel through the Coulomb barrier, which is much higher than the observed kinetic energy of the emitted α-particles. Tunnelling is a quantum mechanical effect.

To estimate B (barrier height of the Coulomb potential), consider the α-particle and the daughter (residual) nucleus to be just at the point of separation (Figure 3.10).

$$V_C = \frac{(Z-2)2e^2}{4\pi\varepsilon_0 r}$$

$$V_C = B \quad \text{when } r = R = r_0(A_d^{1/3} + A_\alpha^{1/3})$$

EXERCISE 3.6:

(a) For a representative case of (${}^{238}_{92}\text{U} \rightarrow {}^{234}_{90}\text{Th} + {}^{4}_{2}\text{He}$), calculate the distance between the α-particle and the ${}^{234}_{90}\text{Th}$ nuclide at the point of separation. Also, estimate the barrier height of the Coulomb potential.

(b) Now using standard conservation laws, work out that the energy of the α-particle emitted by ${}^{238}_{92}\text{U}$ is ~4.2 MeV.

Solution:

(a) At the point of separation,

$$R = 1.2(234^{1/3} + 4^{1/3}) = 9.3 \text{ fm}$$

$$B = \frac{(90)2e^2}{4\pi\varepsilon_0 R} = \frac{180}{9.3 \text{ fm}}(1.44 \text{ MeV fm}) = 27.87 \text{ MeV}$$

(b) From the law of conservation of energy, it follows that

$$Q_\alpha = \frac{1}{2}M_\alpha v_\alpha^2 + \frac{1}{2}M_d v_d^2 = E_\alpha + E_d \tag{3.14}$$

While from the conservation of momentum, it follows that

$$M_\alpha v_\alpha = M_d v_d$$

Squaring and dividing by 2 on both the sides, we get

$$M_\alpha\left(\frac{1}{2}M_\alpha v_\alpha^2\right) = M_d\left(\frac{1}{2}M_d v_d^2\right) \quad \text{or} \quad M_\alpha E_\alpha = M_d E_d \tag{3.15}$$

Combining equations (3.14) and (3.15), we get

$$E_\alpha = Q_\alpha\left(\frac{M_d}{M_d + M_\alpha}\right) \tag{3.16}$$

Using $Q_\alpha = 4.27$ MeV, we get $E_\alpha = 4.198$ MeV and $E_d = 0.072$ MeV (the recoil energy of the daughter nucleus).

Since, $M_d >> M_\alpha$ for most heavy nuclei, $E_\alpha \approx Q_\alpha$, or in other words, most of the decay energy appears as the kinetic energy of the α-particle.

The transmission/tunnelling probability P of a particle of energy E through a rectangular Coulomb potential barrier of height V_C ($V_C > E$) is a standard problem in modern physics[6] with title—'one dimensional potential barrier (tunnel effect)' and is given as

$$P \approx e^{-2Ka}$$

where $K = \sqrt{\dfrac{2m(V_C - E)}{\hbar^2}}$

and a is the width of the barrier.

Close to the nuclear surface, in the region 0–R, α-particle is bound by a strong nuclear potential and requires kinetic energy that depends on V_0 (well depth). In Figure 3.10, the nuclear potential is taken to be a square well of radius R with sharp edge. Actually, the surface will not be sharp and the Coulomb barrier peak will have rounded shape.

Also for the decay of even–even nuclei to the ground states of even–even nuclei, the assumption that no angular momentum is carried away by the α-particle is justified, since both the parent and the daughter nuclei have spin 0. However, in other cases, the orbital angular momentum of the emitted α-particle, $l \neq 0$. In such cases, the potential that the α-particle has to tunnel through becomes bigger by the quantity known as the centrifugal barrier.

EXERCISE 3.7:

(a) Prove that the centrifugal barrier height is given by

$$\frac{l(l+1)\hbar^2}{2mR^2}$$

(b) Illustrate the effect of the centrifugal barrier in increasing the height and the width of the Coulomb barrier in case of ${}^{238}_{94}\text{Pu}$ alpha decay to third excited state (6^+) of ${}^{234}_{92}\text{U}$.

Solution:

(a) If v is the velocity of the particle at the point of separation and the particle is carrying an angular momentum, for a potential barrier of radius R, its angular momentum is given by:

$$L = mvR$$

In case, the emitted particle does not carry any angular momentum, using classical approach, the minimum energy of the particle to come out of the potential well is given by $E_{\min} = B$, where B is the height of the potential barrier. If however, the emitted particle does carry an angular momentum L, then using the classical approach, the minimum energy of the particle needed to come out of the potential well is given by:

6. Arthur Beiser, *Concepts of Modern Physics*, 6th ed., p. 193, McGraw Hill.

$$E_{\text{min}} = B + \frac{1}{2}mv^2 = B + \frac{L^2}{2mR^2}$$

The eigen values of L^2 are given by $l(l + 1)\hbar^2$. Hence, the above equation becomes

$$E_{\text{min}} = B_{\text{effective}} = B + \frac{l(l+1)\hbar^2}{2mR^2}$$

The extra term on the right represents the centrifugal barrier height.

(b) The Coulomb barrier for α-decay is given by

$$V_C = (Z-2)(2)\left(\frac{e^2}{4\pi\varepsilon_0}\right)\left(\frac{1}{r}\right) = 92\times 2(1.44 \text{ MeV fm})\left(\frac{1}{r}\right) = \frac{264.96}{r}$$

The centrifugal barrier is given as follows:

$$\frac{l(l+1)\hbar^2}{2mr^2} = \frac{l(l+1)(\hbar c)^2}{2(mc^2)r^2} = \frac{6\times 7(197.329 \text{ MeV fm})^2}{2(3727 \text{ MeV})}\left(\frac{1}{r^2}\right) = \frac{219.40}{r^2}$$

r will increase from $R = 1.2(234^{1/3} + 4^{1/3}) = 9.3$ fm. Plotting the centrifugal barrier on top of the Coulomb barrier, we get the effective barrier.

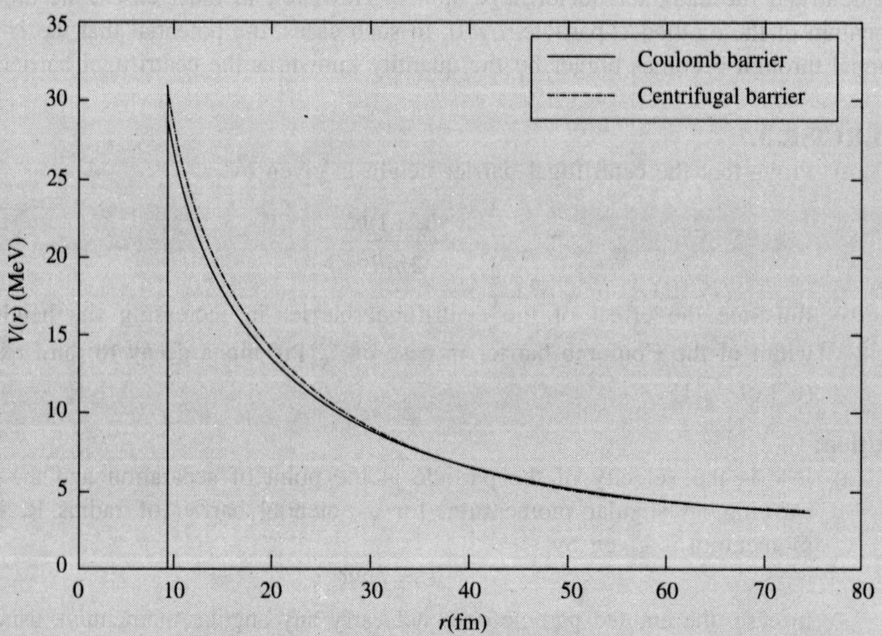

FIGURE 3.11 Effect of the centrifugal barrier to increase the height and width of the original Coulomb barrier.

As can be seen from Figure 3.11, the height of the potential barrier increases slightly and so does the width.

Just for simplicity, we shall continue to use the form as shown in Figure 3.10.

Extending the aforementioned expression for transmission probability to three-dimensional variable potential barrier, because an α-particle inside a nucleus is faced with a barrier of varying height, the transmission probability can be written as follows:

$$P = e^{-2\int K(r)\,dr} = e^{-2G} \tag{3.17}$$

where G is known as the Gamow factor.

Since, $Q_\alpha \approx E_\alpha$, we have

$$K(r) = \sqrt{\frac{2m}{\hbar^2}(V_C(r) - Q_\alpha))} = \sqrt{\frac{2m}{\hbar^2}\frac{(Z-2)2e^2}{4\pi\varepsilon_0}\left(\frac{1}{r} - \frac{1}{b}\right)}$$

$$\therefore \quad G = \sqrt{\frac{2m}{\hbar^2}\frac{(Z-2)\,2e^2}{4\pi\varepsilon_0}} \int_R^b \sqrt{\left(\frac{1}{r} - \frac{1}{b}\right)}\,dr = \sqrt{\frac{2m}{\hbar^2}\frac{(Z-2)2e^2}{4\pi\varepsilon_0}}\, I$$

To evaluate the integral I, we put $r = b\sin^2\theta$. Therefore, $dr = 2b\sin\theta\cos\theta\,d\theta$. Thus,

$$I = \int_{\sin^{-1}\sqrt{R/b}}^{\pi/2} \sqrt{\frac{1}{b}\cot^2\theta}\; 2b\sin\theta\cos\theta\,d\theta$$

$$= \int 2\sqrt{b}\cos^2\theta\,d\theta = \sqrt{b}\int (1 + \cos 2\theta)\,d\theta$$

$$\therefore \quad I = \sqrt{b}\left\{\frac{\pi}{2} - \sin^{-1}\left(\sqrt{\frac{R}{b}}\right) - \sqrt{\frac{R}{b}\left(1 - \frac{R}{b}\right)}\right\}$$

Thus,

$$G = \sqrt{\frac{2m}{\hbar^2}\frac{(Z-2)2e^2}{4\pi\varepsilon_0}}\sqrt{b}\left\{\cos^{-1}\left(\sqrt{\frac{R}{b}}\right) - \sqrt{\frac{R}{b}\left(1 - \frac{R}{b}\right)}\right\}$$

Now, we know that

$$B = \frac{(Z-2)2e^2}{4\pi\varepsilon_0 R} \quad \text{and} \quad Q_\alpha = \frac{(Z-2)2e^2}{4\pi\varepsilon_0 b}$$

or

$$R = \frac{(Z-2)2e^2}{4\pi\varepsilon_0 B} \quad \text{and} \quad b = \frac{(Z-2)2e^2}{4\pi\varepsilon_0 Q_\alpha}$$

Substituting R and b, we get

$$G = \frac{(Z-2)2e^2}{4\pi\varepsilon_0\hbar}\sqrt{\frac{2m}{Q_\alpha}}\left\{\cos^{-1}\left(\sqrt{\frac{Q_\alpha}{B}}\right) - \sqrt{\frac{Q_\alpha}{B}\left(1 - \frac{Q_\alpha}{B}\right)}\right\} \tag{3.18}$$

EXERCISE 3.8:

(a) What is the width of the $^{238}_{92}$U Coulomb barrier seen by the α-particle?
(b) How does the expression for Gamow factor reduce for low energies?
(c) Estimate the value of Gamow factor for $^{238}_{92}$U nuclei.
(d) Determine the probability of α-particle coming out from the Coulomb barrier per attempt for $^{238}_{92}$U nuclei.
(e) Determine the number of attempts of α-particle per unit time trying to come out of the potential barrier for $^{238}_{92}$U nuclei.
(f) Using the above values, estimate the half-life of α-decay for $^{238}_{92}$U nuclei. Comment on how it compares with the measured half-life value.

Solution:

(a) The distance at which the Coulomb potential becomes equal to the energy of the observed decay of $^{238}_{92}$U is b, such that

$$\frac{Q_\alpha}{B} = \frac{R}{b} \quad \text{or} \quad b = \frac{BR}{Q_\alpha}$$

Substituting the values calculated in Exercise 3.6, we get

$$b = \frac{(27.87 \text{ MeV})\,(9.3 \text{ fm})}{4.27 \text{ MeV}} = 60.7 \text{ fm}$$

Thus, the width of the $^{238}_{92}$U Coulomb barrier as seen by the α-particle is

$$b - R = 60.7 - 9.3 = 51.4 \text{ fm}$$

(b) Equation (3.18) can be written as follows:

$$G = \frac{(Z-2)2e^2}{4\pi\varepsilon_0\hbar}\sqrt{\frac{2m}{Q_\alpha}}\, f(x)$$

where $f(x) = [\cos^{-1} x - \sqrt{x(1-x)}]$ and $x = \dfrac{Q_\alpha}{B} = \dfrac{R}{b}$

From Figure 3.10, we know that $b > R$. Therefore, $0 < x < 1$.
At low energies, $x \to 0$ and $f \to \pi/2$ (Figure 3.12). Therefore,

$$G = \frac{(Z-2)2e^2}{4\pi\varepsilon_0\hbar}\sqrt{\frac{2m}{Q_\alpha}}\left(\frac{\pi}{2}\right)$$

(c) Substituting the known values for $^{238}_{92}$U nuclei, m = rest mass of α-particle[7] = 3727 MeV and the fine-structure constant $(e^2/4\pi\varepsilon_0\hbar) = 1/137$, in equation (3.18), we get

7. It is in fact the reduced mass, i.e., $m = \dfrac{(M_\alpha)(M_d)}{(M_\alpha + M_d)}$. The use of the reduced mass takes into account the recoil of the daughter nucleus. For heavier nuclei, the reduced mass is almost equal to the rest mass of α-particle.

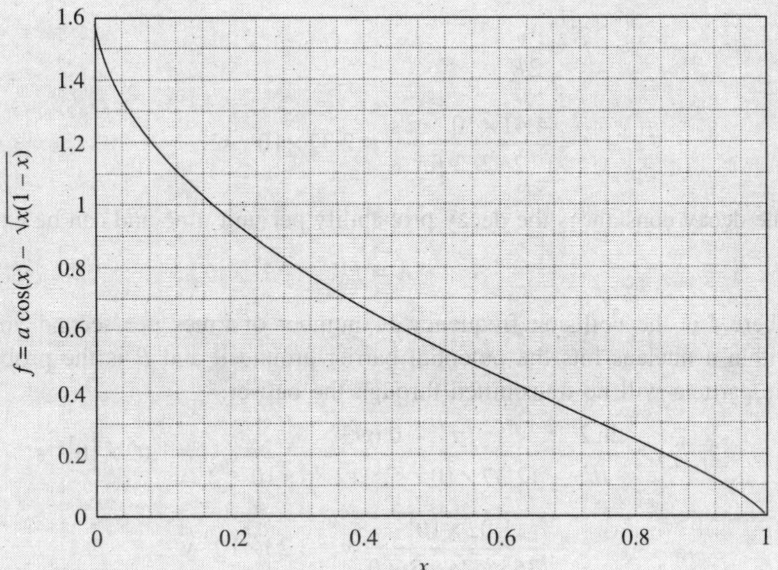

FIGURE 3.12 Plot of $f(x)$ vs. x.

$$G = \frac{90 \times 2}{137}\sqrt{\frac{2 \times 3727\text{ MeV}}{4.27\text{ MeV}}}\left\{\cos^{-1}\left(\sqrt{\frac{4.27}{27.87}}\right) - \sqrt{\frac{4.27}{28.87}}\left(1 - \frac{4.27}{28.87}\right)\right\} = 44.38$$

(d) From equation (3.17), we get

$$P = \exp(-2 \times 44.38) = 2.86 \times 10^{-39}$$

(e) If v_0 is the speed of α-particle inside the nucleus, then its kinetic energy is given by the sum of the kinetic energy of the emitted α-particle, and the well depth of the nuclear potential V_0, which can be taken as 36 MeV. Thus, we get

$$\frac{1}{2}mv_0^2 = Q_\alpha + V_0 \tag{3.19}$$

$$\therefore \qquad v_0 = \sqrt{\frac{2(Q_\alpha + V_0)}{m}} = \sqrt{\frac{2(Q_\alpha + V_0)}{mc^2}}\,c$$

Substituting the values, in the above equation, we get

$$v_0 = \sqrt{\frac{2(4.27 + 36)\text{ MeV}}{3727\text{ MeV}}}\,(3 \times 10^8)\text{ m/s} = 4.41 \times 10^7 \text{ m/s}$$

The frequency with which the α-particle hits the wall of the potential barrier along a nuclear diameter may be estimated as follows:

$$f = \frac{v_0}{2R} \tag{3.20}$$

$$f = \frac{4.41\times 10^7 \text{ m/s}}{2\times 9.3 \text{ fm}} = 2.37\times 10^{21}\text{ s}^{-1}$$

(f) The decay constant is the decay probability per unit time and can be expressed as:

$$\lambda = fP \tag{3.21}$$

where f is the collision frequency or number of times per second an α-particle within a nucleus hits the potential barrier around it and P is the probability that the particle will be transmitted through the barrier.

$$\therefore \qquad T_{1/2} = \frac{\ln 2}{fP} = \frac{0.693}{(2.37\times 10^{21}\text{ s}^{-1})(2.86\times 10^{-39})} = 1.02\times 10^{17}\text{ s}$$

$$= \frac{1.02\times 10^{17}}{365\times 24\times 3600}\text{ y} = 3.23\times 10^{9}\text{ y}$$

The measured half-life of α-decay for ${}^{238}_{92}\text{U}$ ~ 4.5×10^9 y. The agreement between the calculated value and the observed half-life is very good in spite of the crude nature of the calculation. We have taken the pre-formation probability of the α-cluster inside the nucleus to be unity, which is certainly an overestimate. The shape of the potential barrier is also unrealistic. However, this simple treatment undoubtedly illustrates the underlying physics in the alpha decay.

For small x, approximating $\cos^{-1} x = \pi/2 - x$, equation (3.18) can be further reduced to

$$G = \frac{(Z-2)\,2e^2}{4\pi\varepsilon_0\hbar}\sqrt{\frac{2m}{Q_\alpha}}\left\{\frac{\pi}{2} - \sqrt{\frac{Q_\alpha}{B}} - \sqrt{\frac{Q_\alpha}{B}\left(1-\frac{Q_\alpha}{B}\right)}\right\} \approx \frac{(Z-2)\,2e^2}{4\pi\varepsilon_0\hbar}\sqrt{\frac{2m}{Q_\alpha}}\left\{\frac{\pi}{2} - 2\sqrt{\frac{Q_\alpha}{B}}\right\}$$

$V_C(r)$ is equal to B, when $r = R$. Therefore,

$$B = \frac{(Z-2)2e^2}{4\pi\varepsilon_0 R} = \frac{Z_d 2e^2}{4\pi\varepsilon_0 R}$$

Using equation (3.17) and substituting the value of B in the above expression for G, we get

$$\ln P = \frac{4e}{\hbar}\left(\frac{m}{\pi\varepsilon_0}\right)^{1/2} Z_d^{1/2} R^{1/2} - \frac{e^2}{\varepsilon_0\hbar}\left(\frac{m}{2}\right)^{1/2} Z_d Q_\alpha^{-1/2}$$

On evaluating the constants, we get

$$\ln P = 2.97\, Z_d^{1/2} R^{1/2} - 3.95\, Z_d Q_\alpha^{-1/2}$$

Substituting $R = 1.2\, A^{1/3}$,

$$\log_{10} P = 1.42\sqrt{Z_d A^{1/3}} - 1.72\frac{Z_d}{\sqrt{Q_\alpha}} \tag{3.22}$$

We do not expect the α-particles to exist as clusters inside a heavy nucleus. However, there seems to be finite probability of finding two protons and two neutrons at the same place and at the same time inside the nucleus and form a tightly bound Helium nucleus. Taking the pre-formation probability of the α-cluster inside the nucleus into consideration, equation (3.21) can be rewritten as follows:

$$\lambda = p_\alpha f P \tag{3.23}$$

where p_α is the probability of finding an α-cluster inside a nucleus.

Taking logarithm on both the sides, we get

$$\log_{10}\lambda = \log_{10} p_\alpha + \log_{10}\left(\frac{v_0}{2R}\right) + 1.42\sqrt{Z_d A^{1/3}} - 1.72\frac{Z_d}{\sqrt{Q_\alpha}}$$

$$= \log_{10} p_\alpha + \log_{10}\left(\sqrt{\frac{(Q_\alpha + V_0)}{2m\left(r_0 A^{\frac{1}{3}}\right)^2}}\right) + 1.42\sqrt{Z_d A^{1/3}} - 1.72\frac{Z_d}{\sqrt{Q_\alpha}}$$

$$= \log_{10} p_\alpha - \log_{10}\sqrt{2mr_0^2} + \log_{10}\left(\frac{\sqrt{Q_\alpha + V_0}}{A^{1/3}}\right) + 1.42\sqrt{Z_d A^{1/3}} - 1.72\frac{Z_d}{\sqrt{Q_\alpha}}$$

$$= \log_{10} p_\alpha - \frac{1}{2}\log_{10}\left(2\left(\frac{3757}{c^2}\right)(1.2\times 10^{-15})^2\right) + \log_{10}\left(\frac{\sqrt{Q_\alpha + V_0}}{A^{1/3}}\right)$$

$$+ 1.42\sqrt{Z_d A^{1/3}} - 1.72\frac{Z_d}{\sqrt{Q_\alpha}}$$

Therefore,

$$\log_{10}\lambda = (\log_{10} p_\alpha + 21.46) + \log_{10}\left(\frac{\sqrt{Q_\alpha + V_0}}{A^{1/3}}\right) + 1.42\sqrt{Z_d A^{1/3}} - 1.72\frac{Z_d}{\sqrt{Q_\alpha}} \tag{3.24}$$

The above expression agrees with the Geiger–Nuttall law, in which the dominant energy dependence also comes from the last term. In addition, there is also a dependency on Z_d and A.

EXERCISE 3.9: In the previous exercise, what should be the pre-formation probability of the α-cluster inside the nucleus for which the calculated value of half-life of α-decay for ${}^{238}_{92}\mathrm{U}$ nuclei becomes equal to the measured value?

Solution:

$$T_{1/2} = \frac{\ln 2}{p_\alpha f P}$$

$$\therefore \quad p_\alpha = \frac{\left(\dfrac{\ln 2}{fP}\right)}{\text{Measured value of } T_{1/2}} = \frac{3.23\times 10^9 \text{ y}}{4.5\times 10^9 \text{ y}} \approx 0.7$$

3.4.3 Alpha Particle Energy Spectrum and Selection Rules

The energies of the α-particles emitted in the radioactive decay of even–even nuclei often form a group of several discrete peaks corresponding to the transitions from the ground state of the parent nucleus to the ground state and/or to one of the excited states of the daughter nucleus (Figure 3.13). The transitions are designated as α_0, α_1, α_2, … (Figure 3.14).

Nuclear transitions resulting in the emission of α-particles are subjected to the selection rules based on the conservation of angular momentum and parity. For a representative ($X \rightarrow Y + \alpha$) decay, the orbital angular momentum taken by α-particle, which itself has $J^\pi = 0^+$, is given by

$$|J_X - J_Y| \le l \le J_X + J_Y \tag{3.25}$$

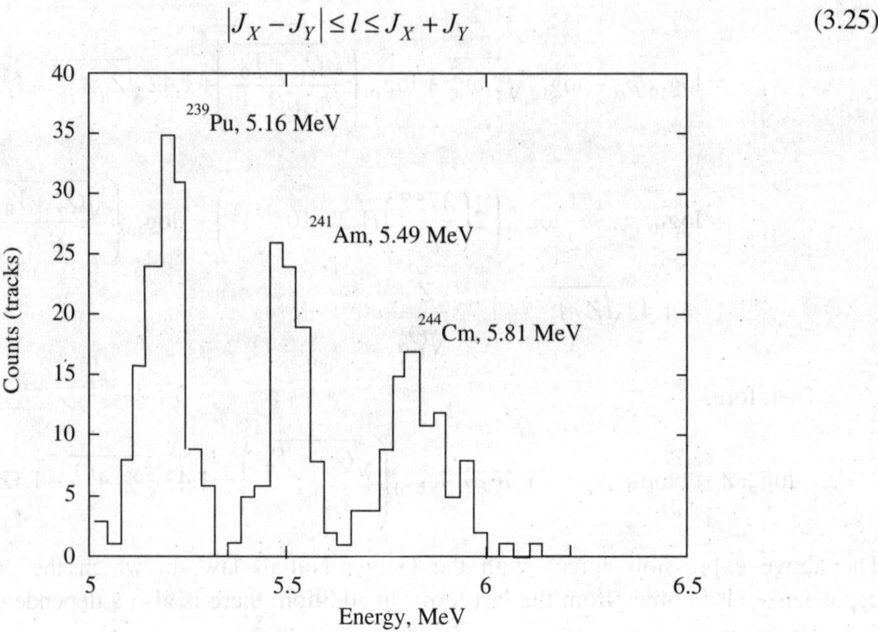

FIGURE 3.13 Alpha particle energy spectrum for a source containing three alpha energy, Pu-239, Am-241 and Cm-244, measured by solid-state nuclear track detector[8].

8. From <http://wiki.urps.info/en/Alpha_spectrometry_using_solid_state_track_detectors>

The selection rule for parity is given by

$$\text{Parity of } X = (\text{Parity of } Y)\ (\text{Even parity of alpha particle})\ (\text{Orbital parity}) \tag{3.26}$$

where orbital parity of α-particle is equal to $(-1)^l$.

A typical example, for the decay of ${}^{238}_{94}\text{Pu}$ to ${}^{234}_{92}\text{U}$ is shown in Figure 3.14, where various groups are observed with rapidly decreasing intensity (branching ratio) as the decay energy falls.

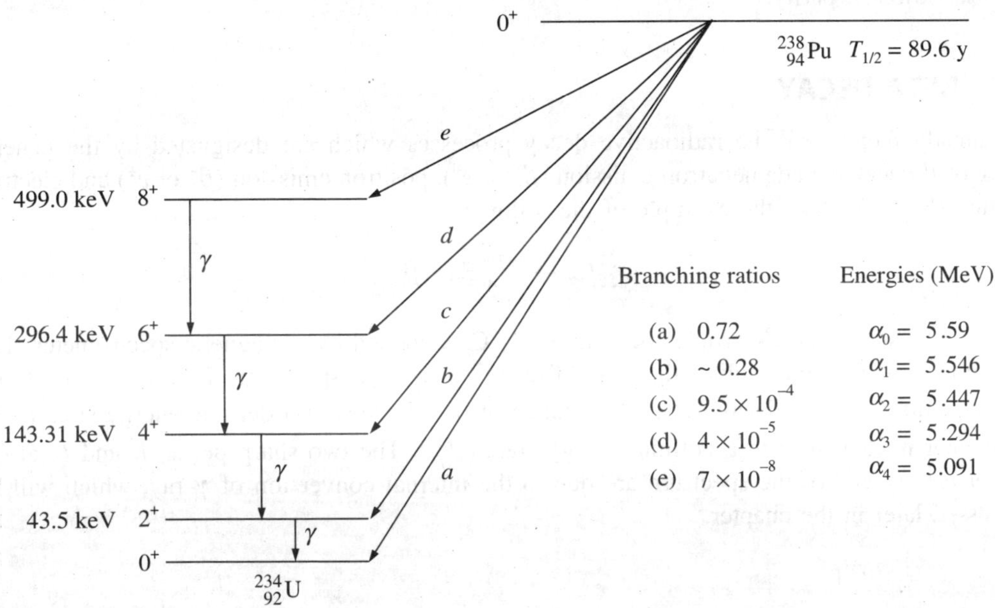

FIGURE 3.14 Alpha decay of Pu-238 into the ground state and excited states of U-234 with the branching ratios. The de-excitation of excited states occurs by gamma transition.

The decreasing intensity can be understood in the light of the following factors:

(i) Strong energy dependence of P (transmission probability), which decreases with increasing energy of the final state.

(ii) Differences in the structure of the initial and final nuclear states which can influence the α-cluster pre-formation probability.

(iii) Large angular momentum is carried away by the alpha particles when the decay proceeds to higher J^π excited states of the daughter nucleus. In our example, $J_i^\pi = 0^+$ and $J_f^\pi = 0^+, 2^+, 4^+, 6^+, 8^+$ (rotational band). The angular momentum carried away by the alpha particles are l = 0, 2, 4, 6, and 8, respectively. The effect of this is an additional centrifugal term in the potential and thus, the height and width of the original Coulomb barrier is increased, which causes a decrease in the corresponding alpha decay probability.

EXERCISE 3.10: Is the α-decay of 1^+ level in $^{20}_{10}$Ne to $0^+\,^{16}_{8}$O ground state possible?

Solution: The decay will proceed via $l = 1$ change. From the parity selection rule, we have

$$+ \neq (+)\ (+)\ (-)$$

The parity is not conserved. Hence, the transition is forbidden because of the non-conservation of parity.

3.5 BETA DECAY

As already mentioned, the radioactive decay processes which are designated by the general name of β-decay include negatron emission (β^- or e^-), positron emission (β^+ or e^+) and electron capture (EC). We take the example of the following:

$$^{137}_{55}\text{Cs} \rightarrow {}^{137m}_{56}\text{Ba} + \beta^-$$

Figure 3.15 shows the β-particle spectrum of $^{137}_{55}$Cs obtained by a magnetic spectrometer. The spectrum is continuous unlike the line (discrete) α-particle spectrum. This was against our understanding that the decay occurs by change of one nucleus in a definite energy state (J_i^π) to another nucleus also in a definite energy state (J_f^π). The two sharp peaks, K and L, at the higher energy end of the spectrum are due to the internal conversion of γ-ray, which will be discussed later in the chapter.

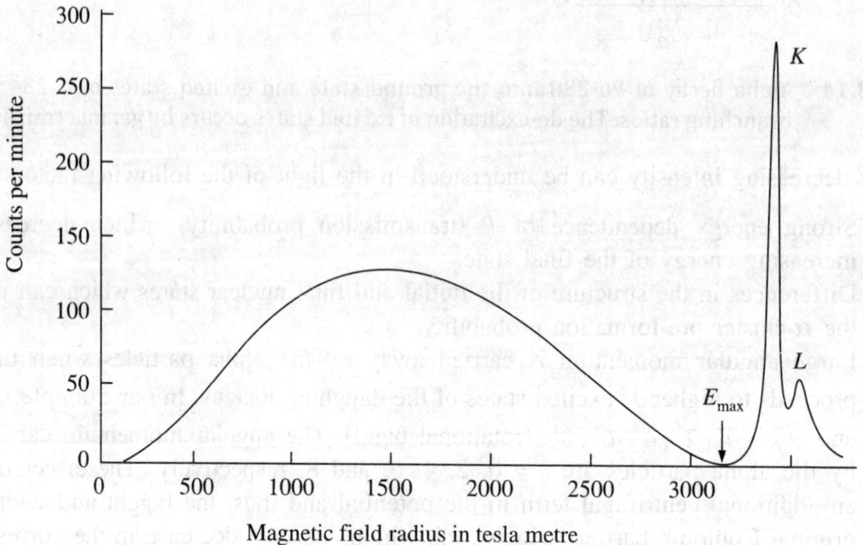

FIGURE 3.15 Spectrum for electrons emitted by Cs-137 as observed with a magnetic spectrometer.

Other puzzling aspects were apparent like non-conservation of linear momentum and angular momentum. When the directions of the emitted electrons and of recoiling nuclei are observed, they were almost never exactly opposite as expected from the point of view of conservation of linear momentum.

$$J_i \text{ (ground state of } {}^{137}_{55}\text{Cs}) = 7/2\hbar$$

$$J_f \text{ (meta-stable state of } {}^{137m}_{56}\text{Ba}) = 11/2\hbar$$

$$\text{Spin of } e^- = 1/2\hbar$$

In a nuclear decay, the angular momentum must be conserved. In our example, the sum of the spin of ${}^{137m}_{56}$Ba and electron is 11/2 + 1/2 = 6, while that of ${}^{137}_{55}$Cs is 7/2. The change in spin (ΔJ) is non-integral and thus, violates the conservation of angular momentum rule.

All these difficulties were sorted out by W. Pauli in 1930, who suggested that a third particle, later called a neutrino/antineutrino, must also be produced in β-decay. This is known as Pauli's neutrino hypothesis.

3.5.1 Pauli's Neutrino Hypothesis

To understand β-decay, Pauli postulated the existence of a new particle—neutrino. According to Pauli, an additional particle, called a neutrino denoted by ν, is emitted in β-decay, and carries away an amount of energy equal to the difference between the observed energy for a particular β-particle and the maximum observed energy $E_{\max}$ of the continuous β-spectrum. The principle of conservation of energy is thus, upheld. To satisfy the law of conservation of angular momentum and charge, the neutrino must have the following properties:

(i) It must have zero charge, because in β-decay the charge is already conserved without the neutrino.

(ii) It must have zero or almost zero mass.

(iii) It must have a spin of 1/2. This will satisfy the law of conservation of angular momentum. In our example, the total spin of the products will be:

$$\overrightarrow{\frac{11}{2}} + \overrightarrow{\frac{1}{2}} + \overrightarrow{\frac{1}{2}} = \frac{13}{2} \text{ or } \frac{11}{2} \text{ or } \frac{9}{2}$$

and when the spin of Cs-137 (= 7/2) is subtracted from it, it gives change in spin ΔJ = 3, 2 or 1, which is an acceptable integral value.

We know that the antiparticle (discussed in Chapter 8) of electron is the positron. Similarly, a neutrino ν has an antiparticle called an antineutrino $\bar{\nu}$. Just like neutrino, an antineutrino too has zero mass, zero charge and a spin of 1/2. The only difference between neutrino and antineutrino is that the neutrino has its spin s_ν always anti-parallel to its momentum p_ν, while the spin $s_{\bar{\nu}}$ of an antineutrino is always parallel to its momentum $p_{\bar{\nu}}$. In other words, ν is a left-handed particle, while $\bar{\nu}$ is a right-handed particle, where the thumb indicates the direction of the momentum and the direction of the curling fingers indicate the spin (upward for counterclockwise, and downward for clockwise) as shown in Figure 3.16. More on the detection and properties of ν and $\bar{\nu}$ will be discussed later in the chapter.

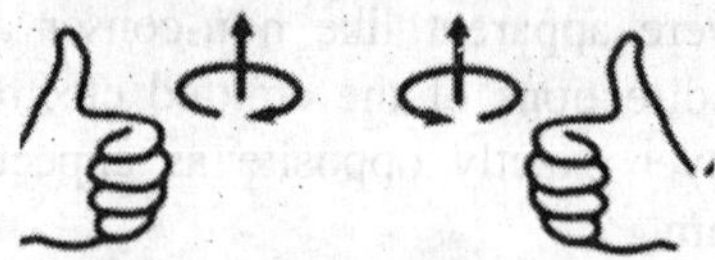

FIGURE 3.16 Neutrino (left-handed particle) and antineutrino (right-handed particle).

Based on Pauli's neutrino hypothesis, the decay in the example given before should be rewritten as follows:

$$^{137}_{55}\text{Cs} \rightarrow \, ^{137m}_{56}\text{Ba} + \beta^- + \overline{\nu}$$

When ν or $\overline{\nu}$ is ejected from the nucleus, it carries away energy which may be considered to be kinetic in nature. In fact, the total decay energy $Q = 0.514$ MeV, is distributed between the e^-, $\overline{\nu}$ and recoil of $^{137m}_{56}\text{Ba}$. Since, the recoil energy is relatively too small, all available energy is more or less shared between e^- and $\overline{\nu}$; there are many different ways that this can be done, so a continuous β-spectrum is expected. The neutrino energy spectrum is complementary to the β-particle energy spectrum. In β-decay, the average value of the β-particle energy ≈ 0.3 E_{max}, where E_{max} is also known as end point energy.

3.5.2 Energy Relations and *Q*-values in Nuclear Beta Decay

Nuclear β-decay is a process that involves nucleon transformation by which an unstable nucleus changes to optimize the N/Z ratio for a given mass number.

Consider three basic β-decay processes:

(i) β^- decay : $n_{\text{bound/free}} \rightarrow p + e^- + \overline{\nu}$

(ii) β^+ decay : $p_{\text{bound}} \rightarrow n + e^+ + \nu$

(iii) EC : $p_{\text{bound}} + e^- \rightarrow n + \nu$

β^- decay: The transformation can be written symbolically as follows:

$$^{A}_{Z}\text{X} \rightarrow \, ^{A}_{Z+1}\text{Y} + \beta^- + \overline{\nu} + Q_{\beta^-}$$

The *Q*-value for this decay,

$$\begin{aligned} Q_{\beta^-} &= \{m_Z c^2 - (m_{Z+1} + m_e)c^2\} \\ &= \{m_Z c^2 + Zm_e c^2 - (m_{Z+1} + (Z+1)m_e)c^2\} \\ &= \{(M_Z - M_{Z+1}\}c^2 \end{aligned}$$

$$\therefore \quad Q_{\beta^-} = \{(M_Z - M_{Z+1}\} \, 931.5 \text{ MeV} \tag{3.27}$$

For example, for the following decay,

$$^{137}_{55}\text{Cs} \rightarrow \, ^{137m}_{56}\text{Ba} + \beta^- + \overline{\nu} + Q_{\beta^-}$$

$$Q_{\beta^-} = \{M(^{137}_{55}\text{Cs}) - M(^{137m}_{56}\text{Ba})\}\ 931.5 \text{ MeV}$$

$$= \{(137 + \text{Mass excess of } ^{137}_{55}\text{Cs}) - (137 + \text{Mass excess of } ^{137m}_{56}\text{Ba})\}\ 931.5 \text{ MeV}$$

$$= -86.550 + 87.064 = 0.514 \text{ MeV}$$

β⁺ decay: The transformation can be written symbolically as follows:

$$^A_Z\text{X} \rightarrow {}^{A}_{Z-1}\text{Y} + \beta^+ + \nu + Q_{\beta^+}$$

The Q-value for this decay,

$$Q_{\beta^+} = \{m_Z c^2 - (m_{Z-1} + m_e)c^2\}$$

$$= \{m_Z c^2 + Zm_e c^2 - (m_{Z-1} + (Z-1)m_e + 2m_e)c^2\}$$

$$= \{(M_Z - M_{Z-1}\}c^2 - 2m_e c^2$$

$$\therefore \quad Q_{\beta^+} = \{(M_Z - M_{Z-1})\}\ 931.5 - 2(0.511) \text{ MeV} \tag{3.28}$$

For example, for the following decay:

$$^{13}_{7}\text{N} \rightarrow {}^{13}_{6}\text{C} + \beta^+ + \nu + Q_{\beta^+}$$

$$Q_{\beta^+} = \{M(^{13}_{7}\text{N}) - M(^{13}_{6}\text{C})\}\ 931.5 - 2(0.511) \text{ MeV}$$

$$= \{13.005738 - 13.003354\}\ 931.5 - 2(0.511) \text{ MeV} = 1.20 \text{ MeV}$$

EXERCISE 3.11:

Daughter Recoil: If the β-particle and $\nu/\bar{\nu}$ are emitted with the same momentum but in opposite direction, the daughter nucleus will experience no recoil. However, if they are both emitted in the same direction or if all the energy is carried away either by the β-particle or $\nu/\bar{\nu}$, the daughter will experience maximum recoil.

(a) Derive the expression for the recoil energy of the daughter nucleus when the daughter experiences maximum recoil. Keep in mind that the β energies are in MeV range, therefore relativistic treatment will be required.

(b) $^{11}_{6}\text{C}$ decays through the emission of positron of maximum energy of 1.0 MeV. Determine the recoil energy of the daughter, using the expression derived in part (a).

Solution:

(a) From the law of conservation of linear momentum, it follows that

$$p_\beta^2 c^2 = p_d^2 c^2 = 2(m_d c^2) E_d \tag{3.29}$$

where E_d is the recoil energy of the daughter and m_d is the rest mass of the daughter. Also the total relativistic energy is given by

$$E^2 = p^2c^2 + m^2c^4 \tag{3.30}$$

where the total energy is the sum of kinetic energy plus rest mass energy. Therefore from equations (3.29) and (3.30), we get the total energy for β-particle

$$(E_{\max} + m_e c^2)^2 = 2(m_d c^2)E_d + m_e^2 c^4$$

where $E_{\max}$ is the maximum kinetic energy of the β-particle and m_e is the rest mass of the electron. Therefore, we get

$$E_d = \frac{m_e E_{\max}}{m_d} + \frac{1}{2}\frac{E_{\max}^2}{m_d c^2} \tag{3.31}$$

(b) The β^+ decay can be written as follows:

$$^{11}_{6}\text{C} \rightarrow {}^{11}_{7}\text{N} + \beta^+ + \nu$$

Substituting the values in the expression for E_d, we get

$$E_d = \frac{(0.0005486u)\,(1\text{ MeV})}{(11.009305\text{ u})} + \frac{1}{2}\frac{(1\text{ MeV})^2}{(11.009305\text{ u})\,(931.5\text{ MeV/u})}$$

$$= 49.83\text{ eV} + 48.76\text{ eV}$$

or $E_d = 98.6$ eV

Electron Capture (EC): A process similar to β^+ decay by which Z decreases by one unit, is called electron capture. An atomic electron is captured by a proton, thus transforming the proton into a bound neutron along with the emission of a neutrino. The transformation can be written symbolically as follows:

$$^{A}_{Z}\text{X} \xrightarrow{EC} {}^{A}_{Z-1}\text{Y} + \nu + Q_{EC}$$

The Q-value for this decay

$$Q_{EC} = \{m_Z c^2 + m_e c^2 - m_{Z-1}c^2\}$$

$$= \{m_Z c^2 + Zm_e c^2 - (m_{Z-1} + (Z-1)m_e)c^2\}$$

$$\therefore \quad Q_{EC} = \{M_Z - M_{Z-1}\}\,931.5\text{ MeV} \tag{3.32}$$

For example, for the following decay

$$^{54}_{25}\text{Mn} \xrightarrow{EC} {}^{54}_{24}\text{Cr} + \nu + Q_{EC}$$

$$Q_{EC} = \{M(^{54}_{25}\text{Mn}) - M(^{54}_{24}\text{Cr})\} \times 931.5\text{ MeV}$$

$$= \{(54 - 0.059638\text{ u}) - (54 - 0.061118\text{ u})\} \times 931.5\text{ MeV} = 1.38\text{ MeV}$$

EXERCISE 3.12:

${}^{57}_{27}Co$, whose decay scheme is given in Figure 3.3, decays almost completely by electron capture. From the obtained Q-value expressions for β^+ and EC decay processes, show that β^+ is always associated with EC, while the reverse is not true?

Solution:

$Q_{\beta^+} = \{M_Z - M_{Z-1}\} \times 931.5 - 2m_ec^2$ MeV, implies that β^+ is possible if

$$M_Z > M_{Z-1} + 2m_e \quad (3.33)$$

$Q_{EC} = \{M_Z - M_{Z-1}\} \times 931.5$ MeV, implies that EC is possible if

$$M_Z > M_{Z-1} \quad (3.34)$$

From equations (3.33) and (3.34), we can conclude that β^+ is always associated with EC while the reverse is not true.

EC mode of decay is dominant for neutron deficient nuclei whose $Z > 80$. Depending on the electron shell from which it is captured, the process is referred to as K-capture, L-capture, etc. The probability for electron capture from the K-shell is larger than that for the capture of an electron from the L-shell, because the wave function of K-electrons is larger at the nucleus than that of L-electrons. The capture of an orbital electron produces no detectable primary radiation from the nucleus, because neutrinos are not easy to detect. There are however, detectable secondary processes.

The electron capture process creates a vacancy in an atomic electron shell, and the rearrangement of atom starts taking place by emitting X-rays and Auger electrons. If the decay proceeds to an excited state of the daughter nucleus, there will be either γ-rays or conversion electrons from the decay of this excited state. These secondary processes are discussed later in this chapter.

In terms of the binding energies, the Q-values are given as follows:

$$Q_{\beta^-} = B(Z+1, N-1) - B(Z, N) + 0.782 \text{ MeV}$$

$$Q_{\beta^+} = B(Z-1, N+1) - B(Z, N) - 2m_ec^2 - 0.782 \text{ MeV}$$

$$Q_{\text{EC}} = B(Z-1, N+1) - B(Z, N) - B_e - 0.782 \text{ MeV}$$

where the amount of energy of 0.782 MeV comes from the mass difference between a neutron and a neutral hydrogen atom. B_e is the binding energy of the electron in $K/L/M$ shell. Since, $B_e \approx 10$ eV, one can ignore it.

EXERCISE 3.13: In the following *K*-capture, the beryllium source is at rest:

$$^{7}_{4}\text{Be} \xrightarrow{\text{EC}} {}^{7}_{3}\text{Li} + \nu$$

The recoil energy of the lithium atoms (mass = 6536 MeV/c^2) was measured to be 55.9 ± 1.0 eV. The mass difference between the two atoms is 0.862 MeV/c^2. Based on the given data, find the range of possible mass of neutrino.

Solution: Just like we used the relativistic equations in Exercise 3.11, here too we will do the same for neutrino. From the law of conservation of energy, it follows that

$$p_\nu^2 c^2 = p_d^2 c^2 = 2(m_d c^2)\, E_d \tag{3.35}$$

where E_d is the recoil energy of the daughter and m_d is the rest mass of the daughter. Also the total relativistic energy is given by

$$E^2 = p^2 c^2 + m^2 c^4 \tag{3.36}$$

From equations (3.35) and (3.36), we get the expression for the rest mass of neutrino,

$$m_\nu c^2 = \sqrt{E^2 - 2(m_d c^2)\, E_d}$$

Substituting the values in the above expression, we get

$$m_\nu c^2 = \sqrt{(0.862\,\text{MeV})^2 - 2(6536\ \text{MeV})(55.9 \pm 1.0\ \text{eV})}$$

$$= \sqrt{0.0123 \pm 0.0131}\ \text{MeV}$$

or $$0 \le m_\nu c^2 \le 159.5\ \text{keV}$$

3.5.3 Fermi Theory of Beta Decay

A simple theory of β-decay was proposed by Fermi in 1934. Although, this theory does not allow for parity violation, which was observed in 1957, it is able to describe the continuous spectra of β-particles and gives a qualitative understanding of the range of β-decay mean-lives.

Fermi assumed that β-decay results from some form of interaction among nucleons, electrons and $\nu/\bar{\nu}$. He anticipated this interaction with a coupling constant $g = 8.8 \times 10^{-5}$ MeV fm^3, later named as weak interaction. The adjective 'weak' indicated that the strength is roughly 10^6 times smaller than the strength of *n-n* (strong) interaction. Fermi further assumed that the electrons and antineutrino are created only at the instant of emission. The act of creation is similar to the process of photon emission in atomic and nuclear decay processes.

In Figure 3.17, we show a diagram representing β-decay of a neutron and bound proton (free proton does not decay) as a 'contact' interaction conceived by Fermi.

It is because of the weakness of the interaction that β-decay does not take place almost instantaneously. In α-decay, it is the Coulomb barrier, otherwise, strong interaction would have resulted in instantaneous α-emission. There is no comparable barrier in β-decay and yet the mean-lives of α and β-decay are of the same order.

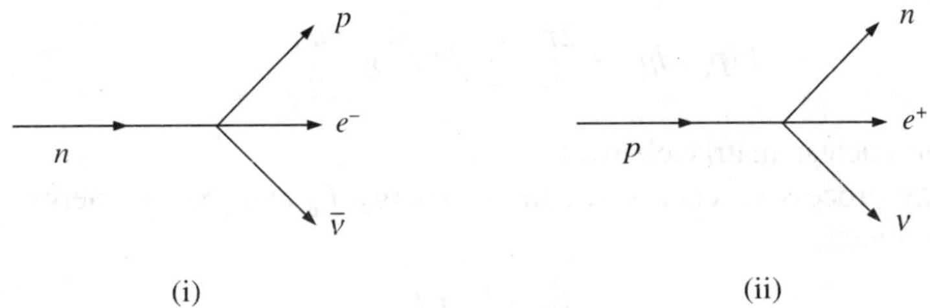

FIGURE 3.17 Beta decay of a (i) neutron and (ii) bound proton as a 'contact' interaction.

Following time-dependent perturbation theory,

$$\text{Decay probability} \propto \left|\langle\psi_f|H'|\psi_i\rangle\right|^2$$

where ψ_i and ψ_f are the initial and final state of nuclear wave functions. H' is perturbation to the total Hamiltonian. The fundamental problem of β-decay theory lies in the choice of a perturbation H' that will produce matrix elements in agreement with the experimental observations.

E_0 (available β-decay energy) can be divided between β and ν in an infinite number of ways, but not all divisions are equally probable, for if they were, then the energy spectrum of β-particles would be perfectly flat. The most probable division occurs when each takes roughly one half. This is why β-energy spectrum has a maximum at about $E_0/2$.

The probability $P(p_e)$ for the emission of an electron of momentum p_e and width of dp_e is,

$$P(p_e)\,dp_e = \frac{2\pi}{\hbar}\left|\langle\psi_{e^-}\psi_{\bar{\nu}}\psi_f|H'|\psi_i\rangle\right|^2 g^2\frac{dn}{dE_0}$$

where ψ_{e^-} and $\psi_{\bar{\nu}}$ are the wave functions of e^-, and $\bar{\nu}$, (dn/dE_0) represents the number of ways in which the available total energy E_0 can be divided between e^- and $\bar{\nu}$.

$\psi_{e^-} = Ae^{i\bar{p}_e\cdot\bar{r}/\hbar}$ and $\psi_{\bar{\nu}} = Be^{i\bar{p}_{\bar{\nu}}\cdot\bar{r}/\hbar}$, since e^- and $\bar{\nu}$ interact weakly with nuclear matter, wave functions are simply the solution of Schrödinger equation for particles in a force-free region. Now, these wave functions must be normalized, i.e.,

$$\int_0^V \psi_e^*\psi_e\,d\tau = \int_0^V \psi_{\bar{\nu}}^*\psi_{\bar{\nu}}\,d\tau = 1$$

where V is the nuclear volume, since the interaction occurs in the nucleus and that is the region we are interested in. Thus, we get $A = B = 1/\sqrt{V}$.

Since, both the particles do not exist inside the nucleus, we can safely assume that the nuclear radius is very small as compared to the de Broglie wavelengths for e^- and $\bar{\nu}$. Therefore, we can put $r = 0$ in the particle wave functions.

$$\psi_e(r=0) = \psi_{\bar{\nu}}(r=0) = 1/\sqrt{V}$$

$$P(p_e)\,dp_e = \frac{2\pi}{\hbar}\frac{1}{V^2}\left|\langle\psi_f|H'|\psi_i\rangle\right|^2 g^2\frac{dn}{dE_0}$$

or
$$P(p_e)\,dp_e = \frac{2\pi}{\hbar}\frac{1}{V^2}\left|M_{if}\right|^2 g^2 \frac{dn}{dE_0} \tag{3.37}$$

where $|M_{if}|$ is the nuclear matrix element.

In a β-decay process of energy E_0, the e^- energy E_e and the $\bar{\nu}$ energy $E_{\bar{\nu}}$ are related by the following equation:

$$E_0 = E_e + E_{\bar{\nu}}$$

Suppose β-particle is found at the point (x, y, z) in the space with (p_x, p_y, p_z) as its momenta along x, y and z axes. From the uncertainty principle, we know

$$\Delta x \Delta p_x \approx h$$

Similarly, for y and z. Hence, we get

$$\Delta x \Delta y \Delta z \Delta p_x \Delta p_y \Delta p_z \approx h^3$$

If we consider an imaginary six-dimensional space (x, y, z, p_x, p_y, p_z), called the phase space, according to the uncertainty principle, one cannot specify the exact point in this space. One could say that the point representing the position and momentum of the particle is somewhere in a volume of size h^3. This volume is known as a unit cell of the phase space.

If we specify that a particle is to be found anywhere within a certain volume of the ordinary space with a momentum lying between certain limits, then it is equivalent to specifying the volume of the phase space corresponding to this situation. The probability of the specified situation is proportional to the number of unit cells within the specified volume of the phase space.

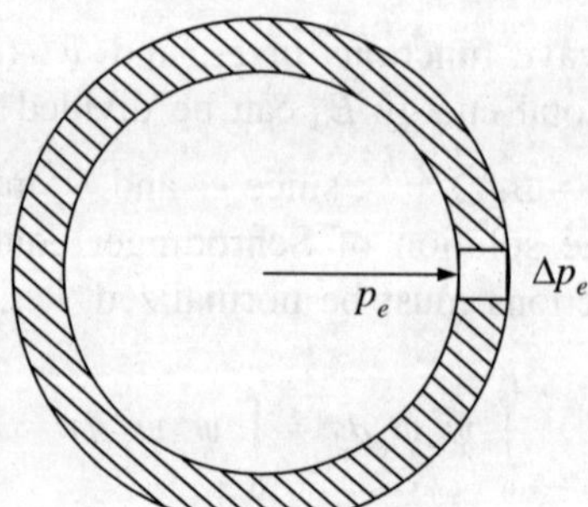

FIGURE 3.18 Cross section of volume in momentum space representing uncertainty in the momentum of an electron.

In Figure 3.18, the volume of the shaded portion is $\Delta p_x \Delta p_y \Delta p_z = 4\pi p_e^2 \Delta p_e$, where p_e is the momentum of the electron emitted in any direction around the nucleus. If we now require that the e^- should be somewhere within a volume V, then the corresponding volume in phase space becomes $4\pi p_e^2 \Delta p_e V$.

Number of different ways in which e^- can have the position and momentum specified by the expression is equal to the number of unit cells within this volume, which is $4\pi p_e^2 \Delta p_e V/h^3$.

Therefore, the number of ways in which the e^- may be in volume V with momentum lying between p_e and $p_e + dp_e$ is given by:

$$dN = 4\pi p_e^2 \Delta p_e V / h^3$$

Similarly, for $\bar{\nu}$

$$dN = 4\pi p_{\bar{\nu}}^2 \Delta p_{\bar{\nu}} V / h^3$$

Hence the number of ways, dn, in which a disintegration can lead to e^- having momentum between p_e and dp_e, while $\bar{\nu}$ has momentum between $p_{\bar{\nu}}$ and $dp_{\bar{\nu}}$ is just the product of the two,

$$dn = \frac{16\pi^2 V^2}{h^6} p_e^2 p_{\bar{\nu}}^2 dp_e dp_{\bar{\nu}}$$

Now, we know that the total relativistic energy of a particle is given by

$$E^2 = p^2c^2 + m^2c^4$$

where the total energy is the sum of the relativistic kinetic energy plus rest mass energy ($E = K + mc^2$). Therefore,

$$p = \frac{1}{c}\sqrt{K(K + 2mc^2)}$$

Since, the total electron energy E_e includes its rest mass (m_ec^2 = 511 keV), the kinetic energy of the electron/positron is $K_e = E_e - m_ec^2$. The maximum electron kinetic energy is at the end point of a β-spectrum.

Taking the $\nu/\bar{\nu}$ rest mass to be zero, the momentum of $\bar{\nu}$ will be

$$p_{\bar{\nu}} = \frac{E_{\bar{\nu}}}{c} = \frac{E_0 - E_e}{c}$$

and for constant E_e,

$$dp_{\bar{\nu}} = \frac{dE_0}{c}$$

therefore,

$$dn = \frac{16\pi^2 V^2}{h^6} p_e^2 \left(\frac{E_0 - E_e}{c}\right)^2 dp_e \left(\frac{dE_0}{c}\right)$$

or

$$\frac{dn}{dE_0} = \frac{16\pi^2 V^2}{h^6 c^3}(E_0 - E_e)^2 p_e^2 dp_e$$

Substituting it in equation (3.37), we get

$$P(p_e)\,dp_e = \frac{2\pi}{\hbar}\frac{1}{V^2}\left|M_{if}\right|^2 g^2 \frac{16\pi^2 V^2}{h^6 c^3}(E_0 - E_e)^2 p_e^2 dp_e$$

Now, replacing $h = 2\pi\hbar$, in the above equation, we get

$$P(p_e)\,dp_e = \frac{1}{2\pi^3 \hbar^7 c^3}\left|M_{if}\right|^2 g^2 (E_0 - E_e)^2 p_e^2 dp_e$$

Until now, we have ignored the effect of nuclear charge of the daughter nucleus on the kinetic energy of the e^-/e^+. The charged electrons/positrons emitted in the decay cannot be treated as free particles, since the decay is in the Coulomb field of the daughter nucleus. With this additional correction factor $F(Z, E_e)$, which is known as the Fermi function, we get

$$P(p_e)\,dp_e = \frac{1}{2\pi^3\hbar^7 c^3}\left|M_{if}\right|^2 g^2 F(Z, E_e)\,(E_0 - E_e)^2 p_e^2\, dp_e \tag{3.38}$$

Extensive tables for $F(Z, E_e)$ are available, but simple approximation is the following non-relativistic formula:

$$F(Z, E_e) = \frac{2\pi\eta}{1 - e^{-2\pi\eta}} \quad \text{where } \eta = \pm\frac{Ze^2}{4\pi\varepsilon_0 \hbar v_e} \tag{3.39}$$

The positive sign holds for electrons, the negative for positrons and v_e is the e^-/e^+ final velocity. As $v_e \to 0$, $F(Z, E_e) \to 2\pi\eta$ for electrons. The decay rate is enhanced at low energies, since the Coulomb field for e^- is attractive. For positrons at low energies, $F(Z, E_e) \to 2\pi|\eta|e^{-2\pi|\eta|}$. The Coulomb field is repulsive for e^+ and one can understand the exponential term as the tunneling factor through the Coulomb barrier, which suppresses positron emission at low energies.

It is also found that $|M_{if}|$ is not a function of the energy of the β-particle. This is especially true for the allowed transitions. Then, we can say

$$P(p_e)\,dp_e = (\text{Constants})\, F(Z, E_e)(E_0 - E_e)^2 p_e^2\, dp_e \tag{3.40}$$

We can see that $P(p_e) = 0$, when $p_e = 0$ or when $E_0 = E_e$. Between zero and the maximum possible β-particle energy, $P(p_e)$ varies smoothly with E_e. The form of momentum spectrum of the β-particles emitted in an allowed β-decay process is given by equation (3.40). By comparing this formula with the experimental β-spectrum, one can:

(i) Test the Fermi theory of β-decay.
(ii) Obtain information if the process is allowed or forbidden.
(iii) Obtain the value of nuclear matrix element $|M_{if}|$.

We can also get the following form:

$$\left[\frac{P(p_e)}{F(Z, E_e)\, p_e^2}\right]^{1/2} \propto (E_0 - E_e) \tag{3.41}$$

If the left hand term is plotted against β-particle energy, a straight line should be obtained with E_0 (end point energy of the β-particles) as the intercept on the energy axis. Such a graph is called Fermi-Kurie plot and provides not only a way to test the theory but also an accurate way of determining E_0. If $m_\nu \neq 0$, but very small, interesting deviations from a straight line appear as shown in Figure 3.19. A straight line curve of Fermi–Kurie plot supports that the β-decay theory is reasonably correct for the allowed transitions. In general, these plots for the forbidden transitions deviate from the straight line.

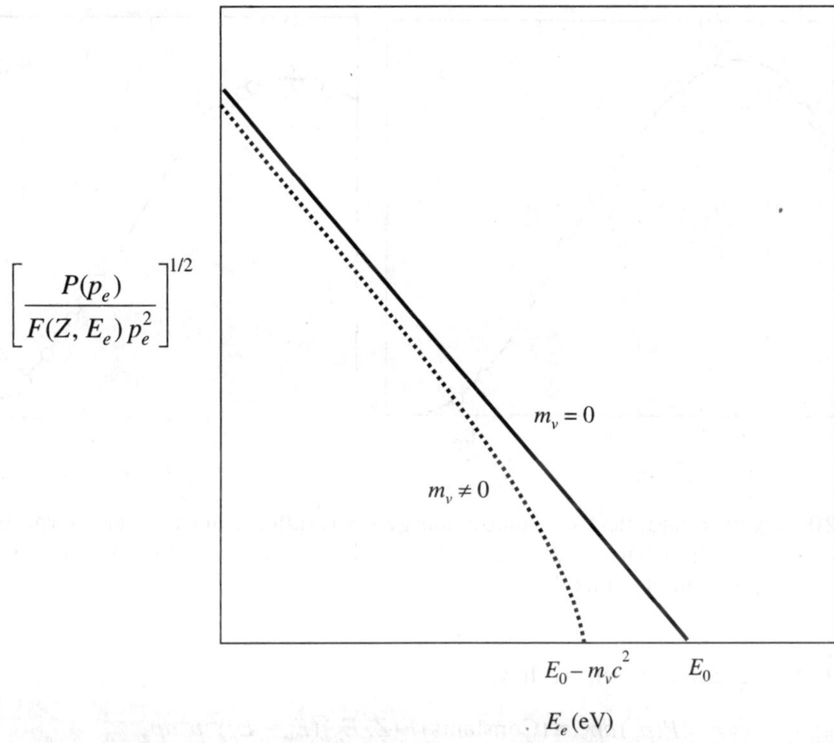

FIGURE 3.19 Fermi–Kurie plot when rest mass of neutrino is zero and non-zero.

EXERCISE 3.14: Figure 3.20 shows fits of the experimental data for the positron and the electron emission on the basis of the Coulomb corrected equation (3.40) for ${}^{64}_{29}\text{Cu}$ nucleus, which is a good example for the energy spectra, since it decays by both β^+ to ${}^{64}_{28}\text{Ni}$ and β^- to ${}^{64}_{30}\text{Zn}$ with comparable transition energies.

(a) Express equation (3.40) in terms of the kinetic energy of the emitted β-particles and obtain the above shapes of the energy spectrum with and without the Coulomb correction factor, $F(Z, E_e)$ using MS Excel. Comment on the need for the Coulomb correction factor.
(b) Discounting the Fermi factor, where does the most intense energy occur?
(c) Where does the most intense energy occur with $F(Z, E_e)$?

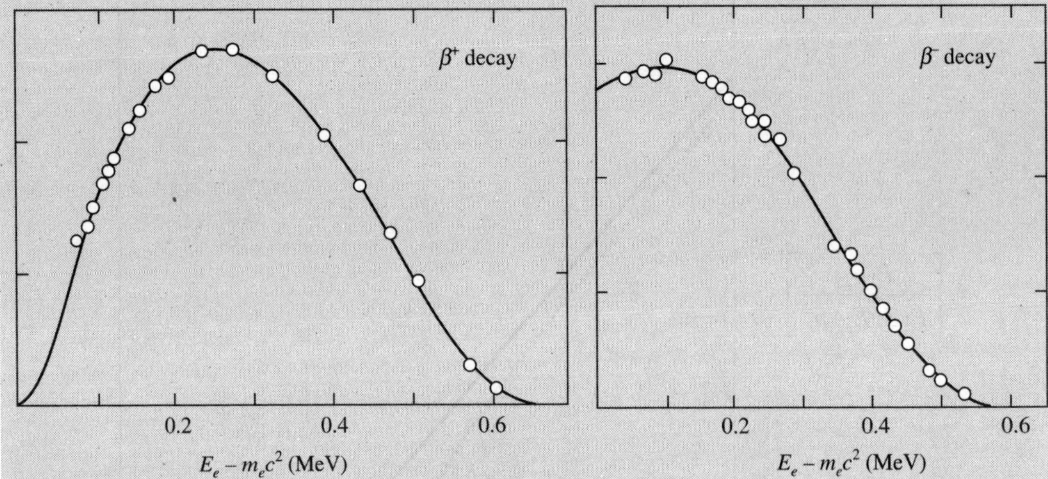

Figure 3.20 Positron and electron kinetic energy distribution from the beta decay of Cu-64, giving the normalized probability distributions in energy from a large sample of experimental points.

Solution:

(a) From equation (3.40), we have

$$P(p_e)\,dp_e = (\text{Constants})\,F(Z, E_e)(E_0 - E_e)^2 p_e^2 dp_e$$

We want the probability distribution in terms of the kinetic energy (K_e) of the β-particle. We know the total relativistic energy is given by:

$$E^2 = p^2c^2 + m^2c^4$$

$$\therefore \qquad p_e^2 = \frac{E_e^2 - m_e^2c^4}{c^2}$$

By substituting p_e^2, we can get the probability distribution as a function of E_e.

$$P(E_e) = (\text{Constants})F(Z, E_e)(E_0 - E_e)^2(E_e^2 - m_e^2c^4)$$

The total energy of the β-particle is the sum of its kinetic energy and its rest mass. Replacing $E_e = K_e + m_ec^2$ in the above expression, we get the probability distribution as a function of K_e.

$$P(K_e) = (\text{Constants})F(Z, K_e)(K_0 - K_e)^2(K_e^2 + 2K_em_ec^2) \qquad (3.42)$$

Here, K_0 is the maximum kinetic energy in the β-spectrum.

***Case* (i)**

$F(Z, K_e)$ is treated as a constant. From Figure 3.20, we can see that $K_0 \approx 0.6$ MeV. Substituting K_0 and the value of the rest mass of electron = 0.511 MeV, we get the following probability distribution:

$$P(K_e) = (\text{Constants})(0.6 - K_e)^2(K_e^2 + 2 \times 0.511\, K_e)$$

As can be seen in Figure 3.21, without the Coulomb correction factor, the shapes of both electron and positron cannot be differentiated and we get a single curve. But we already know from Figure 3.20, that the shapes of the two curves should be different. That is where $F(Z, K_e)$ comes into the picture.

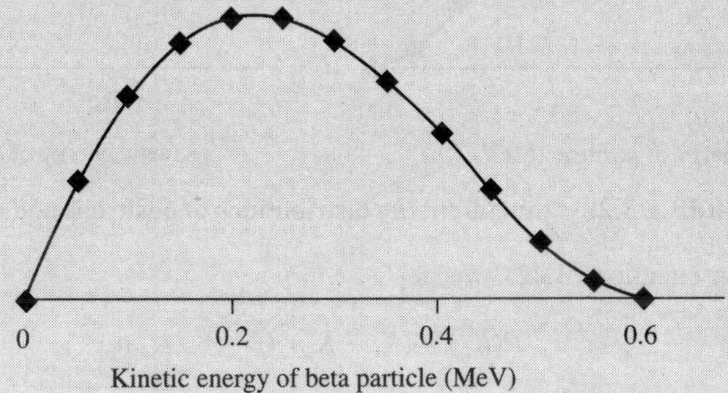

FIGURE 3.21 Kinetic energy distribution of beta particles.

***Case* (ii)**

$F(Z, K_e)$ is incorporated. From equation (3.39), we know

$$\eta = \pm \frac{Ze^2}{4\pi\varepsilon_0 \hbar v_e}$$

Since, the correction factor mainly affects the low energy part of the β-spectrum, where the speeds of the β-particles are small, we should be able to use non-relativistic kinetic energy in our expression for η. Thus, we get

$$\eta = \pm \frac{Ze^2}{4\pi\varepsilon_0 \hbar c \sqrt{\dfrac{2K_e}{m_e c^2}}}$$

Substituting the fine-structure constant and the value of the rest mass of electron, we get

$$\eta = \pm \frac{Ze^2}{4\pi\varepsilon_0 \hbar c \sqrt{\dfrac{2K_e}{m_e c^2}}} = \pm\, 0.00369 \frac{Z}{\sqrt{K_e}}$$

Here Z is the atomic number of the daughter nucleus; 28 in the case of positron emission and 30 for electron emission. We get the following plots (Figure 3.22) for positron and electron emissions with $F(Z, K_e)$.

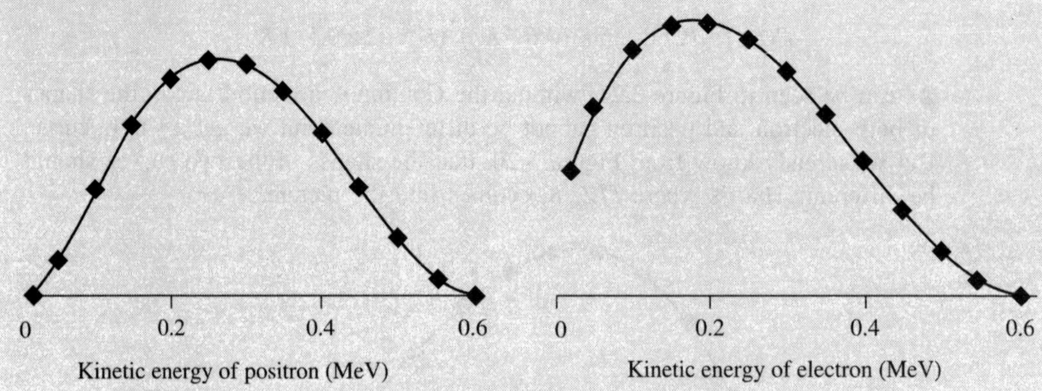

FIGURE 3.22 Kinetic energy distributions of positrons and electrons.

(b) From equation (3.42), we get

$$P(K_e) \propto (K_0 - K_e)^2 (K_e^2 + 2K_e m_e c^2)$$

The maximum intensity occurs at the point of maxima.

$$\frac{d(P)}{dK_e} = 0$$

or $$2K_e^2 + (3m_e c^2 - K_0)K_e - K_0 m_e c^2 = 0$$

or $$K_e = \frac{-(3m_e c^2 - K_0) \pm \sqrt{(3m_e c^2 - K_0)^2 + 8K_0 m_e c^2}}{4}$$

Substituting, $K_0 = 0.6$ MeV and $m_e c^2 = 0.511$ MeV, we get the maximum intensity at 0.22 MeV.

(c) The following Matlab code is used to plot the first derivative (using built-in function *diff*) of $P(K_e)$ for the electron emission.

```
clear all

Z = 30; %Daughter nucleus atomic no.
syms x real positive; %KE
n=0.00369*Z*(x^-0.5); %eta for electron
Fe=(2*pi*n)/(1-exp(-2*pi*n)); %Fermi factor
P=Fe*((0.6-x)^2)*(x^2+2*0.511*x); %Probability distribution
dP = diff(P,x); %First derivative of P
ezplot(dP,[0,0.6])
hold on
plot([0 1],[0 0],'r') %x-axis
grid minor
xlabel('Ke (MeV)')
title('First derivative of P vs beta kinetic energy')
```

Its intersection with the x-axis, or in other words, when its value becomes zero, we should get the maximum intensity of energy. From the following plot (Figure 3.23), we see that at about 0.18 MeV, we get the maximum intensity.

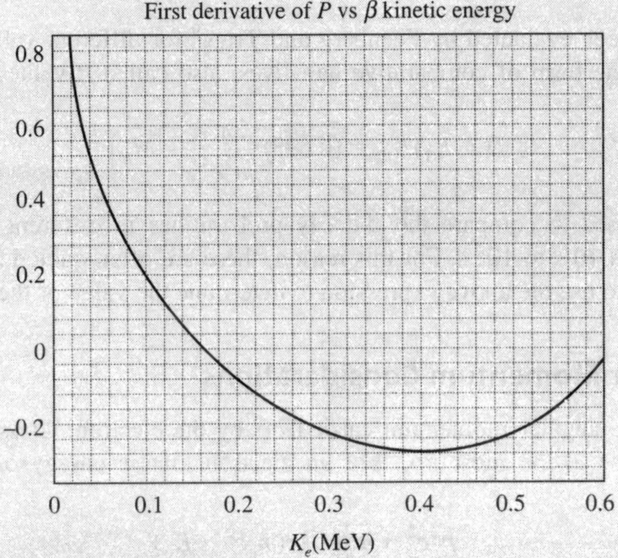

FIGURE 3.23 First derivative of P as a function of the kinetic energy of the beta particles.

Similarly, for positron emission, we can get the maximum intensity at about 0.26 MeV, by using negative sign in η and the atomic number of the daughter nucleus as 28.

3.5.4 Comparative Half-Lives

By integrating the momentum spectrum from 0 to $p_e(\text{max})$, we obtain the total decay probability (λ) as,

$$\lambda = \frac{0.693}{T_{1/2}} = \int_0^{p_e(\text{max})} P(p_e)\, dp_e$$

$$= \frac{1}{2\pi^3 \hbar^7 c^3} \left|M_{if}\right|^2 g^2 \int_0^{p_e(\text{max})} F(Z, E_e)(E_0 - E_e)^2 p_e^2 dp_e$$

The integral in the above expression is a function of p_e, E_0 and Z. It is made dimensionless by multiplying $m_e^5 c^7$, both in the numerator and the denominator, where m_e is the rest mass of the electron.

$$= \frac{m_e^5 c^4 g^2}{2\pi^3 \hbar^7} \left|M_{if}\right|^2 \int_0^{p_e(\text{max})} F(Z, E_e) \frac{(E_0 - E_e)^2}{m_e^2 c^4} \left(\frac{p_e^2}{m_e^2 c^2}\right)\left(\frac{dp_e}{m_e c}\right)$$

This integral is simplified by defining a quantity $f(Z, E_0)$ as follows:

$$f(Z,E_0)=\int_0^{p_e(\max)} F(Z,E_e)\left(\frac{p_e^2}{m_e^2c^2}\right)\left(\frac{dp_e}{m_ec}\right)$$

This relation has been evaluated by Feenberg and Trigg[9] for different values of Z and E_0. The result is given in the form of comparative half-lives, also called *ft* values.

$$f(Z,E_0)T_{1/2}=(0.693)\frac{2\pi^3\hbar^7}{m_e^5c^4g^2}\left|M_{if}\right|^{-2} \tag{3.43}$$

ft values can be used to compare the β-decay probabilities in different nuclei. *ft* values for β-decay range from 10^3s to 10^{20}s. For this reason, these are often quoted as $\log_{10}[f(Z, E_0)T_{1/2}]$ values. One can also use the above expression to determine the value of the coupling constant g.

3.5.5 Angular Momentum Considerations

Let us consider the angular momentum carried off by the electron. Classically, this is $\vec{r}\times\vec{p}_e$ whose value can be at the most Rp_e. For an electron kinetic energy of 1 MeV, p_e can be calculated as

$$p_e^2c^2+m_e^2c^4=(m_ec^2+K_e)^2$$

or
$$p_e^2c^2=(0.511+1)^2-(0.511)^2\ \text{MeV}^2$$

∴
$$p_e=1.4\ \text{MeV}/c$$

For medium sized nucleus, R = 5 fm. Thus, the maximum angular momentum (Rp_e) comes out to be 7 MeVfm/c.

Quantum mechanically, the magnitude of angular momentum $=\sqrt{l(l+1)\hbar^2}$. Thus,

$$\sqrt{l(l+1)}=\frac{7}{\hbar c}=\frac{7\ \text{MeVfm}}{197\ \text{MeVfm}}=0.036$$

or
$$l \ll 1$$

This means that a 1 MeV electron emitted, even with l = 1, is hindered by a large angular momentum barrier. Since, the electron and the neutrino (and their antiparticles) have, on an average, approximately equal momenta, the emission of a neutrino with non-zero angular momentum is also hindered by a large barrier. The rate of emission of β-particles will be the largest when the total orbital angular momentum of the electron and the neutrino, L, is zero. Beta decays with $L = 0$ are referred to as allowed, while those with $L > 0$ are called *forbidden*. Here the term 'forbidden transition' is a misnomer.

9. E. Feenberg, E. and G. Trigg, 'The Interpretation of Comparative Half-Lives in the Fermi Theory of Beta-Decay', *Rev. Mod. Phys.*, 22, 399 (1950).

Alternatively, one can argue that in Fermi's theory, the emitted particles (e, ν) are treated as plane wave.

$$\psi_e \psi_{\bar{\nu}} \sim \exp\left(i \frac{\left(\overrightarrow{p_e} + \overrightarrow{p_\nu}\right) \cdot \vec{r}}{h} \right) \sim e^{\frac{i\vec{p}.\vec{r}}{h}} = e^{i\vec{K}.\vec{r}}$$

where K is the wave vector.

Expanding $e^{i\vec{K}\cdot\vec{r}}$, we get

$$e^{i\vec{K}\cdot\vec{r}} = 1 + i\vec{K}\cdot\vec{r} + \frac{\left(i\vec{K}.\vec{r}\right)^2}{2!} + \cdots$$

In the Fermi theory, only the first term in the expansion is retained and the other terms are neglected. If the 1st term gives vanishing nuclear matrix element, then the 2nd term assumes importance, since this is the only channel by which β-decay can occur. This is known as the first forbidden transition. In the same way, 3rd and higher terms can contribute and these are known as the second and higher order forbidden transitions.

So far, we have ignored the fact that the β-particle and the neutrino are leptons and have spins. Both the electron and the neutrino have (½)$\hbar$ spin, which combine to give the total spin $S = 0$ or $1\hbar$. In the former case ($S = 0$), we have what is known as a Fermi transition, while the latter case ($S = 1$) is called a Gamow–Teller transition. Each involves a different interaction with the weak nuclear field. A single transition can be a mixture of the two (FT and GT), provided that certain selection rules, which are described in the following section, are satisfied.

3.5.6 Selection Rules

Since, the angular momentum and the parity are conserved in any decay, we can immediately deduce how J and π must change in the transitions.

In the Fermi transition, $\vec{S} = 0$, therefore $\overrightarrow{J_f} = \vec{J_i} + \vec{L}$, where J_f and J_i are the angular momentum of the daughter and the parent nucleus, respectively. $\vec{L}$ is the orbital angular momentum carried off by e/e^- and $\bar{\nu}/\nu$.

In the Gamow–Teller transition, $\vec{S} = 1$, therefore $\overrightarrow{J_f} = \vec{J_i} + \vec{L} + \vec{1}$.

Conservation of parity requires $\pi_i = \pi_f (-1)^L$, where π_i and π_f are the parity of the initial and the final nuclear state wave functions and $(-1)^L$ is the parity of the (e–ν) wave function. It is very often important to determine what type of β-transition occurs between two nuclear states. Since, transition rates increase rapidly with decreasing L, this will always be the lowest L. Since, parity switches with each consecutive L, this is either the lowest or the next lowest L. These are determined by $\Delta J = \left|J_f - J_i\right|$, although, there are exceptions where L values are not allowed in cases where either J_f or J_i, are zero.

In Table 3.2 selection rules for various types of transition are shown; those not possible are in parentheses. The yes or no indicate whether the parity changes between the initial and final states.

Table 3.2 Selection Rules for Various Types of Transition

Transition type	*L*	*Fermi*		*Gamow–Teller*		$\log_{10} ft$
		ΔJ	$\Delta\pi$	ΔJ	$\Delta\pi$	
Super-allowed	0	0	No	—	—	~3.5
Allowed	0	0	No	(0), 1	No	5.5 ± 1.5
1st forbidden	1	(0), 1	Yes	0, 1, 2	Yes	7.5 ± 1.5
2nd forbidden	2	(1), 2	No	2, 3	No	~ 12
3rd forbidden	3	(2), 3	Yes	3, 4	Yes	~ 16
4th forbidden	4	(3), 4	No	4, 5	No	~ 21

We have also included the classification super-allowed referring to cases where the matrix element, $M_{if} \approx 1.0$. Examples are the so called mirror transitions, such as $n \to p$, $t \to {}^{3}_{2}\text{He}$ and ${}^{17}_{9}\text{F} \to {}^{17}_{8}\text{O}$, in which the only difference between the initial and the final states is that an unpaired neutron has been changed into a proton or vice-versa.

EXERCISE 3.15: Based on Table 3.2, classify the following β-decay transitions.

(a) ${}^{34}_{17}\text{Cl}\,(0^+) \to {}^{34}_{16}\text{S}\,(0^+)$

(b) ${}^{6}_{2}\text{He}(0^+) \to {}^{6}_{3}\text{Li}\,(1^+)$

(c) $n(1/2^+) \to p(1/2^+)$

(d) ${}^{76}_{35}\text{Br}\,(1^-) \to {}^{76}_{34}\text{Se}(0^+)$

(e) ${}^{22}_{11}\text{Na}(3^+) \to {}^{22}_{10}\text{Ne}\,(0^+)$

(f) ${}^{40}_{19}\text{K}(4^-) \to {}^{40}_{20}\text{Ca}\,(0^+)$

(g) ${}^{115}_{49}\text{In}(9/2^+) \to {}^{115}_{50}\text{Sn}(1/2^+)$

Solution: We know in Fermi transition, $\vec{S} = 0$ and in Gamow–Teller transition, $\vec{S} = 1$. Based on Table 3.2, we can get the following answers:

(a) Allowed transition: Pure Fermi transition
(b) Allowed transition: Pure Gamow–Teller transition
(c) Allowed transition: Both Fermi and Gamow–Teller transitions
(d) 1st forbidden transition
(e) 2nd forbidden transition
(f) 3rd forbidden transition
(g) 4th forbidden transition

3.5.7 Detection of Neutrino and Antineutrino

In order to detect the $\nu/\bar{\nu}$, the following reactions were used. These reactions are sometimes called inverse β-decay.

$$\bar{\nu} + p \rightarrow n + e^{+} \text{ (Threshold energy, } E_{\bar{\nu}} = 18 \text{ MeV)} \tag{3.44}$$

$$\nu + n \rightarrow p + e^{-} \tag{3.45}$$

Reines and Cowan[10] performed the experiments by using high flux of antineutrino available near a 1000 MW nuclear reactor. In nuclear reactors, the fission of ${}^{235}_{92}\text{U}$ produces neutron-rich fission products, which in turn, decay by emitting β^{-} and $\bar{\nu}$. The flux of $\bar{\nu}$ was ~10^{13}/cm^{2}s. Since, the probability that a $\bar{\nu}$ will interact with a proton is extremely small, the experiment required both a high flux of $\bar{\nu}$ and a large number of protons to produce a detectable number of events. The source of the protons was water containing dissolved $CdCl_2$. A positron produced in the capture reaction (3.44), quickly annihilates with an electron within ~10^{-9}s producing two 511 keV γ-rays travelling in opposite directions, in order to conserve the linear momentum. These γ-rays were detected simultaneously in two large tanks containing terphenyl dissolved in triethylbenzene. This solution produces light flash when a γ-ray is absorbed in it. The light flashes were detected by photomultiplier tubes placed in each tank.

Meanwhile, the recoiling neutron slows down through a series of collisions with protons and is then captured in the Cd nucleus by the reaction Cd(n, γ)Cd*, producing few γ-rays of total energy ≈ 8 MeV. This neutron capture process takes about 10^{-5} s. So, a characteristic signature of two simultaneous 511 keV γ-rays, followed by the neutron capture γ-rays, a few microseconds later, indicates a capture reaction. The expected event rate is $FN\sigma_c\varepsilon$, where F is the $\bar{\nu}$ flux, N, the number of protons, σ_c, the capture cross-section and ε is the efficiency of the detector. With the experimental setup developed by Reines and Cowan, they were able to observe about 1 event per hour. To avoid any uncertainty, the experiment was performed with the reactor alternately on and off, and the expected variation in the frequency of neutrino-capture events was observed. After many years of repeated experiments, in 1960, they could confirm the existence of $\bar{\nu}$.

Detection of Neutrino

With the existence of $\bar{\nu}$ confirmed, there could be little doubt about ν, but there was still considerable interest to observe $n + \nu \rightarrow p + e^{-}$ reaction. One cannot use a target of free neutrons, since these are unstable anyway. Soon, it was proposed that the most convenient target to use in this reaction should be stable ${}^{37}_{17}\text{Cl}$. The reaction to look for was

$${}^{37}_{17}\text{Cl} + \nu \rightarrow {}^{37}_{18}\text{Ar} + e^{-} \tag{3.46}$$

The nuclide ${}^{37}_{18}\text{Ar}$ is of course unstable, decaying back to ${}^{37}_{17}\text{Cl}$ through the electron capture, but an equilibrium should be established. The concentration of ${}^{37}_{18}\text{Ar}$ in any sample of ${}^{37}_{17}\text{Cl}$ depends on the neutrino flux to which, it is exposed. The experiment performed by R. Davis

10. Original Papers:
C.L. Cowan, Jr., F. Reines, F.B. Harrison, H.W. Kruse and A.D. McGuire, 'Detection of the Free Neutrino: A Confirmation', Science, 124, 103, 1956.
Frederick Reines and Clyde L. Cowan, Jr., 'The Neutrino' *Nature*, 178, 446, 1956.

consisted of $^{37}_{17}Cl$ in the form of large tanks of C_2Cl_4 and looked for traces of radioactive $^{37}_{18}Ar$ that was there. Davis found no evidence for the above reaction.

This arrangement was first used to show that ν and $\bar{\nu}$ are indeed different particles. If these were identical, then close to a reactor, the above reaction should have proceeded just like $\bar{\nu} + p \rightarrow n + e^+$ reaction.

Since then, various neutrino detection methods have been used. One such neutrino observatory is Super-Kamiokande. It is a large volume of water located deep in the Kamioka mine in Japan, surrounded by photomultiplier tubes that watch for the Cherenkov radiation (see section 7.6.1) emitted when incoming atmospheric neutrinos interact with atomic nuclei in water to produce electrons, muons or tau leptons.

3.5.8 Double Beta Decay

In some even–even nuclei, although β-decay is energetically forbidden, the decay $(A, Z) \rightarrow (A, Z + 2)$ is energetically allowed, and in principle, could occur by emission of two e^- and two $\bar{\nu}$. This is referred to as double β-decay. It is a second order weak interaction and is the rarest type of radioactive decay with life-times of the order of ~ 10^{20} y. It was first observed in 1987 in the decay of

$$^{82}_{34}Se \rightarrow {}^{82}_{36}Kr + 2e^- + 2\bar{\nu}$$

and has subsequently been seen in other isotopes like $^{48}_{20}Ca$, $^{76}_{32}Ge$, $^{96}_{40}Zr$, $^{100}_{42}Mo$, $^{116}_{48}Cd$, $^{128}_{52}Te$ and $^{238}_{92}U$.

EXERCISE 3.16: If in the Fermi–Kurie plot of neutron decay, the electron's end point energy is 0.79 MeV and the rest mass of electron/positron is 0.511 MeV, what is the threshold energy (the theoretical minimum energy at which the reaction takes place such that no kinetic energy is transferred to the final particles that are formed in the centre of mass frame[11]) required by an antineutrino for the following inverse reaction?

$$\bar{\nu} + p \rightarrow n + e^+$$

Solution: In the neutron decay

$$n \rightarrow p + e^- + \bar{\nu} + 0.79 \text{ MeV}$$

$$m_n c^2 = (m_p + m_e)c^2 + 0.79 \text{ MeV}$$

where $m_{\bar{\nu}}$, the rest mass of $\bar{\nu}/\nu$ is taken to be zero.

or
$$(m_n - m_p)c^2 = m_e c^2 + 0.79 \text{ MeV}$$

For the inverse reaction in the CM frame, since, we are interested in calculating the threshold energy, the final particles are assumed to be at rest. Therefore,

$$E_{\bar{\nu}} = (m_n - m_p)c^2 + m_e c^2$$

or
$$E_{\bar{\nu}} = 0.79 + 2(0.511) = 1.81 \text{ MeV}$$

11. See Appendix B.

We conclude this discussion by looking once more at the question of distinguishing between neutrino and antineutrino. If they were identical, then the following reaction could be visualized where the emitted neutrino again combines with the neutron to form a proton and an electron:

$$n \rightarrow p + e^- + \{[\nu] + n \rightarrow p + e^-\}$$

or

$$2n \rightarrow 2p + 2e^-$$

Thus, we should have the possibility of neutrino-less β-decay with Z changing by two units at a time. Such neutrino-less β-decay has so far not been observed.

3.6 GAMMA DECAY

Frequently α and β-decay leave the daughter nuclei in an excited state. This excitation is removed either by γ-photon emissions or by a process called internal conversion (IC). Since, these two modes of decay are very closely related, we will consider them together in this section, but we will first discuss in more detail the γ-ray emission process.

Gamma rays were discovered by Paul Villard in 1900. They are similar to the X-rays, but are emitted from the nucleus and generally have shorter wavelengths and high penetrability. In majority of cases, the emission of γ-rays occurs immediately after α or β-decay, i.e., within $< \sim 10^{-12}$ s, but in some instances, the nucleus may remain in the higher energy states for a measurable length of time. The longer-lived excited nuclei are called *isomers*. An example is ${}^{60m}_{27}\text{Co}$ (Figure 3.3) which decays with a $T_{1/2} \sim 10.5$ m to the ground state of ${}^{60}_{27}\text{Co}$. This transition is often known as *isomeric transition* (IT). In γ-decay, a nucleus rearranges its constituent protons and neutrons in order to make a transition from higher mass (energy) state to lower mass (energy) state through the emission of electromagnetic radiation (γ-photon). The emitted γ-ray is mono-energetic, having energy equal to the energy level difference, minus the small fraction transferred to the recoil nucleus.

When the excited energy of a nucleus is higher than the separation energy of the nucleon, the nucleus will decay by particle emission. This decay is very fast with $T_{1/2} \sim 10^{-20}$ s. Below the separation energy, de-excitation by γ-emission to either the ground state directly or via cascades of γ-rays to the ground state takes place. $T_{1/2}$ of such decays is longer and can be experimentally determined by measurement of Γ-value or the line width of γ-peak and using the relation $T_{1/2} = \hbar/\Gamma$. Gamma ray energies cover a wide range up to tens of MeV, but typically, are of the order of 1 MeV.

3.6.1 Kinematics of Gamma Decay

In γ-decay of a nucleus, the decay energy is distributed between the γ-ray energy (E_γ) and the kinetic energy of the recoiling product nuclide (E_d). We can therefore write

$$E_{\text{final}} - E_{\text{initial}} = E_\gamma + E_d = Q_\gamma$$

The linear momentum conservation requires that the recoil daughter momentum to be equal to the forward γ-ray momentum, i.e.,

$$p_d = E_\gamma / c$$

EXERCISE 3.17: From Figure 3.3, for cesium-137 decay, $E_\gamma = 662$ keV

$$^{137}_{55}\text{Cs} \xrightarrow{\beta} {}^{137m}_{56}\text{Ba} \xrightarrow{\gamma} {}^{137}_{56}\text{Ba}$$

Calculate the value of E_d and prove that $Q_\gamma \approx E_\gamma$, which is true in most of the cases when only the γ-ray energy is considered.

Solution:

$$p_d c = E_\gamma$$

Squaring both the sides, we get

$$m_d^2 v_d^2 c^2 = E_\gamma^2$$

or

$$2m_d c^2 \left(\frac{1}{2} m_d v_d^2\right) = 2m_d c^2 E_d = E_\gamma^2$$

$\therefore$

$$E_d = \frac{E_\gamma^2}{2m_d c^2}$$

Substituting the values,

$$E_d = \frac{(662\ \text{keV})^2}{2(137\,\text{u})\,(931.5\times10^3\ \text{keV})} \approx 17\ \text{eV}$$

As E_d is quite small and it may be neglected. Hence, $Q_\gamma \approx E_\gamma$.

Even though E_d is quite small, the recoil shifts the γ-radiation out of resonance condition, since the natural line width Γ of the radiation is even smaller. The emission of recoilless γ-photons is possible if one implants the nucleus in a lattice. This idea was found by Mössbauer and gave rise to a nuclear technique, which proved to be very useful in various domains of physics. The Mössbauer effect is described in detail in Chapter 9.

3.6.2 Mechanism of Gamma Decay

Gamma radiation refers to high frequency electromagnetic radiation, in which just like any other EM radiation, a changing electric field induces a magnetic field and vice versa. Such EM radiation can be generated by an oscillating electric charge, which sets up an external oscillating electric field, or by an oscillating electric current or varying magnetic moment, which sets up an oscillating magnetic field. Nucleus consists of protons, which are charged particles with spin and intrinsic magnetic moments, and neutrons, which though uncharged, carry spin and intrinsic magnetic moments, in motion. So in principle, a nucleus has a charge distribution and a current distribution. Charge distributions gives rise to the static electric multipole moments and the current distributions to the static magnetic multipole moments, when the nucleus is in the ground state as discussed in Chapter 1. However, in an excited state, the electric and magnetic multipoles will vary with time. These oscillatory electric and magnetic multipole moments will radiate energy in the form of EM waves (γ-radiation) and the nucleus will make a transition to the lower energy levels to ultimately arrive at the ground state. Correspondingly, γ-rays emitted by the former mechanism are called *electric multipole (EL) radiation* and by the latter process are said to give rise to *magnetic multipole (ML) radiation.*

A photon carries away angular momentum of magnitude given by a quantum number L, which can have an integer value greater than zero. The value $L = 0$ is excluded as photons with zero angular momentum do not exist. The γ-rays carrying away L values of 1, 2, 3, ..., units of $\hbar$ are called dipole, quadrupole, octupole ..., radiations.

A proper treatment using a quantized electromagnetic radiation field is beyond the scope of this book and we refer the interested reader to the book by Blatt and Weisskopf listed in the bibliography. Here, we will just outline the results, without proof.

Selection Rules

The total angular momentum of a photon must include the fact that the photon is a spin one vector boson. The minimum value of L for a photon is 1. The angular momentum of the photon is related by the angular momentum conservation to the spins of the initial and the final nuclear states. Thus, we have the following vector equation:

$$\vec{J_i} = \vec{J_f} + \vec{L}$$

which leads to a selection rule that L can have any integer value between

$$\left|\vec{J_i} - \vec{J_f}\right| \leq L \leq \vec{J_i} + \vec{J_f}$$

It is also necessary to take account of the parity. In classical physics, an electric dipole $q\vec{r}$ is formed by having two equal and opposite charge q separated by a distance $\vec{r}$. Under the parity operation of inversion, $q\vec{r}$ transforms to $-q\vec{r}$ and therefore, electric dipole operator has odd or negative parity. A magnetic dipole is equivalent to a charge circulating with velocity $\vec{v}$ to form a current loop of radius $\vec{r}$ (Figure 3.24).

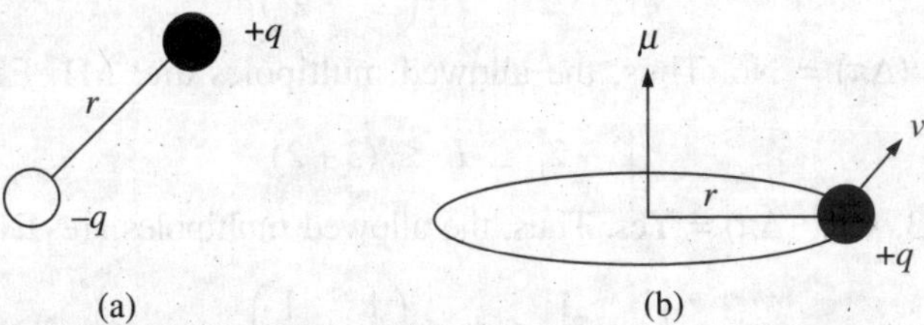

FIGURE 3.24 (a) Electric dipole moment formed by positive and negative charges separated by a distance r, (b) magnetic dipole moment created by a charge q moving in a circle of radius r with speed v.

The magnetic dipole is of the form $q\vec{r} \times \vec{v}$, which does not change sign under an inversion and thus, has even or positive parity. The general result, which we state without proof is that the electric multipole radiation has parity $(-1)^L$, whereas magnetic multipole radiation has a parity of $(-1)^{L+1}$. We thus, are led to the selection rules for the γ-emission given in Table 3.3.

Table 3.3 Selection Rules and Multipolarities in Gamma Decay

Type	*Symbol*	*L*	*Parity change*
Electric dipole	$E1$	1	Yes
Magnetic dipole	$M1$	1	No
Electric quadrupole	$E2$	2	No
Magnetic quadrupole	$M2$	2	Yes
Electric octupole	$E3$	3	Yes
Magnetic octupole	$M3$	3	No

EXERCISE 3.18: List all the possible multipole γ-ray transitions between the following pair of nuclear states.

(a) $5/2^+ \rightarrow 3/2^+$
(b) $3/2^- \rightarrow 1/2^-$
(c) $3^- \rightarrow 2^+$
(d) $1/2^- \rightarrow 1/2^+$

Solution:

(a) Allowed values of L,

$$\left|\frac{5}{2}-\frac{3}{2}\right| \le L \le \left(\frac{5}{2}+\frac{3}{2}\right)$$

Thus, L = 1, 2, 3, 4. Parity change ($\Delta\pi$) = No. Therefore, the allowed electric and magnetic multipole transitions are: M1, E2, M3, E4.

(b)
$$\left|\frac{3}{2}-\frac{1}{2}\right| \le L \le \left(\frac{3}{2}+\frac{1}{2}\right)$$

L = 1, 2. ($\Delta\pi$) = No. Thus, the allowed multipoles are: M1, E2.

(c)
$$|3-2| \le L \le (3+2)$$

L = 1, 2, 3, 4, 5. ($\Delta\pi$) = Yes. Thus, the allowed multipoles are: E1, M2, E3, M4, E5.

(d)
$$\left|\frac{1}{2}-\frac{1}{2}\right| \le L \le \left(\frac{1}{2}+\frac{1}{2}\right)$$

Thus, L = 0, 1. ($\Delta\pi$) = Yes. Thus, the allowed multipole orders are: E1.

Transition Rates

In semi-classical radiation theory, the transition probability per unit time, i.e., the emission rate, is given by

$$T^{E,M}(L) = \frac{1}{4\pi\varepsilon_0}\frac{8\pi(L+1)}{L[(2L+1)!!]^2}\frac{1}{\hbar}\left(\frac{E_\gamma}{\hbar c}\right)^{2L+1} B^{E,M}(L) \tag{3.47}$$

where E_γ is the photon energy, E and M refer to the electric and the magnetic radiation, and the double factorial stands for

$$(2L+1)!! = (2L+1)(2L-1)(2L-3)\ldots 3\cdot 1$$

The function $B^{E,M}(L)$ is the so called reduced transition probability, and contains all the nuclear information. It is basically the square of the matrix element of the appropriate operator causing the transition involving complicated nuclear wave functions and producing photons with multipolarity L. For electric transitions, B is measured in units of $e^2\text{fm}^{2L}$ and for magnetic transition, in units of $(\mu_N/c^2)\, fm^{2L-2}$, where μ_N is the nuclear magneton.

An estimate, referred to as a single particle estimate, can be obtained by assuming that the transition is due to a single proton making a transition between two shell model states. Weisskopf has shown that for an electric transition, a reasonable estimation for the single particle reduced transition probability is given by

$$B_{sp}(EL) = \frac{e^2}{4\pi}\left(\frac{3R^L}{L+3}\right)^2$$

and for a magnetic transition

$$B_{sp}(ML) = 10\left(\frac{\hbar}{m_p cR}\right)^2 B_{sp}(EL)$$

where R is the nuclear radius and m_p is the mass of proton. Substituting $R = r_0A^{1/3}$, where r_0 = 1.21 fm, we obtain the following expressions for the electric and the magnetic single-particle reduced transition rates:

$$B_{sp}(EL) = \frac{e^2}{4\pi}\left(\frac{3}{L+3}\right)^2 (r_0)^{2L} A^{2L/3} \tag{3.48}$$

$$B_{sp}(ML) = \frac{10}{\pi}\left(\frac{e\hbar}{2m_p c}\right)^2\left(\frac{3}{L+3}\right)^2 (r_0)^{2L-2} A^{(2L-2)/3} \tag{3.49}$$

Using the above approximate expressions in equation (3.47), the following simplified transition rates have been obtained, which are known as Weisskopf estimates:

$$T(E1) = 1.0\times 10^{14} A^{2/3} E_\gamma^3 \qquad T(M1) = 3.1\times 10^{13} E_\gamma^3$$

$$T(E2) = 7.3\times 10^{7} A^{4/3} E_\gamma^5 \qquad T(M2) = 2.2\times 10^{7} A^{2/3} E_\gamma^5$$

$$T(E3) = 34 A^2 E_\gamma^7 \qquad T(M3) = 10 A^{4/3} E_\gamma^7$$

$$T(E4) = 1.1\times 10^{-5} A^{8/3} E_\gamma^9 \qquad T(M4) = 3.3\times 10^{-6} A^2 E_\gamma^9$$

For a medium mass nucleus with $A = 100$, the plot based on the aforementioned expressions is shown in Figure 3.25. It is clear from the figure that there is a very strong dependence of the emission rate on the photon energy as well as on the multipolarity. One can see that for a given transition, there is a very substantial decrease in the decay rates with increasing L. Also, the electric transitions have decay rates about two orders of magnitude larger than the corresponding magnetic transitions.

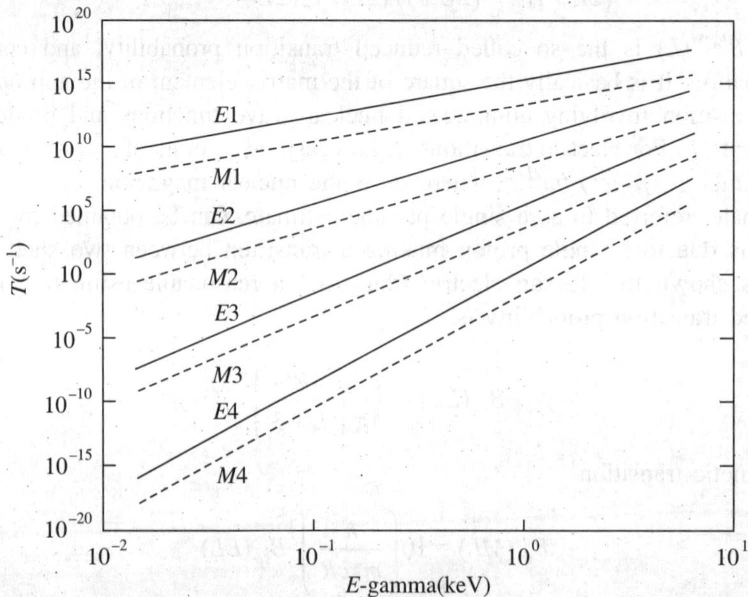

FIGURE 3.25 Weisskopf single-particle gamma transition rates for E and M transitions of different multipolarities calculated as a function of photon energy, for a medium mass nucleus with $A = 100$.

3.6.3 Internal Conversion (IC)

A transition between two nuclear states in the same nucleus can be induced by electric or magnetic interaction with the atomic electrons. The wave function of an orbital electron may overlap that of the excited nucleus, the excitation energy of the nucleus may be transferred directly to the orbital electron, which escapes from the atom with certain kinetic energy $E_e = \Delta E - B$ (neglecting nuclear recoil), where ΔE is the transition energy $(E_f - E_i)$ and B is the binding energy of the electron in the atom. Obviously, electrons emitted from different shells K, L_I, L_{II}, L_{III}, etc. will appear at different energies. Nuclear recoil effects are same in internal conversion as they are in the γ-ray emission.

No γ-ray is emitted in an internal conversion process; it is an alternate mode to the γ-ray emission of de-excitation of nuclei. Conversion electrons are mono-energetic. In Figure 3.15, two sharp peaks are observed just beyond E_{max}. The first peak, designated as K, is due to the conversion electrons originating in K-atomic shell, while the peak marked as L is due to the conversion electrons originating in L-atomic shell. The decay of $^{137m}_{56}$Ba (Figure 3.3) proceeds both, by the emission of 662 keV γ-ray and by the competitive process of IC. The ratio of the number of conversion electrons and the number of γ-rays emitted in this competition is known as the internal conversion coefficient (α). It may be written as a sum of the partial coefficients corresponding to *K, L, M,* …, shells.

$$\alpha = \alpha_K + \alpha_L + \alpha_M + \ldots$$

where $\alpha_K = I_{eK}/I_\gamma$, $\alpha_L = I_{eL}/I_\gamma$ and so on, I being the intensity.

Internal conversion is accompanied by the emission of X-rays and/or Auger electrons as the vacancy in the atomic shell is filled in the rearrangement process of the atom.

Extensive tabulations of calculated conversion coefficients are available[12]. A few observations that one can make are as follows:

(i) Internal conversion coefficient increases with the atomic number as Z^3.
(ii) $\alpha_K > \alpha_L > \alpha_M$
(iii) For higher L, γ-decay probabilities decreases and IC probability increases.
(iv) In de-excitation process between $0^+ \rightarrow 0^+$ nuclear states, the γ-decay is strictly forbidden; the only possible decay-channel is that of IC.

The Auger Effect

In the internal conversion and orbital electron capture processes, the atom will be left with a vacancy in one of its electron shells. This vacancy is usually filled with an electron from a higher shell and two competing processes take place—either emission of an X-ray, whose energy is equal to the difference in the electron binding energy between the electron shells (outer shells to a shell closer to nucleus) or the energy difference between the orbitals is used to knock another electron from the orbit. For example, if a vacancy in the *K*-shell is filled with an electron from the *L*-shell, *K* X-rays may be emitted from the atom. Alternatively, this energy difference is enough to remove another electron from the *L* or *M*-shell. Pierre Auger discovered this process, which is now known by his name as Auger effect. The transitions are written as $[KL_IL_{III}]$ with an understanding that the primary vacancy is in the *K*-shell, L_I sub-shell electron fills this vacancy and the energy difference between the *K*-shell and the L_I sub-shell is used to knock out electron of the L_{III} sub-shell. The Auger effect creates an additional electron shell vacancy which will lead to further X-rays or Auger electron emission. The complete process of atomic rearrangement may thus, be very complex.

Auger process is more probable in light nuclei, while the emission of X-rays is more probable in heavy nuclei. A term fluorescent yield (ω_K) is defined as:

$$\omega_K = \frac{\text{Number of } K \text{ X–rays emitted}}{\text{Number of primary vacancies in the } K \text{ shell}}$$

For example, if $\omega_K = 0.4$, it means that there is a 40% chance that *K* X-rays are emitted and Auger electrons are emitted in the rest (i.e., 60%) of the cases.

12. M.E. Rose, G.H. Goertzel and C.L. Perry, *K-shell Internal conversion coefficients*, Oak Ridge National Laboratory (ORNL) 1023.
Hager-Seltzer Internal conversion coefficients (HSICC) using <www.nndc.bnl.gov/hsicc>

Auger electrons have energy in the range of 100 eV to few keV. The kinetic energy of the Auger electrons corresponds to the energy difference between the initial electronic transition and the ionization energy of the shell from which the Auger electrons were ejected.

3.7 APPLICATIONS OF RADIOACTIVITY

Radioactivity has found wide range of applications, since it was discovered in the late nineteenth century. Today, it is used in diverse fields such as medicine, industry, agriculture, archaeology, geology, art, crime detection, space exploration, etc. A brief overview of some of them is presented in the following sections. There are various resources available for more detailed information in this area[13].

Medicine

Radioisotopes are extensively used in diagnosis and therapy. Radiography, which means making an image on a film using radiation, popularly known as X-ray, is the most common example of the diagnostic application. It is used to locate any fractured bone inside the body, cavities in the teeth and study the gastrointestinal tract.

Tracers, which are physically and chemically identical with stable isotopes of the same element and hence, they can take the place of their stable isotopes, and because they are radioactive, they can be readily traced and used for diagnostic and therapeutic purposes. Iodine-131 being used for the treatment of an overactive thyroid gland is perhaps the most common therapeutic procedure. Another common therapeutic use of radioisotopes is in treating cancer. Cobalt-60 is widely used to treat cancer, where the γ-rays emitted from cobalt-60 is used to selectively kill the cancerous cells.

EXERCISE 3.19: For a diagnostic procedure, a patient is administered Tc-99m radionuclide in a pharmaceutical, which is preferably taken by the tissue under study. The radionuclide has a half-life of 6 h and gives off gamma rays of 0.14 MeV energy. Although the half-life of the radionuclide is short, its activity will remain in the body of the patient for some time after the procedure and even very small activity can be detected. A gamma camera is used to observe how the tissue behaves and how quickly the radionuclide moves.

If the initial activity that was measured in the tissue on the day of the administration was 1 mCi, what is the expected count rate after one week?

Solution: From equation (3.5), we have

$$A = A_0 e^{-\lambda t} = 1 \times \exp\left(-\frac{0.693}{6} \times 7 \times 24\right) = 3.74 \times 10^{-9} \text{ mCi}$$

Industry and Agriculture

Radiography is used to detect cracks and determine the structural integrity in the metal structure, like in airplanes, bridges, boilers, pipes and the industrial equipment. Radioisotopes are also

13. G.C. Lowenthal and P.L. Airey, *Practical Applications of Radioactivity and Nuclear Radiations*, Cambridge University Press, 2001.

commonly used to detect leakages and blockages in buried pipelines by monitoring the activity of the tracer element, to measure and control the thickness of metal and plastic sheets and for process optimization. Gamma irradiation, which involves exposing an item to intense controlled amounts of γ-radiations, commonly from cobalt-60 or cesium-137, is used for the sterilization of food products in order to increase their shelf life and disposable medical supplies, such as syringes and other medical instruments. It is especially advantageous in the case of plastic equipment as compared to heating because it avoids heat damage.

In the field of agriculture, radioisotopes are used to increase the food production by producing improved varieties of crops, optimizing the use of fertilizers, insect control and food preservation. For example, induced mutations being used to genetically improve crop variety or tracers being used to study the mechanism of how plants utilize a particular element to grow and reproduce.

One of the most important applications of radioactivity is for the production of electricity. Nuclear reactors make use of the controlled fission reaction, where a heavy nucleus is broken into two or more fragments and as a result, enormous amount of energy is produced. Today, there are several commercial nuclear power plants in operation around the world.

EXERCISE 3.20: When uranium-235 is bombarded with a thermal neutron, the following fission reaction takes place and about 200 MeV of energy is released from each fissioning uranium nucleus.

$$^{235}_{92}\text{U} + {}^{1}_{0}n \rightarrow \text{fission products} + \text{neutrons} + \text{energy} \;(\sim 200 \text{ MeV})$$

Calculate the amount of energy released if 1 g of uranium-235 was to fission completely in a controlled fashion (if it were uncontrolled, then it will be what we know as atom bomb).

Now, in order to get an idea about the extent of energy that a nuclear power plant is capable of producing vis-à-vis conventional coal-powered power plant, estimate the amount of coal that will have to burnt, to produce the equivalent amount of energy. (The amount of heat content depends on the quality of the coal. For the purpose of calculation, assume the average heat content of the coal to be 16 kJ/g.)

Solution: Number of nuclei present in 1 g of uranium-235

$$N_0 = \frac{1}{235}\,[6.023\times 10^{23}] = 2.563\times 10^{21} \text{ nuclei}$$

Therefore, the amount of energy released if 1 g of uranium-235 was to fission completely

$$= (2.563\times 10^{21})(200 \text{ MeV})\left(1.6\times 10^{-16}\,\frac{\text{kJ}}{\text{MeV}}\right) = 8.2\times 10^{7} \text{ kJ or kWs}$$

$$= \frac{(8.2\times 10^{7})(10^{-3})}{3600\times 24} \approx 1 \text{ MWd}$$

Thus, 1 g of uranium-235 would produce about 1 MW of power for one day. The amount of coal needed to produce the same amount of energy would be

$$= \frac{8.2\times 10^{7} \text{ kJ}}{16 \text{ kJ/g}} = 5.1\times 10^{6}\text{g} \approx 5 \text{ tons}$$

Space Exploration

Today, more and more countries are joining the space race and developing their own outer space missions. One of the main motivations of unmanned space missions to Moon and the Mars is to understand the elemental composition of the soil found there. RBS (Rutherford Back Scattering) probe, discussed in detail in Chapter 9, is used to get the elemental composition. For this purpose, these space flights are equipped with an α-emitter to record the spectra of the back scattered α-particles. Space flights themselves are powered by radioisotope generators (RTGs), which make use of the heat released in the α-decay of plutonium-238 to produce electricity.

EXERCISE 3.21:

(a) Plutonium-238 has been used as the power source in the space flights. It has an α-decay half-life of 90 y. Each of the α-particles is emitted with 5.5 MeV of energy. What is the power released if there are 238 g of plutonium-238?

(b) If the power source in the previous part produces 8 times the minimum required to run a piece of apparatus, for how much time will the source produce sufficient power for the apparatus to function?

Solution:

(a) The initial activity,

$$A_0 = N_0\lambda = (6.023 \times 10^{23})\left(\frac{0.693}{90 \times 365 \times 24 \times 3600}\right) = 1.54 \times 10^{14} \text{ dps}$$

From equation (3.16), we get

$$Q_\alpha = E_\alpha\left(\frac{M_d + M_\alpha}{M_d}\right) = 5.5\left(\frac{234+4}{234}\right) = 5.6 \text{ MeV}$$

Therefore, the energy released per second initially = A_0(5.6 MeV) = 8.63×10^{14} MeV/s

(b) The power source can continue to successfully power the apparatus as long as it produces the minimum power required to run it. Therefore, $A(t) = A_0/8$. From equation (3.6), we have

$$\frac{A(t)}{A_0} = \left(\frac{1}{2}\right)^{t/T_{1/2}} \quad \text{or} \quad \left(\frac{1}{2}\right)^3 = \left(\frac{1}{2}\right)^{t/T_{1/2}}$$

Hence, the apparatus can be powered normally for 3 half-lives of ${}^{238}_{94}\text{Pu}$ or 270 y.

Radioactive Dating

Radioactive dating is the method of determining the age of organic matter or geological samples by using the radioactive technique. This process is widely used by researchers to determine the age of materials and artifacts. Some of the commonly used radioactive dating methods have been discussed in the following sections:

Radiocarbon Dating[14]**:** Radiocarbon or carbon-14 dating method is most commonly used to date biological samples (bones, wood, paper, wool, etc.). Carbon-14 is continuously formed in the upper atmosphere when ${}^{14}_{7}N$ is broken down to form the radiocarbon following cosmic ray neutron bombardment. Since, all living matter is made up of carbon atoms and it continuously absorbs carbon dioxide from the atmosphere, it is natural that it also contains ${}^{14}_{6}C$, which is present in trace amount, and it stays in equilibrium with the atmospheric carbon as long as the living matter is alive. But, when the living matter is dead, it stops acquiring any new carbon. The unstable ${}^{14}_{6}C$ isotope within the organism naturally decays and over the time there is a gradual decrease in the ratio of ${}^{14}_{6}C$ to other stable isotopes of carbon, and the specific activity of the carbon content declines. The half-life of radiocarbon is 5730 y and the specific activity of carbon is about 14 disintegrations per minute. This dating method breaks down for periods beyond about ten half-lives of ${}^{14}_{6}C$.

There is an inherent assumption involved here that the production rate of radiocarbon in the upper atmosphere has been relatively constant over thousands of years, which might not be completely true especially now, since the burning of the fossil fuels and nuclear weapons must have altered the atmospheric balance.

EXERCISE 3.22:

(a) What is the relative abundance of ${}^{14}_{6}C$ to ${}^{12}_{6}C$ in the atmosphere?

(b) The radiocarbon activity in a piece of charcoal from the remains of a medieval campfire is found to be 13 dpm per gram. How long ago did the fire occur?

(c) A nuclear explosion has taken place leading to increase in concentration of ${}^{14}_{6}C$ in nearby areas. Its concentration in nearby areas is C_1 and C_2 in areas faraway. If the age of the fossil determined at the two places is t_1 and t_2, respectively, at which of the two places will the age of the fossil increase and by how much?

Solution:

(a) We know that 1 g of carbon content has an activity of 14 dpm, which comes entirely from carbon-14. From equation (3.4), we know $A = \lambda N$. Therefore,

$$14 \text{ dpm} = \frac{0.693}{5730 \times 365 \times 24 \times 60 \text{ m}} N({}^{14}_{6}C)$$

$$\therefore \qquad \text{Relative abundance of } {}^{14}_{6}C = \frac{N({}^{14}_{6}C)}{\text{Number of carbon atoms in 1 g}}$$

$$= \frac{6.084 \times 10^{10}}{\frac{1}{12}(6.023 \times 10^{23})} = 1.2 \times 10^{-12}$$

14. The method was invented by Willard Libby, an American physical chemist, in the late 1940s for which, he received the Nobel Prize (Chemistry) in 1960. He also discovered that tritium could be used for dating water. Unfortunately, Libby was also responsible for the gaseous diffusion separation and enrichment of the uranium-235, which was used in the atomic bomb on Hiroshima.

(b) From equation (3.5), we have

$$t = \frac{1}{\lambda}\ln\left(\frac{A_0}{A}\right) = \frac{5730\text{ y}}{0.693}\ln\left(\frac{14}{13}\right) \approx 600\text{ y}$$

(c) The activity of the fossil at the two places is the same. Since, the concentration of ${}^{14}_{6}C$ at the two places differs, which is indicative of the initial activity of the fossil, when it was living, the age of the fossil calculated at the two sites will vary accordingly. Since,

$$t = \frac{1}{\lambda}\ln\left(\frac{A_0}{A}\right)$$

$$\Rightarrow t_1 - t_2 = \frac{1}{\lambda}\ln\left(\frac{C_1}{C_2}\right)$$

The age of the fossil will increase at the site where explosion has taken place by an amount equal to $t_1 - t_2$.

Geological Dating: Geological dating using the lead isotopes, is one of the methods to estimate the age of rocks, ores, etc. Because lead is found to be stable and is at the end of the decay chain of ${}^{238}_{92}U$ ($T_{1/2} = 4.47 \times 10^9$ y), the ratio of ${}^{206}_{82}Pb$ and ${}^{238}_{92}U$ slowly increases over time and can be used as a measure of the age. Similarly, the ratio of ${}^{206}_{82}Pb$ and ${}^{204}_{82}Pb$ isotopes, is also a function of time, since the abundance of ${}^{204}_{82}Pb$ isotope can be presumed to be constant as it is stable and no other nuclide decays to it, and the ${}^{206}_{82}Pb$ isotope though stable is at the end of the decay chain of ${}^{238}_{92}U$ isotope, hence its abundance is slowly increasing over time.

On earth, mostly iron is found as ${}^{56}_{26}Fe$, which is a stable isotope. Elsewhere in the universe, however, with four additional neutrons, a radioactive isotope ${}^{60}_{26}Fe$ is found. Recently, half-life of ${}^{60}Fe$ has been measured precisely to be equal to 2.6 million years. It is believed that using ${}^{60}Fe$ as a chronometer, the scientists should be able to date the events such as supernova and some other stars.[15]

EXERCISE 3.23: Assuming that the entire ${}^{206}_{82}Pb$ in a given sample of uranium ore is resulted from decay of ${}^{238}_{92}U$, find out how old is the ore if the ratio of ${}^{238}_{92}U$ to ${}^{206}_{82}Pb$ isotopes is 0.5.

Solution: Initially the ore consists no ${}^{206}_{82}Pb$. Thereafter, the number of atoms of ${}^{206}_{82}Pb$ in the ore is equal to the atoms of ${}^{238}_{92}U$ decayed over time. Therefore, if N_0 is the total number of atoms of ${}^{206}_{82}Pb$ and ${}^{238}_{92}U$, which remains constant, then using equation (3.2), we have

15. Phys. Rev. Lett. 114, 041101, 2015.

$$\frac{\text{Number of } {}^{238}_{92}\text{U atoms}}{\text{Number of } {}^{206}_{82}\text{Pb atoms}} = \frac{N_0 - N_0 e^{-\lambda t}}{N_0 e^{-\lambda t}} = 0.5$$

$$\therefore \quad t = \frac{4.47 \times 10^9 \text{ y}}{0.693} \ln(1.5) = 2.6 \times 10^9 \text{ y}$$

The sample of uranium ore is about 2.6 billion years old.

PROBLEMS

3.1 Calculate the time required for 10% of a pure sample of ${}^{232}_{90}$Th ($T_{1/2} = 1.4 \times 10^{10}$ y) to disintegrate. **[Ans:** 2.1×10^9 y]

3.2 Nuclei of a certain radioactive element A are being produced at a constant rate K. The element has a decay constant λ. At time $t = 0$, there are N_0 nuclei of the element. Calculate the number of nuclei (N) of A at time t. If $K = 2N_0\lambda$, calculate the number of nuclei of A after one half-life of A, and also the limiting value of N.
[*Hint:* As $t \to \infty$, $N = 2N_0$]

3.3 Certain nuclei A decay into daughter nuclei B, which are also unstable. The respective decay constants are λ_A and $\lambda_B = 2\lambda_A$; calculate the time (in terms of λ_A) and the number of nuclides of B (in terms of N_{A0}), when the number of B nuclides are maximum. Assume initially we have a pure sample of A.
[*Hint:* For N_B at the point of maxima, its derivative should be zero, which will give $N_B(\text{max}) = 0.25\ N_{A0}$]

3.4 ${}^{118}_{48}$Cd radioisotope goes through the following decay chain:

$${}^{118}_{48}\text{Cd} \xrightarrow{T_{1/2} = 30\text{ m}} {}^{118}_{49}\text{In} \xrightarrow{T_{1/2} = 4.5\text{ m}} {}^{118}_{50}\text{Sn (stable)}$$

Assuming that initially, the sample consists of Cd only; find the fraction of nuclei transformed into stable ones after one hour. In what proportion, does the activity of the sample diminish at this point? **[Ans:** 0.71; 0.544]

3.5 Calculate the fractions of the parent, daughter and granddaughter nuclei as a function of time that are given in the previous example, by numerical solution of the Bateman equations. Also indicate in the plot, the time for the maximum daughter concentration.

3.6 For the following decay scheme, prove that the total number of atoms remain constant

$$A \to B \to C \to D$$

3.7 It is found that a solution containing 1 g of α-emitter ${}^{226}_{88}$Ra isotope never accumulate more than 6.4×10^{-6} g of its daughter element radon ($T_{1/2} = 3.825$ d). Calculate the half-life of radium radionuclide, based on the above information. **[Ans:** 1637 y]

3.8 Calculate the ratios of λ_α probability for l angular momentum compared to $\lambda_\alpha(l = 0)$. Take the value of $Q_\alpha = 4.88$ MeV, $R = 9.87$ fm and $Z_d = 86$.
[Ans: 1, 0.7, 0.37, 0.137, 0.037 and 0.0071 for $l = 0$–5, respectively]

3.9 Determine the classically forbidden regions for α-particles of 4.27 MeV and 6.81 MeV emitting from ${}^{238}_{92}U$ and ${}^{228}_{92}U$, respectively. Also, compare their tunneling probabilities P and justify your answer quantitatively.

[**Ans:** 51 and 26 fm, $P = E(-38)$ for U-238 and $E(-25)$ for U-228]

3.10 Estimate the Z dependence of the α-decay for nuclei close to ${}^{238}_{92}U$ by calculating the effect on the half-life of changing Z by 2 at an α-particle energy of 5 MeV.

3.11 In the following decay scheme (Figure 3.26), determine the energy of the orbital electron capture decay mode. The K-shell binding energy of ${}^{22}_{11}Na$ is 1.0721 keV and the atomic masses of ${}^{22}_{11}Na$ and ${}^{22}_{10}Ne$ are 21.9944368 u and 21.9913855 u, respectively. Also determine the energy carried away by the neutrino.

[**Ans:** –2.841 MeV, 1.567 MeV]

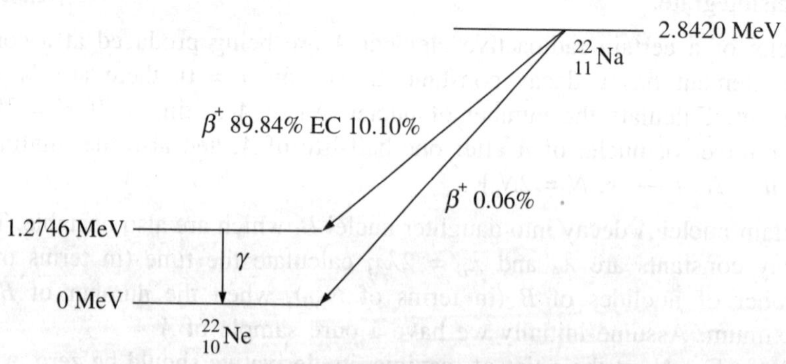

FIGURE 3.26 Decay scheme of Na-22.

3.12 The total half-life of ${}^{232}_{94}Pu$ isotope is 34.1 m. Its decay by α-emission is 20% and the rest is by electron capture. Calculate the half-life for both the decay modes.

[**Ans:** 170.5 m, 42.6 m]

3.13 A free neutron undergoes β-decay with a half-life of 10.6 m. Calculate the Q-value of the decay. [**Ans:** 0.782 MeV]

3.14 A neutron at rest decays into a proton with a decay energy of 0.782 MeV. What will be the maximum kinetic energy of the residual proton? [**Ans:** 7.5×10^{-4} MeV]

3.15 Calculate the energy released (E_{max}) for the following β-decay processes:

$${}^{32}_{15}P(31.9720707\,u) \to {}^{32}_{16}S(31.9739072\,u) + \beta^- + \bar{\nu}$$

$${}^{13}_{7}N(13.0057386\,u) \to {}^{13}_{6}C(13.0033548\,u) + \beta^+ + \nu$$

[**Ans:** 1.7107 MeV, 1.196 MeV]

3.16 The isotope ${}^{194}_{79}Au$ undergoes β-decay and has a mean-life of 56 h. One mode of decay is as follows:

$${}^{194}_{79}Au \to {}^{194}_{78}Pt + \beta^+ + \nu + 1.5\text{ MeV}$$

In this decay, the positron is created inside the nucleus, which must tunnel through a Coulomb barrier to escape. If the positron has an energy of about 1 MeV, calculate the tunneling probability (suppression factor).
[*Hint:* Expression (3.18) derived for α-decay, for the positron use: $Z = 1$, $Z_d = Z - 1$, $Q_\alpha = Q_\beta$ and in the reduced mass, use M_β instead of M_α]

3.17 The maximum kinetic energy K_0 of the electrons emitted in the decay of carbon-14 is 0.156 MeV. If the number of electrons with kinetic energy K and $K + dK$ is assumed to have the approximate form

$$N(K)\,dK \propto \sqrt{K}(K_0 - K)^2 dK$$

(a) Find the mean energy of the electrons.
(b) Find the rate of evolution of heat by a radiocarbon source emitting 3.7×10^7 electrons per second.

$$\left[Hint\text{: } \langle K \rangle = \frac{\int_0^{K_0} KN(K)\,dK}{\int_0^{K_0} N(K)\,dK} = \frac{K_0}{3} \right]$$

What is the degree of forbiddenness in the β-decay for the following nuclear transitions?

(a) $1/2^+ \rightarrow 3/2^+$
(b) $1/2^+ \rightarrow 1/2^-$
(c) $2^+ \rightarrow 1^+$
(d) $2^+ \rightarrow 5^+$

3.18 In the β-decay, if a $5/2^+$ nuclear state decays by 1st forbidden transition, what are the possible spin-parity states for the final nucleus? [**Ans:** $1/2^-$, $3/2^-$, $5/2^-$, $7/2^-$, $9/2^-$]

3.19 Determine the radioactivity in the human body, if there are 140 g of K (atomic mass = 39.0983 u) in a typical person's body. The abundance of ${}^{40}_{19}$K radioisotope is 0.0117% and its half-life is 1.277×10^9 y. [**Ans:** 116 nCi]

3.20 Calculate the specific activity (Ci/μg) of pure ${}^{239}_{94}$Pu ($T_{1/2} = 2.4 \times 10^4$ y). [**Ans:** 6.24×10^4]

3.21 A piece of wood from the ruins of an ancient Japanese temple (Figure 3.27) was found to have a radiocarbon activity of 12 disintegrations per minute per gram of its carbon content.

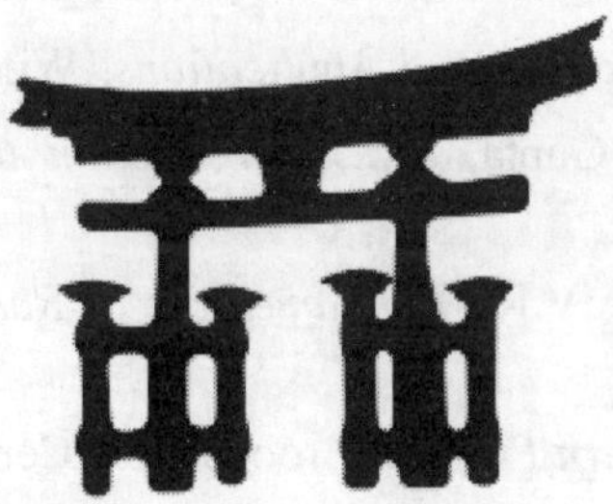

FIGURE 3.27 An ancient Japanese temple.

Approximately, how long ago did the tree die from which the wood sample came? **[Ans:** 1250 y]

3.22 Radionuclide A undergoes simultaneous decay into two different nuclei P and Q. The half-lives for the decay of A to P and Q are 9 h and 4.5 h, respectively. Calculate the time after which the amount of Q doubles that of A. Assume that initially neither P nor Q were present. **[Ans:** 6 h]

3.23 A sample consists of a mixture containing ${}^{239}_{94}\text{Pu}$ and ${}^{240}_{94}\text{Pu}$. If the specific activity is 3.42×10^8 dpm/mg, what is the proportion of plutonium in the sample?

3.24 A dose of 5 mCi of ${}^{32}_{15}\text{P}$ (half-life from Table 3.1) is injected into a patient's bloodstream whose blood volume is of 3.5 litres. What would be the count rate per ml of withdrawn blood after 28 days, if the counter has an efficiency of only 10%.

[Ans: 1.3×10^3]

3.25 If the age of earth is 4.5 billion years, what should be the ratio of ${}^{206}_{82}\text{Pb}$ to ${}^{238}_{92}\text{U}$ in a uranium-bearing rock as old as the earth?

BIBLIOGRAPHY

Beiser, A., *Concepts of Modern Physics*, McGraw-Hill, New York, 2003.

Blatt, J.M. and Weisskopf, V.F., *Theoretical Nuclear Physics*, Wiley, New York, 1952.

Choppin, G.R. and Rydberg, J., *Nuclear Chemistry: Theory and Applications*, Pergamon Press, Oxford, 1980.

Cottingham, W.N. and Greenwood, D.A., *An Introduction to Nuclear Physics*, Cambridge University Press, Cambridge, 1986.

Devanathan, V., *Nuclear Physics*, Narosa, New Delhi, 2006.

Garg, J.B., *Nuclear Physics: Basic Concepts*, Macmillan, New Delhi, 2011.

Harvey, B.G., *Introduction to Nuclear Physics and Chemistry*, Prentice-Hall, New Jersey, 1970.

Heyde, K., *Basic Ideas and Concepts in Nuclear Physics*, Overseas Press India, New Delhi, 2005.

Jevremovic, T., *Nuclear Principles in Engineering*, Springer, New York, 2009.

Lilley, J., *Nuclear Physics: Principles and Applications*, Wiley, New York, 2002.

Mittal, V.K., Verma, R.C. and Gupta, S.C., *Introduction to Nuclear and Particle Physics*, PHI Learning, Delhi, 2013.

Sood, D.D. and Ramamoorthy, A.V.R., *Fundamentals of Radiochemistry*, IANCAS, Mumbai, 2000.

Thornton, S.T. and Rex, A., *Modern Physics*, Brooks/Cole Cengage Learning, New Delhi, 2007.

4

Nuclear Force

"A Thousand Paper Cranes. Peace on Earth and in the Heavens."

[Inscribed, in the handwriting of Yukawa, on the surface of the bell inside the Children's Peace Monument at Hiroshima, Japan. The crane is a symbol of longevity and happiness in Japan. The monument to mourn all the children whose death was caused by the atomic bomb was inspired by 12-year-old Sadako Sasaki, who believed that if she could fold 1000 paper cranes she would be cured of the leukemia that resulted from her exposure to the radiation of the atomic bomb when two years old. She died before completing them.]

—Hideki Yukawa

Nuclear forces (also known as strong forces) are the forces that act between two or more nucleons. They bind the nucleons into atomic nuclei. The nuclear force is about ten million times stronger than the chemical binding that holds the atoms together in molecules. Exact understanding of the nuclear forces between the nucleons has been the central question in nuclear physics and during the last seven decades, many nuclear physicists have devoted a lot of their time in this pursuit. The original idea was that the force is mediated by a particle (known as π-meson) and this resulted in the birth of 'elementary particle physics,' which is discussed in Chapter 8. It is now common knowledge that the nucleons are not elementary particles and the force acting between quarks that make them up is responsible for their interactions. This force is mediated by the exchange of gluons and holds the quarks together inside a nucleon.

In this chapter, with a historical perspective, we concentrate on the simplest two-nucleon system. There are two experimentally achievable situations:

(i) When a neutron and a proton are bound together, as a deuteron[1]. It is the simplest bound state of nucleons, where the two nucleons are held together by the nuclear force alone, and therefore gives an ideal system for studying the nucleon–nucleon interaction.

(ii) In collisions between two nucleons, usually referred to as scattering processes.

1. A ${}^2_1\text{H}$ nucleus is called a deuteron, while a neutral atom of ${}^2_1\text{H}$ is called deuterium.

4.1 HISTORICAL PERSPECTIVE

After the discovery of the neutron by Chadwick in 1932, it was clear that the nucleus contains neutrons and protons. Two forces—gravitational and electric forces were known by that time. In order to understand the nature of the force that holds the nucleus, the scientists started comparing the nucleus with the atom. The radius of the nucleus is ~10^{-15} m, which is much smaller than the size of the atom (~10^{-10} m). Binding energy of the nucleus is in the MeV range, while electronic binding energy in the atom is in the ~eV range. The repulsive electric force between protons should blow the nucleus apart. Therefore, the concept of a new attractive strong force between nucleons, i.e. nuclear force, was proposed, which must be much stronger than the Coulomb force. It cannot be attractive at all separations otherwise nuclei will just collapse in on themselves. So, at very short ranges, there must be a repulsive core. Studies on the B/A curve between A = 20 to A = 160 indicated a constancy, reflecting the saturation property of the nuclear force. Another important property is the charge symmetry of the nuclear forces. The evidence of charge symmetry comes from the similarity of the energy levels of the mirror nuclei such as ($^3_1$H, $^3_2$He), ($^5_2$He, $^5_3$Li), ($^{13}_6$C, $^{13}_7$N) and ($^{14}_6$C, $^{14}_8$O). The principle of charge symmetry was extended to imply charge independence of nuclear force.

In 1935, the first theory for this new force was developed by the physicist H. Yukawa, who proposed that the nucleons exchange particles between each other as shown in Figure 4.1.

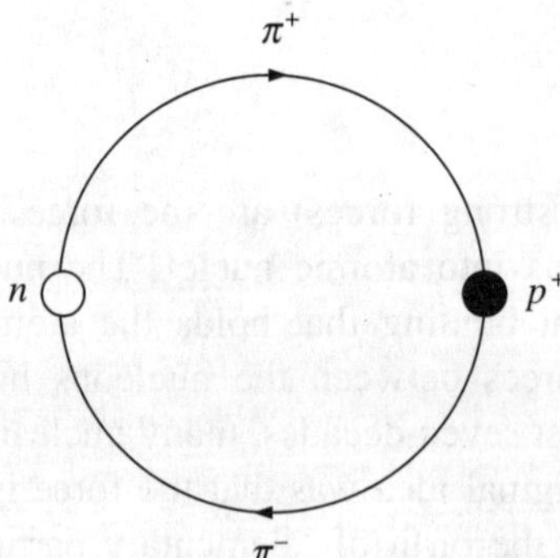

FIGURE 4.1 Mechanism of particle exchange between the pair of nucleons.

Thus, each nucleon acts as a source and a sink for π-mesons. When two nucleons are brought together, a π-meson emitted by one particle is absorbed by the other, resulting in lowering of the energy of the system and the attractive force between nucleons is generated. Yukawa formulated his theory in analogy to electromagnetic interactions, where the exchange of a massless photons is taken to be the cause of the force. Detailed Yukawa theory of nuclear force is given in the later section of the chapter.

These meson models were qualitatively quite successful in understanding the nuclear forces until 1970, when Gell-Mann[2] proposed that all hadrons and mesons are not elementary particles but are made up of quarks. Thus, a proton is made up of 3 quarks (*uud*) and the neutron is also made up of 3 quarks (*udd*). The force between nucleons is a facet of strong

2. See Chapter 8 of this book on 'Particle Physics.'

interaction between quarks, and the fundamental theory of strong interaction is quantum chromodynamics (QCD). The problem with a derivation of nuclear forces from QCD is two-fold. First, each nucleon consists of 3 quarks and therefore, a system of two nucleons is already a six-body problem. Second, the force between quarks, which is due to the exchange of gluons, must satisfy the condition of confinement and has the feature of being very strong at the low-energy scale. This extraordinary strength makes it difficult to find 'converging' mathematical solutions. A proper derivation of the nuclear force from QCD is still in developing stage.

4.2 THE DEUTERON (D)—BOUND STATE OF TWO NUCLEONS

The only bound system of two nucleons found in nature is the deuteron, which consists of a neutron and a proton, such that

$$m_d(2.013553\,\text{u}) < [m_p(1.007276\,\text{u}) + m_n(1.008665\,\text{u})]$$

The other possibilities consistent with Pauli exclusion principle are di-neutron and di-proton, which do not exist in nature.

The experiments have provided the following facts about deuteron:

(i) Binding energy, $B_d = -2.2244$ MeV.
(ii) Total angular momentum and parity $J^\pi = 1^+$.
(iii) Root mean square radius (size) of deuteron $\langle r_d \rangle = 2.1$ fm.
(iv) Magnetic moment, $\mu_d = 0.857393$ nm.
(v) Electric quadrupole moment, $Q_d = 0.00282$ b.
(vi) Deuteron does not have any excited states.

The two body problem of a neutron and a proton interacting with a force represented by central potential $V(\vec{r})$, where $\vec{r} = \overrightarrow{r_p} - \overrightarrow{r_n}$, describes the relative position of the proton with respect to the neutron, reduces by the well-known centre-of-mass transformation of classical mechanics to the problem of a single particle in a potential well $V(\vec{r})$, and mass of the 'single particle' is the reduced mass,

$$\mu = \frac{m_n m_p}{m_n + m_p} \approx \frac{1}{2} m_p$$

The time-independent Schrödinger equation is given by

$$\left[-\frac{\hbar^2}{2\mu}\nabla^2 + V(\vec{r})\right]\psi = E\psi \tag{4.1}$$

Since, the nucleus is spherical, we consider a spherical symmetric potential, i.e., central potential, in which case the wave function ψ can be separated as a radial function $R(r)$ and the spherical part $Y_l^m(\theta, \varphi)$, known as spherical harmonics.

$$\psi = R(r) Y_l^m(\theta, \varphi)$$

From equation (4.1)[3], one can see that the radial function $R(r)$ satisfies the following one-dimensional wave equation

$$-\frac{\hbar^2}{2\mu r^2}\frac{d\left(r^2\frac{dR(r)}{dr}\right)}{dr}+\left[V(r)+\frac{l(l+1)\hbar^2}{2\mu r^2}\right]R(r)=ER(r)$$

where the potential in the Schrödinger equation is replaced by the effective potential, which is the sum of the nuclear potential $V(r)$ and the centrifugal potential $l(l+1)\hbar^2/2\mu r^2$ that arises from the rotation of the two nucleons about their centre of mass. Experimentally, it has been found that the deuteron has only one energy (lowest energy) corresponding to the ground state and thus, is obtained when $l = 0$.

Substituting $R(r) = u(r)/r$ to further simplify the above equation, we get

$$-\frac{\hbar^2}{2\mu}\frac{d^2u(r)}{dr^2}+\left[V(r)+\frac{l(l+1)\hbar^2}{2\mu r^2}\right]u(r)=Eu(r)$$

To obtain a solution for the above equation, we assume a hardcore square well potential as shown in Figure 4.2. There are three separate regions, such that

$$V=\infty \quad \text{for } r<c$$
$$V=-V_0 \quad \text{for } c<r<b+c$$
$$V=0 \quad \text{for } r>b+c$$

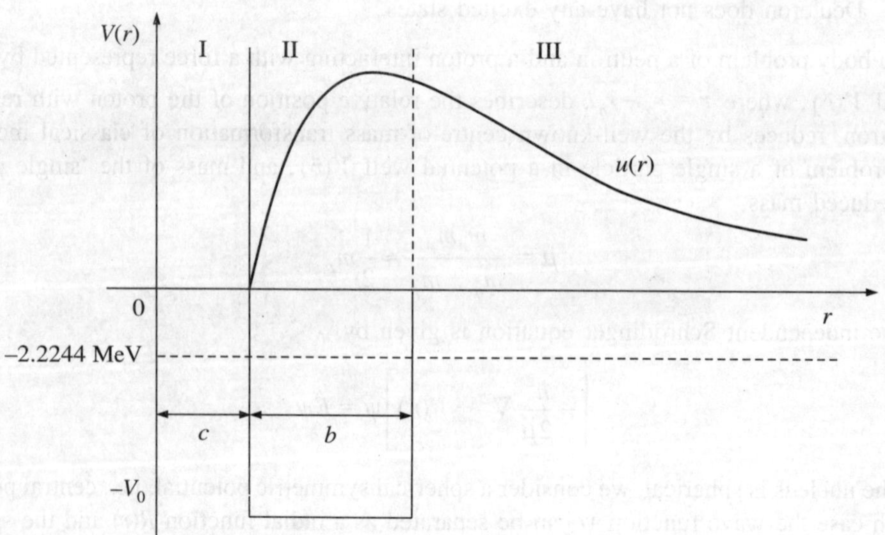

FIGURE 4.2 Simplified neutron-proton potential and deuteron radial wave function $u(r)$.

3. The Laplace operator in the polar coordinates (r, θ, φ) is

$$\nabla^2=\frac{1}{r^2}\frac{\partial}{\partial r}\left(r^2\frac{\partial}{\partial r}\right)+\frac{1}{r^2\sin\theta}\frac{\partial}{\partial\theta}\left(\sin\theta\frac{\partial}{\partial\theta}\right)+\frac{1}{r^2\sin^2\theta}\frac{\partial}{\partial\theta}\left(\frac{\partial^2}{\partial\varphi^2}\right)$$

It should be duly noted that a square well potential may not be the actual potential but it is used here just for simplicity.

Now, there is no interest of solving the above equation in the 1st region, because the two nucleons cannot come closer than a certain distance owing to a repulsive force. Two nucleons, thus, make a bound system only in the 2nd and the 3rd region, where the nuclear attractive force is present.

In the 2nd region,

$$-\frac{\hbar^2}{2\mu}\frac{d^2u}{dr^2} = (V_0 + E)u$$

or

$$\frac{d^2u}{dr^2} = -K^2u$$

where $K = \sqrt{\frac{2\mu(V_0 + E)}{\hbar^2}}$.

The general solution of the above equation is

$$u(r) = A \sin K(r - c) + B \cos K(r - c) \qquad \text{for } c < r < (b + c)$$

Applying the boundary condition at $r = c$, $u(r)$ should be zero. Thus, B must be zero. Hence, the possible solution is

$$u_{\text{II}}(r) = A \sin K(r - c)$$

In the 3rd region,

$$-\frac{\hbar^2}{2\mu}\frac{d^2u}{dr^2} = Eu$$

or

$$\frac{d^2u}{dr^2} = k^2u$$

where $k = \sqrt{\frac{-2\mu E}{\hbar^2}}$ [k is positive since the binding energy E is negative].

The general solution of the above equation is

$$u(r) = Ce^{kr} + De^{-kr} \qquad \text{for } r > (b + c)$$

Applying the boundary condition at $r \to \infty$, $u(r) \to 0$. Thus, C must be zero. Hence, the possible solution is

$$u_{\text{III}}(r) = De^{-kr}$$

Since, the radial wave function has to be continuous at the boundary at $r = (b + c)$, it follows that

$$u_{\text{II}}(r)\big|_{r=b+c} = u_{\text{III}}(r)\big|_{r=b+c}$$

$$A \sin Kb = De^{-k(b+c)} \tag{4.2}$$

also

$$\frac{du_{II}(r)}{dr}\bigg|_{r=b+c} = \frac{du_{III}(r)}{dr}\bigg|_{r=b+c}$$

$$AK\cos Kb = -kDe^{-k(b+c)} \tag{4.3}$$

Dividing equation (4.3) by equation (4.2), we get

$$\frac{AK\cos Kb}{A\sin Kb} = \frac{-kDe^{-k(b+c)}}{De^{-k(b+c)}}$$

or

$$\cot Kb = -\frac{k}{K} \tag{4.4}$$

i.e.,

$$\cot\sqrt{\frac{2\mu(V_0+E)}{\hbar^2}}\,b = -\frac{\sqrt{\dfrac{-2\mu E}{\hbar^2}}}{\sqrt{\dfrac{2\mu(V_0+E)}{\hbar^2}}}$$

EXERCISE 4.1: The size of the deuteron has been determined from electron scattering experiments. The mean square distance between the neutron and proton r_d was found to be 4.2 fm^2. The radius of deuteron is found to be approximately equal to the range of the nuclear force b = 2.1 fm.

Determine the theoretical root mean square radius of the deuteron and compare it with its experimentally determined value. Take K = 0.877 fm^{-1}, k = 0.232 fm^{-1} and c = 0.4 fm. The mean square radius of proton $\langle r_p^2\rangle = 0.8$ fm^2.

Solution:

$$\psi = \frac{1}{r}u(r)Y_{l,m}(\theta,\varphi)$$

The total probability for the deuteron to be in an element of volume $d\tau$ is given as (spherical shell of radius r and thickness dr)

$$\int |\psi|^2\, d\tau \quad \text{where } d\tau = 4\pi r^2 dr$$

For $l = 0$,

$$Y_{l,m}(\theta,\varphi) = \frac{1}{\sqrt{4\pi}}$$

or

$$\int |\psi|^2\, d\tau = \int_0^\infty \left|\frac{u(r)}{r}\frac{1}{\sqrt{4\pi}}\right|^2 d\tau = \int_0^\infty |u(r)|^2\, dr$$

Substituting the wave function in 2nd and the 3rd region, we get

$$\int_0^\infty |u(r)|^2\, dr = \int_c^{b+c}\left[A^2\sin^2 K(r-c)\right]dr + \int_{b+c}^\infty D^2 e^{-2kr}\, dr$$

The total probability of finding the deuteron in 2nd and 3rd regions is 1. Therefore, the above expression must be equal to 1.

$$\therefore \qquad \frac{A^2}{2}\left[b - \frac{\sin 2Kb}{2K}\right] + \left[\frac{D^2}{2k}\, e^{-2k(b+c)}\right] = 1 \qquad (4.5)$$

A and D are constants, whose values can be obtained by combining equation (4.5) with equations (4.2) and (4.3). Thus,

$$A^2 = \frac{2k}{1+kb} \text{ and } B^2 = \frac{2k\sin^2 Kb}{1+kb}\, e^{2k(b+c)}$$

Now, if the proton and the neutron are assumed to be point masses, the mean square radius of the deuteron will be equal to the average of the square of the distance between the centre of mass to each nucleon, i.e.,

$$\langle r_d^2 \rangle = \int \left(\frac{r}{2}\right)^2 u(r)^2 dr$$

$$= \frac{1}{4}\int_c^{b+c} [A^2 r^2 \sin^2 K(r-c)]\, dr + \frac{1}{4}\int_{b+c}^{\infty} D^2 r^2 e^{-2kr}\, dr$$

Part-wise integration gives

$$\langle r_d^2 \rangle = \frac{1}{8k^2} - \frac{1}{8K^2} + \frac{(2c+b)(1+kb)}{8k} + \frac{c^2}{4} - \frac{kb^3}{24(1+kb)}$$

$$= \frac{1}{8(0.232)^2} - \frac{1}{8(0.877)^2} + \frac{(2\times 0.4 + 2.1)(1 + 0.232\times 2.1)}{8(0.232)} + \frac{(0.4)^2}{4} - \frac{0.232(2.1)^3}{24(1+0.232\times 2.1)}$$

or

$$\langle r_d^2 \rangle = 4.39 \text{ fm}^2$$

Since, proton/neutron is actually not a point mass, the corrected value of the deuteron rms radius should take the nucleon size into account, i.e.,

$$\langle r_d^2 \rangle = 4.39 + \langle r_p^2 \rangle = 5.19 \text{ fm}^2$$

$$\therefore \qquad \sqrt{\langle r_d^2 \rangle} = 2.28 \text{ fm}$$

The theoretical value of the deuteron radius $\langle r_d \rangle \approx 2.28$ fm is found to be close to the experimentally determined value of 2.1 fm.

The binding energy of the deuteron (E) is found to be –2.2244 MeV and can be experimentally determined by various methods, like

(i) Allowing slow neutrons to be captured by the protons in a material containing hydrogen and measuring the energy of γ-photon that is emitted in the reaction,

$$n + {}_1^1\text{H} \rightarrow {}_1^2\text{H} + \gamma$$

(ii) Another method uses the reverse reaction called photodisintegration, in which a γ-ray photon breaks down the deuteron.

$${}_1^2\text{H} + \gamma \rightarrow n + {}_1^1\text{H}$$

The wave function is maximum at $r = 1.79$ fm, when $\sin Kr = 1$, and the fall is about 4% at the edge ($r = b$) of the potential. In the 3rd region, the wave function decreases slowly and at $r = 4.31$ fm, it has decreased by 37%. The exponential tail represents the important part of the wave function which is outside the well and indicates that the deuteron is a weakly-bound system.

EXERCISE 4.2:

(a) Substituting the values of b, E and $\hbar^2/2\mu$ for the deuteron in equation (4.4), determine the well-depth (V_0).

(b) Derive the expression for the kinetic energy in terms of K which was defined earlier.

(c) Show that the deuteron cannot have an excited state (but still be bound) with $L > 0$.

(d) Can the deuteron have any bound excited states with $L = 0$?

Solution:

(a) The reduced mass (μ) is given by

$$\mu = \frac{m_n m_p}{m_n + m_p} \approx \frac{1}{2} m_p$$

Thus, $\hbar^2/2\mu = \hbar^2/m_p = 41.3$ MeVfm^2. From equation (4.4), we get

$$f(V_0) = \cot\sqrt{\frac{(V_0 - 2.2244)}{41.3}}\,2.1 + \frac{1.4914}{\sqrt{(V_0 - 2.2244)}} = 0$$

Plotting $f(V_0)$ vs. V_0 (Figure 4.3), we get the well depth V_0 = ~34 MeV when $f(V_0) = 0$.

The nuclear potential that binds the deuteron is thus, to some approximation, a 34 MeV deep square well and has a width of 2.1 fm. Thus, the depth is much larger than the Coulomb force at these distances; the Coulomb force between two protons corresponds to a potential of only 0.6 MeV at 2.1 fm. It can also be noted that in comparison with the depth of the well, the 2.2244 MeV binding energy of the deuteron is very small; the deuteron is loosely bound.

(b) We know that

$$K = \sqrt{\frac{2\mu(V_0 + E)}{\hbar^2}}$$

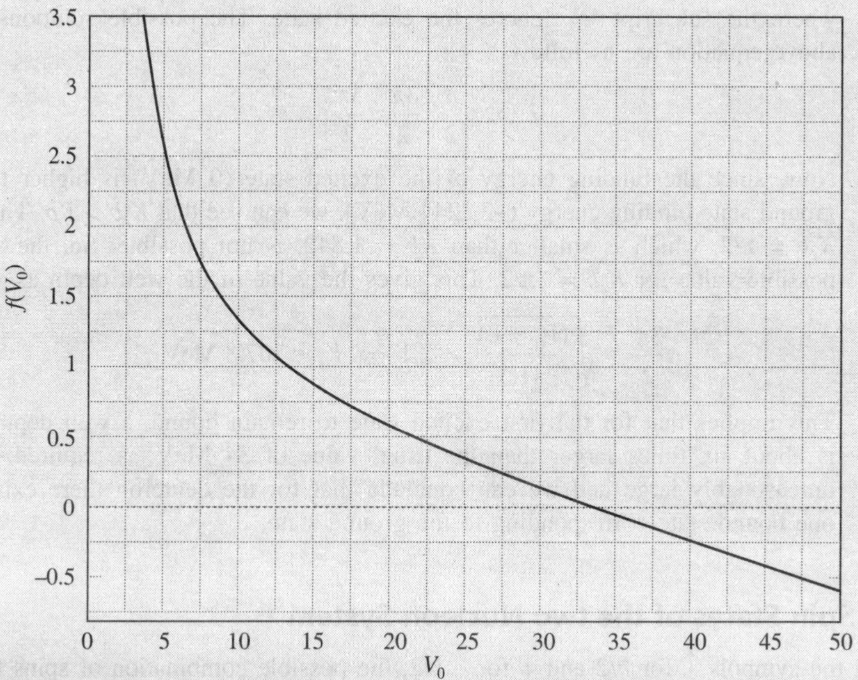

FIGURE 4.3 Plot of $f(V_0)$ vs. V_0.

Since, the well depth should be equal to the sum of the kinetic energy and the potential energy (binding energy), we get

$$\text{KE} = V_0 + E = \frac{\hbar^2 K^2}{2\mu}$$

(c) If there are any bound excited states, they must be between the ground state ($E = -2.2244$ MeV) and the top of the well ($E = 0$ MeV). Now, if the centrifugal barrier height is found to be greater than the top of the well, then we can easily infer that the bound states with $L > 0$ do not exist.

$$\frac{\hbar^2 l(l+1)}{2\mu r^2} \geq \frac{\hbar^2}{2\mu}\frac{2}{b^2} = 18.7 \text{ MeV}$$

(d) For the deuteron, $V_0 = \sim 34$ MeV, $b = 2.1$ fm and the ground state binding energy is equal to -2.2244 MeV. Therefore, for the ground state,

$$Kb = \sqrt{\frac{2\mu(V_0 + E)}{\hbar^2}}\, b = \sqrt{\frac{(34 - 2.2244)}{41.3}} \times 2.1 = 1.842$$

The largest possible binding energy for any excited state is 0. Let us assume that the deuteron has a first excited state, and its binding energy is 0. Then from equation (4.4), we have

$$\cot K_e b = 0$$

where the subscript 'e' denotes the excited state. The possible solutions of the above equation are as follows:

$$K_e b = \frac{\pi}{2}, \frac{3\pi}{2}, \frac{5\pi}{2}, \ldots$$

Now, since the binding energy of the excited state (0 MeV) is higher than the ground state binding energy (–2.2244 MeV), we can see that $K_e b > Kb$. Therefore, $K_e b = \pi/2$, which is smaller than $Kb = 1.842$, is not possible. So, the smallest possible value for $K_e b = 3\pi/2$. This gives the value of the well depth as

$$\frac{3\pi}{2} = \sqrt{\frac{(V_0 + 0)}{41.3}} \times 2.1 \Rightarrow V_0 = 207.8 \text{ MeV}$$

This implies that for the first excited state to remain bound, a well depth which is about six times larger than the usual value of 34 MeV, is required. This is unreasonably large and we can conclude that for the deuteron there exists only one bound state corresponding to the ground state.

4.2.1 Spin States of the two Nucleon System

Adopting the symbols ↑ for $\hbar/2$ and ↓ for $-\hbar/2$, the possible combination of spins for a two nucleon system are as follows:

$$|\uparrow\uparrow\rangle\ |\uparrow\downarrow\rangle\ |\downarrow\uparrow\rangle\ |\downarrow\downarrow\rangle$$

If the spin angular momenta of the neutron and the proton couple, then the spin angular momentum of the composite system,

$$\vec{S} = \vec{s_p} + \vec{s_n} = \vec{\frac{1}{2}} + \vec{\frac{1}{2}} = 1, 0$$

When $S = 0$, then the physically measurable quantity—the z-component of S, $S_z = 0$, where $S_z = s_{pz} + s_{nz}$. The state with $S = 0$, $S_z = 0$ (singlet state) will be

$$\frac{1}{\sqrt{2}}\left[|\uparrow\downarrow\rangle - |\downarrow\uparrow\rangle\right]$$

where $\sqrt{2}$ is the normalization factor.

Similarly, if $S = 1$ (triplet state) then $S_z = \hbar, 0, -\hbar$ and the states will be

$$S = 1, S_z = \hbar \Rightarrow |\uparrow\uparrow\rangle$$

$$S = 1, S_z = -\hbar \Rightarrow |\downarrow\downarrow\rangle$$

$$S = 1, S_z = 0 \Rightarrow \frac{1}{\sqrt{2}}\left[|\uparrow\downarrow\rangle + |\downarrow\uparrow\rangle\right]$$

As can be seen, there are four nuclear states for an (n, p) system (not necessarily deuteron), if the angular momenta couple. Out of these four possibilities, the probability of $S = 1$ is 75%.

EXERCISE 4.3: For the deuteron, the total angular momentum, which is the vector sum of the orbital and spin angular momentum, i.e., $\vec{I} = \vec{L} + \vec{S}$, was experimentally measured as $I = 1$. Parity is also an experimentally measured quantity and for the deuteron, it is even (or positive). Thus, we know that for the deuteron $I^{\pi} = 1^{+}$.

What are the possible values of the orbital angular momentum L? Can we say, that the nuclear interaction between a neutron and a proton is spin dependent?

Solution:

$$\vec{I} = \vec{L} + \vec{S}$$

There are three options, where the vector sum of the spin and the orbital angular momenta will be 1.

$$L = 0 \quad \text{and} \quad S = 1$$
$$L = 1 \quad \text{and} \quad S = 0$$
$$L = 2 \quad \text{and} \quad S = 1$$

Since, the parity $\pi = (-1)^L = +1$, $L = 1$ is not possible. Therefore, only $L = 2$ can be present with $L = 0$. This also means that S is equal to 1, which implies that the nuclear interaction between a neutron and a proton must be spin dependent.

A state with (S, L, I) is usually labelled as ${}^{2S+1}L_I$, where $L = 0, 1, 2 \ldots$, are usually called the $S, P, D \ldots$ states. Thus, the superimposition of the states ${}^3S_1$ and ${}^3D_1$ forms the ground state of the deuteron, which means that the ground state wave function is a combination of ${}^3S_1$ and ${}^3D_1$ states, given as:

$$\psi = a({}^3S_1) + b({}^3D_1)$$

Because of the spin-dependence, the potential describing the interaction between a neutron and a proton must contain a spin-dependent term. If a spin-dependent central potential term is added to the central potential with proper sign and functional form of r, r being the inter-nucleon distance, the resultant (modified) potential well has to be deeper for the triplet state, since, it explains the deuteron bound state. Let us see if this indeed is the case.

EXERCISE 4.4: Show that the energy of the triplet state ($S = 1$) is not equal to the energy of the singlet state ($S = 0$).

Solution:

$$\vec{S} = \overrightarrow{s_p} + \overrightarrow{s_n}$$

$$\vec{S}\cdot\vec{S} = (\overrightarrow{s_p} + \overrightarrow{s_n})\cdot(\overrightarrow{s_p} + \overrightarrow{s_n}) = \overrightarrow{s_p}\cdot\overrightarrow{s_p} + 2\overrightarrow{s_p}\cdot\overrightarrow{s_n} + \overrightarrow{s_n}\cdot\overrightarrow{s_n}$$

$$\therefore \qquad \overrightarrow{s_p}\cdot\overrightarrow{s_n} = \frac{1}{2}\left[\vec{S}\cdot\vec{S} - \overrightarrow{s_p}\cdot\overrightarrow{s_p} - \overrightarrow{s_n}\cdot\overrightarrow{s_n}\right] = \frac{1}{2}[S(S+1) - s_p(s_p+1) - s_n(s_n+1)]$$

The expectation value of $\vec{s_p} \cdot \vec{s_n}$ may be obtained from the above identity is:

For the triplet state ($S = 1$)

$$\left\langle \vec{s_p} \cdot \vec{s_n} \right\rangle = \frac{\hbar^2}{2}\left[1(1+1) - \frac{1}{2}\left(\frac{1}{2}+1\right) - \frac{1}{2}\left(\frac{1}{2}+1\right)\right] = \frac{\hbar^2}{4}$$

For the singlet state ($S = 0$)

$$\left\langle \vec{s_p} \cdot \vec{s_n} \right\rangle = \frac{\hbar^2}{2}\left[0(0+1) - \frac{1}{2}\left(\frac{1}{2}+1\right) - \frac{1}{2}\left(\frac{1}{2}+1\right)\right] = -\frac{3\hbar^2}{4}$$

Thus, the energy of the triplet state is not equal to the energy of the singlet state.

The spin-dependent central potential term can be constructed with the help of a spin exchange operator, such that the singlet and the triplet states have eigen values –1 and +1.

$$P_\sigma = \frac{1}{2}\left(1 + \frac{4}{\hbar^2}\vec{s_p} \cdot \vec{s_n}\right)$$

Using the values of $\left\langle \vec{s_p} \cdot \vec{s_n} \right\rangle$ for the triplet and the singlet states, we can see that

$$P_\sigma = 1 \text{ for } S = 1 \text{ and } P_\sigma = -1 \text{ for } S = 0$$

In terms of the Pauli spin matrices $\vec{\sigma}$ [4]

$$P_\sigma = \frac{1}{2}(1 + \vec{\sigma_n} \cdot \vec{\sigma_p}) \quad \text{where } \vec{s} = \frac{\hbar}{2}\vec{\sigma} \tag{4.6}$$

The nuclear potential, representing the nuclear force can now be written in terms of the spin-independent central potential and the spin-dependent central potential as follows:

$$V(r) = V_c(r) + V_\sigma(r)P_\sigma$$

EXERCISE 4.5: Why is $^3P_1$ not a component of the ground state of the deuteron?

Solution: $^3P_1$ implies $L = 1$. Therefore, $^3P_1$ state cannot be a component of ($L = 0$) $^3S_1$ ground state of deuteron. Both even and odd values of L cannot be present in the same wave function due to definite parity $\pi = (-1)^L$ of a nuclear state.

4.2.2 Magnetic Dipole Moment of Deuteron

The orbital motion of the proton inside the nucleus of the deuteron gives rise to currents, which in turn, produce the magnetic field or the magnetic moment. The neutron, on the other hand, will not have any magnetic moment due to the orbital motion because it has no charge. The circulating protons are not the only source of magnetic fields in the nucleus. Besides the

4. Verma, H.C., *Quantum Physics*, Section 19.5, Surya Publications, 2006.

orbital angular momentum, both protons and neutrons possess intrinsic spin due to their internal structures and consequently magnetic moments too.

From equation (1.20) in Chapter 1, the net magnetic moment of the deuteron can be written as follows:

$$\mu_z = g_l \frac{\mu_N}{\hbar} l_z + g_{sp} \frac{\mu_N}{\hbar} s_{pz} + g_{sn} \frac{\mu_N}{\hbar} s_{nz}$$

where g is a factor called the gyromatic ratio. $g_l = 1$ and $g_{sp} = 5.585691$ for the proton, while $g_{sn} = -3.826084$ for the neutron.

The observed value (expected value) is given by $\langle\psi|\mu_z|\psi\rangle$, where ψ is the bound state wave function of the deuteron. If a magnetic field is applied along the z-axis, then μ_z takes the maximum value along the z-axis and that is what is measured experimentally.

If we assume central potential and $L = 0$, so that

$$g_l \frac{\mu_N}{\hbar} l_z = 0$$

If $I = 1$ (total angular momentum), its z-component $M_I = +1, 0, -1$ in units of $\hbar$. In an experiment to measure the magnetic moment, a magnetic field is applied, say in the z-direction. Quantum-mechanically the maximum value of μ_z is compared with the experimental value. This maximum value of μ_z is obtained by using the wave function corresponding to $I = 1$ and $M_I = 1$. Therefore,

$$\langle\mu_z\rangle = \langle I, M_I = I|\mu_z|I, M_I = I\rangle$$

The state will be $|\uparrow\uparrow\rangle$ as $s_{pz} = \hbar/2$ and $s_{nz} = \hbar/2$. Therefore,

$$\langle\mu_z\rangle = \langle\uparrow\uparrow\left| g_{sp} \frac{\mu_N}{\hbar} s_{pz} + g_{sn} \frac{\mu_N}{\hbar} s_{nz} \right|\uparrow\uparrow\rangle$$

$$= g_{sp} \frac{\mu_N}{\hbar} \langle\uparrow\uparrow|s_{pz}|\uparrow\uparrow\rangle + g_{sn} \frac{\mu_N}{\hbar} \langle\uparrow\uparrow|s_{nz}|\uparrow\uparrow\rangle$$

The operator s_{pz} operates on state $|\uparrow\uparrow\rangle$, where the first arrow represents the proton eigen state and the second arrow represents the neutron eigen state. We know that when an operator operates on the eigen function, it gives rise to the eigen value and the eigen function, therefore, we get

$$\langle\mu_z\rangle = g_{sp} \frac{\mu_N}{\hbar}\left(\frac{\hbar}{2}\right)\langle\uparrow\uparrow|\uparrow\uparrow\rangle + g_{sn} \frac{\mu_N}{\hbar}\left(\frac{\hbar}{2}\right)\langle\uparrow\uparrow|\uparrow\uparrow\rangle$$

$\langle\uparrow\uparrow|\uparrow\uparrow\rangle = 1$ for the normalized wave function. Therefore,

$$\langle\mu_z\rangle = \frac{\mu_N}{2}(g_{sp} + g_{sn})$$

$$= \frac{\mu_N}{2}(5.585691 - 3.826084)$$

or $$\langle\mu_z\rangle_d = 0.8798\mu_N$$

We have obtained this expected (theoretical) value of the magnetic moment by assuming central nuclear potential and $^3S_1$ ($L = 0$) state, along with the experimental value of $I^\pi = 1^+$ for the deuteron. However, the experimental value of $\langle \mu_z \rangle_d = 0.8574\ \mu_N$. So there is a small discrepancy of ($-0.0224\ \mu_N$) that should be considered more carefully.

One can think of at least three possible interpretations for this small departure of the measured value of the deuteron magnetic dipole moment from the expectation value in $^3S_1$ state.

(i) The internal structures of the proton and the neutron are modified by the fact that the two nucleons are in a bound state. As a result, g_{sp} and g_{sn} values given for free nucleons may be different.

(ii) There are contributions from the charged mesons exchanged between the proton and the neutron, which have not been taken into account.

(iii) There is some orbital motion in the ground state of the deuteron, i.e., other $L = 2$ value should be allowed leading to a small admixture of the $^3D_1$ state in the ground state of the deuteron.

The first point is ruled out, since the deuteron is a loosely bound system. The binding energy of –2.22 MeV can hardly be expected to affect the motion of quarks inside a nucleon that are bound by energies of the order of hundreds of MeV. The second point is possible; however, it is known that the mesonic currents are important only in odd-mass nuclei. For simplicity, we shall consider only the third interpretation, and concentrate on the admixture of the $^3D_1$ state with $^3S_1$ state as a source for the small discrepancy between the experimental and the calculated value of the deuteron's magnetic moment.

EXERCISE 4.6: The deuteron ground state wave function is a combination of $^3S_1$ and $^3D_1$ states,

$$\psi = a(^3S_1) + b(^3D_1)$$

For the deuteron, $\langle \mu_z \rangle_{L=0} = 0.8798\mu_N$ and by taking the appropriate wave function for the $^3D_1$ state, the expectation value of μ_z comes out to be equal to 0.3101 μ_N. For the deuteron, $\langle \mu_z \rangle_{\text{experimental}} = 0.8574\ \mu_N$.

Based on the magnetic moment of the deuteron, suggest the percentage admixture of the $^3D_1$ state in the ground state of the deuteron.

Solution: If $\psi = a(^3S_1) + b(^3D_1)$

$$\langle \mu_z \rangle_{\text{observed}} = |a|^2\, 0.8798\,\mu_N + |b|^2\, 0.3101\mu_N = 0.8574\mu_N$$

Also, as the wave function is normalized,

$$|a|^2 + |b|^2 = 1$$

From the above two equations, we get $|a|^2 = 0.96$ and $|b|^2 = 0.4$. Thus, the magnetic moment of deuteron suggests that there is a 4% admixture of the $^3D_1$ state in the ground state of the deuteron.

4.2.3 Electric Quadrupole Moment of Deuteron

In electrostatics, it is a well-known fact that the potential due to an arbitrary charge distribution at points far away from the source is characterized by the distance and the moments of a multipole expansion of the source. For a nuclear charge distribution, the lowest non-vanishing multipole moment is the quadrupole moment, since, the expectation value of the electric dipole operator and all the other odd multipole electric operators vanishes due to the fact that the operators change sign under space inversion (see section 1.8.5 in Chapter 1).

If the deuteron is in an $L = 0$ state, the wave function has no (θ, φ) dependence, and hence is a function of r only. It must, therefore, exhibit spherical symmetric charge distribution leading to a zero quadrupole moment. However, the observed value of the quadrupole moment of the deuteron was found to be

$$Q = +2.88 \text{ mb}$$

This is relatively small quadrupole moment, nonetheless, it is not zero. This indicates that the wave function of the deuteron ground state is not a simple $L = 0$, which is in agreement with the conclusion drawn from the magnetic moment. Non-zero electric quadrupole moment of the deuteron provides further evidence for an admixture of the $L = 0$ and $L = 2$ states and shows that the interaction potential must have a non-central (tensor) component. The ground state of the deuteron, $\{0.96(^3S_1) + 0.04(^3D_1)\}$, is consistent with the observed values of the deuteron magnetic and quadrupole moments.

Once the mixture of L-values are included in the wave function, it means that we simultaneously have $L = 2$ and $L = 0$. This puts the conservation of the orbital angular momentum in jeopardy. At this point, a torque must be considered to act for changing of the orbital angular momentum. The torque is related to the potential by the following relation:

$$\vec{\tau} = \vec{r} \times \vec{F} = rF_\theta = -\left(\frac{\partial V}{\partial \theta}\right)$$

A changing orbital angular momentum, therefore, implies that the potential V is a function of θ and not merely a function of r, as was assumed earlier. Since, a central force is defined as one for which V is a function of r, therefore, this is a non-central force; it is called the *tensor force*. Thus, the nuclear potential describing the interaction between a proton and a neutron should also contain a non-central potential term.

The only fixed direction in space for the deuteron is the spin vector $\vec{S}$. The angle θ between the radius vector $\vec{r}$ and the tensor force $\vec{F}$ upon which the tensor force depends must, therefore, be measured from this direction. Thus, the tensor force is a function of $\vec{S} \cdot \vec{r}$. When the two nucleons are in $\vec{S} = 0$ state, there is no preferred direction in space, so there can be no tensor force. Two bar magnets provide a classical example of a tensor force, where the force between the two magnets depends on the way they are oriented (Figure 4.4).

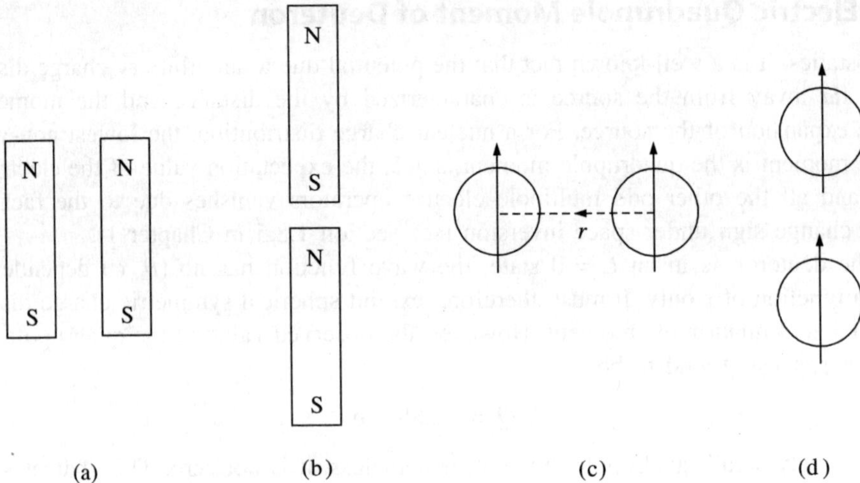

FIGURE 4.4 Illustration of the tensor force. (a) For two magnets kept as shown the force is repulsive, (b) the force is attractive, (c) for the deuteron, the force is repulsive when the spins are parallel and perpendicular to the line joining the nucleons, (d) when the parallel spins are also parallel to the line joining the nucleons, the force is attractive.

Taking a cue from electromagnetism, where the potential energy of an electric dipole $\overrightarrow{d_1}$ in the field of another dipole $\overrightarrow{d_2}$ at a distance $\vec{r}$ is

$$-\frac{1}{r^3}\left\{\frac{3\left(\overrightarrow{d_1}\cdot\vec{r}\right)\left(\overrightarrow{d_2}\cdot\vec{r}\right)}{r^2}-\overrightarrow{d_1}\cdot\overrightarrow{d_2}\right\}$$

We extend the idea to the interaction between two nucleons, in order to construct the tensor force term. The r^{-3} dependence is replaced by an arbitrary function of r, and we write the simplest form of the non-central term of the nuclear potential as follows:

$$V_T(r)\, S_{pn}$$

where $V_T(r)$ is some r dependence part, and S_{pn} is the tensor force operator defined as follows:

$$S_{pn}=2\left\{\frac{3(\vec{S}\cdot\vec{r})^2}{r^2}-\vec{S}\cdot\vec{S}\right\}$$

In terms of the Pauli spin matrices $\vec{\sigma}$, it can be written as:

$$S_{pn}=\left\{\frac{3\left(\overrightarrow{\sigma_p}\cdot\vec{r}\right)\left(\overrightarrow{\sigma_n}\cdot\vec{r}\right)}{r^2}-\overrightarrow{\sigma_p}\cdot\overrightarrow{\sigma_n}\right\} \tag{4.7}$$

Just like the dipole-dipole potential, S_{pn} averages to zero over the solid angle.

Hence, the general potential for nuclear two body forces depending only on the position of the nucleons and their spins is given by:

$$V(r) = V_c(r) + V_\sigma(r)P_\sigma + V_T(r)S_{pn} \tag{4.8}$$

EXERCISE 4.7: Show that the angular average of the tensor force S_{pn} is zero.

Solution: We consider the orientations of the spin vectors and the radius vector as shown in Figure 4.5, where α is the angle between the two spin vectors.

$$\overrightarrow{\sigma_p} = \sigma_p \hat{z}$$

$$\overrightarrow{\sigma_n} = \sigma_n \cos \alpha\hat{z} + \sigma_n \sin \alpha\hat{x}$$

$$\hat{r} = \sin\theta \cos\varphi\hat{x} + \sin\theta \sin\varphi\hat{y} + \cos\theta\hat{z}$$

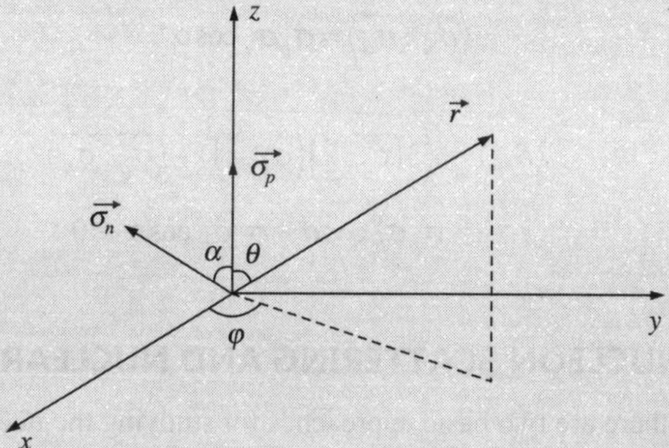

FIGURE 4.5 Orientations of the spin vectors in the spherical coordinate system.

Now,

$$\left(\overrightarrow{\sigma_p}.\hat{r}\right)\left(\overrightarrow{\sigma_n}.\hat{r}\right) = \sigma_p \cos\theta(\sigma_n \sin\alpha \sin\theta \cos\varphi + \sigma_n \cos\alpha \cos\theta)$$

$$= \sigma_p\sigma_n \sin\theta\cos\theta\sin\alpha\cos\varphi + \sigma_p\sigma_n \cos^2\theta\cos\alpha)$$

Therefore, its average value is given as (φ varies from 0 to 2π and θ from 0 to π to cover all the directions)

$$\langle\overrightarrow{\sigma_p}\cdot\hat{r}\rangle\langle\overrightarrow{\sigma_n}\cdot\hat{r}\rangle = \frac{1}{4\pi}\iint \left(\overrightarrow{\sigma_p}.\hat{r}\right)\left(\overrightarrow{\sigma_n}.\hat{r}\right)\sin\theta\, d\theta d\varphi$$

$$= \frac{1}{4\pi}\iint \sigma_p\sigma_n \sin^2\theta\cos\theta\sin\alpha\cos\varphi\, d\theta\, d\varphi + \frac{1}{4\pi}\iint \sigma_p\sigma_n \cos^2\theta \sin\theta \cos\alpha\, d\theta\, d\varphi$$

Now, since $\int_0^{2\pi} \cos\varphi\, d\varphi = 0$, the first term will be zero. The second term reduces to

$$= \frac{1}{4\pi} 2\pi \int \sigma_p \sigma_n \cos^2\theta \sin\theta \cos\alpha\, d\theta$$

$$= \frac{1}{2} \sigma_p \sigma_n \cos\alpha \int_0^{\pi} \cos^2\theta \sin\theta\, d\theta$$

Substituting $\cos\theta = t \Rightarrow -\sin\theta\, d\theta = dt$, in the above integral, we get

$$\langle \overrightarrow{\sigma_p} \cdot \hat{r} \rangle \langle \overrightarrow{\sigma_n} \cdot \hat{r} \rangle = \frac{1}{2} \sigma_p \sigma_n \cos\alpha \left(\frac{2}{3}\right) = \frac{1}{3} \sigma_p \sigma_n \cos\alpha$$

Now, the second term in the tensor force operator is

$$\langle \overrightarrow{\sigma_p} \cdot \overrightarrow{\sigma_n} \rangle = \sigma_p \sigma_n \cos\alpha$$

Therefore,

$$\langle S_{pn} \rangle = \langle 3(\overrightarrow{\sigma_p} \cdot \hat{r})(\overrightarrow{\sigma_n} \cdot \hat{r}) \rangle - \langle \overrightarrow{\sigma_p} \cdot \overrightarrow{\sigma_n} \rangle$$

$$= \sigma_p \sigma_n \cos\alpha - \sigma_p \sigma_n \cos\alpha = 0$$

4.3 NUCLEON–NUCLEON SCATTERING AND NUCLEAR FORCE

As pointed out earlier, there are two basic approaches for studying the nuclear force, by studying the bound state—the deuteron; and by studying the scattering of nucleons by nucleons. From the properties of the deuteron, we gained some vital insights about the nuclear potential. However, without a doubt, the most straightforward method for studying the nuclear interaction is the nucleon–nucleon scattering. Scattering of nucleons is produced as a result of the nuclear force between them. A simple scattering experiment consists of a projectile beam of the neutrons or the protons at a target of hydrogen and observing the scattered particles at various angles of deflection (Figure 4.6). The distribution of the scattered particles can be studied by putting the detector at different angles.

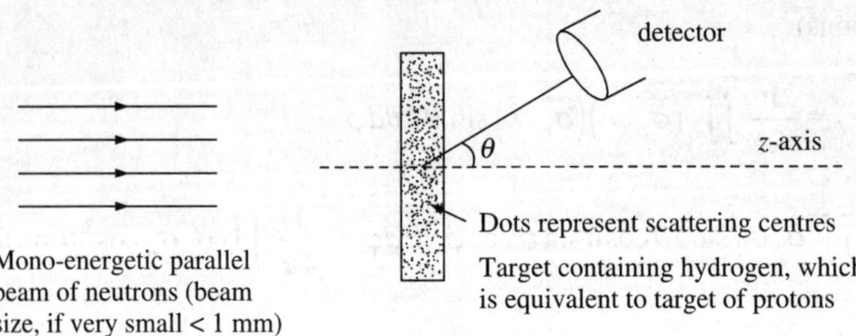

FIGURE 4.6 Low energy *n–p* scattering.

The number of particles collected in the detector

$$dN \propto t\,d\Omega I N_t \tag{4.9}$$

where t is the time of collection, $d\Omega$ is the solid angle subtended by the detector, I is the intensity of the projectile beam (number of particles/cm^2-time) and N_t is the number of target particles or scattering centres.

The proportionality constant in the above expression depends upon the type of interaction between the neutron and the proton. The constant of proportionality $\sigma(\theta, \varphi)$, also written as $d\sigma(\theta, \varphi)/d\Omega$, is known as the differential cross-section. Thus,

$$\begin{aligned} dN &= \sigma(\theta, \varphi)\, t\, d\Omega\, I\, N_t \\ &= \sigma(\theta, \varphi)\, t\, d\Omega\, I\, n_a A \end{aligned}$$

where n_a is the areal density and A is the beam area.

$$dN = \sigma(\theta, \varphi)\, d\Omega\, n_a\, N_p$$

where $N_p = IAt$, is the number of projectiles hitting the target. Therefore,

$$\sigma(\theta, \varphi) = \frac{dN}{d\Omega\, n_a N_p} \tag{4.10}$$

The total scattering cross-section σ (in barn) is obtained by summing the scattered particles over all the angles which is given as follows:

$$\sigma = 2\pi \int_0^{\pi} \frac{d\sigma(\theta)}{d\Omega} \sin\theta\, d\theta \tag{4.11}$$

4.3.1 Wave Mechanical Treatment of *n–p* Scattering

Similar to the wave mechanical treatment of the deuteron problem, we study the neutron–proton scattering using wave mechanics. The Schrödinger equation for the two body problem is given by:

$$\left[-\frac{\hbar^2}{2\mu}\nabla^2 + V(\vec{r})\right]\psi = E\psi$$

The potential energy is zero for a free particle, therefore, the above equation reduces to

$$-\frac{\hbar^2}{2\mu}\nabla^2\psi = E\psi$$

The solution of the above equation gives the free particle wave function, describing a beam of particles moving in the z-direction with constant momentum as:

$$\psi = e^{ikz} = e^{ikr\cos\theta} \tag{4.12}$$

where $k = \sqrt{2\mu E/\hbar^2}$ is its wave number, E is the kinetic energy of the incident particle and μ is the reduced mass of the system.

The linear momentum is given as

$$\vec{p} = \hbar\vec{k} = \hbar k\hat{z}$$

The wave function in equation (4.12) is a simultaneous eigen function of the Hamiltonian and the linear momentum operator; but it is not the eigen function of the angular momentum operator. All $l = 0, 1, 2, 3, \ldots,$ values get superimposed in this wave function. The wave function can be expanded into a series of partial waves,

$$\psi = e^{ikz} = \sum_{l=0}^{\infty} (2l+1)i^l j_l(kr) P_l(\cos\theta) \tag{4.13}$$

where $j_l(kr)$ are the spherical Bessel functions [e.g. $j_0(kr) = \sin kr/kr$, $j_1(kr) = \sin kr/(kr)^2 - \cos kr/kr$, $j_2(kr) = (3/(kr)^3 - 1/kr)\sin kr - 3\cos kr/(kr)^2$] and $P_l(\cos\theta)$ are the Legendre polynomials of degree l [e.g. $P_0(\cos\theta) = 1$, $P_1(\cos\theta) = \cos\theta$, $P_2(\cos\theta) = (3\cos^2\theta - 1)/2$]. It must be noted that this is a purely mathematical rearrangement.

In a scattering experiment, the detector is kept at a large distance compared to the fm range of the nuclear interaction, therefore, we have to write the above equation when $r \to \infty$. Replacing the asymptotic form of the Bessel function,

$$j_l(kr) \sim \frac{\sin\left(kr - \frac{\pi l}{2}\right)}{kr}$$

the wave function becomes

$$\psi = \sum_{l=0}^{\infty} (2l+1)\, i^l \frac{\sin\left(kr - \frac{\pi l}{2}\right)}{kr} P_l(\cos\theta) \tag{4.14}$$

Since, the intensity of the incoming wave is to remain same in the scattering (because no absorption takes place), a phase shift is introduced in the wave. When the nuclear potential is switched on, i.e., the scattering event has taken place, then

$$\psi = \sum_{l=0}^{\infty} (2l+1) i^l \frac{\sin\left(kr - \frac{\pi l}{2} - \delta_l\right)}{kr} P_l(\cos\theta) \tag{4.15}$$

where δ_t is the phase shift in presence of the potential. Following equations (4.14) and (4.15), the expression for the differential scattering cross-section turns out to be

$$\sigma(\theta) = \frac{1}{k^2}\left|\sum_{l=0}^{\infty} (2l+1)\, e^{i\delta_l} \sin\delta_l P_l(\cos\theta)\right|^2 \tag{4.16}$$

EXERCISE 4.8: Using a simple semi-classical treatment, show that only $l = 0$ will contribute to the scattering cross-section. Also, calculate the total *n-p* scattering cross-section.

Solution: The projectile (neutron) approaches towards the target particle, as shown in Figure 4.7.

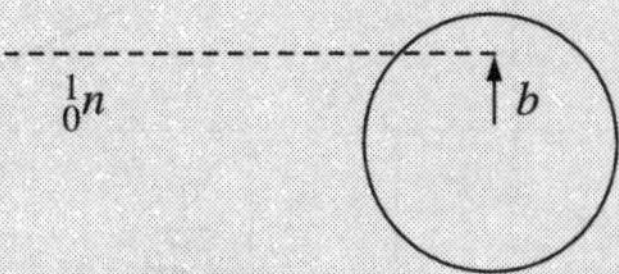

FIGURE 4.7 A neutron approaching the target particle.

Then the angular momentum is given by

$$l = mvb = \left(\sqrt{2m_p E}\right)b$$

where b is the nuclear force range, and E is the kinetic energy of the incident neutron.

Suppose $E = 1$ MeV,

$$l = \left\{\sqrt{2\left(938\ \frac{\text{MeV}}{c^2}\right)(1\ \text{MeV})}\right\}(2.1\ \text{fm})$$

or
$$\frac{l^2}{\hbar^2} = \frac{8273}{\hbar^2 c^2} = \frac{8273\ \text{MeV}^2\ \text{fm}^2}{(200\ \text{MeV fm})^2} = 0.2$$

Hence, the only present angular momentum value is $l = 0$.

From equation (4.16), for low energy ($E < 1$ MeV) *n-p* scattering

$$\sigma(\theta) = \frac{1}{k^2}\sin^2\delta_0 \cdot 1$$

From equation (4.11), thus the total *n-p* scattering cross-section is

$$\sigma_{\text{total}} = 2\pi \int_0^\pi \sigma(\theta) \sin\theta\, d\theta$$

or
$$\sigma_{\text{total}} = \frac{4\pi}{k^2}\sin^2\delta_0 \tag{4.17}$$

4.3.2 Determination of the Phase Shift

One can determine the phase shift S_l for low energy *n-p* scattering by solving the Schrödinger equation in the region of interaction. We again assume the hard core square well potential as shown in Figure 4.8.

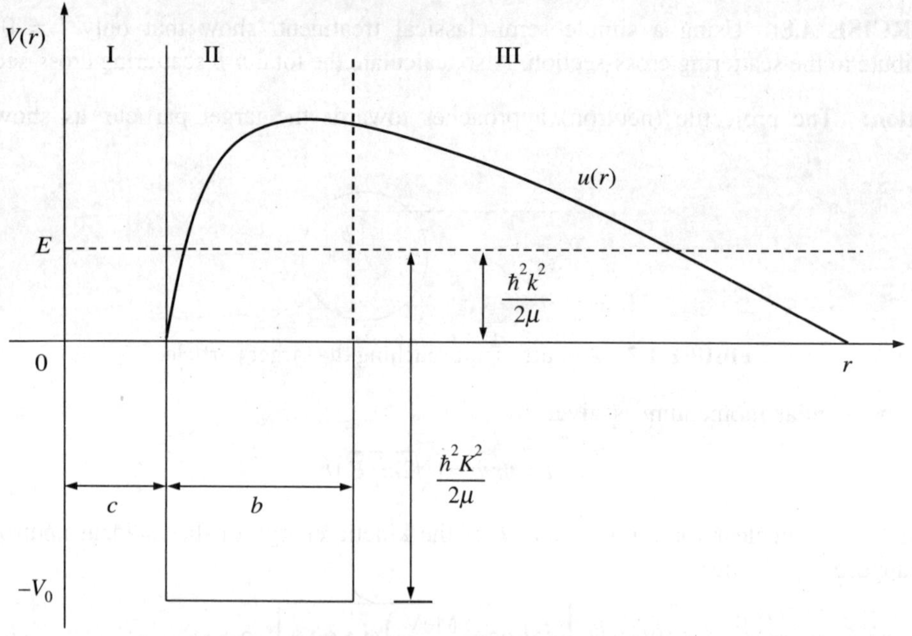

Figure 4.8 Square-well *n-p* potential and the radial function $u(r)$ for *n-p* scattering.

In the 2nd region, the radial part of the wave function for $l = 0$ must satisfy the radial wave function.

$$-\frac{\hbar^2}{2\mu}\frac{d^2u}{dr^2} - V_0 u = Eu \tag{4.18}$$

or
$$-\frac{\hbar^2}{2\mu}\frac{d^2u}{dr^2} = -(V_0 + E)u$$

or
$$\frac{d^2u}{dr^2} = -K^2u \quad \text{where } K = \sqrt{\frac{2\mu(V_0 + E)}{\hbar^2}} \tag{4.19}$$

The general solution of the above equation is

$$u(r) = A \sin K(r - c) + B \cos K(r - c) \qquad \text{for } r < (b + c)$$

Applying the boundary condition at $r = c$, $u(r)$ should be zero. Thus, B must be zero. The physically acceptable solution is

$$u_{\text{II}}(r) = A \sin K(r - c)$$

In the 3rd region, equation (4.18) becomes

$$\frac{d^2u}{dr^2} = -k^2u \quad \text{where } k = \sqrt{\frac{2\mu E}{\hbar^2}} \tag{4.20}$$

The general solution of the above equation is

$$u(r) = C \sin kr + D \cos kr$$

where $V_0 = 0$ for $r > (b + c)$. Since, this region does not contain $r = b + c$, so both the terms may be present. Physically acceptable solution is (after rewriting the above equation)

$$u_{III}(r) = C' \sin(kr + \delta_0)$$

where $C = C' \cos \delta_0$, $D = C' \sin \delta_0$ and δ_0 is the phase shift suffered by the incident neutron. Matching the wave functions and their derivatives at the boundary $r_0 = (b + c)$, the following equations are obtained:

$$A \sin Kb = C' \sin(kr_0 + \delta_0) \tag{4.21}$$

$$A \cos Kd = C' \cos(kr_0 + \delta_0) \tag{4.22}$$

Dividing equation (4.22) by equation (4.21), we get

$$K \cot Kb = k \cot(kr_0 + \delta_0) \tag{4.23}$$

From the earlier section of deuteron, we had obtained $V_0 = 34$ MeV and $b = 2.1$ fm. Using these values in equation (4.23), one can obtain the values of δ_0 as a function of neutron energy E. Once we know the values of δ_0 and k, the scattering cross-section σ_0 can be obtained from equation (4.17),

$$\sigma_0 = \frac{4\pi}{k^2} \sin^2 \delta_0$$

For low energy neutrons (E in keV range), $\sigma_{\text{calculated}} \approx 5\,\text{b}$).

Figure 4.9 shows the experimental cross-section for *n-p* scattering. The cross-section is indeed constant at low energy, and it decreases with E at large energy, consistent with equation (4.17). However, the low energy cross-section ~20 b is not in agreement with our calculated value of ~5 b. We have taken the nuclear potential as was taken for deuteron. This discrepancy is due to the fact that the deuteron was a triplet state ($S = 1$) formed with the parallel spins of the neutron and the proton, whereas *n-p* scattering takes place both in the triplet spin ($S = 1$) and the singlet spin ($S = 0$) states.

As the neutron approaches the target, the probability of being in a triplet state is 75% and the probability of being in a singlet state is 25% (see section 4.2.1). So, the *s*-wave ($l = 0$) scattering cross-section can be written as a sum of the triplet spin and the singlet spin scattering cross-sections, i.e.,

$$\sigma = \frac{3}{4}\sigma_t + \frac{1}{4}\sigma_s$$

Using the measured value of $\sigma \approx 20\,\text{b}$ at low energy and $\sigma_t \approx 5\,\text{b}$, in the above equation, we can deduce that $\sigma_s = 65\,\text{b}$. The vast difference between σ_t and σ_s indicates that the nuclear force must be spin dependent.

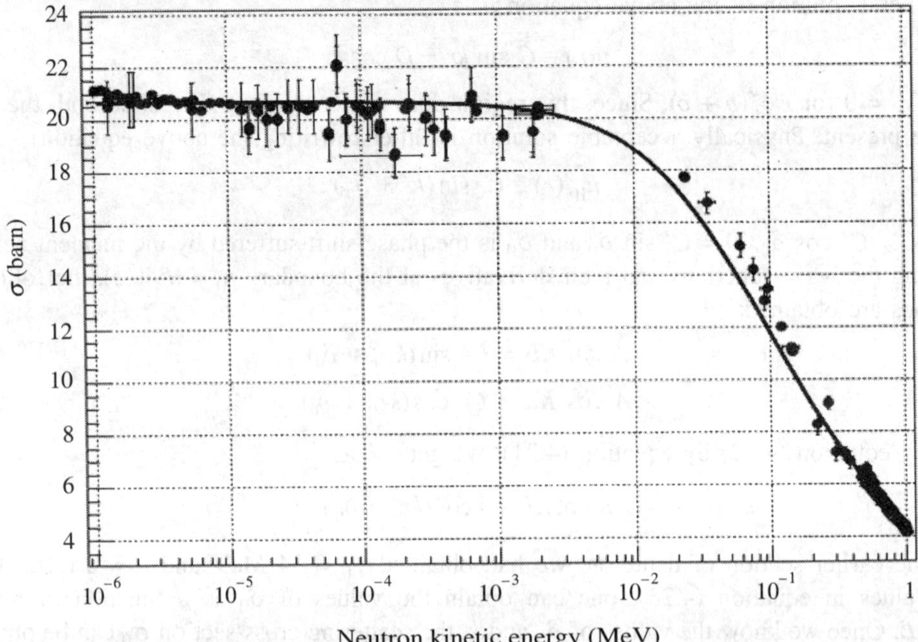

FIGURE 4.9 Neutron–Proton scattering cross-section at low energy. [Abramovsky, V.A. and N.V. Radchenko, "Total cross-section of neutron–proton scattering at low energies in quark-gluon model." arXiv:1108.0090 [hep-ph] (2011)].

EXERCISE 4.9: A beam of particles of energy $\hbar^2k^2/2m$, moving in the $+z$ direction, is scattered by a short-range central potential $V(r)$. One looks for the stationary solution of the Schrödinger equation which is of the asymptotic form,

$$\psi \approx e^{ikz} + \frac{f(\theta)e^{ikr}}{r}$$

Using the method of partial wave analysis, derive the partial-wave decomposition

$$f(\theta) = \frac{1}{2ik}\sum_{l=0}^{\infty}(2l+1)\,(e^{2i\delta_l}-1)\,P_l(\cos\theta)$$

where $f(\theta)$ is the scattering amplitude. The scattering can be assumed to be azimuthally symmetric.

Solution: Let the total wave function be $\psi = \psi_i + \psi_s$, where ψ_i represents the incident wave and ψ_s, the scattered wave.

In the absence of the scattering potential, the incident plane wave is the most general solution of the wave equation: $\nabla^2\psi_i + k^2\psi_i = 0$ and is given by (by choosing unit amplitude)

$$\psi_i = Ae^{ikz} = e^{ikz}$$

Assuming the scattered wave as

$$\psi_s = \frac{f(\theta)e^{ikr}}{r}$$

such that there is an inverse square r dependence of the scattered wave from the scattering centre. This $1/r$ dependence provides the conservation of particles in the outgoing wave, i.e., no absorption of the incident particles.

$$\sigma(\theta) = |f(\theta)|^2$$

Therefore, the total wave function can be written as follows:

$$\psi = e^{ikr\cos\theta} + \frac{f(\theta)e^{ikr}}{r}$$

or
$$f(\theta) = re^{-ikr}(\psi - e^{ikr\cos\theta}) \tag{4.24}$$

Now ψ_i can be expanded as the sum of partial waves

$$\psi_i = e^{ikr\cos\theta} = \sum_{l=0}^{\infty} A_l j_l(kr) P_l(\cos\theta)$$

where $j_l(kr)$ are the spherical Bessel functions and $P_l(\cos\theta)$ are the Legendre polynomials of degree l. For $r \to \infty$, $j_l(kr) \approx (1/kr)\sin(kr - \pi l/2)$. A_l are some constants which can be evaluated by multiplying the above equation by $P_l(\cos\theta)\sin\theta\, d\theta$ on both sides and integrating by substituting $\cos\theta = t$

$$A_l j_l(kr)\frac{2}{2l+1} = \int_{-1}^{+1} e^{ikrt} P_l(t)\, dt$$

where we have used the orthonormal property of the Legendre polynomials.

Integrating the RHS by parts, we get

$$\frac{i}{ikr}\left[e^{ikrt}P_l(t)\right]_{-1}^{+1} - \frac{1}{ikr}\int e^{ikrt}\frac{dP_l(t)}{dt}\,dt$$

The second term is of the order of $1/r^2$ which can be neglected. Therefore, using $P_l(1) = 1$ and $P_l(-1) = (-1)^l$,

$$A_l j_l(kr)\frac{2}{2l+1} \approx \frac{i}{ikr}[e^{ikr} - e^{-ikr}(-1)^l]$$

Using the identity $e^{i\pi l/2} = i^l$, we get

$$A_l j_l(kr)\frac{2}{2l+1} \approx \frac{\left[\frac{2i^l}{kr}\right]\left[e^{i\left(kr-\frac{\pi l}{2}\right)} - e^{-i\left(kr-\frac{\pi l}{2}\right)}\right]}{2i}$$

$$= \frac{2i^l \sin\left(kr - \frac{\pi l}{2}\right)}{kr}$$

Thus,

$$A_l j_l(kr) = \frac{(2l+1)i^l \sin\left(kr - \frac{\pi l}{2}\right)}{kr}$$

If the potential energy between the interacting particles $V(r)$ is spherically symmetric, then the Schrödinger equation becomes

$$\nabla^2\psi + \left\{K^2 - \frac{2\mu}{\hbar^2} V(r)\right\}\psi = 0$$

The general solution of the above equation can be written as the product of the radial function and the Legendre polynomials.

$$\psi = \sum_{l=0}^{\infty} B_l R_l(r) P_l(\cos\theta)$$

We can expand the total wave function into following components

$$\sum_{r\to\infty} \left(\frac{B_l}{kr}\right) \sin\left(kr - \frac{\pi l}{2} + \delta_l\right) P_l(\cos\theta)$$

where B_l are arbitrary coefficients and δ_l is the phase-shift of the lth wave. From equation (4.24),

$$f(\theta) = re^{-ikr}\left[\Sigma\left(\frac{B_l}{kr}\right)\sin\left(kr - \frac{\pi l}{2} + \delta_l\right)P_l(\cos\theta) - \frac{\Sigma(2l+1)i^l}{kr}\sin\left(kr - \frac{\pi l}{2}\right)P_l(\cos\theta)\right]$$

or

$$e^{ikr} f(\theta) = \frac{1}{2ik}\left\{\Sigma B_l P_l(\cos\theta)\left[e^{i\left(kr - \frac{\pi l}{2}\right)} - e^{-i\left(kr - \frac{\pi l}{2} + \delta_l\right)}\right]\right.$$

$$\left. - \Sigma(2l+1)\, i^l P_l(\cos\theta)\left[e^{i\left(kr - \frac{\pi l}{2}\right)} - e^{-i\left(kr - \frac{\pi l}{2}\right)}\right]\right\}$$

Equating the coefficients of e^{-ikr}, we get

$$0 = -\Sigma\frac{1}{2ik} B_l P_l(\cos\theta)\left[e^{-i\left(-\frac{\pi l}{2} + \delta_l\right)} + \frac{\Sigma(2l+1)i^l P_l(\cos\theta)}{2ik} e^{\frac{i\pi l}{2}}\right]$$

Therefore,

$$B_l = (2l + 1)i^l e^{i\delta_l}$$

Equating the coefficients of e^{ikr} and using the value of B_l, we get

$$f(\theta) = \frac{1}{2ik}\left[\Sigma\,(2l+1)i^l e^{i\delta_l} P_l(\cos\,\theta)\,e^{i\left(-\frac{\pi l}{2}+\delta_l\right)} - \Sigma\,(2l+1)i^l P_l(\cos\,\theta)\,e^{-\left(\frac{i\pi l}{2}\right)}\right]$$

or

$$f(\theta) = \frac{1}{2ik}\,\Sigma(2l+1)i^l P_l(\cos\,\theta)\,e^{-\frac{i\pi l}{2}}\,[e^{2i\delta_l} - 1]$$

Using the identity $e^{i\pi l/2} = i^l$, we finally get

$$f(\theta) = \frac{1}{2ik}\sum_{l=0}^{\infty}(2l+1)\,(e^{2i\delta_l} - 1)\,P_l(\cos\,\theta)$$

4.3.3 Relation of Phase Shift and Scattering Length

The scattering length is defined as the intercept of the radial wave function with the r-axis outside the range of the nuclear force. In the measurement of *n-p* cross-section, we obtain values of the phase shift, which is related to the scattering cross-section. The phase shift and the scattering length are also related if the potential well is deeper than 50 MeV and the range is short such that the radial wave function $u(r)$ at the boundary $r = b + c$ has a negative slope, and the straight line of the slope would intersect r-axis at a positive distance a_t from the origin (Figure 4.10).

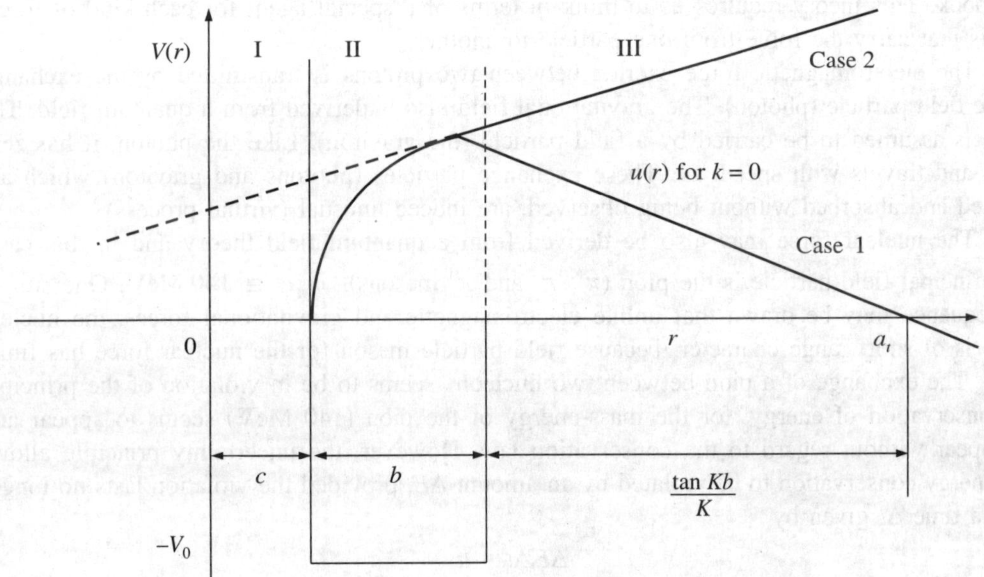

FIGURE 4.10 Radial wave function $u(r)$ in the limit of zero neutron energy.

Case 1: Bound state of n-p system

The radial wave function outside the range of nuclear potential in the 3rd region is given by:

$$u_{\text{III}}(r) = C' \sin(kr + \delta_0)$$

$u(r) = 0$ at $r = a_t$. Hence, $C' \sin(ka_t + \delta_0) = 0$

or
$$\delta_0 = \pi - ka_t$$

Case 2: Unbound state of n-p system

If the nuclear potential is not deep enough as is the case for (*n*, *n*) and (*p*, *p*) pairs, it will not form a bound system. Then the slope at the boundary is positive and it would intercept *r*-axis at a negative value of $-a_s$. The scattering length is then given by:

$$a_s = -\frac{\delta_0}{k}$$

Substituting the value of the phase shift and the scattering length in equation (4.17), we get the scattering cross-section $\sigma_0 = 4\pi a_t^2$ for the bound system, which is a triplet state ($S = 1$) with parallel spins and $\sigma_0 = 4\pi a_s^2$ for the unbound system which is a singlet state ($S = 0$) with anti-parallel spins.

4.4 QUANTUM FIELD THEORY (MESON THEORY OF NUCLEAR FORCES)

The exact description of the quantum field theory is quite complex and is beyond the scope of this book. This theory requires us to think in terms of a special agent for each kind of force; agents that carry the force from one particle to another.

The electromagnetic force exerted between two protons is transmitted by the exchange of the field particle (photon). The gravitational field also is derived from a quantum field. The force is assumed to be carried by a field particle (the graviton). Like the photon, it has zero mass and travels with speed '*c*'. These exchange particles (photons and graviton) which are emitted and absorbed without being observed, are indeed unusual (virtual process).

The nuclear force may also be derived from a quantum field theory and in this case, the principal field particle is the pion (π^+, π^- and π^0 mesons), $m_\pi c^2 \cong 140$ MeV. One simple consequence may be drawn that unlike electromagnetic and gravitational forces, the nuclear force is of short range character, because field particle meson for the nuclear force has finite mass. The exchange of a pion between two nucleons seems to be in violation of the principle of conservation of energy, for the mass-energy of the pion (140 MeV) seems to appear and disappear without regard to the conservation law. However, the uncertainty principle allows the energy conservation to be violated by an amount ΔE, provided the violation lasts no longer than a time Δt given by

$$\Delta E \Delta t \sim \hbar$$

Thus, the uncertainty principle restricts the existence of the pion within a time $\Delta t \le \hbar/mc^2$, during which it could cover at most a distance

$$R \approx c\Delta t$$

$$= \frac{c\hbar}{m_\pi c^2} = \frac{\hbar}{m_\pi c} \ (\approx 1.4 \text{ fm})$$

which is equal to the nuclear force range. For electromagnetic and gravitational field, the range is infinite because $m \approx 0$. Thus, the range of the field is given by the Compton wavelength $(\hbar/mc)$ of the associated quantum. According to quantum electrodynamics, the mechanism of electromagnetic interaction involves the transfer of a photon from one charge to the other. The equation of motion for a freely moving photon can be written in the following form:

$$E^2 = p^2c^2$$

In order to obtain the equation for the potential field of a unit charge, we must make the substitution

$$E \to -\frac{\hbar}{i}\frac{\partial}{\partial t} \quad \text{and} \quad p \to \frac{\hbar}{i}\nabla$$

The equation for the potential in the empty space will be of the form

$$\nabla^2\psi - \frac{1}{c^2}\frac{\partial^2\psi}{\partial t^2} = 0$$

For the time-independent case, $\partial\psi/\partial t$ is zero. Thus,

$$\nabla^2\psi = 0$$

The Laplace equation in spherical coordinate is given by:

$$\nabla^2\psi = \frac{1}{r^2}\frac{\partial}{\partial r}\left(r^2\frac{\partial\psi}{\partial r}\right) + \frac{1}{r^2\sin\theta}\frac{\partial}{\partial\theta}\left(\sin\theta\frac{\partial\psi}{\partial\theta}\right) + \frac{1}{r^2\sin^2\theta}\frac{\partial^2\psi}{\partial\varphi^2} = 0$$

Due to the spherical symmetry, the potential (ψ) will be independent of (θ, φ) and will depend only on r (radial distance). Therefore, the above equation reduces to

$$\nabla^2\psi = \frac{1}{r^2}\frac{\partial}{\partial r}\left(r^2\frac{\partial\psi}{\partial r}\right) = 0$$

The solution of the above equation is the following wave function, which is the expression for the interaction potential energy of a unit charge ($-e$) in the potential $V = e/4\pi\varepsilon_0 r$,

$$\psi = -\frac{e^2}{4\pi\varepsilon_0}\frac{1}{r}$$

According to the meson theories, the transfer of interaction takes place through a π-meson, a particle with a nonzero mass ($m \neq 0$). The equation for a freely moving particle with $m \neq 0$ is

$$E^2 = p^2c^2 + m^2c^4$$

After substituting the energy and the momentum operators, the equation for the meson potential field of a nucleon in empty space assumes the following form:

$$\nabla^2\varphi - \frac{1}{c^2}\frac{\partial^2\varphi}{\partial t^2} - \frac{m^2c^2}{\hbar^2}\varphi = 0$$

Just like before, for the time-independent case, $\partial\varphi/\partial t$ is zero. Thus, the above equation reduces to

$$\nabla^2\varphi = \frac{m^2c^2}{\hbar^2}\varphi$$

or
$$\frac{1}{r^2}\frac{\partial}{\partial r}\left(r^2\frac{\partial\varphi}{\partial r}\right) = \frac{m^2c^2}{\hbar^2}\varphi$$

or
$$\frac{1}{r^2}\left(2r\frac{\partial\varphi}{\partial r} + r^2\frac{\partial^2\varphi}{\partial r^2}\right) = \frac{m^2c^2}{\hbar^2}\varphi$$

or
$$\frac{\partial^2\varphi}{\partial r^2} + \frac{2}{r}\frac{\partial\varphi}{\partial r} - \frac{1}{R^2}\varphi = 0 \qquad \text{where } R = \frac{\hbar}{mc}$$

If $r \to \infty$
$$\frac{\partial^2\varphi}{\partial r^2} = \frac{1}{R^2}\varphi$$

or
$$\varphi \propto e^{-r/R}$$

If $r \to 0$, $\varphi \to \infty$ means $\varphi \sim 1/r$ type. The first and the second terms will be dominating.

$$\frac{\partial^2\varphi}{\partial r^2} + \frac{2}{r}\frac{\partial\varphi}{\partial r} = 0$$

and if $\varphi \sim f(r) \approx r^\alpha$, then

$$\alpha(\alpha-1)r^{\alpha-2} + \frac{2}{r}\alpha r^{\alpha-1} = 0$$

or $\alpha^2 + \alpha = 0$, i.e., $\alpha = -10$. Therefore, φ should also be inversely proportional to r. Thus, the solution for the meson potential, also known as the Yukawa potential, is

$$\varphi(r) = \frac{g}{4\pi r}e^{-r/R}$$

where g is the strength of the Yukawa potential. In the Yukawa theory, it plays the same role as the charge in electrostatics and measures the 'strong nuclear charge'.

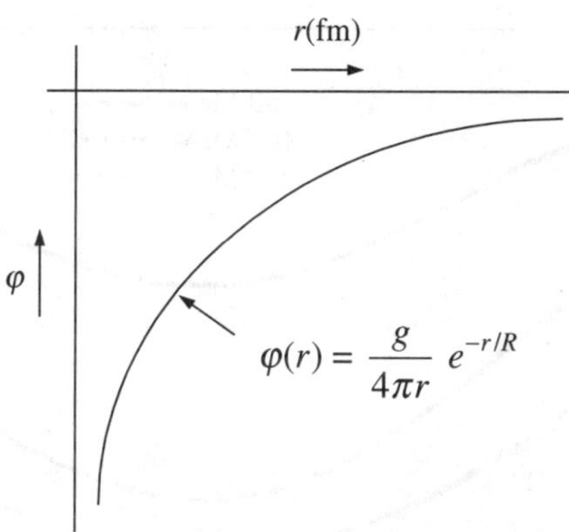

FIGURE 4.11 Yukawa potential.

In the historical context of the nuclear forces with range R = 10^{-15} m, the Yukawa hypothesis predicted a spinless quantum of mass $mc^2 = \hbar c/R \cong 100$ MeV. The meson predicted by Yukawa[5] was finally found in 1947 in cosmic ray and in 1948 in the laboratory and was called the pion with mass of about 140 MeV and spin 0.

The exchange of π-meson for a free nucleon is classically forbidden for a proton at rest by the law of conservation of energy, $p \rightarrow n + \pi^+$. According to quantum mechanics, however, the energy $\Delta E = m_\pi c^2$ necessary may be borrowed, provided that it is returned within a time given by $\Delta E \Delta t \sim \hbar$ and the meson exchange can therefore be considered as a virtual process. The strong interaction which takes place between the constituent quarks is mediated via bosons called gluon.

4.5 EXPERIMENTAL VERIFICATIONS

The experimental evidence for the exchange of pions between two interacting nucleons is found in the *n-p* scattering. The first high energy experiment was performed with incident neutrons of energy 90 MeV. The measurements show that the differential cross-section (Figure 4.12) is approximately symmetric about a scattering angle of 90°. There is a strong peak in the cross-section at forward angles near 0° corresponding to a small momentum transfer between the neutron and the proton. This is understandable; but a 'backward' peak, i.e., a peak at ~180° as a result of the head-on collision between the neutron and the proton in which the incident neutron has its motion reversed is quite unlikely. A reasonable successful explanation can be found based on the exchange model. During the collision, the neutron and the proton exchange take place, i.e., the forward moving neutron becomes a proton and the backward moving proton becomes a neutron (as viewed in the CM frame). The incident neutron then reappears in the laboratory as a forward-moving nucleon (now a proton). When the two nucleons are very close, the following set of reactions take place.

Neutron emits a negatively charged π^- meson into its field, becoming a proton. Thereafter, the π^- meson joins the field of the proton and is absorbed by the proton to become a neutron.

$$n \rightarrow p + \pi^- \quad \text{then} \quad \pi^- + p \rightarrow n$$

5. Yukawa was awarded the Nobel Prize (Physics) in 1949.

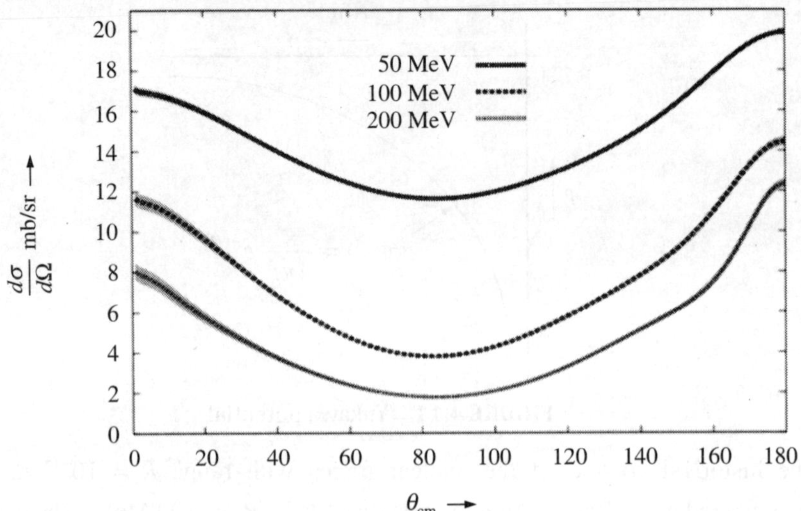

FIGURE 4.12 Differential cross-section for *n-p* scattering at different energies. [A. Calle Cordon, E. Ruiz Arriola, "Serber symmetry, Large Nc and Yukawa-like One Boson Exchange Potentials," arXiv:0904.0421 [nucl-th] (2009)]

In case the proton emits a π^+ meson, it is subsequently absorbed by the neutron.

$$p \to n + \pi^+ \quad \text{then} \quad \pi^+ + n \to p$$

Thus, in about half the *n-p* scatterings, a meson transfers charge as well as momentum between the two interacting nucleons. In about half the scatterings though, neutrons and protons do not exchange identities when they interact but a meson is still exchanged as it carries the transferred momentum. The two sets of reactions that occur are as follows:

$$n \to p + \pi^0 \quad \text{then} \quad \pi^0 + p \to n$$

$$p \to n + \pi^0 \quad \text{then} \quad \pi^0 + n \to p$$

The situation means that an isolated proton or a neutron should be surrounded by a meson field which contains different mesons and the nucleon must reabsorb the meson it has emitted within a very small interval. Experimental verification of these predictions is provided by the electron scattering measurements of the charge distribution of the proton and the neutron (Figure 4.13).

The charge density of the proton is everywhere positive and extends out to a distance *r* of about 2 fm, For the neutron, the charge density is positive for smaller *r* and at larger *r*, it is negative. The volume integral of the charge density is however, zero, since the neutron has no net charge. At values of *r* close to 2 fm, the nucleon charge densities are to some degree proportional to the intensity of their meson field. Both the proton and the neutron charge densities decrease gradually with *r*. The nucleon force that acts between the two nucleons when their meson fields overlap, also decreases gradually as their separation increases. When the two nucleons are close enough to interact, the attractive part of the nuclear potential gradually starts to show up.

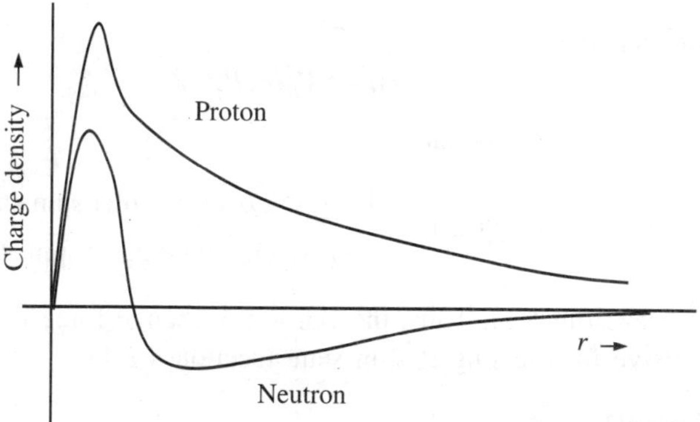

FIGURE 4.13 Electron scattering measurements of radial distributions of charge for a proton and a neutron.

The meson theory also explains how the neutron can have an intrinsic magnetic dipole moment, even though its net charge is zero. The neutron sometimes becomes a proton and a π^- meson. The proton has an intrinsic magnetic moment and π^- meson can produce a current that makes an additional contribution to the magnetic dipole moment.

The concept of exchange forces thus, leads us to the most general two-nucleon potential with the exchange character and tensor forces.

$$V(r) = V_W(r) + V_M(r)P_r + V_B(r)P_\sigma + V_H(r)P_r P_\sigma + V_{WT}(r)S_{pn} + V_{MT}(r)S_{pn}P_r \tag{4.25}$$

where $V_W(r)$, $V_M(r)$, $V_B(r)$ and $V_H(r)$ are of the Yukawa form $e^{-\mu r}/\mu r$ with different values for the constant μ and are known as Wigner force, Majorana force, Bartlett force and Heisenberg force. They are discussed briefly in the following section. A detailed discussion is beyond the scope of this book.

Wigner Force

It is a no exchange force. It is attractive for all states of the two-nucleon system.

$$V(r) = V_W(r)$$

Majorana Force

It arises from spatial exchange.

$$V(r) = V_M(r)\ P_r$$

where P_r is the spatial exchange operator.

$$P_r\varphi(1, 2) = \varphi(2, 1) = \begin{cases} +\varphi(1, 2), & \text{for even parity states} \\ -\varphi(1, 2), & \text{for odd parity states} \end{cases}$$

$\varphi(1, 2)$ is the spatial wave function. Thus, the Majorana exchange force is attractive for even parity states and repulsive for odd parity states.

Bartlett Force

It arises from spin-exchange.

$$V(r) = V_B(r)P_\sigma$$

where P_σ is the spin exchange operator.

$$P_\sigma \chi(1, 2) = \chi(2, 1) = \begin{cases} +\chi(1, 2), & \text{for triplet spin state} \\ -\chi(1, 2), & \text{for singlet spin state} \end{cases}$$

$\chi(1, 2)$ is the spin wave function. Thus, the Bartlett exchange force is attractive for the triplet spin state and repulsive for the singlet spin state (section 4.2.1).

Heisenberg Force

It arises from space-spin exchange.

$$V(r) = V_H(r)P_rP_\sigma$$

$$P_rP_\sigma\varphi(1, 2)\,\chi(1, 2) = \varphi(2, 1)\,\chi(2, 1) = \begin{cases} +\varphi(1, 2)\,\chi(1, 2), & \# \\ -\varphi(1, 2)\,\chi(1, 2), & \$ \end{cases}$$

#: for even parity triplet spin state or odd parity singlet spin state
$: for odd parity triplet spin state or even parity singlet spin state

Thus, the Heisenberg exchange force is attractive for even parity triplet spin state and odd parity singlet spin state but repulsive for odd parity triplet spin state and even parity singlet spin state.

PROBLEMS

4.1 Briefly explain the properties of the nuclear forces.

4.2 If the scattering experiments determine the separation of the neutron and the proton to be about b = 1.5 fm in the deuteron, the binding energy determined from mass measurements is found to be E = –2.226 MeV and a spherical square well potential $V(r) = -V_0$ for $r < b$ and $V(r) = 0$ for $r > b$ is assumed, what is the value of V_0?
[Ans: 59.7 MeV]

4.3 Explain why a stable system of di-neutron has not been observed?

4.4 The following normalized radial wave function is a useful approximation to describe the ground state of the deuteron,

$$R(r) = \left(\frac{1}{r}\right)\sqrt{\frac{\alpha}{2\pi}}\,e^{-\alpha r}$$

where $1/\alpha$ = 4.3 fm. Find the root mean square separation of the neutron and the proton in this nucleus. **[Ans:** 3.0 fm]

4.5 If $E_n = 10$ MeV, what are the values of l that will contribute to the scattering cross-sections in *n-p* scattering?

4.6 Show that for low energy hard sphere scattering, the cross-section is equal to $4\pi R^2$, where R is the radius of the potential well.

4.7 1 MeV neutrons are scattered by a target. The angular distribution of the neutrons is isotropic in centre-of-mass frame. The total cross-section is measured to be 0.1 b. Using the partial wave representation, calculate the phase shifts of the partial waves involved. [**Ans:** $\pm 11.3°$]

4.8 For a system of one proton and one neutron (not necessarily a deuteron), what are the various possible states based on isospin, spin and orbital quantum numbers? [*Hint:* The total wave function has to be antisymmetric.]

4.9 When an electron is scattered elastically by another electron, a photon is exchanged. The photon is said to be virtual. What does this mean and why is this the case?

4.10 Is it possible to detect the Yukawa mesons in actual practice? Under what conditions can they be detected?

4.11 The measured *n-p* scattering cross-section at low energies is about 20 b, whereas the theoretical value based upon the scattering length obtained from the deuteron is about 5 b. How do you explain this discrepancy?

4.12 A π^0 meson decays into two γ-rays. If the π^0 is at rest, what is the energy of each γ-ray?

4.13 If the binding energy of the deuteron were 12 MeV, what would be roughly the depth of the potential well, assuming it to be square?

BIBLIOGRAPHY

Cohen, B.L., *Concepts of Nuclear Physics*, McGraw-Hill, New York, 1971.

Enge, H.A., *Introduction to Nuclear Physics,* Addison Wesley, London, 1966.

Garg, J.B., *Nuclear Physics: Basic Concepts*, Macmillan, New Delhi, 2011.

Krane, K.S., *Introductory Nuclear Physics*, Wiley, New York, 2008.

Patel, S.B., *Nuclear Physics: An Introduction,* Wiley Eastern, New Delhi, 1991.

Roy, R.R. and Nigam, B.P., *Nuclear Physics: Theory and Experiment*, New Age International, New Delhi, 2012.

5 Nuclear Reactions

"Everyone thinks of changing the world, but no one thinks of changing himself."
—Leo Tolstoy

Nuclear reactions have played an important role in the investigation of size, shape and energy levels of nuclei. In fact, it was Rutherford who first performed an experiment in 1919 with α-particles from $^{210}_{84}\text{Po}$ radioactive source on nitrogen and demonstrated the following nuclear transmutation (nuclear reaction):

$$^{14}_{7}\text{N} + ^{4}_{2}\text{He} \rightarrow ^{17}_{8}\text{O} + ^{1}_{1}\text{H}$$

Nuclear reaction is a process in which two nuclides or else a nucleus of an atom and subatomic particles (such as neutron, proton, high energy electrons) or ions (such as deuterons, $^{3}_{2}\text{He}$ ions, $^{4}_{2}\text{He}$ ions and heavy ions) collide to produce one or more nuclides that are different from the nuclides that began the process. Nuclear reactions are generally produced by exposing the nuclei of a target to a beam of nuclear projectiles or γ-rays. We shall consider the nuclear reaction of the following type:

$$a + A \rightarrow B + b \tag{5.1}$$

This reaction represents the most general nuclear reactions at low energy; where A is the target supposed to be at rest in the laboratory frame, a are the projectiles, b are the ejectiles and B is called the residual (product) nucleus. a and b may be elementary particles or γ-rays, or they may themselves be nuclei, e.g. α-particles or deuterons. The reaction shown above is also represented as $A(a, b)B$. There are a few exceptions to this:

(i) Radiative capture process: $a + A \rightarrow C + \gamma$

(ii) Elastic scattering: $a + A \rightarrow a + A$, in elastic scattering, the interaction may be simple Coulomb repulsion or more complicated nuclear interactions can take place. In the former case, the process is called Rutherford scattering.

(iii) Inelastic scattering: $a + A \rightarrow a + A^*$, where A^* indicates that the nucleus A is in the excited state.

Thus, the two sides of equation (5.1) represent the physical situation before and after the interaction. The left hand side of the equation is known as the entrance channel, while the right hand side is termed as the exit channel. If more than one particle, say b_1, b_2, ..., are emitted, then the reaction is written as

$$a + A \rightarrow B + b_1 + b_2 + \cdots$$

For example,

$$^{238}_{92}\text{U} + {}^{14}_{7}\text{N} \rightarrow {}^{243}_{97}\text{Bk} + {}^{4}_{2}\text{He} + 5\,{}^{1}_{0}n$$

We can also write the above equation as:

$$^{238}_{92}\text{U}({}^{14}_{7}\text{N}, \alpha\, 5n)\ {}^{243}_{97}\text{Bk}$$

With the advent of particle accelerators, it has become possible to observe variety of nuclear reactions. All these reactions must obey certain conservation laws:

(i) Conservation of total mass and energy
(ii) Conservation of linear momentum
(iii) Conservation of angular momentum
(iv) Conservation of total charge
(v) Conservation of parity in the strong interaction but not in the weak interaction
(vi) Conservation of baryon number
(vii) Conservation of lepton number
(viii) Conservation of isobaric spin

Apart from these quantities, statistics also follow conservation laws, whereas the magnetic dipole moment and electric quadrupole moment are not conserved as these are determined by the internal structure of nuclei.

Conservation of total mass and energy in equation (5.1) gives

$$(M_A + M_a) \times 931.478 + E_a = (M_B + M_b) \times 931.478 + E_B + E_b$$

or

$$(E_B + E_b - E_a)\,\text{MeV} = (M_A + M_a - M_B - M_b) \times 931.478\ \text{MeV} \tag{5.2}$$

where E_a, E_b and E_B are the kinetic energies of projectiles, ejectiles and product nuclei. M_A, M_a, M_B and M_b are the rest mass energies.

The Q-value of a reaction is defined as

$$Q = E_B + E_b - E_a$$

i.e., it is the change in the total kinetic energy.

From equation (5.2), one can also write

$$Q = (M_A + M_a - M_B - M_b) \times 931.478\ \text{MeV}$$

When $Q > 0$, the reaction is *exoergic,* and the liberated energy appears in the form of kinetic energy of the reaction products. When $Q < 0$, the reaction is *endoergic* and it can only occur if the incident particle has threshold energy. This threshold energy is always greater than $|Q|$.

EXERCISE 5.1: For the reaction $A(a, b)B$, show that

$$Q = E_b\left(\frac{M_b + M_B}{M_B}\right) - E_a\left(\frac{M_B - M_a}{M_B}\right) - \frac{2\sqrt{M_a M_b}}{M_B}\sqrt{E_a E_b}\cos\theta$$

where the E's and M's are the laboratory system kinetic energies and the rest mass energies of particles a, b and the nuclei A, B and θ is the angle between the directions of particles a and b as shown in Figure 5.1.

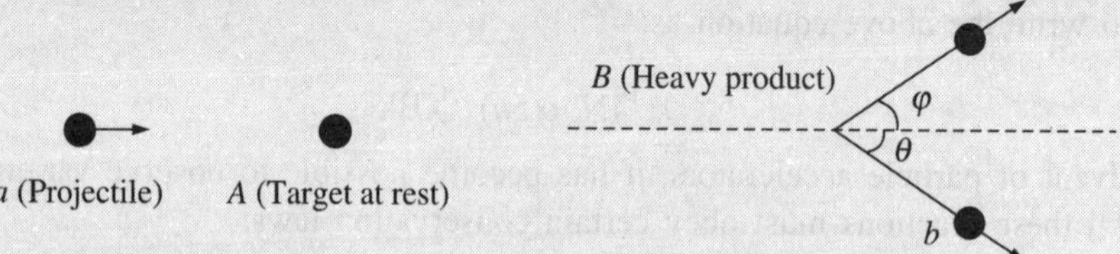

FIGURE 5.1 $A(a, b)B$ reaction in the laboratory frame.

Solution: Applying the law of conservation of linear momentum, for the x and the y components, we get

$$M_a v_a = M_b v_b \cos\theta + M_B v_B \cos\varphi$$

$$M_B v_B \sin\varphi = M_b v_b \sin\theta$$

Squaring and adding the two equations, we get

$$M_B^2 v_B^2(\sin^2\varphi + \cos^2\varphi) = M_a^2 v_a^2 + M_b^2 v_b^2(\sin^2\theta + \cos^2\theta) - 2M_a M_b v_a v_b \cos\theta$$

Since,

$$E_a = \frac{1}{2}M_a v_a^2,\ E_b = \frac{1}{2}M_b v_b^2,\ E_B = \frac{1}{2}M_B v_B^2$$

$$2E_B M_B = 2E_a M_a + 2E_b M_b - 4(E_a E_b M_a M_b)^{1/2}\cos\theta$$

or

$$E_B = E_a\frac{M_a}{M_B} + E_b\frac{M_b}{M_B} - \frac{2}{M_B}(E_a E_b M_a M_b)^{1/2}\cos\theta$$

The Q-value of a reaction is defined as $Q = E_B + E_b - E_a$. Replacing E_B from the above expression, we get

$$Q = E_b\left(\frac{M_b + M_B}{M_B}\right) - E_a\left(\frac{M_B - M_a}{M_B}\right) - \frac{2\sqrt{M_a M_b}}{M_B}\sqrt{E_a E_b}\cos\theta \qquad (5.3)$$

5.1 GENERAL SOLUTION OF *Q*-EQUATION

Equation (5.3) is quadratic in $\sqrt{E_b}$ for a fixed Q-value and varying E_a (projectile energy), when written in the following form:

$$\left[\left(\sqrt{E_b}\right)^2 (M_b + M_B) - 2\sqrt{M_a M_b E_a}\ \cos\theta\left(\sqrt{E_b}\right) + \{E_a(M_a - M_B) - QM_B\}\right] = 0$$

$$\therefore \quad \sqrt{E_b} = \frac{\sqrt{M_a M_b E_a}\ \cos\theta}{(M_b + M_B)} \pm \frac{\sqrt{M_a M_b E_a \cos^2\theta + (M_b + M_B)\{QM_B - E_a(M_a - M_B)\}}}{(M_b + M_B)}$$

It is clear that the physical solutions would only correspond to the positive and real values of $\sqrt{E_b}$.

For Exoergic Reaction

$Q > 0$ with $M_B > M_a$, therefore, + sign is to be chosen to get real value of E_b. Thus, E_b is single valued function of θ. E_b depends on cos θ and therefore, E_b is the smallest in the backward direction, i.e., if $\theta = 180°$.

For Endoergic Reaction

$Q < 0$, the reaction cannot proceed at all unless, $E_a >> E_{th}$ (threshold energy). This condition is essential for the expression under the square root to be positive, i.e.,

$$M_a M_b E_a \cos^2\theta + (M_b + M_B)\{QM_B - E_a(M_a - M_B)\} \geq 0$$

or

$$E_a = -Q\left[\frac{M_b + M_B}{M_b + M_B - M_a - \left(\dfrac{M_a M_b}{M_B}\right)\sin^2\theta}\right]$$

At $\theta = 0°$

$$E_{th} = -Q\left[\frac{M_b + M_B}{M_b + M_B - M_a}\right] = -Q\left[\frac{M_a + M_A}{M_A}\right] \tag{5.4}$$

Below the threshold ($E_a < E_{th}$), the reaction is impossible. At the threshold, the particles emerge at 0°. When the incident energy is increased, they (outgoing particles) are observed in a cone which gradually widens (θ has a maximum value θ_{max}) until the cone angle ($2\theta_{max}$) reaches 180°.

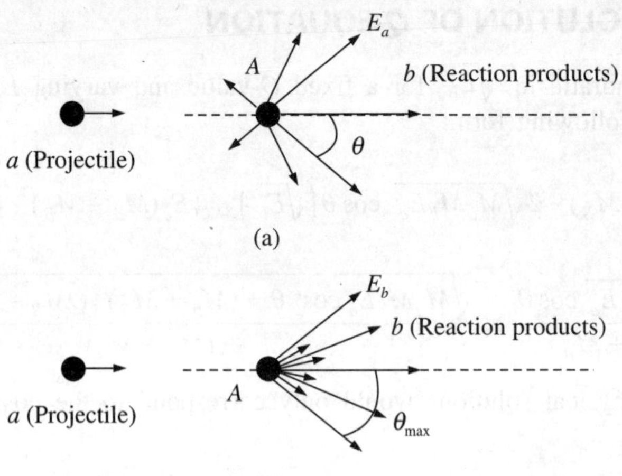

FIGURE 5.2 Kinematics of nuclear reactions: the energy of product b is proportional to the length of arrows. (a) the reaction products are emitted in all directions, when $Q > 0$ ($M_B > M_a$) or $Q < 0$ ($E_a > E_{max}$). The energy is maximum at forward angles and minimum at backward angles; (b) the reaction products are emitted in forward direction only, when $Q < 0$ ($M_B < M_a$) or $Q > 0$ ($M_B < M_a$) and ($E_{th} < E_a < E_{max}$); two energy groups are observed at each angle.

$\sqrt{E_b}$ may have two possible values corresponding to '±' signs before the bracket. Both signs can be accepted only if

$$QM_B - E_a(M_a - M_B) > 0$$

or

$$E_a < \frac{QM_B}{(M_a - M_B)} = E_{max} \tag{5.5}$$

Two groups of particles of different energies are observed (as shown in Figure 5.2) when

$$E_{th} < E_a < E_{max}$$

and the angle of emission is limited by

$$0 < \theta < \theta_{max}$$

θ_{max} may be calculated from

$$M_a M_b E_a \cos^2 \theta_{max} + (M_b + M_B)\{QM_B - E_a(M_a - M_B)\} = 0$$

or

$$\cos^2 \theta_{max} = -\frac{(M_b + M_B)\{QM_B - E_a(M_a - M_B)\}}{M_a M_b E_a} \tag{5.6}$$

i.e.,

$$\theta_{max} < 90°$$

If $E_a > E_{max}$, the energy is single valued as for exoergic reactions.

EXERCISE 5.2: Calculate the values of E_{th}, $E_{\max}$, $\theta_{\max}$ and the energy of the product neutrons, for the following endoergic reaction, which is used for lithium analysis and for neutron production,

$$^7_3\text{Li} + p \rightarrow {}^7_4\text{Be} + n - 1.643 \text{ MeV}$$

What will be the threshold energy for the appearance of neutrons in 90° direction?

Solution: $M_a = 1$, $M_A = 7$, $M_b = 1$, $M_B = 7$. The threshold energy is given by:

$$E_{th} = -Q\left[\frac{M_b + M_B}{M_b + M_B - M_a}\right] = +1.643\left(\frac{8}{7}\right) = 1.878 \text{ MeV}$$

The two neutron groups of different energies E_{n1} and E_{n2} are emitted at forward angles when the proton energy $E_a < E_{\max}$, where $E_{\max}$ is given by

$$E_{\max} = -\frac{7}{6}Q = 1.917 \text{ MeV}$$

If the incident proton energy is 1.9 MeV, i.e., $E_a < E_{\max}$, the neutron emission angle is limited, and from equation (5.6), the maximum angle $\theta_{\max}$ is given by

$$\cos^2\theta_{\max} = -\frac{(1+7)\{(-1.643)(7) - 1.9(1-7)\}}{1.9} = 0.425$$

$$\therefore \qquad \theta_{\max} = 49.3°$$

At zero degree (i.e., forward direction), the neutron energies E_{n1} and E_{n2} are given as (for $E_a = 1.9$ MeV)

$$\sqrt{E_n} = \frac{\sqrt{M_a M_b E_a}\cos\theta}{(M_b + M_B)} \pm \frac{\sqrt{M_a M_b E_a \cos^2\theta + (M_b + M_B)\{QM_B - E_a(M_a - M_B)\}}}{(M_b + M_B)}$$

or

$$\sqrt{E_{n1}} = \frac{1.378 + \sqrt{1.9 - 0.808}}{8} = \frac{2.422}{8}$$

$$\therefore \qquad E_{n1} = 0.0918 \text{ MeV}$$

and

$$\sqrt{E_{n2}} = \frac{1.378 - \sqrt{1.9 - 0.808}}{8} = \frac{0.3331}{8}$$

or

$$E_{n2} = 0.0018 \text{ MeV}$$

For $\theta = \theta_{\max} = 49.3°$,

$$E_{n1} = E_{n2} = \left[\frac{\sqrt{1.9}\cos 49.3°}{8}\right]^2 = 0.0126 \text{ MeV}$$

The threshold for the appearance of neutrons in the 90° direction is

$$E_{th} = -Q\left[\frac{M_b + M_B}{M_b + M_B - M_a - \left(\frac{M_a M_b}{M_B}\right)\sin^2 90°}\right] = +1.643\left(\frac{8}{7 - \frac{1}{7}}\right) = 1.917 \text{ MeV}$$

Besides the threshold energy, another condition that pertains to the reactions involving charged projectiles only, is that the energy of the incident projectile should be sufficient to overcome the electrostatic repulsion between the positively charged projectile and the positively charged nucleus. This requirement is to ensure that the incident projectile can get close enough after tunneling through the Coulomb barrier to the target nucleus to interact with it. The Coulomb barrier height is given by:

$$V_c(\text{Coulomb barrier}) = 1.44 \frac{Z_p Z_T}{R_p + R_T} \text{ MeV} \tag{5.7}$$

where Z_p, R_p and Z_T, R_T are the atomic number and radii of the incoming projectile and target, respectively. The radii in this expression are to be taken in Fermi (fm) and can be calculated by using the equation, $R = r_0 A^{1/3}$ fm, where A is the mass number and r_0 is the nuclear unit radius.

EXERCISE 5.3:

(a) Consider the possible reaction ${}^{195}_{78}\text{Pt}(p, n){}^{195}_{79}\text{Au}$ to be carried out with variable energy of protons. At what energy would the reaction occur? [Given: r_0 = 1.4 fm]

(b) Calculate the minimum energy of α-particle to investigate the reaction ${}^{14}_{7}\text{N}(\alpha, p){}^{17}_{8}\text{O}$.

(c) What should be the kinetic energy of the projectile so that the following reaction takes place?

$$\alpha + {}^{232}_{96}\text{Th} \rightarrow {}^{236}_{98}\text{U}^*$$

Take mass/mass excesses from the standard compilation—Nuclear Wallet Cards: <http://www.nndc.bnl.gov/wallet/>

Solution:

(a) Recalling that the mass excess Δ_A is related to M_A and A by $M_A = \Delta_A + A$, we can write

$$\begin{aligned} Q &= \text{Mass excess of } {}^{195}_{78}\text{Pt} + \text{Mass excess of } p - \text{Mass excess of } {}^{195}_{79}\text{Au} \\ &\quad - \text{Mass excess of } n \\ &= (-32.796 + 7.289) - (-32.569 + 8.071) = -1.009 \text{ MeV} \end{aligned}$$

So, the reaction can occur if $E_p > 1.009$ MeV (theoretically), but in this case

$$V_C(\text{Coulomb barrier}) = 1.44 \frac{(1)(78)}{(1.4 \times 195^{1/3} + 1.4)} = 12 \text{ MeV}$$

Therefore, we conclude that the reaction would occur with protons of energies greater than 12 MeV.

(b) For the reaction $\alpha + {}^{14}_{7}\text{N} \rightarrow {}^{17}_{8}\text{O} + p$, the Q-value is given by

$$\begin{aligned} Q &= \Delta_\alpha + \Delta_N - \Delta_O - \Delta_p \\ &= 2.425 + 2.863 + 0.809 - 7.289 \\ &= -1.192 \text{ MeV} \end{aligned}$$

Q is negative, therefore, the required threshold energy for this reaction to occur is (taking mass numbers just for good approximation)

$$E_{th} = -Q\left[\frac{M_a + M_A}{M_A}\right] = 1.192\left(\frac{4+14}{14}\right) = 1.53 \text{ MeV}$$

Minimum energy of α-particles should be 1.53 MeV.

(c) Based on the mass excesses, the calculated value of $Q = -4.57$ MeV. Since, Q is negative, it is an endoergic reaction and therefore, the threshold energy is

$$E_{th} = 4.57\left(\frac{4+232}{232}\right) = 4.65 \text{ MeV}$$

The reaction also has a Coulomb barrier

$$V_C = 1.44\frac{(2)(92)}{1.4\,(232^{1/3} + 4^{1/3})} = 24.47 \text{ MeV}$$

Thus, the minimum energy required to overcome this Coulomb barrier is [See *Exercise* 5.7 to understand how much of the projectile's kinetic energy in the laboratory frame is spent as translational energy of the compound nucleus, in order to conserve the linear momentum.]

$$V_C\left(\frac{232+4}{232}\right) = 24.89 \text{ MeV}$$

Therefore, we have to supply an energy equal to 24.89 MeV in the form of kinetic energy of α-particles for the reaction to occur. However, out of this, only 4.65 MeV will be spent to take care of the potential threshold. The remaining energy will appear as excitation energy of uranium. This excess energy is dissipated in the form of γ-emission and as a result, the $^{236}_{98}U$ nucleus would attain ground state configuration.

5.2 CONCEPT OF CROSS-SECTION

The concept of cross-section was briefly touched upon in section 1.1.2 of Chapter 1, while discussing the Rutherford scattering. The cross-section (denoted by σ) of a target material for any given reaction represents the probability of a particular interaction. It is a property of the nucleus and the projectile energy. One can visualize this probability in terms of the effective area offered by the nucleus to the incident particle. This effective area should not be confused with the geometrical area of the nucleus (πR^2). This effective area could be larger, smaller or equal to the geometrical area of the nucleus, depending upon the likelihood of a particular nuclear interaction.

In general, for a given energy and a target-projectile combination, several nuclear reactions such as elastic scattering, inelastic scattering, capture reaction, many body reactions, etc. are possible simultaneously. The probability of occurrence of each of these interactions/reactions can be separately measured and the cross-section for each reaction is known as the partial reaction cross-sections denoted by σ_1, σ_2, σ_3, ... The sum of all the partial cross-sections is known as the total reaction cross-section σ_t.

The unit for nuclear cross-section is the barn, where

$$1\,\text{b} = 10^{-28}\ \text{m}^2 = 100\ \text{fm}^2$$

Let I_0 be the beam intensity of the mono-energetic projectiles falling on a thin target of thickness t and area a. For the purpose of the interaction/reaction, each nucleus is represented by a disc of area σ normal to the direction of the incident particles. If the incident particles strike a disc, the interaction/reaction occurs, otherwise it does not. We assume that the target is sufficiently thin so that I_0 does not get attenuated. If the target contains n nuclei per unit volume, then the total number of nuclei in the target is (nat). The total effective area available for the reaction, is thus equal to $(nat\sigma)$. Therefore,

$$\text{Fractional effective area} = \frac{nat\sigma}{a} = nt\sigma \tag{5.8}$$

And, the number of interactions occurring per unit area per second is given by:

$$N = I_0 nt\sigma \tag{5.9}$$

EXERCISE 5.4: A 0.01 mm thick ${}^7_3\text{Li}$ target is bombarded with 10^{13} protons/s. As a result, 10^8 neutrons/s are produced. Evaluate σ for this reaction. (Density of Li = 500 kg/m^3)

Solution: Neutrons are produced by the reaction: ${}^7_3\text{Li} + p \rightarrow {}^7_4\text{Be} + n$

$$n = \frac{N_a \rho}{A}$$

where N_a is the Avogadro number and A is the atomic mass number.

The number of neutrons produced per second is $nt\sigma I_0$. Therefore,

$$\frac{(6.023\times10^{23}\ \text{mol}^{-1})\,(500\times10^3\,\text{g/m}^3)}{(7\ \text{g/mol})}\,(10^{-5}\ \text{m})\,(\sigma)\,(10^{13}\ \text{s}^{-1}) = (10^8\ \text{s}^{-1})$$

On solving, we get σ = 0.23 b.

Let us now consider the same beam of particles incident on a thick slab. As the beam passes through, the particles in the beam interact with the target nuclei and are absorbed by the target or get scattered in some direction. Thus, the beam loses these particles, resulting in a decrease in its intensity.

The decrease ($-dI$) in the intensity as the beam intensity I passes through a thickness dt of the target is the number of incident particles per unit area per unit time interacting with the target nuclei. Thus, from equation (5.9), we get

$$-dI = (I\sigma n)dt$$

Integrating the above equation, we get

$$\int_{I_0}^{I} \frac{dI}{I} = -n\sigma \int_{0}^{t} dt$$

or
$$\ln I - \ln I_0 = -n\sigma t$$

or
$$I = I_0 e^{-n\sigma t} \tag{5.10}$$

As can be seen, the number of surviving particles in the beam, i.e., I, decreases exponentially with increasing slab thickness t.

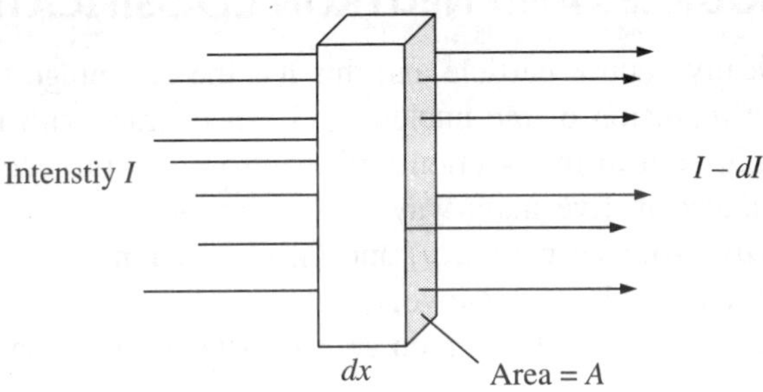

FIGURE 5.3 Relationship between cross-section and beam intensity.

Mean Free Path

The mean free path λ of a particle in a material is the average distance it can travel in the material before any interaction takes place. Since, at any thickness t, the probability that a particle interacts in the thickness dt is $e^{-n\sigma t}\, dt$, we have

$$\lambda = \frac{\int_0^\infty te^{-n\sigma t}\, dt}{\int_0^\infty e^{-n\sigma t}\, dt}$$

or
$$\lambda = \frac{1}{n\sigma} \tag{5.11}$$

EXERCISE 5.5: Consider the reaction ${}^{197}_{79}\text{Au}(n, \gamma)\,{}^{198}_{79}\text{Au}$. If the ${}^{197}_{79}\text{Au}$ target of area 5 cm^2 and thickness 0.3 mm is bombarded with a beam of 2×10^{12} neutron/m^2 s, what would be the number of ${}^{198}_{79}\text{Au}$ nuclei produced per second, assuming the cross-section of this reaction to be 94 b? (Density of Au = 19.3×10^3 kg/m^3)

Solution: The number of ${}^{197}_{79}\text{Au}$ nuclei per unit volume is given by

$$n = \frac{N_a \rho}{A} = \frac{(6.023\times10^{23}\ \text{mol}^{-1})\,(19.3\times10^{6}\ \text{g/m}^3)}{(197\ \text{g/mol})} = 5.89\times10^{28}\ \text{m}^{-3}$$

The number of neutrons hitting the target per second is equal to target area multiplied by the number of neutrons per unit area per second.

$$I_0 = (2\times10^{12} \text{ neutrons/m}^2\text{s})\,(5\times10^{-4}\,\text{m}^2) = 10^9 \text{ neutrons/s}$$

The number of ${}^{198}_{79}\text{Au}$ nuclei produced per second is equal to the number of reaction events taking place,

$$= (I_0 - I) = I_0(1 - e^{-n\sigma t})$$

$$= 10^9\{1 - \exp[-(5.89\times10^{28}\text{ m}^{-3})\,(94\times10^{-28}\text{ m}^2)\,(0.3\times10^{-3}\text{ m})]\} = 1.53\times10^8$$

5.3 NEUTRON SOURCES AND NEUTRON CLASSIFICATION

A neutron is an electrically neutral particle and this has the advantage that it does not have to overcome the Coulomb repulsion of the nucleus and can interact with nuclei at low energies. We shall focus our discussion in this section with neutron as a projectile.

Neutrons are produced in five main ways:

(i) From radioactive sources via (α, n) and (γ, n) reactions.

(ii) From accelerated based sources involving mainly protons and deuterons as projectiles on suitable target—the most common are the *D-D* and *D-T* reactions as shown below:

$${}^3_1\text{H}({}^2_1\text{H}, n){}^4_2\text{He}, \qquad Q = 17.58 \text{ MeV}$$

$${}^2_1\text{H}({}^2_1\text{H}, n){}^3_2\text{He}, \qquad Q = 3.27 \text{ MeV}$$

These reactions produce 14.1 MeV and 2.5 MeV neutrons, respectively.

(iii) From neutron reactors (both prompt and delayed-fission neutrons), e.g., Dhruva reactor at BARC, India.

(iv) Photo-nuclear neutrons production, e.g., white neutron sources at Oak Ridge Electron Linear Accelerator (ORELA), USA.

(v) Large accelerator based neutron source like the Spallation Neutron Source (SNS) at Oak Ridge National Laboratory (ORNL), USA. It provides neutron beams having extremely high energy and high flux.

Neutrons are classified according to their energies since their interactions with matter are energy dependent. The most common classification is given in Table 5.1.

Table 5.1 Classification of Neutrons

Neutron Energy	*Name*
0 – 0.025 eV	Cold
0.025 eV	Thermal
0.025 – 0.4 eV	Epithermal
0.4 – 1 eV	Cadmium
1 – 10 eV	Slow
10 – 300 eV	Resonance
300 eV – 1 MeV	Intermediate
1 MeV – 20 MeV	Fast
> 20 MeV	Relativistic

EXERCISE 5.6: Are thermal neutrons mono-energetic as shown in Table 5.1? Justify your answer.

Solution: No.

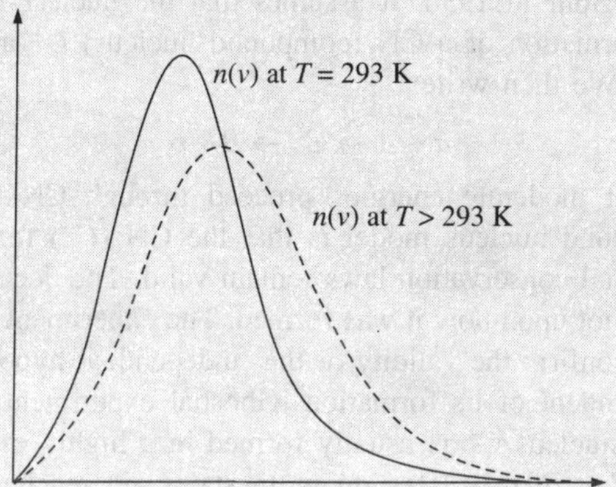

FIGURE 5.4 Maxwellian velocity distribution of neutrons.

Neutrons, while traversing matter, suffer collisions in which they lose energy until their energy distribution is the same as that of the atoms and molecules of the surrounding medium. The neutrons are then in thermal equilibrium with the surrounding medium at ordinary room temperature and are, therefore, termed thermal neutrons. At equilibrium, the thermal neutrons follow the Maxwellian distribution (Figure 5.4).

$$\frac{n(v)}{n} = \frac{4\pi v^2}{\left(2\pi k_B T/m\right)^{3/2}} e^{-mv^2/2k_BT}$$

where n = thermal neutron population per unit volume, m = neutron rest mass, T = temperature in K, $n(v)$ = Maxwellian velocity distribution of neutrons per unit volume and unit velocity interval.

The most probable neutron velocity v_p is found by setting

$$\frac{dn(v)}{dv} = 0$$

or
$$\frac{8\pi vn}{(2\pi k_BT/m)^{3/2}} e^{-mv^2/2k_BT} - \frac{4\pi v^2 n}{(2\pi k_BT/m)^{3/2}} \frac{2mv}{2k_BT} e^{-mv^2/2k_BT} = 0$$

or
$$v_p = \sqrt{\frac{2k_BT}{m}} = \sqrt{\frac{2(1.38\times10^{-23}\ \text{J/K})(293\text{K})}{1.66\times10^{-27}\ \text{K}}} = 2200\ \text{m/s}$$

Now,
$$E_p = \frac{1}{2}mv_p^2 = \frac{m}{2}\frac{2k_BT}{m} = k_BT$$
$$= (1.38\times10^{-23}\ \text{J/K})\,(293\text{K}) = 4.043\times10^{-21}\ \text{J}$$

or
$$E_p = 0.025\ \text{eV}$$

5.4 MECHANISM OF NUCLEAR REACTIONS

The exact mechanism of nuclear reactions is not yet completely understood but simple physical models exist, which give a simplified picture of the more complicated processes. The first model was proposed by Bohr in 1936. It assumes that the nuclear reaction proceeds in two independent steps: the formation of a CN (compound nucleus) C^* and the decomposition of this nucleus into $B + b$. We then write:

$$a + A \rightarrow C^* \rightarrow B + b$$

Most of the reactions at moderate energies proceed through CN formation. The essential hypothesis of the compound nucleus model is that the CN (C^*) forgets how it was formed except that the fundamental conservation laws remain valid. The decay of C^* depends only on the properties of C^* and not upon how it was formed. The experiment of S.N. Ghoshal in 1950, was the first attempt to confirm the validity of the 'independent hypothesis', i.e., de-excitation (decay) of CN is independent of its formation. Ghoshal experiment is discussed later in the chapter. The compound nucleus C^* is usually formed in a highly excited state. For reactions involving light nuclei ($Z < 20$), isolated quantum states are reached if projectile energy E_a is not too high and in this case, there are resonance effects in the excitation function (σ vs. E_a curve) of the reaction. With increasing energy, the density of the excited states increases rapidly and the resonances overlap. When the projectile energy E_a is so high that the continuum region (level spacing is smaller than the line width, i.e., $D_{CN} < \Gamma_{CN}$) of the compound nucleus is reached, the statistical models are used to calculate the reaction cross-sections. Experiments show that the neutron emission from a highly excited CN is very similar to the evaporation of molecules from a heated liquid and therefore, ejectiles follow Maxwell-Boltzmann distribution.

Another mechanism is suggested by the direct reaction model in which no intermediate quantum state is present. Rather the incoming projectile interacts with one or few nucleons of the target nucleus as it traverses the latter. The direct reaction cross-section is slowly dependent on the energy and does not exhibit narrow resonances, in contrast with the reactions that proceed through CN formation. In direct interaction process, a high momentum transfer between a and b takes place and consequently only low lying states of B are produced in the outgoing channel.

There is experimental evidence that a third type of mechanism, pre-equilibrium emission process is important in nuclear reactions induced by light projectiles of energies (> 10 MeV). Pre-equilibrium process is an intermediate between the two extremes, i.e., the compound nucleus mechanism and direct reaction mechanism. In the process of fusion of a target and projectile, if particles are emitted before statistical equilibrium is attained, then the third type of mechanism is possible, called pre-equilibrium process.

These reaction processes can be subdivided according to the time scales or equivalently, the number of intra-nuclear collisions taking place before the equilibrated state (CN) is reached. The direct reactions (one step process) are characterized by their short interaction time ($\sim 10^{-22}$ s) and discrete peaks at higher energies in the energy level spectra. A compound nucleus process (multistep process) involves long reaction time ($\sim 10^{-18}$ s) and is predominant at low energies.

It is often possible to distinguish between different reactions owing to the different exit channels. However, this is not always the case, e.g., ejectile b emerging from a nuclear collision at a particular angle may be the result of a direct reaction, pre-compound (pre-equilibrium) or a compound-nucleus reaction. The data analysis gets quite complicated due to the interference of these competing processes. Different reaction mechanisms can, however, be distinguished from each other on the basis of the energy spectra and angular distribution of the emitted particle *b*.

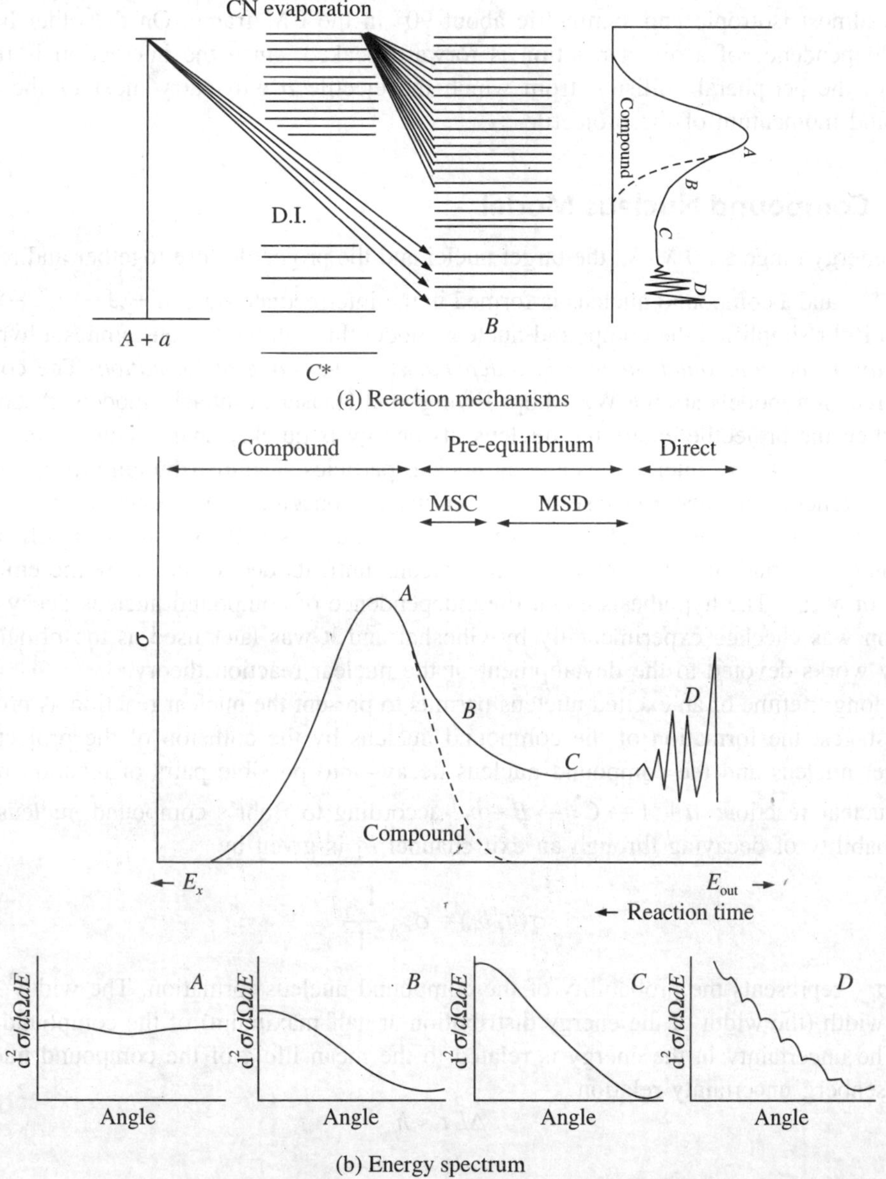

FIGURE 5.5 (a) Reaction mechanisms; (b) typical energy spectrum of a reaction *A*(*a*, *b*) *B* with incident energy of several tens of MeV. The direct interaction (DI) mechanism is responsible for high energy particle emission. The evaporation process (CN) yields low energy particles with a typical Maxwellian energy distribution. The associated angular distributions show a gradual transition to isotropy for decreasing outgoing energies. [A.J. Koning and J.M. Akkermans, *Proceedings of the Workshop*: *Nuclear Reaction Data and Nuclear Reactors*, World Scientific (1998)]

Figure 5.5 depicts a typical resultant particle spectrum. Each mechanism preferentially excites certain parts of the nuclear level spectrum and is characterized by different types of angular distributions. The characteristic angular distribution of the particles evaporated from a CN is almost isotropic and symmetric about 90° in the CM frame. On the other hand, the angular dependence of a direct reaction is forward peaked, since the interaction is restricted mainly to the peripheral collision from which the ejectile b will carry most of the incident energy and momentum of the projectile.

5.4.1 Compound Nucleus Model

For the energy range ≤ 20 MeV, the target nuclei and the projectile fuse together and remain so for ~10^{-14} s and a compound nucleus is formed in the intermediate state $a + A \rightarrow C^* \rightarrow B + b$. In 1936, Bohr simplified the compound nucleus model through his famous amnesia hypothesis: *The decay of the compound nucleus is independent of the mode of formation.* The compound nuclear reaction models are the Weisskopf Ewing[1] and Hauser Feshbach[2] models. According to Bohr, when the projectile enters the nucleus, its energy is quickly shared among the nucleons in the CN as a result of interactions among nuclear particles. The distribution presumably takes place in a random manner. Because the probability of concentration of enough energy on one nucleon to escape is rather small, the formed excited nucleus will exist for a significant time, undergoing a myriad of collisions between nucleons until its decay occurs by the emission of nucleon or γ-ray. The hypothesis about the independence of compound nucleus decay from its formation was checked experimentally by Ghoshal and it was later used as the principal idea in many works devoted to the development of the nuclear reaction theory.

A long lifetime of an excited nucleus permits to present the nuclear reaction as proceeding in two stages: the formation of the compound nucleus by the collision of the projectile with the target nucleus and the compound nucleus decays into possible pairs of reaction products. For a nuclear reaction: $a + A \rightarrow C^* \rightarrow B + b_1$, according to Bohr's compound nucleus theory, the probability of decaying through an exit channel b_1 is given by

$$\sigma(a, b_1) = \sigma_{CN} \frac{\Gamma_{b_1}}{\Gamma}$$

where σ_{CN} represents the probability of the compound nucleus formation. The width[3] Γ is the natural width (the width of the energy distribution at half-maximum) of the compound nucleus level. The uncertainty in the energy is related to the mean life τ of the compound nucleus by the Heisenberg uncertainty relation

$$\Delta E \tau \approx \hbar$$

1. Weisskopf, V.F. and Ewing, D.H., *Phys. Rev.* 57, 472, 1940.
2. Hauser, W. and Feshbach, H., The Inelastic Scattering of Neutrons, *Phys. Rev.*, 87, 366, 1952.
3. The level width is a measure of the spread in energy of an unstable state, i.e., the energy value is not exact, but there is an energy distribution peaking at the most probable energy value. For the ground state, $\tau \rightarrow \infty$ and $\Gamma = 0$, since it is stable. For all other nuclear levels, there are finite mean lives and therefore, finite Γs, indicating various modes of decay.

In case of a nuclear state, the above equation can be written as follows:

$$\Gamma\tau \approx \hbar$$

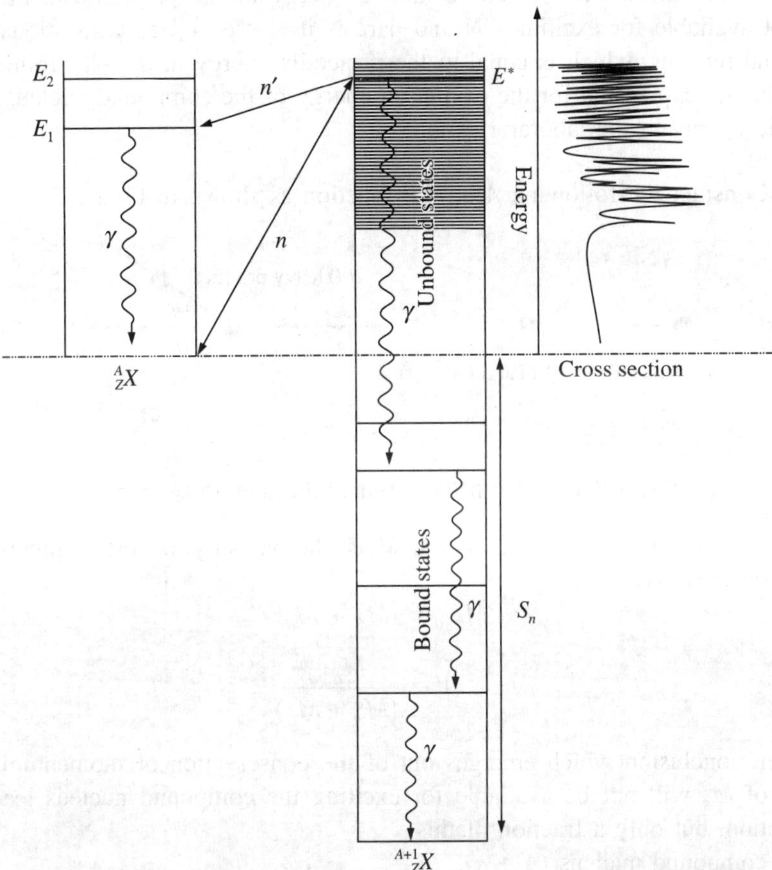

FIGURE 5.6 Neutron capture, elastic and inelastic level scheme.

If Γ_{b_1}, Γ_{b_2}, Γ_{b_3},..., are the partial widths of the modes of decay b_1, b_2, b_3,..., then the decay rate through channel b_1 can be expressed by Γ_{b_1}/Γ, where the total decay rate probability has to be one and $\sum_{i=1}^{n} \Gamma_{b_i} = \Gamma$.

Figure 5.6 depicts the neutron capture, elastic and inelastic exit channels. After the neutron capture, the CN ${}^{A+1}_{Z}\text{X}$ is formed, whose energy levels (bound as well as virtual) have been shown. Compound nucleus with excitation energy E^* can decay by the following processes:

(i) Neutron emission with the same energy with which it was captured (elastic channel).
(ii) Neutron emission with less energy and associated γ-ray of E_1 energy corresponding to the first excited level of ${}^{A}_{Z}\text{X}$ nucleus (inelastic channel).
(iii) Gamma ray emission.

The resolved and the unresolved resonances have been shown on the right.

EXERCISE 5.7: Show that the entire kinetic energy of the projectile in the laboratory frame is not available for exciting CN and part of it is used up as translational energy of the compound nucleus, which is equal to the projectile energy in the CM frame.
Hence, derive the expression for the excitation energy of the compound nucleus in terms of the projectile energy in the laboratory frame.

Solution: Consider the following $A(a, b)B$ reaction as shown in Figure 5.7.

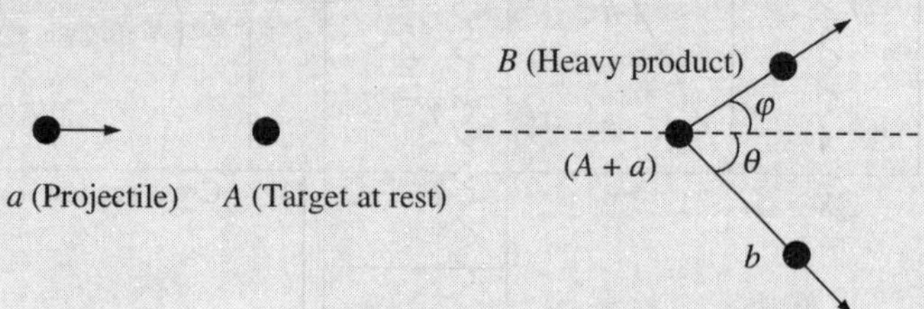

FIGURE 5.7 $A(a, b)B$ reaction in the laboratory frame.

Conservation of linear momentum, where V is the velocity of the compound nucleus (composite system) gives:

$$M_a v_a = (M_A + M_a)V$$

or

$$V = \frac{M_a v_a}{(M_A + M_a)}$$

An important conclusion which emerges out of the conservation of momentum is that the entire K.E. of M_a will not be available for exciting the compound nucleus leading to the nuclear reaction, but only a fraction of this.
K.E. of the compound nucleus $(A + a)$,

$$E_{CN} = \frac{1}{2}(M_A + M_a)V^2$$

$$= \frac{M_a}{(M_A + M_a)}\left(\frac{1}{2}M_a v_a^2\right)$$

$$\therefore \qquad E_{CN} = \frac{M_a}{(M_A + M_a)}E_a$$

Since, E_{CN} is used up in translation, the energy available for exciting the target nucleus is only

$$(E_a - E_{CN}) = \frac{M_A}{(M_A + M_a)}E_a$$

The left side $(E_a - E_{CN})$ is also referred as the K.E. of the (projectile + target) in the CM frame.

or
$$E_{a,\text{CMS}} = \frac{M_A}{(M_A + M_a)} E_a$$

where E_a is the projectile energy in the laboratory system and $E_{a,\text{CMS}}$ represents the projectile kinetic energy in the centre of mass system.

Thus, the excitation energy of CN is given by

$$E_{CN} = B_a + E_{a,\text{CMS}} = Q + E_{a,\text{CMS}}$$

$$\therefore \qquad E_{CN} = Q + \frac{M_A}{(M_A + M_a)} E_a \qquad (5.12)$$

where B_a is the binding energy of the projectile in the compound nucleus.

5.4.2 Direct Reaction Model

The direct reactions are characterized by their short interaction time (10^{-22} s), which is roughly equal to the time taken by the incident particle to transverse the target nucleus. In this type of reaction mechanism, an intermediate energy projectile transfers energy or picks up or looses nucleon to the nucleus in a single (10^{-21} s) event. The energy and the momentum transfers are relatively small. Direct reactions occur most probably for reactions in which the nucleus is left in one of its low lying excited states. For incident energies above ~10 MeV, the low lying, discrete states of the residual nucleus are nearly completely excited by the direct reaction processes. The associated angular distributions are strongly peaked in the forward direction. They also exhibit markedly oscillatory behaviour. This specific oscillatory shape enables one to determine the spin and the parity of the residual nuclear state. The analysis of direct reactions is almost always performed with either a DWBA (Distorted Wave Born Approximation) or a coupled channel approach[4].

The direct reaction mainly includes:

(i) Stripping reaction, (d, n), (d, p) reactions
(ii) Pickup reaction, (n, d) and (n, t) reactions
(iii) Inelastic scattering, e.g. (n, n') reactions

The stripping reactions refer to those reactions in which the incident projectile looses one or more of its constituent while passing in the vicinity of the target nucleus such as the (d, n), (d, p) or the (α, n), (α, p) reactions. The reverse of the stripping reaction is the pickup reaction, in which the incident projectile such as the neutron picks up a proton from the target nucleus while passing through it and the ejectile is a deuteron.

In direct interaction processes involving inelastic scattering such as the (n, n') reactions, the incident particle suffers an inelastic collision with the target nucleus without the formation of the compound nucleus.

4. G.R. Satchler, *Direct Nuclear Reactions*, Oxford University Press (NY, 1983); N. Austern, *Direct Nuclear Reaction Theories*, Wiley-Interscience (1970).

Table 5.2 shows a comparison between compound nucleus reactions and direct reactions.

Table 5.2 Comparison of Compound Nucleus Reactions and Direct Reactions

Feature	*CN reactions*	*Direct reactions*
Times involved	$10^{-14} - 10^{-16}$ s	~$10^{-20} - 10^{-21}$ s
Dominance of reaction	Low energy	High energy
Nature of reaction	Surface phenomenon	Nuclear interior
Cross-section	~barn	~millibarn
Angular distribution	Isotropic in CM frame	Peaked in the forward hemisphere
Location of peaks	Energy dependent	Orbital angular momentum dependent

5.4.3 Pre-equilibrium Model

In the process of fusion of a target and a projectile, if the particles are emitted before statistical equilibrium is attained, then the third type of the mechanism is possible, called pre-equilibrium processes, which is the intermediate state between the direct and the compound processes. Many experimental results at moderate energies are best explained by a mixture of contributions from both the mechanisms. During 1950–60, the evidence that was accumulated, suggested that in some nuclear reactions, it is not possible to understand all the emission processes in terms of the compound nucleus (slow process, $\tau \approx 10^{-18}$ s) and direct processes ($\tau \approx 10^{-22}$ s). Deviations from the Maxwellian shape for the emission spectra were observed for intermediate to high emission energies. Pre-equilibrium processes are important in nuclear reactions induced by light projectiles of energies > 10 MeV.

Pre-equilibrium emission takes place after the first stage of the reaction, but much before the attainment of the statistical equilibrium of the compound nucleus. It is imagined that the incident particle creates more complex states of the compound system step-by-step and gradually loses its memory of the incident energy and the angular momentum. Pre-equilibrium processes provide a significant part of the reaction cross-section for incident energies between 10 MeV and 200 MeV. Multistep compound (MSC) and multistep direct (MSD) depicted in Figure 5.6, are the competing processes within the domain of pre-equilibrium reactions.

In MSC, the emission of ejectiles takes place from the bound configuration of the composite system (target nucleus + projectile) before the attainment of statistical equilibrium. The treatment of MSC models are based on quantum mechanical theory of composite system with statistical postulates[5]. The MSC mechanism yields symmetrical angular distribution analogous to the CN mechanism. In MSD reactions, the reaction proceeds by the unbound composite system and gives rise to smooth forward peaked angular distributions that are the signature of pre-equilibrium decay. The detailed treatment is beyond the scope of this book[6].

The commonly used pre-equilibrium models are the exciton model and the hybrid model. Both of these are semi-classical models and originate from the paper by Griffin[7], 1966 and later development by Kalbach,[8] 1977.

---

5. D. Agassi, H.A Weidenmueller and G. Mantzouranis, Phys. Rep. 22(1975) 145; K. W. McVoy and X. T. Tang, Phys. Rep. 94, 139, 1983.
6. Further details can be found in H. Feshbach, A. Kerman and S. Koonin, Ann. Physc. (NY) 125, 429, 1980.
7. Griffin J.J., 'Statistical model of intermediate structure', Phys. Rev. Lett. 17 478, 1966.
8. Kalbach C., 'The Griffin model, complex particles and direct nuclear reaction', Z. Physik A283 401, 1977.

Griffin, proposed the pre-equilibrium 'exciton model.' According to the exciton model, after the initial interactions between the incident particle and the target nucleus, the excited system can pass through a series of stages of increasing complexity before equilibrium is reached; and the emission may occur from these stages giving the pre-equilibrium (PE) particles. The different stages of complexity are classified according to the number of particles and holes excited, and the exciton model calculations involve solving a number of master equations that describe the equilibrium of an excited system through a series of two body collisions producing more complex configurations of particle-hole pairs. The nuclear state is characterized by the excitation energy E and the total number of particles and holes above and below the Fermi surface. Particles and holes are indiscriminately referred to as excitons. Equilibrium process of the excited nucleus is imagined to proceed as shown in Figure 5.8.

After entering the target nucleus, the incident particle collides with one of the nucleons of the Fermi sea. Thus, an initial state with $n = 3$ is formed (in the case of nucleon-induced reactions). Subsequent interaction results in changes in the number of excitons. At any stage, there is a non-zero probability of particle emission, and intuitively it is clear that the emitted particle retains some 'memory' of the incident energy and direction: Bohr's amnesia hypothesis is not valid. This phase is called the pre-equilibrium phase, which is expected to be responsible for the experimentally observed high energy tails and forward peaked angular distributions. If an emission does not occur at an early stage, the system eventually reaches a (quasi) equilibrium. The equilibrium situation, corresponding to high exciton number, is established after a large number of interactions, i.e., after a long lapse of time, and the system has 'forgotten' about the initial stage. Accordingly, this stage may be called the compound or evaporation stage. Hence, in principle, the exciton model enables one to compute the emission cross-section in a unified way, without introducing arbitrary adjustments between equilibrium and pre-equilibrium contributions.

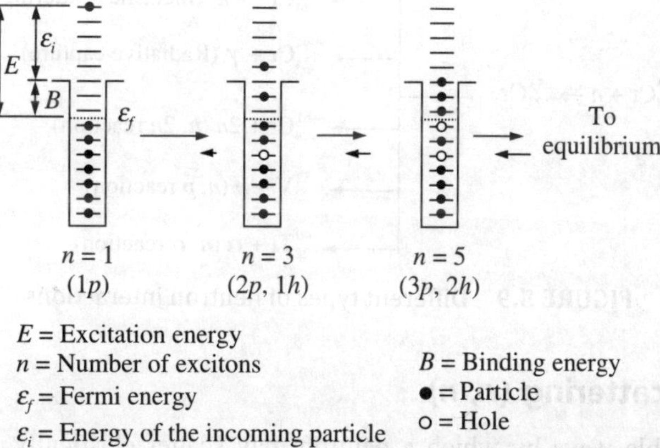

FIGURE 5.8 Schematic representation of the exciton model. The relative probability and the direction of the transitions are roughly indicated by the length and the direction of the arrows. [A.J. Koning and J.M. Akkermans, *Proceedings of the Workshop: Nuclear Reaction Data and Nuclear Reactors,* World Scientific (1998)].

5.5 NEUTRON INTERACTIONS THROUGH CN FORMATION

One of the characteristic features of the neutron interactions with matter that proceed through CN formation is that, at certain incident neutron energies, the cross-sections have maximum values. These maximum values at different energies are called resonances, and the particular energies at which the maxima occur are called resonance energies. The experimental results obtained at the resonance energies are often very useful in getting precise information about the excited levels of CN.

The different types of neutron interactions are depicted in Figure 5.9.

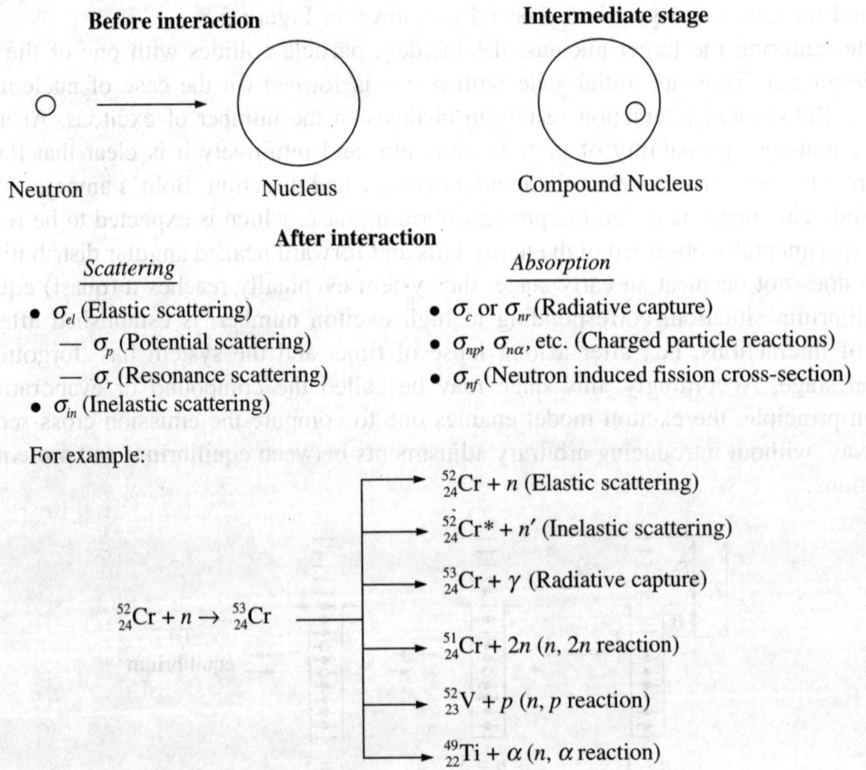

FIGURE 5.9 Different types of neutron interactions.

5.5.1 Elastic Scattering (*n, n*)

There are two possible ways by which a neutron may scatter elastically.

(i) Resonance or compound elastic scattering: CN is formed and the re-emission of the neutron takes place.

(ii) Potential (shape) elastic scattering: The neutron is scattered away from the nucleus by the short range nuclear force. It is a more common form of neutron elastic scattering. The cross-section is approximately constant as is expressed by the relation

$$\sigma_{\text{pot.elastic}} = 4\pi R^2$$

where R is the nuclear radius.

In elastic scattering, the momentum and the kinetic energies are conserved and some kinetic energy is transferred from the neutron to the target nucleus. The largest energy transfer occurs in the head-on collision of neutron with light elements such as hydrogen.

5.5.2 Inelastic Scattering (*n*, *n*′)

If the energy of the neutron in the CM frame exceeds the energy of the first excited state in the target, inelastic scattering becomes possible. In the case of inelastic scattering, the neutron is emitted with lower energy, leaving the target nucleus in an excited state.

The inelastic scattering proceeds in two steps as depicted in Figure 5.10. Thus,

$$E_n' + E_\gamma = E_n$$

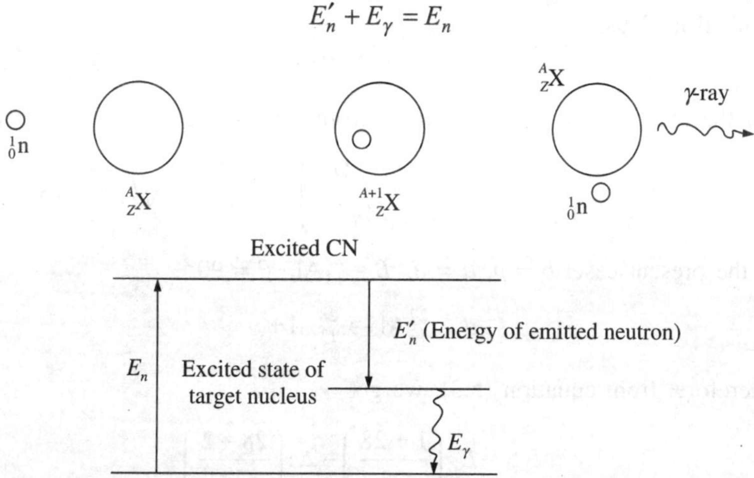

FIGURE 5.10 Schematics of (*n*, *n*′) reaction.

5.5.3 Radiative Capture (*n*, γ)

Neutron capture is often called radiative capture, because γ-rays are produced in these reactions.

$${}^1_0\text{n} + {}^A_Z\text{X} \rightarrow {}^{A+1}_{Z}\text{X}^* \rightarrow {}^{A+1}_{Z}\text{X} + \gamma$$

In the resonance region, $\sigma(n, \gamma)$ follows $1/v$ law [see equation (5.17)]. Above the resonance region (≈ 1 keV in heavy nuclei), $\sigma(n, \gamma)$ drops rapidly.

5.5.4 Charged Particle Emission (*n*, α) (*n*, *p*)

They are represented as follows:

$${}^1_0\text{n} + {}^A_Z\text{X} \rightarrow {}^{A+1}_{Z}\text{X}^* \rightarrow {}^{A-3}_{Z-2}\text{Y} + {}_{Z-2}{}^{4}\text{He}^{++}$$

$${}^1_0\text{n} + {}^A_Z\text{X} \rightarrow {}^{A+1}_{Z}\text{X}^* \rightarrow {}_{Z-1}^{A}\text{Y} + p$$

For example:

$${}^{1}_{0}\text{n} + {}^{10}_{5}\text{B} \rightarrow {}^{11}_{5}\text{B}^* \rightarrow {}^{7}_{3}\text{Li} + \alpha$$

$${}^{1}_{0}\text{n} + {}^{6}_{3}\text{Li} \rightarrow {}^{7}_{3}\text{Li}^* \rightarrow {}^{3}_{1}\text{H} + \alpha$$

$${}^{1}_{0}\text{n} + {}^{16}_{8}\text{O} \rightarrow {}^{17}_{8}\text{O}^* \rightarrow {}^{16}_{7}\text{N} + p$$

EXERCISE 5.8:

(a) Calculate the energy of protons detected at 90° when 2.1 MeV deuterons are incident on ${}^{27}_{13}\text{Al}$ to produce ${}^{28}_{13}\text{Al}$ (Q = 5.5 MeV).

(b) An aluminium target is bombarded by α-particles of energy 7.68 MeV, and the resultant proton groups at 90° were found to possess energies 8.63, 6.41, 5.15 and 3.98 MeV. Using the above information, draw an energy level diagram of the residual nucleus.

Solution:

(a) For the reaction $A(a, b)B$, we know from equation (5.3)

$$Q = E_b\left(\frac{M_b + M_B}{M_B}\right) - E_a\left(\frac{M_B - M_a}{M_B}\right) - \frac{2\sqrt{M_a M_b}}{M_B}\sqrt{E_a E_b}\cos\theta$$

In the present case, $b = p$, $a = d$, $B = {}^{28}_{13}\text{Al}$, $\theta = 90°$

$$d + {}^{27}_{13}\text{Al} \rightarrow {}^{28}_{13}\text{Al} + p$$

Therefore, from equation (5.3), we get

$$5.5 = E_p\left(\frac{1+28}{28}\right) - 2.1\left(\frac{28-2}{28}\right)$$

or

$$E_p = 7.19 \text{ MeV}$$

(b) For the following reaction:

$${}^{4}_{2}\text{He} + {}^{27}_{13}\text{Al} \rightarrow {}^{30}_{14}\text{Si} + p$$

$$Q = E_p\left(\frac{1+30}{30}\right) - 7.68\left(\frac{30-4}{30}\right)$$

Thus, we have

$$E_p = 8.63 \text{ MeV}, \; Q_0 = 2.262 \text{ MeV}$$

$$E_p = 6.41 \text{ MeV}, \; Q_0 = -0.032 \text{ MeV}$$

$$E_p = 5.15 \text{ MeV}, \; Q_0 = -1.334 \text{ MeV}$$

$$E_p = 3.98 \text{ MeV}, \; Q_0 = -2.543 \text{ MeV}$$

The energy levels are determined by

$$E_0 = Q_0 - Q_0 = 0 \text{ (Ground state)}$$

$$E_1 = Q_0 - Q_1 = 2.294 \text{ MeV}$$

$$E_2 = Q_0 - Q_2 = 3.596 \text{ MeV}$$

$$E_3 = Q_0 - Q_3 = 4.805 \text{ MeV}$$

The energy levels of ${}^{30}_{14}\text{Si}$ are shown in Figure 5.11.

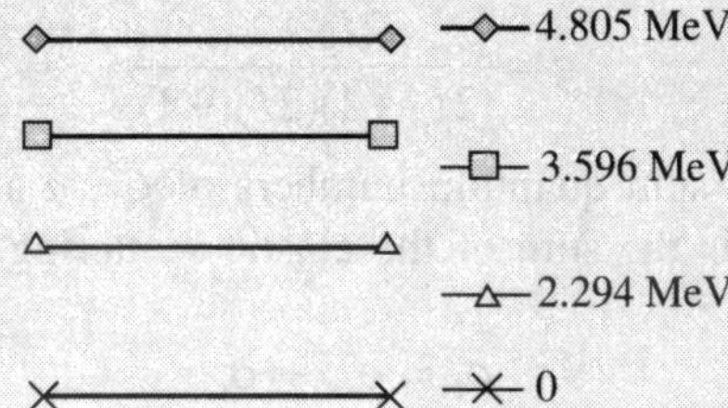

FIGURE 5.11 Energy level diagram of the residual nucleus (Si–30).

5.6 RESONANCES IN A COMPOUND NUCLEAR REACTION

When the projectile energy E_a in the entrance channel is varied in such a way that the excitation energy matches with one of the levels of the compound nucleus, then the excitation function of the nuclear reaction shows a peak, which is called resonance. Thus, the excited states of the CN, C^* are responsible for the existence of resonances in nuclear reaction cross-sections (Figure 5.12).

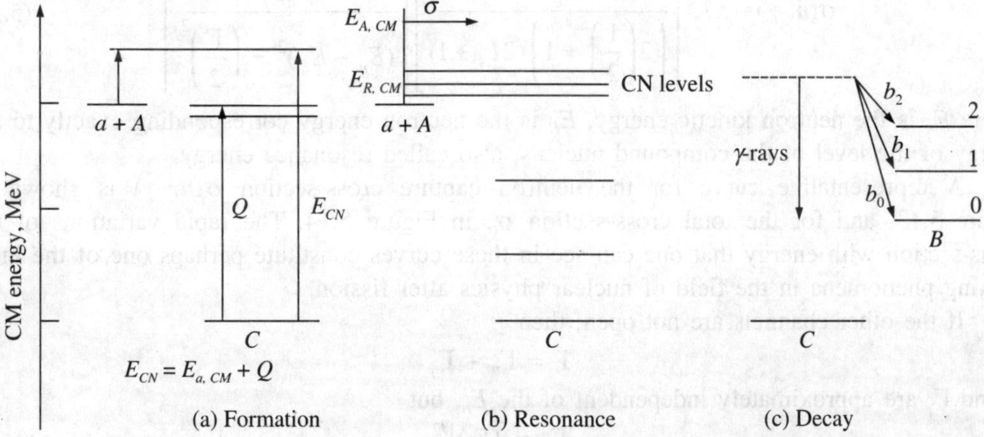

Figure 5.12 Energy diagrams (always drawn in the CM system) for compound nucleus reactions (single resonances): (a) compound nucleus energy; (b) compound nucleus levels corresponding to resonances in the reaction cross-section, the level width is the resonance width Γ'; (c) compound nucleus decaying by particle or photon emission.

The resonant reaction cross-section is given by the Breit–Wigner equation:

$$\sigma_{sc} = \frac{\pi \lambdabar^2 \Gamma_s^2 g}{(E_a - E_R)^2 + \left(\frac{\Gamma}{2}\right)^2} \tag{5.13}$$

$$\sigma_r = \frac{\pi \lambdabar^2 \Gamma_r \Gamma_s g}{(E_a - E_R)^2 + \left(\frac{\Gamma}{2}\right)^2} \tag{5.14}$$

where g is the statistical factor,

$$g = \frac{2I_c + 1}{(2I_a + 1)(2I_A + 1)}$$

in which I_c, I_a and I_A are the spin quantum numbers of C^*, a and A, respectively.

The total cross-section is the sum of the elastic scattering cross-section and the reaction cross-section, i.e.,

$$\sigma_t = \sigma_{sc} + \sigma_r$$

$$\therefore \qquad \sigma_t = \frac{\pi \lambdabar^2 \Gamma_s g}{(E_a - E_R)^2 + \left(\frac{\Gamma}{2}\right)^2} \tag{5.15}$$

The energy width Γ of a level is related to the mean-life by the uncertainty principle:

$$\Gamma \tau \approx \hbar$$

where $\Gamma = \sum_i \Gamma_i$ in which Γ_i represents partial widths corresponding to each of the possible modes of decay.

From equation (5.15), Breit–Wigner expression for neutron capture reaction is thus,

$$\sigma(n, \gamma) = \pi \lambdabar^2 \left[\frac{2I_c + 1}{\left(2\left(\frac{1}{2}\right) + 1\right)(2I_A + 1)}\right]\left[\frac{\Gamma_n \Gamma_r}{(E_n - E_R)^2 + \left(\frac{\Gamma}{2}\right)^2}\right] \tag{5.16}$$

where E_n is the neutron kinetic energy, E_R is the neutron energy corresponding exactly to the energy of the level of the compound nucleus, also called resonance energy.

A representative curve for the neutron capture cross-section $\sigma(n, \gamma)$ is shown in Figure 5.13, and for the total cross-section σ_t, in Figure 5.14. The rapid variations of the cross-section with energy that one can see in these curves constitute perhaps one of the most striking phenomena in the field of nuclear physics after fission.

If the other channels are not open, then

$$\Gamma = \Gamma_n + \Gamma_\gamma$$

Γ and Γ_γ are approximately independent of the E_n, but

$$\Gamma_n \propto (E_n)^{1/2}$$

For neutron energies well below E_R, so that $(E_n - E_R)^2 \approx E_R^2$, the de-Broglie wavelength is related to the neutron velocity as follows:

$$\lambda = \frac{h}{mv}$$

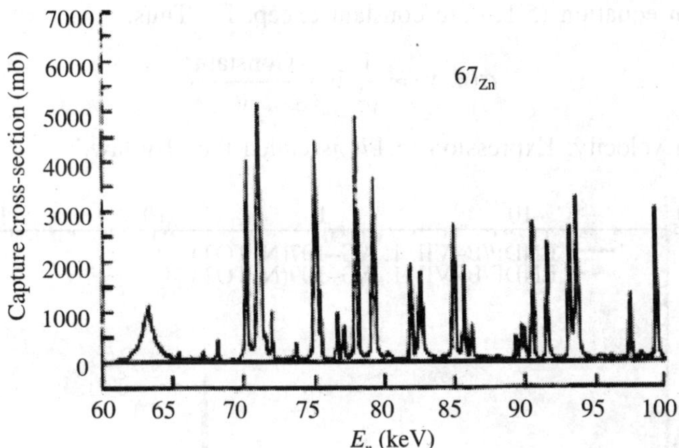

FIGURE 5.13 The neutron capture cross-section data for a thin oxide sample of Zn-67 nucleus in the neutron energy range 6 to 10 keV. The full curves merely connect the data points. [H.M. Agrawal, J.B. Garg, V.K. Tikku, J.A. Harvey and R.L. Macklin, High-resolution neutron total and capture cross sections in 67Zn', *J. Phys G: Nucl. Phys.*, 18, 1069–1087 (1992)].

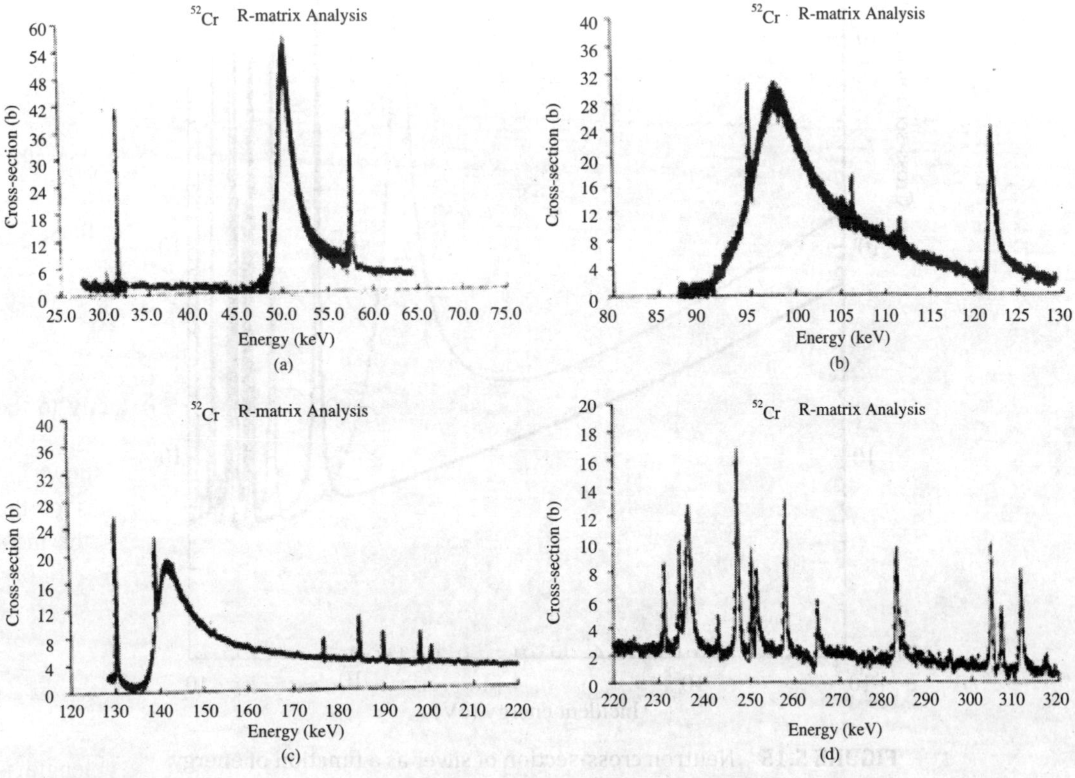

FIGURE 5.14 R-matrix multilevel fit (solid lines) to the total cross-section data (points) for Cr-52 in the neutron energy range from 25 to 320 keV. [H.M. Agrawal, J.B. Garg and J.A. Harvey, *Phys. Rev.* C30, 1880 (1984)].

All other factors in equation (5.16) are constant except Γ_n. Thus,

$$\sigma(n, \gamma) \propto \frac{1}{v^2} v = \frac{\text{Constant}}{v} \tag{5.17}$$

where v is neutron velocity. Expression (5.17) is called the '1/v law.'

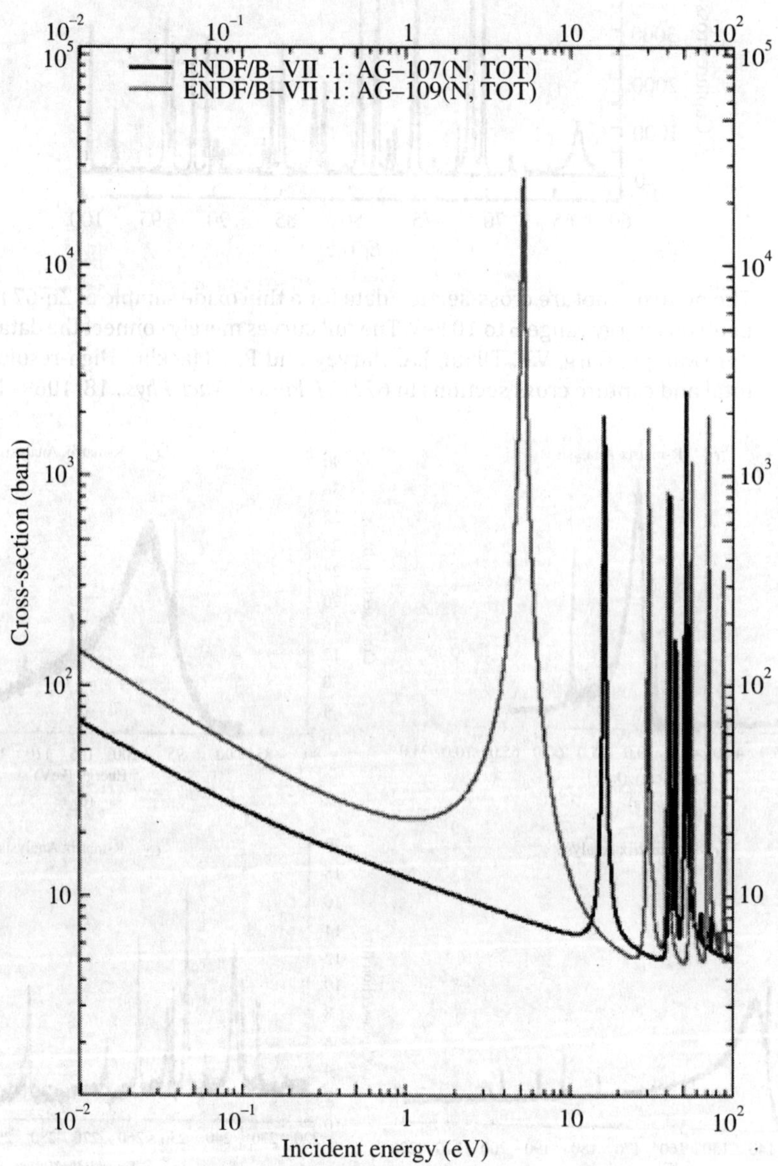

FIGURE 5.15 Neutron cross-section of silver as a function of energy. [National Nuclear Data Center: ENDF/B-VII.1 (USA, 2011)]

The $1/v$ behaviour of the cross-section in slow energy neutron induced reactions is shown in Figure 5.15, by the straight portion of the curves. Resonances have been shown for Ag-107 and Ag-109 isotopes. The spacing between the resonances is large (few eV to keV) as expected for silver isotopes.

EXERCISE 5.9: A nucleus has a neutron resonance at 65 eV with no other resonances nearby. For this particular resonance, $\Gamma_n = 4.2$ eV, $\Gamma_\gamma = 1.3$ eV, and $\Gamma_\alpha = 2.7$ eV. All other partial widths can be neglected. What is the cross-section for (n, γ) and (n, α) reactions at 70 eV?

Solution:

$$\Gamma = \Gamma_n + \Gamma_\gamma + \Gamma_\alpha$$
$$= 4.2 + 1.3 + 2.7 = 8.2 \text{ eV}$$

From equation (5.14), we have

$$\sigma(n, \gamma) = \pi \bar{\lambda}^2 (g) \left[\frac{\Gamma_n \Gamma_r}{(E - E_R)^2 + \left(\frac{\Gamma}{2}\right)^2} \right]$$

Since, we do not have information about the target nucleus, we ignore the g-factor. The de-Broglie wavelength λ is given by

$$\lambda = \frac{h}{\sqrt{2mE}} = \frac{2\pi(\hbar c)}{\sqrt{2mc^2}\sqrt{E}}$$

$$= \frac{2\pi(197 \times 10^6 \text{ eVfm})}{\sqrt{2 \times 940 \times 10^6 \text{ eV}}\sqrt{70 \text{ eV}}} = 3.42 \times 10^3 \text{ fm}$$

Hence, the neutron capture cross-section is

$$\sigma(n, \gamma) = \pi \left(\frac{3.42 \times 10^3 \text{ fm}}{2\pi} \right)^2 \left[\frac{(4.2 \text{ eV})(1.3 \text{ eV})}{(70 - 65)^2 + \left(\frac{8.2}{2}\right)^2} \right] = 1215 \text{ b}$$

and,

$$\sigma(n, \alpha) = \sigma(n, \gamma) \frac{\Gamma_\alpha}{\Gamma_\gamma} = 1215 \left(\frac{2.7}{1.3} \right) = 2523 \text{ b}$$

EXERCISE 5.10: The nuclide ${}^{19}_{9}\text{F}$ was bombarded with protons of varying energy and the yields of neutrons from the (p, n) reaction were measured. Resonances were observed at the following proton energies in MeV: 4.29, 4.46, 4.49, 4.57, 4.62, 4.71, 4.78, 4.99, 5.07 and 5.20. The threshold energy for the reaction was found experimentally to be 4.253 MeV. Calculate from the data:

(a) the mass difference ${}^{19}_{10}\text{Ne} - {}^{19}_{9}\text{F}$,

(b) the endpoint energy of the β^+-decay of ${}^{19}_{10}\text{Ne}$,

(c) the excitation energies of the states of ${}^{20}_{10}\text{Ne}$, which correspond to the resonances.

Solution:

(a) In the following reaction,

$${}^{19}_{9}\text{F}_{10} + p \to {}^{19}_{10}\text{Ne}_9 + n$$

$$E_{th} = -Q\frac{19+1}{19+1-1} = -Q\frac{20}{19}$$

$$\therefore \qquad Q = -\frac{19}{20}E_{th} = -\frac{19}{20} \times 4.253 = -4.040 \text{ MeV}$$

$$Q = [M({}^{19}_{9}\text{F}) + M(p) - M({}^{19}_{10}\text{Ne}) - M(n)]$$

or

$$M({}^{19}_{10}\text{Ne}) - M({}^{19}_{9}\text{F}) = M(p) - M(n) - Q$$

$$= 1.008144 - 1.008983 - \left(-\frac{4.040}{931.478}\right) = 0.003498 \text{ u}$$

(b) In the following β^+-decay

$${}^{19}_{10}\text{Ne}_9 \to {}^{19}_{9}\text{F}_{10} + \beta^+ + \nu$$

$$E_{\beta^+} = 0.003498 \times 931.478 \text{ MeV} - 2m_ec^2$$

$$= 3.25 - 1.02 = 2.23 \text{ MeV}$$

(c) The binding energy of proton as obtained from the mass difference $\{{}^{19}_{9}\text{F} + p - {}^{20}_{10}\text{Ne}\}$

$= \{(19 + \text{Mass excess of } {}^{19}_{9}\text{F}) + (1 + \text{Mass excess of } p) - (20 + \text{Mass excess of } {}^{20}_{10}\text{Ne})\}$

$= -1486 + 7288.99 - (-7041.5) = 12844.4$ keV

$= 12.85$ MeV

Table 5.3 Excitation energies of the states of CN.

Proton resonance energy (MeV)	*Energy to be added to binding energy* (MeV)	*Excited level of* $^{20}_{10}Ne$ (MeV)
4.29	$4.29 \times \frac{19}{19+1} = 4.075$	12.85 + 4.075 = 16.925
4.46	$4.46 \times \frac{19}{19+1} = 4.237$	12.85 + 4.237 = 17.087
4.49	$4.49 \times \frac{19}{19+1} = 4.266$	12.85 + 4.266 = 17.116
4.57	$4.57 \times \frac{19}{19+1} = 4.342$	12.85 + 4.342 = 17.192
4.62	$4.62 \times \frac{19}{19+1} = 4.389$	12.85 + 4.389 = 17.239
4.71	$4.71 \times \frac{19}{19+1} = 4.475$	12.85 + 4.475 = 17.325
4.78	$4.78 \times \frac{19}{19+1} = 4.541$	12.85 + 4.541 = 17.391
4.99	$4.99 \times \frac{19}{19+1} = 4.741$	12.85 + 4.471 + 17.591
5.07	$5.07 \times \frac{19}{19+1} = 4.817$	12.85 + 4.817 = 17.667
5.20	$5.20 \times \frac{19}{19+1} = 4.94$	12.85 + 4.94 = 17.79

5.7 GHOSHAL EXPERIMENT

The validity of Bohr's hypothesis of the compound nucleus has been elegantly demonstrated by several experiments. We now discuss the Ghoshal experiment, which proved that the decay of CN is independent of the way it was formed. In this experiment, the same compound nucleus $^{64}_{30}Zn^*$ at a specific energy was produced with the bombardment of $^{60}_{28}Ni$ by α-particles and $^{63}_{29}Cu$ by protons. The energies of α-particles were taken between 10 and 40 MeV, while those of protons were kept between 3 and 33 MeV, and the fusion excitation functions were measured. The energies of the α-particles were kept 7 MeV higher to produce the same excitation in $^{64}_{30}Zn^*$.

The reactions observed were:

(i) ${}^{60}_{28}\text{Ni}(\alpha, n)\,{}^{63}_{30}\text{Zn}$ (ii) ${}^{60}_{28}\text{Ni}(\alpha, 2n)\,{}^{64}_{30}\text{Zn}$

(iii) ${}^{60}_{28}\text{Ni}(\alpha, pn)\,{}^{63}_{29}\text{Cu}$ (iv) ${}^{63}_{29}\text{Cu}(p, n)\,{}^{63}_{30}\text{Zn}$

(v) ${}^{63}_{29}\text{Cu}(p, 2n)\,{}^{62}_{30}\text{Zn}$ (vi) ${}^{63}_{29}\text{Cu}(p, pn)\,{}^{62}_{29}\text{Cu}$

Figure 5.16 depicts all these reactions.

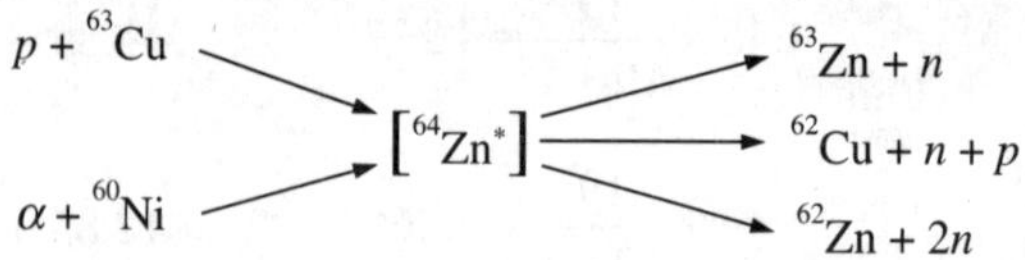

FIGURE 5.16 Reactions observed in Ghoshal experiment.

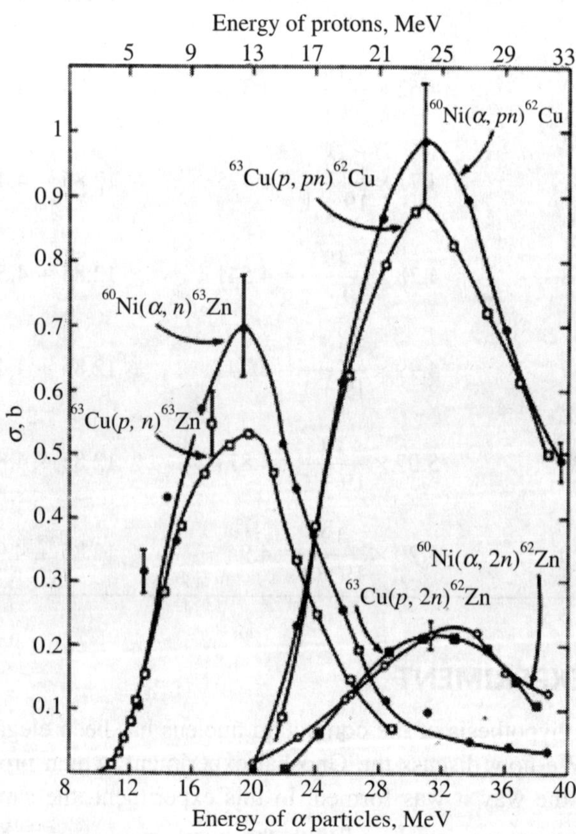

FIGURE 5.17 Yields of the decay products of the compound nucleus Zn-64 according to Ghoshal. The proton energy scale has been shifted 7 MeV to the right. [S.N. Ghoshal, *Phys. Rev.*, 80, 939 (1950)].

The excitation functions were experimentally determined by measuring the evaporation residues at various projectile energies. For these reactions, stacked foil arrangement for irradiation of several target foils was used. The measured cross-sections as obtained by Ghoshal are given in Figure 5.17. The figure shows clearly that the competitive *ratio of the yields is independent of the mode of formation of the compound nucleus*; i.e.,

$$\sigma(\alpha, n): \sigma(\alpha, 2n): \sigma(\alpha, pn) = \sigma(p, n): \sigma(p, 2n): \sigma(p, pn)$$

The results are accurate within 10%.

EXERCISE 5.11: Suppose the α-particles of energy 14 MeV incident on a $^{60}_{28}\text{Ni}$ nucleus form the compound nucleus $^{64}_{30}\text{Zn}^*$ via the following reaction:

$$^{4}_{2}\text{He} + ^{60}_{28}\text{Ni} \rightarrow ^{64}_{30}\text{Zn}^*$$

(a) What is the excited energy of $^{64}_{30}\text{Zn}^*$ relative to the ground state $^{64}_{30}\text{Zn}$?

(b) The same compound nucleus can also be formed via the following reaction:

$$^{1}_{1}\text{H} + ^{63}_{29}\text{Cu} \rightarrow ^{64}_{30}\text{Zn}^*$$

If we want to attain the same excited energy in $^{64}_{30}\text{Zn}^*$, what must be the energy of the incident proton? [Ignore the recoil energy of $^{64}_{30}\text{Zn}^*$ nucleus]

Given: $M(^{60}_{28}\text{Ni}) = 59.93078$ u, $M(^{63}_{29}\text{Cu}) = 62.92959$ u, $M(^{64}_{30}\text{Zn}) = 63.92914$ u, $M(^{4}_{2}\text{He}) = 4.00260$ u, $M(^{1}_{1}\text{H}) = 1.007825$ u

Solution:

$$^{4}_{2}\text{He} + ^{60}_{28}\text{Ni} \rightarrow ^{64}_{30}\text{Zn}^*$$

(a) From equation (5.12), the excitation energy of CN ($^{64}_{30}\text{Zn}^*$) is given by

$$E_{\text{CN}} = Q + \frac{M_A}{(M_A + M_a)} E_{\text{He}}$$

$$= \{(59.93078 + 4.00260 - 63.92914) \times 931.5\} + \left\{\frac{59.93078}{(59.93078 + 4.00260)}\right\} \times 14$$

$= 17.08$ MeV

(b) The other mode of formation of the same CN is

$$^{1}_{1}\text{H} + ^{63}_{29}\text{Cu} \rightarrow ^{64}_{30}\text{Zn}^*$$

Now,

$$E_{\text{CN}} = \{(62.92959 + 1.007825 - 63.92914) \times 931.5\} + \left\{\frac{62.92959}{(62.92959 + 1.007825)}\right\} E_{\text{H}}$$

We have to find out the energy of the incident proton so as to get the same excitation energy of CN. i.e.,

$$17.08\ \text{MeV} = \left\{\frac{62.92959}{(62.92959 + 1.007825)}\right\} E_H + 7.708\ \text{MeV}$$

or

$$E_H = 9.522\ \text{MeV}$$

The kinetic energy of the incident proton is thus, equal to 9.522 MeV.

Note: The shift between the kinetic energies of the proton and the α-particle must be $(14.0 - 9.52) \approx 4.5$ MeV. In Figure 5.17, the shift between the proton and the α-particle energy scales was chosen as 7 MeV to achieve the best match of the two sets of curves (at that time, the mass excesses were not precisely known). If the shift between the scale is taken as 4.5 MeV, we will not get the best fit, however, the curves will still match within the error bars.

5.8 CROSS-SECTIONS FOR CN REACTIONS AVERAGED OVER MANY RESONANCES

If the incident particle is of low energy (less than 50 MeV), one of the reactions that can occur is the capturing of the incident particle in the nucleus to form a compound nucleus. The compound nucleus thus formed, is in a high state of excitation energy, which is equal to the binding energy of the captured particle to the nucleus plus the kinetic energy of the captured particle in the centre of mass system. Many nucleons are excited with energy levels well above the ground state. The energy levels become so closely spaced that it is no longer possible to recognize the discrete energy levels and the nucleus no longer exists in a single, well defined quantum state. Even though the energy and the angular momentum should remain constant, the nucleus itself will oscillate very rapidly between the different possible states. In such a scenario, we may be able to treat the energy levels statistically and the properties of the excited nucleus can be, therefore, described by means of the level density parameter (a) and the nuclear temperature (T).

Immediately after the formation of highly excited compound nucleus, it will begin to decay. In most cases, CN is energetically capable of emitting one of more particles such as neutrons, protons, deuterons and helium ions apart from the energy which is emitted in the form of γ-radiation. The emission of the γ-radiation is, however, a slow process in comparison to the emission of the particles; hence, whenever the nuclear excitation energy exceeds the binding energy of a particle, the emission of a particle will nearly always take place in preference to the emission of the γ-radiation. Since, there is no Coulomb barrier, the emission of the neutrons is easier than the emission of the charged particles, particularly from the nuclei of high atomic number. In order for a nucleon to escape, the quantity of energy at least equal to its binding energy should by chance become concentrated on that nucleon. The difficulty of concentrating a large amount of energy on a single nucleon suggests that the evaporated nucleons will be emitted with rather low kinetic energies, which is experimentally the case.

For neutrons, the energy spectra is approximately of the form

$$P(E) = Ee^{-E/T} \tag{5.18}$$

Here, $P(E)$ is the probability of the emission of a neutron of energy E and T is the nuclear temperature expressed in MeV. The shape of the spectrum for a typical temperature, $T = 2$ MeV, is shown in Figure 5.18.

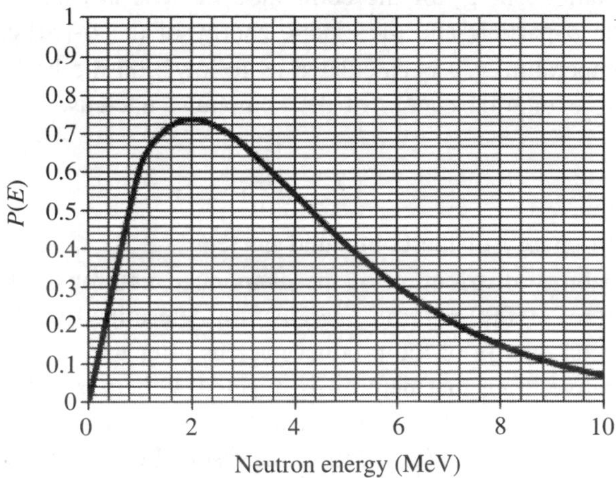

FIGURE 5.18 Energy spectrum of evaporated neutrons for nuclear temperature of 2 MeV.

EXERCISE 5.12: *Monte Carlo Method:* Since, the calculation of the cross-section ratios will involve the kinetic energies of the evaporated neutrons, they need to be determined in some manner. As already mentioned, the energy levels can be treated statistically. The domain of the possible inputs has already been defined with the help of the probability curve shown in Figure 5.18. Now, a set of random values will have to be generated, such that they follow the probability distribution. In order to do this, we make use of the Monte Carlo Method.

Here, first we choose a random number between 0 and 10, because it is understood that the range of energies for neutrons mostly lie between 0 and 10 MeV. Thereafter, a second random number between 0 and 1 is chosen, corresponding to the probability. If this number is found to be less than the probability for the emission of the neutron whose energy is taken as the first random number, then we reject this neutron. Thus, the neutrons are weighed according to their probability of evaporation. This process is repeated until a set of values is generated. Needless to say, with increasing number of values, the accuracy of the method should also increase, thus making the data more reliable.

Now, using Monte Carlo Method, calculate the ratio of the (α, 2n) to (α, 3n) cross-section for a $^{209}_{83}$Bi target if the incident α-particle is of low energy, and then compare it with the following experimental data[1]. Take the level density parameter, $a = A/22$.

1. Library of Experimental Nuclear Reaction Data, 1999, IAEA-NDS-CD-05 (Austria).

E_α (MeV)	29.8	30.4	31.0	33.0	34.2	35.7	37.3	38.6	40.1	41.0
$\dfrac{\sigma(\alpha, 2n)}{\sigma(\alpha, 3n)}$	79	25.14	11.209	1.80	0.69	0.38	0.20	0.14	0.11	0.094

Solution: The excitation energy of the compound nucleus is calculated by summing the binding energy of the α-particle and the kinetic energy of the α-particle in the CM frame.

The nuclear temperature (T) is calculated as follows, on the basis of the level density parameter (a) and the excitation energy of the compound nucleus.

$$T = \sqrt{\frac{E_x}{a}}$$

where E_x = excitation energy of the compound nucleus.

Using the probability distribution of the evaporated neutrons, which is dependent on the nuclear temperature, 10,000 neutron energies are determined by Monte Carlo Method. If the residual excitation energy of the product nucleus after the evaporation of one neutron comes out to be negative, then that neutron is rejected. Otherwise, it is compared with the binding energy of the second evaporating neutron. If it is found to be less than the binding energy, then the nucleus is labelled (α, n). Else, the residual excitation energy of the resultant nucleus after two neutrons have been emitted is compared with the binding energy of the third evaporating neutron. Depending on it being less than the binding energy of the neutron or not, the nucleus is labelled (α, $2n$) and (α, $3n$), respectively. The cross-section ratio $\{\sigma(\alpha, 2n)/\sigma(\alpha, 3n)\}$ is just the ratio of the number of (α, $2n$) cases to the number of (α, $3n$) cases.

The following Matlab code is used to arrive at the analytical solution.

```
%Monte Carlo Method for Calculation of Crosssection Ratio
clear all

for k=1:10
meN = 8.071; %Mass excess of neutron MeV
me4He = 2.424; %Mass excess of 4-He MeV
meA = -6.460; %Mass excess of 213-At MeV
meA2 = -8.641; %Mass excess of 212-At MeV
meA3 = -11.679; %Mass excess of 211-At MeV
meA4 = -12.075; %Mass excess of 210-At MeV
meB = -18.262; %Mass excess of 209-Bi MeV

BEalpha = me4He + meB - meA; %Binding energy of alpha particle, MeV

%Kinetic Energy of alpha particle in lab frame, MeV
KEalpha_lab = [29.8 30.4 31.0 33.0 34.2 35.7 37.3 38.6 40.1 41.0];

Mb = 209; %Mass number of 209-Bi
Ma = 213; %Mass number of 213-At
```

```
%Kinetic Energy of alpha particle in C.M. frame, MeV
KEalpha_CM = KEalpha_lab(k)*Mb/Ma;

Ex = BEalpha + KEalpha_CM; %Excitation energy of compound nucleus, MeV
a = Ma/22; %Level density parameter MeV^-1
T = (Ex/a)^0.5; %Nuclear temperature MeV

BE1 = (meN + meA2 - meA); %B.E. of 1st neutron evaporating, MeV
BE2 = (meN + meA3 - meA2); %B.E. of 2nd neutron evaporating, MeV
BE3 = (meN + meA4 - meA3); %B.E. of 3rd neutron evaporating, MeV

%Monte Carlo method
for i=1:10000

    n1(i) = 0; %Counter for evaporation of 1st neutron only
    n2(i) = 0; %Counter for evaporation of 1st and 2nd neutron
    n3(i) = 0; %Counter for evaporation of 1st, 2nd and 3rd neutron

    %For 1st neutron
    e = 10*rand; %Random number between 0 and 10
    p = e*exp(-e/T); %Probability of evaporation
    r = rand; %Random number between 0 and 1

        while p<r
        e = 10*rand;
        r = rand;
        p = e*exp(-e/T);
        end

    %For 2nd neutron
    e2 = 10*rand; %Random number between 0 and 10
    p2 = e2*exp(-e2/T); %Probability of evaporation
    r2 = rand; %Random number between 0 and 1

        while p2<r2
        e2 = 10*rand;
        r2 = rand;
        p2 = e2*exp(-e2/T);
        end

    P(i) = p;
    E(i) = e; %1st neutron kinetic energy
    E2(i) = e2; %2nd neutron kinetic energy

    %Excitation energy for 212-At
    Ex_2(i) = Ex - (BE1 + E(i));

    %Excitation energy for 211-At
    Ex_3(i) = Ex_2(i) - (BE2 + E2(i));
```

```
        if Ex_2(i)>=0
            if Ex_2(i)<BE2
                n1(i)=1;
            elseif Ex_3(i)<BE3
                n2(i)=1;
            else
                n3(i)=1;
            end
        end
    end

    %To calculate the ratio of crosssection
    N2 = 0; N3 = 0;
    for j=1:10000
        N2 = N2 + n2(j);
        N3 = N3 + n3(j);
    end
    Crosssection_Ratio23(k) = N2/N3; %Ratio of crosssection
    end

    Crosssection_Ratio23
```

The results are listed in Table 5.4. A reasonably good agreement between the theoretical and the experimental values is found.

Table 5.4

E_α (MeV)	29.8	30.4	31.0	33.0	34.2	35.7	37.3	38.6	40.1	41.0
Experimental Values	79	25.14	11.209	1.80	0.69	0.38	0.20	0.14	0.11	0.094
Theoretical Values	78.41	21.83	9.78	2.03	1.08	0.53	0.27	0.16	0.08	0.05

5.9 NUCLEAR FISSION

Soon after the discovery of the neutron, the possibility of new types of nuclear reactions became apparent. Fermi et al., irradiated the heavy elements, notably uranium, with slow neutrons and analyzed the products by radiochemical methods. Hahn and Strassmann carried out the detailed work of identifying the activities of products by radiochemical techniques and finally concluded that the radioactive products were isotopes of much lighter elements formed by the splitting of the ^{239}U nucleus into two parts of comparable size. Frisch and Meitner in 1939 used the word 'fission' to describe the process in which the nucleus undergoes division to form two products of comparable mass, which fly off with high velocity due to mutual Coulomb repulsion. The total kinetic energy was estimated to be $\approx$ 200 MeV. This was an entirely new type of reaction and top nuclear scientists from all over the world got involved in research to exploit this reaction for large energy production either to supply abundant power for all the people or to destroy civilization through the use of nuclear weapons.

By 1940, the following facts were established:

(i) Natural uranium (0.7% ^{235}U and 99.3% ^{238}U) undergoes fission reaction either by slow neutrons or by fast neutrons, but ^{238}U always required fast neutrons. Eventually, it was ascertained that ^{235}U was fissile to slow neutrons.

(ii) The elements thorium and protactinium could also undergo fission with fast neutrons.

(iii) In all the cases, very large disintegration energies were released, equal to about ten times the order of energies previously experienced.

(iv) Fission fragments were all radioactive and decayed to stable nuclides by a series of β^- emissions.

(v) The atomic weights of the fission products ranged from about 70 to 160, though, needless to say, one parent uranium nucleus could only produce two fragments. These were eventually identified by radioactive methods.

Later research showed that the fission could also be initiated by energetic deuterons and α-particles from the accelerators and even 'photofission' was possible using incident γ-rays[9]. A heavy nucleus (with $A > 220$) fissions by itself. This type of fission is known as spontaneous fission. We will focus only on the neutron-induced fission in the subsequent sections.

5.9.1 Mechanism of the Fission Process

In nuclear fission, a neutron interacts with the target nucleus leading to the formation of an unstable compound nucleus that splits into smaller nuclei, releasing two or more neutrons and energy. The CN thus, temporality contains all the charge and mass involved in the reaction and exists in the excited state. Various stages of the fission process are depicted in Figure 5.19, for neutron interaction with ^{235}U.

In stages (g) and (h), the fragments are in excited state and the excess neutrons and γ-rays are emitted. These neutrons and γ-rays are known as prompt neutrons and prompt γ-rays. In stage (i), the fission fragments decay by a successive emission of β-particles to attain stability with stable end products. In the process, several γ-rays, called the delayed γ-rays, are also emitted. However, we may have a situation where β^- decay is forbidden due to the selection rules and fragments have no option but to become stable after emitting a neutron. The neutrons so emitted are known as the delayed neutrons. Direct neutron measurements reveal that on an average 2.5 neutrons are emitted in fission. The *n*/*p* ratio for $^{235}_{92}$U is 1.6, while the ratio essential for stability is (1.2–1.4) in the nuclides produced in the fission. The fission fragments always have too large a value for the *n*/*p* ratio. Consequently, prompt neutrons are emitted. However, the number of neutrons emitted in the fission process is not sufficient to lower the *n*/*p* ratios to stable values. Further successive β-decay lowers the *n*/*p* ratio until the products are stable.

9. *Phys. Rev. Lett.*, 84 (2000), 5740–43.

The probabilities of the fission processes taking place vary widely as shown in Figure 5.20. It is clear from the figure that ${}^{235}_{92}\text{U}$ is fissionable by both slow and fast neutrons, but ${}^{238}_{92}\text{U}$ is fissionable by fast neutrons only.

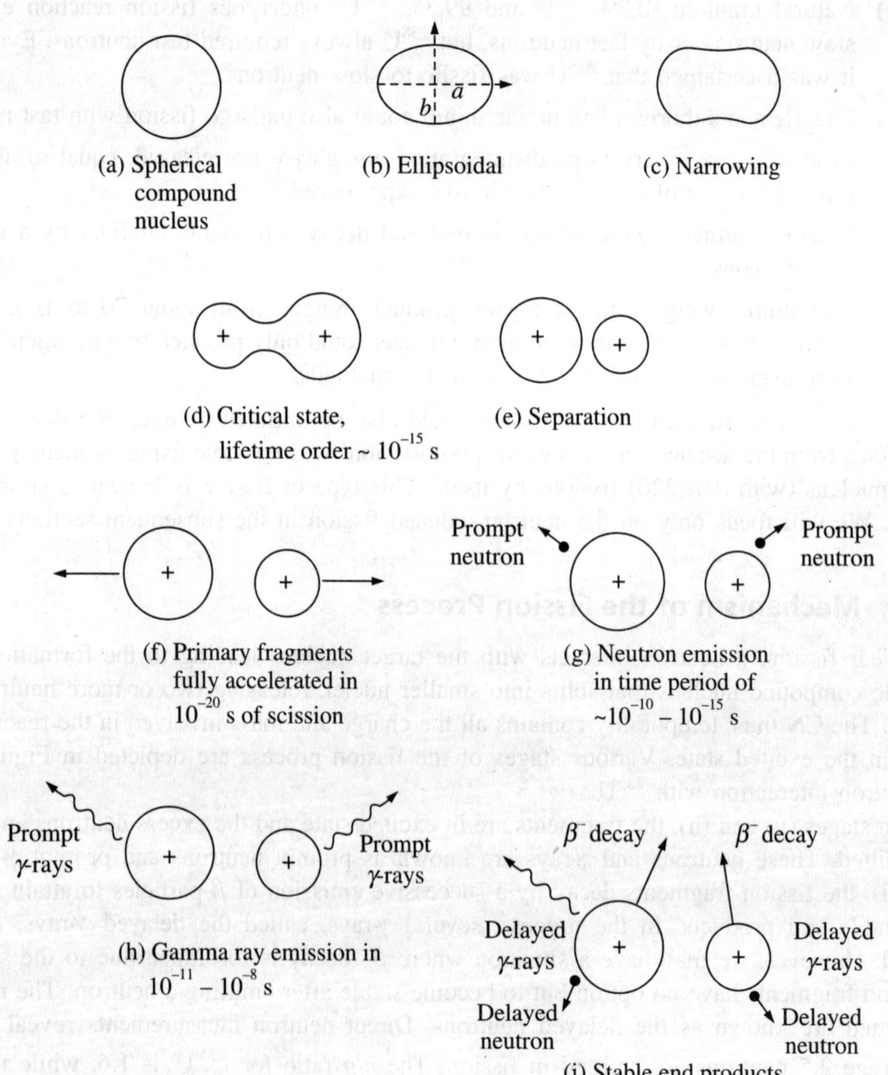

FIGURE 5.19 Schematics of a fission process.

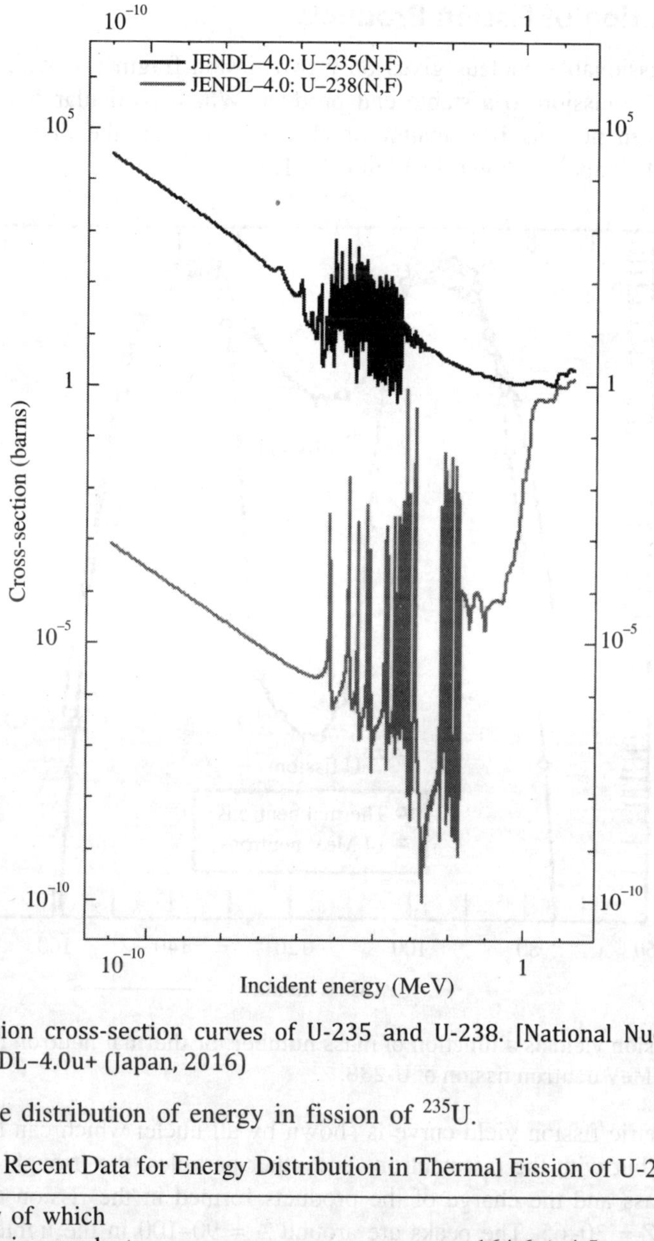

FIGURE 5.20 Fission cross-section curves of U-235 and U-238. [National Nuclear Data Center: JENDL–4.0u+ (Japan, 2016)

Table 5.5 gives the distribution of energy in fission of ^{235}U.

Table 5.5 Recent Data for Energy Distribution in Thermal Fission of U-235 in MeV

Prompt energy of which		
K.E. of the fission products	164.6 ± 4.5	
K.E. of 2.5 neutrons (prompt)	4.9 ± 0.5	176.5 ± 5.5
γ-energy (prompt)	7.0 ± 0.5	
Delayed energy from fission products decay of which		
K.E. of β-particles	6.5 ± 1.5	
neutrino radiation	10.5 ± 2	23.5 ± 5
γ-energy	6.5 ± 1.5	
Total		200 ± 6

5.9.2 Distribution of Fission Products

We know that a fissionable nucleus gives only two fission fragments, which thereafter decay successively by β^- emission to a stable end product. Which particular fragment nuclides are produced by a given nucleus is a matter of chance, however, the distribution curve has a peculiar saddleback shape as shown in Figure 5.21.

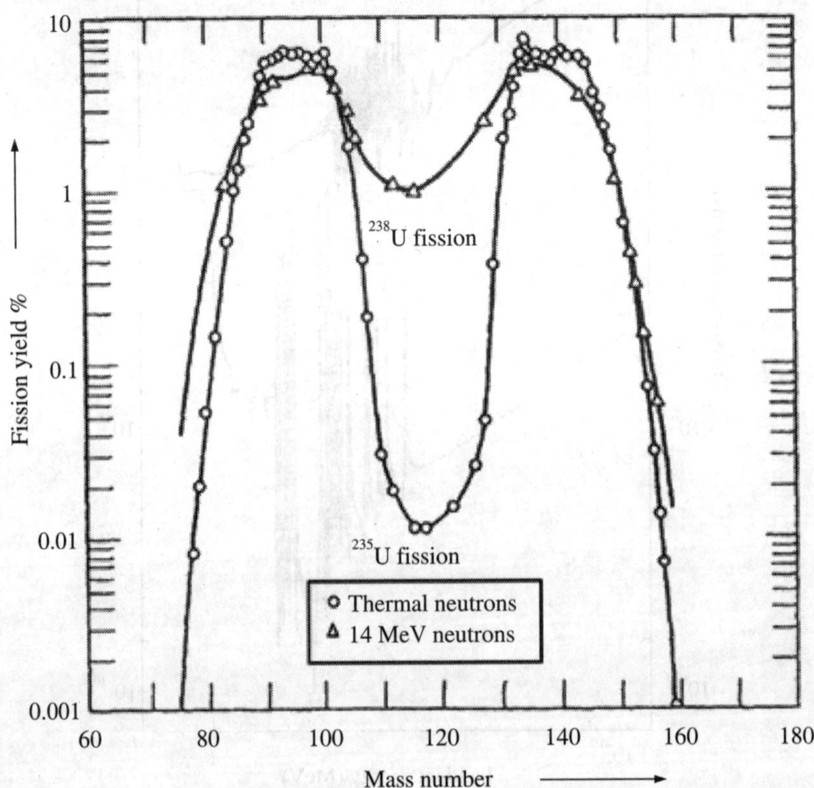

FIGURE 5.21 Fission yield as a function of mass number for thermal-neutron fission of U-235 and 14 MeV neutron fission of U-238.

The asymmetric fission yield curve is shown by all nuclei which can be fissioned by the thermal neutrons, but with the fast neutrons and other particles, the 'trough' in the curve tends to fill up. The mass and the charge of the products formed in the fission are in the range of A = 70–160 and Z = 30–65. The peaks are around A = 90–100 in the lighter mass region and around A = 134–144 in the heavier mass region, which is attributed to the influence of double magic configuration in the products. The energy distribution of the fission fragments can be obtained by assuming that the two fragments are ejected with equal and opposite momenta so that

$$m_1v_1 = m_2v_2$$

or
$$\frac{E_2}{E_1} = \frac{\frac{1}{2}m_2\left(\frac{m_1 v_1}{m_2}\right)^2}{\frac{1}{2}m_1 v_1^2}$$

$$= \frac{m_1}{m_2} \cong \frac{95}{140}$$

$\therefore$
$$\frac{E_2}{E_1} \approx \frac{2}{3}$$

Thus, the heavier fragments have lower kinetic energy with peak at ~70 MeV and the lighter fragments have higher kinetic energy with peak at ~100 MeV. Appreciable change in the kinetic energy has not been observed experimentally if the projectile energy is increased.

Bohr and Wheeler explained the fission process based on the liquid drop model and obtained a semi-quantitative expression for the neutron energy required to initiate fission in a given nucleus. The excitation energy added to the compound nucleus (e.g., $n + {}^{235}_{92}\text{U} \rightarrow {}^{236}_{92}\text{U}$) is equal to the sum of the binding energy of the incident neutron and its kinetic energy. This energy appears to initiate a series of rapid oscillations in the drop (CN), which at times assumes the ellipsoidal shape as shown in Figure 5.19. The restoring force (surface tension) arises from the short range intra-nuclear forces. If the oscillations become so violent that the critical state is reached, and as each of the two fragments is positively charged, the final splitting and separation is inevitable. Thus, there is a threshold energy or a critical energy required to produce the critical state (d) in Figure 5.19, after which the CN cannot return to the beginning state (a), because of the large Coulomb repulsion of the two parts.

The critical energy, which must be supplied to the neutron, is shown in Figure 5.22. In this figure, one can note how the energy E_{crit} must be added to the system to enable the energy of the nucleus to become greater than E_b (zero deformation ground state energy). Once the maximum barrier height has been overcome, the CN descends to the state of lowest potential energy and the fragments separate with the release of energy equal to

$$\Delta E = (\text{Mass of CN} - \text{Mass of fragments})\, c^2$$

The value of the critical deformation energy E_{crit} was calculated by Bohr and Wheeler on the basis of LDM. They obtained

$$E_{\text{crit}} = 0.89A^{2/3} - 0.02\frac{Z(Z-1)}{A^{1/3}} \text{ MeV} \tag{5.19}$$

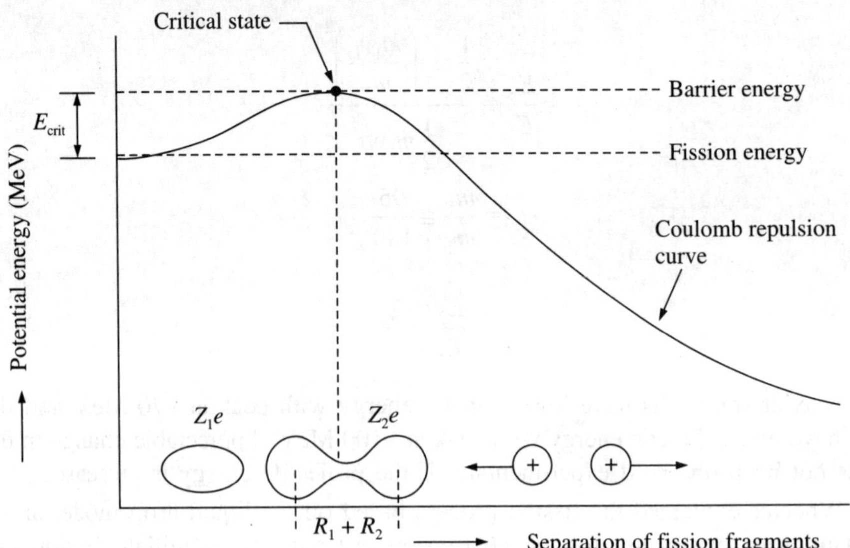

FIGURE 5.22 Potential energy curve for fission.

EXERCISE 5.13: What is the minimum *A* for fission to be energetically possible? [No external energy is supplied]

Solution: We have already solved this problem in Chapter 2, Exercise 2.3(c), when we were discussing SEMF. By considering a binary and symmetric fission, we had arrived at the following condition for Q = positive

$$\frac{A}{Z^2} < 0.06$$

or

$$\frac{Z^2}{A} > 16.67 \approx 17 \tag{5.20}$$

Z^2/A is known as the *fissionability parameter*, because the liquid drop model predicts that the probability of the fission should increase with the increase in Z^2/A.

When we put *A* and *Z* values for different elements from the periodic table in the above equation, it turns out that Q = positive for about $A \geq 100$, which means SF (spontaneous fission) is possible.

Nuclei for which $Z^2/A > 17$, are meta-stable with respect to fission. This condition is satisfied by β-stable nuclei heavier than $^{98}_{42}$Mo. However, even nuclei well beyond this limit do not fall apart but have very long lifetimes against SF. A plot of half-lives against (Z^2/A) is shown in Figure 5.23, for some heavier isotopes.

The process of fission is strongly inhibited by the tunneling factor, and SF is only observed in the heaviest of the elements. At large separations, the Coulomb potential between the charges of the two nuclei determines the barrier shape. At small distortions, however, there is a competition between the surface tension, which pulls the nucleus back to a more spherical shape and the Coulomb repulsion, which pushes them apart.

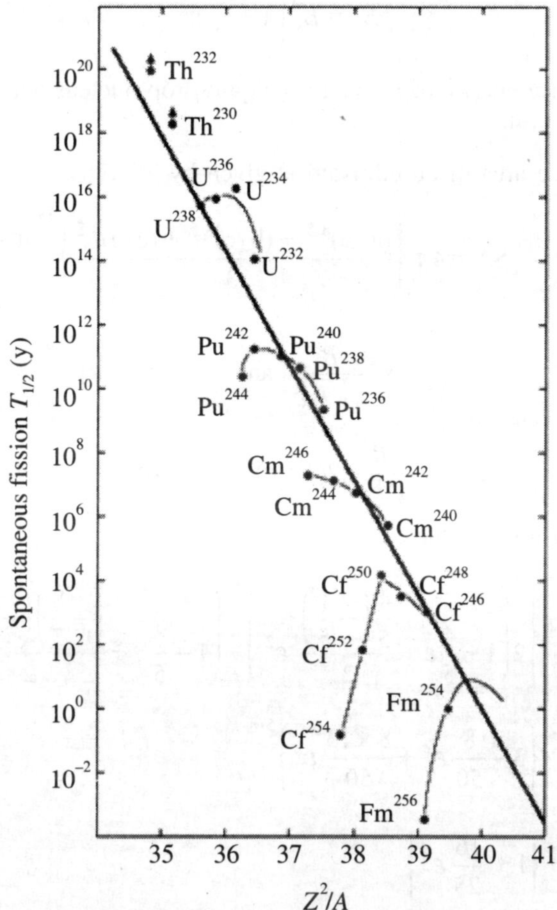

FIGURE 5.23 Spontaneous fission life-times.

In Figure 5.19, which gives a schematic representation of the fission process, the nucleus is treated as a liquid drop. It is reasonable to consider that the spherical nucleus of radius R deforms into an ellipsoid (stage b). If we introduce a deformation parameter ε such that

$$a = R(1 + \varepsilon)$$

where a is the major semi-axis of the ellipsoid.

If the volume of the drop (nucleus) remains the same, then

$$\frac{4}{3}\pi R^3 = \frac{4}{3}\pi ab^2$$

where $b = R/(1 + \varepsilon)^{1/2}$

EXERCISE 5.14: For small deformation, prove that the surface energy of a deformed nucleus

$$E_s = E_s^0\left(1+\frac{2}{5}\varepsilon^2\right)$$

where E_s^0 is the surface energy of a spherical liquid drop nucleus and ε is the deformation parameter.

Solution: The surface area of an ellipsoid is given by

$$\text{SA} = 4\pi\left\{\frac{(\text{a}\cdot\text{b})^{8/5} + (\text{b}\cdot\text{c})^{8/5} + (\text{c}\cdot\text{a})^{8/5}}{3}\right\}^{5/8}$$

In our case,

$$b = c = \frac{R}{\sqrt{1+\varepsilon}} \quad \text{and} \quad a = R(1+\varepsilon)$$

or

$$\text{SA} = \frac{4\pi R^2}{3^{5/8}}\{2(1+\varepsilon)^{4/5} + (1+\varepsilon)^{-8/5}\}^{5/8}$$

Using binomial expansion, we have

$$\text{SA} = \frac{4\pi R^2}{3^{5/8}}\left\{2\left(1+\frac{4}{5}\varepsilon+\frac{\frac{4}{5}\left(-\frac{1}{5}\right)}{2}\varepsilon^2\right)+\left(1-\frac{8}{5}\varepsilon+\frac{\left(-\frac{8}{5}\right)\left(-\frac{13}{5}\right)}{2}\varepsilon^2\right)\right\}^{5/8}$$

$$= \frac{4\pi R^2}{3^{5/8}}\left\{3-\frac{8}{50}\varepsilon^2+\frac{8\times 13}{50}\varepsilon^2\right\}^{5/8}$$

$$= 4\pi R^2\left\{1+\frac{16}{25}\varepsilon^2\right\}^{5/8}$$

Again using binomial expansion, we have

$$\text{SA} = 4\pi R^2\left\{1+\frac{5}{8}.\frac{16}{25}\varepsilon^2\right\}$$

$$= 4\pi R^2\left(1+\frac{2}{5}\varepsilon^2\right)$$

Thus, the modified surface energy of a deformed nucleus, which is a function of its surface area is

$$E_s = E_s^0\left(1+\frac{2}{5}\varepsilon^2\right) \tag{5.21}$$

where E_s^0 is the surface energy of a spherical liquid drop nucleus.

Thus, we can see that the surface energy increases with deformation. On the other hand, the Coulomb energy decreases on deformation. A fairly lengthy calculation gives the following result for the Coulomb energy for a uniformly charged ellipsoid,

$$E_c = \frac{3}{5}\frac{(Z_e)^2}{4\pi\varepsilon_0 R}\left(1-\frac{\varepsilon^2}{5}\right)$$

or

$$E_c = E_c^0\left(1-\frac{\varepsilon^2}{5}\right) \tag{5.22}$$

where E_c^0 is the Coulomb energy of a spherical liquid drop nucleus of charge Z and mass A. The change in E_s and E_c are given by

$$\Delta S = E_s - E_s^0 = \frac{2}{5}\varepsilon^2 E_s^0 \tag{5.23}$$

$$\Delta C = E_c - E_c^0 = -\frac{1}{5}\varepsilon^2 E_c^0 \tag{5.24}$$

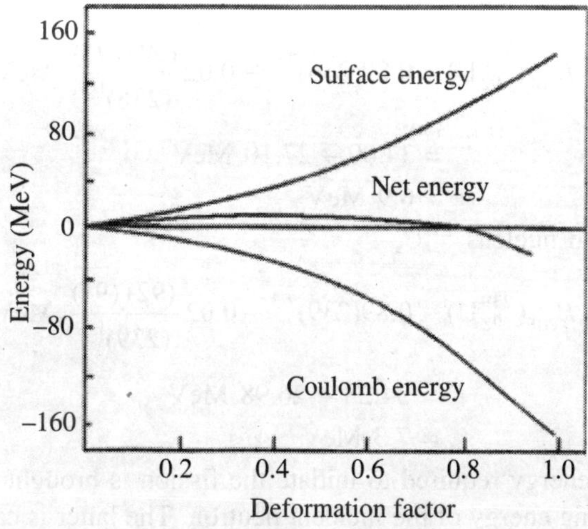

FIGURE 5.24 Surface, Coulombic and net deformation energies vs. deformation factor ε.

These changes reflect in the potential energy of the nucleus. With increase in deformation, the potential energy of the nucleus increases due to ΔS and decreases due to ΔC. The potential energy variation as a function of deformation is shown in Figure 5.24. The transition point of balance of magnitudes of ΔS and ΔC corresponds to the maximum potential energy, known as the saddle point or the critical state, which the nucleus must pass to undergo fission.

From equations (5.23) and (5.24), it is expected that the nucleus undergoes fission when

$$|\Delta S| = 2|\Delta C|$$

Fissibility parameter (χ) is a measure of the tendency of a nucleus to undergo fission,

$$\chi = \frac{E_c^0}{2E_s^0} = \frac{Z^2}{50.13\,A} \tag{5.25}$$

If $\chi < 1$, the nucleus is stable.
If $\chi > 1$, there will be no potential barrier to inhibit SF.

Thus, equation (5.25) suggests that there is an absolute upper limit of $Z \cong 144$ for the chemical elements and the periodic table cannot be extended infinitely.

EXERCISE 5.15: (a) On the basis of the liquid drop model, explain why ${}^{235}_{92}\text{U}$ is fissionable by thermal neutrons while ${}^{238}_{92}\text{U}$ is fissionable only by fast neutrons.

(b) On the basis of SEMF, show that the energy released in spontaneous fission is at maximum, when the two fragments are of equal mass. For simplicity, assume that in splitting up the nucleus, the protons and the neutrons divide in the same ratio.

Solution: Using equation (5.19), we have

(a) For compound nucleus ${}^{236}\text{U}$,

$$E_{\text{crit}}({}^{236}_{92}\text{U}) = 0.89(236)^{2/3} - 0.02\frac{(92)(91)}{(236)^{1/3}} \text{ MeV}$$

$$= 34.00 - 27.10 \text{ MeV}$$

$$= 6.9 \text{ MeV}$$

For compound nucleus ${}^{239}\text{U}$,

$$E_{\text{crit}}({}^{239}_{92}\text{U}) = 0.89(239)^{2/3} - 0.02\frac{(92)(91)}{(239)^{1/3}} \text{ MeV}$$

$$= 34.28 - 26.98 \text{ MeV}$$

$$= 7.3 \text{ MeV}$$

This critical energy required to initiate the fission is brought by the kinetic energy and the binding energy of the incident neutron. The latter is calculated from SEMF. In the case of ${}^{235}\text{U}$ fission, $[B({}^{236}_{92}\text{U}) - B({}^{235}_{92}\text{U})]$ and in the case of ${}^{238}\text{U}$ fission, $[B({}^{239}_{92}\text{U}) - B({}^{238}_{92}\text{U})]$. For ${}^{235}\text{U}$ fission, the binding energy of the added neutron is 6.8 MeV, very close to 6.9 MeV required, so that this particular isotope of uranium is fissionable with low (0.025 eV) energy neutrons. However, for ${}^{238}\text{U}$ fission, the binding energy of the added neutron is 5.9 MeV which is 1.4 MeV less than the required critical energy and therefore, the ${}^{238}\text{U}$ nucleus is fissionable by neutrons of energy greater than 1.4 MeV. The experimental value of this threshold energy is found to be 1.1 MeV. The difference of 0.3 MeV is attributed to the choice of constants in SEMF.

Note: Nuclei having odd N ($^{235}_{92}U_{143}$, $^{239}_{94}Pu_{145}$, etc.) are more fissile compared to those having even N ($^{238}_{92}U_{146}$, $^{240}_{94}Pu_{146}$, etc.). This can be explained by considering the excitation energy of the compound nuclei and the fission barrier associated with them; e.g., the absorbed neutron pairs with the unpaired neutron of ^{235}U and the resulting even-even nucleus is more tightly bound in the ground state than the emitting even-odd nucleus.

(b) We are assuming that the protons and the neutrons divide in the same ratio; i.e., if A and Z are the mass number and the atomic number of the nucleus that is fissioning then,

$$\frac{A_1}{A} = \frac{Z_1}{Z} = x \quad \text{and} \quad \frac{A_2}{A} = \frac{Z_2}{Z} = y$$

such that $x + y = 1$.
The expression for Q in terms of the binding energy is

$$Q = -B(A, Z) + B(A_1, Z_1) + B(A_2, Z_2)$$

Therefore, from SEMF and ignoring the pairing energy term, we get

$$Q = a_s\{A^{2/3} - A_1^{2/3} - A_2^{2/3}\} + a_c\left\{\frac{Z(Z-1)}{A^{1/3}} - \frac{Z_1(Z_1-1)}{A_1^{1/3}} - \frac{Z_2(Z_2-1)}{A_2^{1/3}}\right\}$$

Substituting A_1, Z_1 and A_2, Z_2 in terms of x and y, and using the approximation, $Z(Z - 1) \to Z^2$, we have

$$Q = a_s A^{2/3}(1 - x^{2/3} - y^{2/3}) + a_c \frac{Z^2}{A^{1/3}}(1 - x^{5/3} - y^{5/3})$$

We also know that $x + y = 1$ or $y = 1 - x$. Therefore,

$$Q = a_s A^{2/3}\{1 - x^{2/3} - (1-x)^{2/3}\} + a_c \frac{Z^2}{A^{1/3}}\{1 - x^{5/3} - (1-x)^{5/3}\}$$

Now for $Q(x)$ to be maximum, $dQ/dx = 0$, i.e.,

$$\frac{2}{3} a_s A^{2/3}\{-x^{-1/3} + (1-x)^{-1/3}\} + \frac{5}{3} a_c \frac{Z^2}{A^{1/3}}\{-x^{2/3} + (1-x)^{2/3}\} = 0$$

Clearly, we can see that the above expression is zero when $x = 1/2$. Now, to confirm that this indeed is the point of maxima and not minima, we see that the slope dQ/dx should be positive as we approach $x = 1/2$ from the left. If it is a point of minima, then the slope should be negative as we approach $x = 1/2$ from the left.

We check at point $x = 0.49$, and we see that dQ/dx turns out to be positive. Hence, we can safely say that $x = 1/2$ is the point of maxima. Thus, the energy released in a spontaneous fission is at maximum when the two fragments are of equal mass.

5.10 NUCLEAR FUSION

Power from nuclear fission is now a reality globally. On a long-term basis, an infinite supply of uranium and plutonium is required. As the uranium bearing minerals are diminished in time, we shall have to depend on the breeder reactor for utilizing plutonium.

An alternative to the fission reaction as a source of energy is the fusion reaction. We have seen that the plot of $\bar{B}$ vs. A has a maximum at $A \approx 56$ (Fe, Ni) and slowly decreases for heavier nuclei. For lighter nuclei, $\bar{B}$ drops quickly, so that with the exception of magic nuclei, lighter nuclei are less tightly bound than medium sized nuclei. Thus, in principle, if lighter nuclei are forced together, they will fuse to produce a more tightly bound nucleus and the energy will be released in the form of kinetic energy of the products and their excitation energy. This process is called nuclear fusion. Research into controlled fusion, with the aim of producing electricity, is being conducted for over six decades now.

In 1939, H.A. Bethe (USA) suggested that the production of energy in the sun and the stars is by thermonuclear reactions in which the protons are continuously transformed into helium nuclei. For comparatively lower stellar temperatures, he proposed the following cycle:

$$ {}_1^1\text{H} + {}_1^1\text{H} \rightarrow {}_1^2\text{D} + \beta^+ + \nu $$

$$ {}_1^2\text{D} + {}_1^1\text{H} \rightarrow {}_2^3\text{He} $$

$$ {}_2^3\text{He} + {}_1^1\text{H} \rightarrow {}_2^4\text{He} + \beta^+ + \nu $$

Hence, by adding these reactions together, we get

$$ 4\,{}_1^1\text{H} \rightarrow {}_2^4\text{He} + 2\beta^+ + 2\nu $$

The total energy released is about 27 MeV.

The energy released in the fusion of light elements is due to the interplay of the two opposing forces, the nuclear force which combines together the protons and the neutrons, and the Coulomb force, which causes the protons to repel each other. The practical problem for fusion to occur is the Coulomb repulsive barrier, which inhibits two nuclei from getting close enough within the range of nuclear force, so that they may fuse. This Coulomb barrier is given by

$$ V_c = \frac{Z_1 Z_2 e^2}{4\pi\varepsilon_0 (R_1 + R_2)} \tag{5.26} $$

where Z_1 and Z_2 are the atomic numbers of the two nuclei and R_1 and R_2 are their effective radii given by $r_0 A^{1/3}$, where $r_0 = 1.2$ fm.

If $A_1 = A_2 = 2Z_1 = 2Z_2$, we get $V_c \approx 0.15A^{5/3}$ MeV. If $A = 2$, $V_c \approx 0.5$ MeV and this energy needs to be supplied to overcome the Coulomb barrier. This is relatively small amount of energy to supply, and one would expect it to be achieved by simply colliding two accelerated beams of light nuclei. But in practice, nearly all the particles would be elastically scattered. Another practical way is to heat a confined mixture of the nuclei to supply enough thermal energy to overcome the Coulomb barrier. This is known as *thermonuclear fusion.*

The temperature required may be estimated from the relation

$$ E = k_B\, T $$

where k_B is the Boltzmann constant. For an energy of ~0.5 MeV, T comes out to be $\approx 5 \times 10^9$ K. This is above the typical temperature ~10^8 K found inside the stars. Fusion actually occurs at a lower temperature than this estimate due to a combination of two reasons.

(i) Phenomenon of quantum tunnelling, which means that the full height of the Coulomb barrier does not have to be overcome. The probability of barrier penetration depends on the Gamow factor $\sim \exp\left[-\sqrt{E_G/E}\right]$, where $E_G = 2mc^2(\pi\alpha Z_1 Z_2)^2$. Here m is the reduced mass of the two fusing nuclei of electric charges $Z_1 e$ and $Z_2 e$, and α is the fine structure constant. Thus, the probability of barrier penetration increases as E increases. Nonetheless, the probability of fusion is still very small. For example, if two protons are to be fused, say at 10^7 K, we find that $E_G = 490$ keV and $E \approx 1$ keV, and the probability of fusion is proportional to $\exp\left[-\sqrt{E_G/E}\right] = \exp[-22] \approx 10^{-96}$, which certainly is a huge suppression factor. Therefore, the actual fusion rate is still extremely slow.

(ii) Another reason that the fusion occurs at a lower temperature than expected is that a collection of nuclei at a given temperature will have a Maxwellian distribution of energies about the mean, so there will be some nuclei with energies substantially higher than the mean energy. Nevertheless, even the temperature inside a star of ~10^8 K corresponds to an energy ~10 keV. So, the fraction of nuclei with energies of the order of ~1 MeV in such a star would only be of the order $\exp[-E/k_B T] = \exp[-100] \approx 10^{-43}$, a minute amount.

We now examine the interplay of these two factors, i.e., increasing barrier penetration factor with energy and the Maxwellian decreasing exponential function. To understand this, consider the fusion of two nuclei x and y having number densities n_x and n_y (i.e., the number of particles per unit volume) and at a temperature T high enough so that all atoms are ionized and these ions and free electrons are moving about very rapidly. The mixture is still electrically neutral, of course, and the whole state is called the *plasma state*. We assume this plasma with uniform values of number densities and temperature. We also assume that the velocities of the two nuclei are given by the Maxwell-Boltzmann distribution, so that the probability of having two nuclei with relative speed v in the range v to $v + dv$ is

$$P(v)\,dv = \left(\frac{2}{\pi}\right)^{1/2}\left(\frac{m}{k_B T}\right)^{3/2} \exp\left[-\frac{mv^2}{2k_B T}\right] v^2\,dv \tag{5.27}$$

where m is the reduced mass of the pair.

An important parameter for the actual realization of fusion energy is the reaction rate. The fusion reaction rate per unit volume is given by

$$R_{xy} = n_x n_y \langle \sigma_{xy} v \rangle \tag{5.28}$$

where n_x and n_y are the concentrations of x and y particles, and $\langle \sigma_{xy} v \rangle$ is the average value of the fusion cross-section and velocity produced. This average is also called the *reactivity*.

$$\langle \sigma_{xy} v \rangle = \int_0^\infty \sigma_{xy} v P(v)\,dv \tag{5.29}$$

For fusion reaction, one must take into account the repulsive Coulomb barrier between the positively charged nuclei. The fusion cross-section may be approximated as follows:

$$\sigma_{xy}(E) = \frac{S(E)}{E} \exp\left[-\sqrt{\frac{E_G}{E}}\right] \tag{5.30}$$

where $S(E)$ is known as the astrophysical factor, and contains all the constants and terms related to the nuclei involved.

Using equations (5.27) and (5.30) in equation (5.28), we get

$$R_{xy} = n_x n_y \left(\frac{2}{\pi}\right)^{1/2} \left(\frac{m}{k_B T}\right)^{3/2} \int_0^\infty \frac{S(E)}{E} \exp(-\varphi)\, dE \tag{5.31}$$

where

$$\varphi(E) = \left\{\frac{E}{k_B T} + \sqrt{\frac{E_G}{E}}\right\}$$

Equation (5.31) represents the overlap between the Maxwell–Boltzmann distribution, which is peaked at low energies and the Gamow factor, which increases with increasing energy. The product of these two terms produces a peak in the overlap region of these two functions, known as *Gamow peak* (Figure 5.25). This peak occurs at an energy $E = E_0$, where

$$E_0 = \left[\frac{1}{4} E_G (k_B T)^2\right]^{1/3} \tag{5.32}$$

and the fusion takes place over a relatively narrow range of energies $E_0 \pm \Delta E_0$, where

$$\Delta E_0 \cong \frac{4}{3^{1/2} 2^{1/3}} E_G^{1/6} (k_B T)^{5/6} \tag{5.33}$$

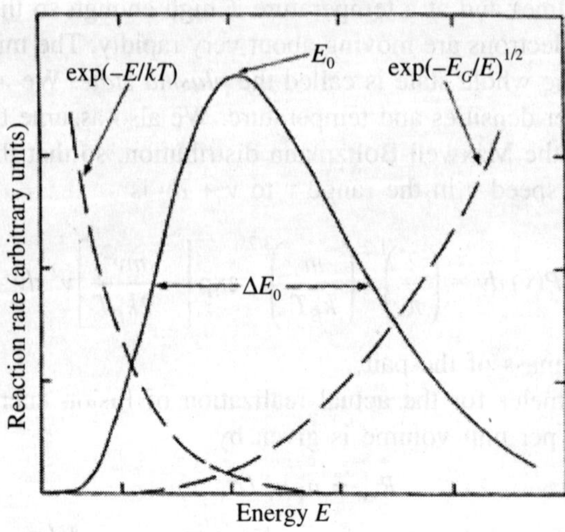

FIGURE 5.25 The right-hand curve is proportional to the barrier penetration factor and the left-hand curve is proportional to the Maxwell–Boltzmann distribution. The solid curve is the combined effect and is proportional to the overall probability of fusion with a peak at E_0 and a width of ΔE_0.

EXERCISE 5.16:

(a) Prove that the Gamow peak occurs at

$$E_0 = \left[\frac{1}{4} E_G (k_B T)^2\right]^{1/3}$$

(b) Determine the Gamow peak and its width in case of *p-p* fusion at temperature $T \sim 2 \times 10^7$ K. Plot the function $\exp(-\varphi)$ for the above *p-p* fusion reaction and show the Gamow peak.

Solution:

(a) In equation (5.31), the function $\exp(-\varphi)$ is sharply peaked. It falls off rapidly at high energies because of the Boltzmann factor and at low energies because of the Gamow factor. The peak lies at $E = E_0$, where the function $\varphi(E)$ is minimum, i.e.,

$$\frac{d\varphi}{dE} = 0$$

or

$$\frac{1}{k_B T} - \frac{\sqrt{E_G}}{2E^{3/2}} = 0$$

or

$$E_0 = \left[\frac{1}{4} E_G (k_B T)^2\right]^{1/3}$$

(b)

$$E_G = 2mc^2 (\pi \alpha Z_1 Z_2)^2$$

$$= 2\left(\frac{938.3 \text{ MeV}}{2}\right)\left(\frac{\pi}{137}\right)^2 \cong 493 \text{ keV}$$

From equation (5.32), we get

$$E_0 = \left[\frac{1}{4} (493 \text{ keV})\, (8.62 \times 10^{-8} \text{ keV/K})^2 (2 \times 10^7 \text{K})^2\right]^{1/3}$$

$$\cong 7.2 \text{ keV}$$

From equation (5.33), we get

$$\Delta E_0 = \frac{4}{3^{1/2} 2^{1/3}} E_G^{1/6} (k_B T)^{5/6}$$

$$= \frac{4}{3^{1/2} 2^{1/3}} (493 \text{ keV})^{1/6} (8.62 \times 10^{-8} \text{ keV/K})^{5/6} (2 \times 10^7 \text{K})^{5/6} \cong 8.2 \text{ keV}$$

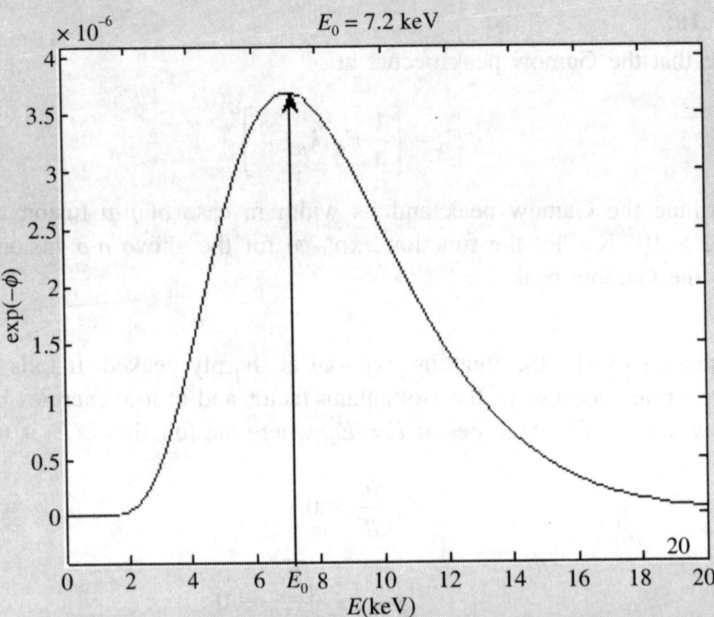

FIGURE 5.26 Gamow peak in case of *p-p* fusion at $T \sim 2 \times 10^7$ K.

The following Matlab code is used to get the above plot (Figure 5.26).

```
clear all
kT = 1.7; %keV Boltzmann constant * Temperature
Eg = 493; %keV

syms x real positive;
f=exp(-(x/kT + sqrt(Eg/x)));
ezplot(f,[0,20])
xlabel('E (keV)')
ylabel('exp(-phi)')
title('E0 = 7.2 keV')
annotation('textarrow',[0.41 0.41],[0.1,0.85],'String','E0');
```

5.10.1 Conditions for Fusion

Research into using controlled thermonuclear fusion for power production began way back in 1950 and it continues to this day. Various reaction candidates have been proposed for getting fusion energy. For instance,

$$^2_1\text{D} + {}^2_1\text{D} \rightarrow {}^3_1\text{T}\,(1.01\text{ MeV}) + p(3.02\text{ MeV})$$

$$^2_1\text{D} + {}^2_1\text{D} \rightarrow {}^3_2\text{He}(0.82\text{ MeV}) + n(2.45\text{ MeV})$$

$$^2_1D + ^3_1T \rightarrow ^4_2He\ (3.5\ MeV) + n(14.1\ MeV)$$

$$^2_1D + ^3_2He \rightarrow ^4_2He\ (3.6\ MeV) + p(14.7\ MeV)$$

To assess the usefulness of these reactions, apart from the knowledge of the reactants, the products and the energy released, one also needs to know about the cross-section. For any given fusion device, there is a maximum plasma pressure that it can sustain, and an economical device would always operate near this maximum value. At this pressure, the largest fusion output is obtained when the temperature is chosen such that $\langle \sigma_{xy}\ v\rangle/T^2$ is the maximum. At this temperature, ($nT\tau$), where n is the ion density and τ is the confinement time, the ignition is minimum and it is inversely proportional to $\langle \sigma_{xy}\ v\rangle/T^2$. Plasma is ignited if the fusion reactions produce enough power so as to maintain a steady temperature without the need for any external heating. The optimum temperature and the value of $\langle \sigma_{xy}\ v\rangle/T^2$ at that temperature for the reactions given above, is listed in Table 5.6.

Even with a high enough temperature to overcome the Coulomb barrier, a critical density of ions needs to be maintained to make the collision probability high enough to achieve a net yield of energy from the fusion reaction. Confinement time in nuclear fusion devices is defined as the time during which the plasma is maintained at a temperature above the critical ignition temperature. In order to generate more energy from the fusion than has been invested to heat the plasma, the plasma needs to be held up to this temperature for some minimum length of time. The ion density multiplied by the confinement time required for the fusion is called the Lawson's Criterion.

Table 5.6 Optimum Temperature and $\langle \sigma_{xy}\ v\rangle/T^2$ at that Temperature for Deuteron Reactions

Fuel	T (keV)	$\langle \sigma_{xy}\ v\rangle/T^2$
$^2_1D + ^2_1D$	15	1.28×10^{-26}
$^2_1D + ^3_1T$	13.6	1.24×10^{-24}
$^2_1D + ^3_2He$	58	2.24×10^{-26}

Lawson's Criterion

In 1957, J.D. Lawson demonstrated how the product of the ion density and the confinement time determined the minimum conditions for productive fusion. When the critical temperature for nuclear fusion has been achieved, it must be maintained at that temperature for a long enough confinement time τ and at a high enough ion density n to obtain a net yield of energy. Lawson's criteria for *d-d* fusion is taken as $n\tau \geq 10^{14}$ s.cm^{-3} and for *d-t* fusion it is taken as $n\tau \geq 10^{16}$ s.cm^{-3}.

The values of $\langle \sigma_{xy}\ v\rangle$ for *d-d* and *d-t* reactions are shown in Figure 5.27. It can be seen that the deuterium-tritium (*d-t*) reaction has the advantage over deuterium-deuterium (*d-d*) reaction of a much higher cross-section. The heat of the reaction is also greater. However, the main disadvantage is that tritium does not occur naturally ($T_{1/2}$ ~ 17.7 y) and is expensive to manufacture, which increases the overall cost. From Figure 5.27, it can be seen that the rate for the *d-t* reaction peaks at $E = k_B T$ = 30–40 keV and a working energy where σ_{xy} is considered reasonable is about 20 keV, i.e., 3×10^8 K.

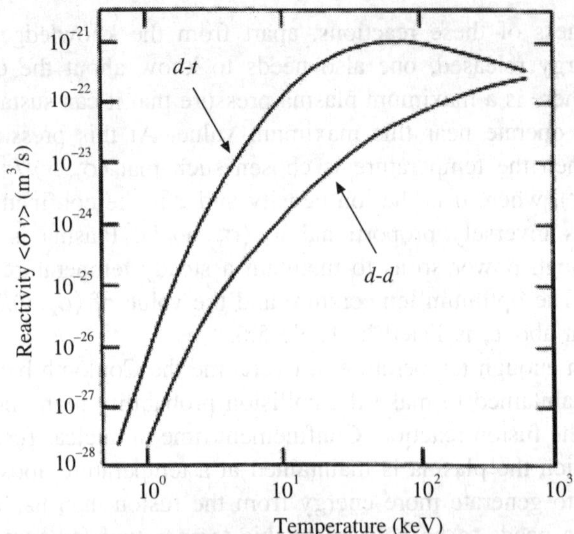

FIGURE 5.27 $\langle \sigma_{xy} v \rangle$ for *d-d* and *d-t* fusion reactions. [Data from the NRL Plasma Formulary, Revised 2006]

The effective energy produced by the fusion process will be reduced by the heat radiated by the hot plasma. The distortions have a temperature comparable or greater than that of the ions, so they will collide with the ions and emit X-rays of 10–30 keV energy (Bremsstrahlung). Thus, for a plasma with given constituents and at a fixed ion density, there will be a minimum temperature below which the radiation losses will exceed the power produced by fusion. For example, for *d-t* reaction with an ion density $n \geq 10^{21}$ m^{-3}, $k_B T_{min} \cong 4$ keV.

At the temperature required for fusion, any material container will vapourize and so the main problem is how to contain the plasma for sufficiently long times for the reaction to take place. The two main techniques are magnetic confinement and inertial confinement. Both these techniques present enormous technical challenges. In the magnetic confinement approach, the hot plasma is kept out of contact with the walls of its container by keeping it moving in circular or helical paths by means of the magnetic force on the charged particles. Research is also underway on a completely different approach to plasma confinement. In the inertial confinement approach, no attempt is made to confine the plasma. Instead, the idea is to use intense pulsed lasers or electron or ion beams to compress the target to the fusion-producing temperatures and densities, and to fuse nuclei so quickly that they do not have time to move apart appreciably because of their inertia.

Workable designs for a reactor that theoretically can deliver ten times more fusion energy than the amount needed to heat up plasma to the required temperature are in development phase. A collaborative project ITER (International Thermonuclear Experimental Reactor) is a step in the direction to produce clean fusion energy by the year, 2027.[10]

10. S. Atzeni, J. Meyer-ter-Vehn, The Physics of Internal Fusion, Oxford University Press, Oxford, (2004).

EXERCISE 5.17: The following reaction has been used to test the reciprocity theorem. It will take place for low energy deuterons (l = 0) leaving the $^{12}_{6}$C nucleus in the ground state.

$$d + {}^{14}_{7}\text{N} \rightarrow \alpha + {}^{12}_{6}\text{C}$$

Given I^π (deuteron) = 1^+, $I^\pi(\alpha) = 0^+$; estimate the spin of $^{14}_{7}$N in the ground state.

If the incident kinetic energy of d is 20 MeV in the laboratory frame, calculate the laboratory kinetic energy at which α-particles should scatter from $^{12}_{6}$C to test the reciprocity theorem. What is the expected ratio between the cross-sections for the direct and the inverse reactions? Can the α-particle be emitted with an orbital angular momentum, l = 1?

(Atomic mass of $^{14}_{7}$N, d and α are 14.003074, 2.014102 and 4.002603 u, respectively. 1 u = 931.44 MeV)

Solution: According to the inverse reaction and the reciprocity theorem (reciprocity is a consequence of time reversal symmetry; see section 8.7.2 in Chapter 8),

$$\frac{\sigma(b \rightarrow a)}{\sigma(a \rightarrow b)} = \frac{(2I_A + 1)(2I_a + 1)}{(2I_B + 1)(2I_b + 1)} \frac{p_a^2}{p_b^2}$$

In the reaction,

$$d + {}^{14}_{7}\text{N} \rightarrow \alpha + {}^{12}_{6}\text{C}$$

$$a = d, A = {}^{14}_{7}\text{N}, b = \alpha, B = {}^{12}_{6}\text{C}$$

Therefore,

$$Q = [(M_d + M_N) - (M_\alpha + M_C)] \times 931.44 \text{ MeV}$$

$$= [(2.014102 + 14.003074) - (14.002603 + 12.0)] \times 931.44 \text{ MeV}$$

$$= 13.75 \text{ MeV}$$

For the forward reaction, the energy available in the CMS is

$$E^* = Q + \frac{E_d M_N}{M_N + M_d} \qquad (\text{* represents CMS})$$

$$= (13.57 \text{ MeV}) + \frac{(20 \text{ MeV})(14)}{(14 + 2)} = 31.07 \text{ MeV}$$

This energy (= 31.07 MeV) is shared between α and $^{12}_{6}$C in tune with the conservation of energy and momentum. From the conservation of energy, we have

$$\frac{p_\alpha^{*2}}{2m_\alpha} + \frac{p_C^{*2}}{2m_C} = 31.07$$

And the conservation of momentum gives

$$p_\alpha^* = p_C^*$$

From the two equations, we get

$$p_\alpha^* = 416.8 \text{ MeV/c}$$

The inverse reaction

$$\alpha + {}^{12}_{6}\text{C} \rightarrow d + {}^{14}_{7}\text{N}$$

is endothermic with $Q = -13.57$ MeV. So, energy of 31.07 + 13.57 = 44.64 MeV must be provided. This corresponds to the laboratory kinetic energy for the α-particle of

$$E_\alpha = 44.64 \times \frac{(12+4)}{12} = 59.52 \text{ MeV}$$

In the CMS, the energy of 31.07 MeV is shared between d and ${}^{14}_{7}\text{N}$. Just like before, from the conservation of energy and the conservation of momentum, we have

$$\frac{p_d^{*2}}{2m_d} + \frac{p_N^{*2}}{2m_N} = 31.07$$

$$p_d^* = p_N^*$$

From the two equations, we get

$$p_d^* = 318.3 \text{ MeV/c}$$

Since, $I_C = 0$, $I_\alpha = 0$ and $I_d = 1$, the ratio of the cross-sections is

$$\frac{\sigma_{dN}}{\sigma_{\alpha C}} = \frac{(1)(1)}{(2I_N + 1)(3)}\left(\frac{416.8}{318.6}\right)^2$$

If the ratio ($\sigma_{dN}/\sigma_{\alpha C}$) is known, then one can determine the nuclear spin of ${}^{14}_{7}\text{N}$ nucleus. If $I^\pi({}^{14}_{7}\text{N}) = 1^+$, then

$$\frac{\sigma_{dN}}{\sigma_{\alpha C}} = \frac{(1)(1)}{(3)(3)}\left(\frac{416.8}{318.6}\right)^2 = 0.19$$

Alpha particle cannot be emitted with $l = 1$, because of the parity violation. Intrinsic parity of all the four particles is even (positive). Additional component to parity due to the orbital motion of the α-particle is $(-1)^L$; therefore $l = 0$ is the only possibility.

PROBLEMS

5.1 A neutron passing through a body of matter and not absorbed in a nuclear reaction undergoes frequent elastic collisions in which some of its kinetic energy is given up to nuclei in its path. Very soon the neutron reaches thermal equilibrium, which means that it is equally likely to gain or to lose energy in further collisions. At room temperature a thermal neutron has an average energy of 1.5 kT = 0.04 eV and a most probable energy of kT = 0.025 eV. The cross-section of ${}^{113}_{48}\text{Cd}$ for capturing thermal neutrons is 2×10^4 b, the mean atomic mass of natural cadmium is 112 u, and its density is 8.64 g/cm^3.

(a) What fraction of an incident beam of thermal neutrons is absorbed by a cadmium sheet of 0.1 mm thickness? **[Ans:** 0.67]

(b) Cadmium is observed to be very efficient absorber of the thermal neutrons. What thickness of cadmium is needed to absorb 99% of an incident beam of thermal neutrons? **[Ans:** 0.41 mm]

(c) What is the mean free path of the thermal neutrons in ${}^{113}_{48}\text{Cd}$? **[Ans:** 0.0893 mm]

5.2 The following reaction is used for iron identification.

$$ {}^{56}_{26}\text{Fe} + p \rightarrow {}^{56}_{26}\text{Fe}^* + p' $$

In this reaction, ${}^{56}_{26}\text{Fe}$ nucleus is left in its first excited state (ΔE = 0.847 MeV) which decays by γ-ray emission. Determine the theoretical reaction threshold. **[Ans:** 0.862 MeV]

5.3 The reaction ${}^{3}_{1}\text{H}(d, n)\,{}^{4}_{2}\text{He}$ has a Q-value of 17.6 MeV. What is the range of the neutron energies that may be obtained from this reaction for an incident deuteron beam of 300 keV? **[Ans:** 13.08–15.41 MeV]

5.4 Cadmium has a resonance for the neutrons of energy 0.178 eV and the peak value of the total cross-section is about 7000 b. Estimate the contribution of the scattering to this resonance. **[Ans:** 3.35 b]

5.5 Neutrons incident on a heavy nucleus with spin zero, show a resonance at an incident energy of 250 eV in the total cross-section with a peak magnitude of 1300 barn, the observed width of the peak being Γ = 20 eV. Find the elastic partial width of the resonance. **[Ans:** 2.5 eV]

5.6 The nucleus ${}^{5}_{3}\text{Li}$ is apparent as a resonance in the elastic scattering of protons from ${}^{4}_{2}\text{He}$ at a proton energy of 2 MeV. The resonance has a width of 0.5 MeV and spin 3/2. What is the lifetime of ${}^{5}_{3}\text{Li}$? Estimate the cross-section at the resonance energy.

[Ans: 1.32×10^{-21}s, 3.2 b]

5.7 A neutron with kinetic energy T_0 (non-relativistic) elastically collides with a stationary nucleus of mass M. In the CM frame, the scattering is isotropic. Prove that on average, the energy of the neutron after the collision is

$$ T_1 = \frac{M^2 + m_n^2}{(M + m_n)^2} T_0 $$

5.8 Show that in the scattering of a particle of mass M_0 of a target nucleus of mass M, the momentum transfer q from a to b has the same form in both the CMS and the laboratory frame.

5.9 Consider the following two reaction branches proceeding via a compound nucleus ${}^{8}_{4}\text{Be}^*$.

$$p + {}^{7}_{3}\text{Li} \rightarrow {}^{8}_{4}\text{Be}^* \rightarrow \begin{cases} 2\,{}^{4}_{2}\text{He} \\ {}^{8}_{4}\text{Be} + \gamma \end{cases}$$

The spin and parity of ${}^{7}_{3}\text{Li}$ and ${}^{8}_{4}\text{Be}$ nuclei in the ground state are equal to $(3/2)^-$ and 0^+, respectively; the spin of the α-particle is 0^+ and the internal parity of the proton is to be assumed positive. Using the laws of conservation of angular momentum and parity, for the cases when the orbital angular momentum of the proton is equal to 0 and 1, find the possible values of the spin and parity of the compound nucleus. Also, find the states (spin and parity) of the compound nucleus in both the reaction branches.

[**Ans:** For the CN, the possible states will be 1^-, 2^- for $l = 0$ and 0^+, 1^+, 2^+, 3^+ for $l = 1$. In the first reaction branch, $I_C^\pi = 2^+, 0^+$ and in the second reaction branch, $I_C^\pi = 2^+$]

5.10 Estimate the number of partial waves, which will be important in nuclear collisions occurring between 9 MeV neutrons and ${}^{125}_{50}\text{Sn}$ nuclei. Assume that the interaction radius is given by $R = 1.2A^{1/3}$, where A is the mass number of the target. [**Ans:** 4]

5.11 The energy released in a (*d, p*) reaction to a certain excited state of an even–even nucleus is 8 MeV. The target nucleus can be assumed to remain at rest. Estimate the bombarding energy for which the outgoing protons at forward angles will have roughly the same momenta as the incident deuterons. If the angular distribution at this bombarding energy peaks at an angle of 40° and the target nucleus has a radius of about 4.1 fm, calculate the probable l transfer. What are the spin and parity assignments of the excited state. [**Ans:** $l = 2$ and hence excited state of 2^+]

5.12 What is the half-life of the resonance in ${}^{113}_{48}\text{Cd}$, if it has a total width of 0.133 eV? Compare this with the collision time for a 0.17 eV neutron, which is approximately the time it takes for the neutron to pass the nucleus. [**Ans:** 2.04×10^{-18} s]

5.13 An *s*-wave ($l = 0$) resonance in the total cross-section for the neutrons incident on ${}^{238}_{92}\text{U}$ is obtained at the centre of mass energy of $E_n = 115$ eV. The total width of the resonance is 0.094 eV and the peak cross-section is 19200 b. Calculate the partial widths Γ_n and Γ_r for neutron and γ emission (assuming that these are the only final channels populated) and the peak cross-section for the (n, γ) capture reaction. Assume that $\Gamma_n > \Gamma_\gamma$. [**Ans:** $\Gamma_n = 0.0797$ eV, $\Gamma_\gamma = 0.0143$ eV]

5.14 Repeat Exercise 5.12 to calculate the ratio of the (α, n) to $(\alpha, 2n)$ cross-section for a ${}^{86}_{38}\text{Sr}$ target. Compare the analytical results with the following experimental data [IAEA-NDA-CD-05 (Austria)]. Take the level density parameter, $a = A/13$.

$E\alpha$ (MeV)	16.7	17.6	19.2	23.2
$\dfrac{\sigma(\alpha, n)}{\sigma(\alpha, 2n)}$	3.639	2.076	0.6935	0.1303

5.15 Show that ${}^{8}_{4}\text{Be}$ can decay into two α-particles with an energy release of 0.1 MeV, but ${}^{12}_{6}\text{C}$ cannot decay into three α-particles. Show that the energy released (including the energy carried by the photon) in the reaction ${}^{1}_{1}\text{H} + {}^{4}_{2}\text{He} \rightarrow {}^{6}_{3}\text{Li} + \gamma$ is 1.5 MeV.

5.16 A proton with kinetic energy K_E collides with another stationary proton, and a proton-antiproton pair is produced.

$$p + p \rightarrow p + p + p + \overline{p}$$

If the momentum of the bombarding proton is shared equally by the four particles that emerge from the collision, find the minimum value of the K_E.

[**Ans:** 5628 MeV]

5.17 Calculate the fission energy released in the fission of ${}^{235}_{92}\text{U}$ based on

(a) $\overline{B}$ vs. A curve

(b) the fact that fission is purely Coulomb repulsive energy once the scission state is achieved.

(c) the exact mass differences and assuming the reaction as

$${}^{235}_{92}\text{U} + n \rightarrow {}^{98}_{42}\text{Mo (stable)} + {}^{136}_{54}\text{Xe (stable)} + 2n$$

[**Ans:** (a) 212 MeV; (b) 227 MeV (if $r_0 = 1.37$); (c) 207 MeV]

5.18 A beryllium target was bombarded with α-particles accelerated to a kinetic energy of 21.7 MeV in a cyclotron. It was reported that four groups of protons were observed corresponding to Q-values of –6.92, –7.87, –8.57 and –10.74 MeV, respectively.

(a) What were the energies of the proton groups observed at an angle of 90° with the incident beam?

(b) Which Q-value corresponds to the ground state of the product nucleus, and how does it compare with the value calculated from the atomic masses?

(c) To which levels of the product nucleus do the different proton groups correspond?

(d) What are the threshold energies for the excitation of the different states of the product nucleus?

[**Ans:** (a) Q-value (MeV), proton energy (MeV): (–6.92, 6.97), (–7.87, 6.09), (–8.57, 5.45), (–10.74, 3.44); (b) The Q-value (–6.92 MeV) corresponds to the ground state. The value calculated from the masses is (–6.88 MeV); (c) The different proton groups correspond to the ground state and two excited levels in ${}^{12}\text{B}$ at 0.95, 1.65 and 3.82 MeV, respectively; (d) The threshold energies (MeV) are 9.99, 11.36, 12.38 and 15.51, for the ground state and the three excited states, respectively.]

5.19 When the nuclide ${}^{12}_{6}$C is bombarded with slow neutrons, γ-rays with an energy of 4.947 MeV are produced by the reaction ${}^{12}_{6}\text{C}(n, \gamma)\,{}^{13}_{6}\text{C}$. Compare this energy with the binding energy of the last neutron in ${}^{13}_{6}$C:

(a) as calculated from the atomic masses of the nuclei involved,

(b) as deduced from the Q-value, 2.723 MeV, of the reaction ${}^{12}_{6}\text{C}(d, p)\,{}^{13}_{6}\text{C}$,

(c) as deduced from the Q-value, –0.281 MeV, of the reaction ${}^{12}_{6}\text{C}(d, n)\,{}^{13}_{7}\text{N}$ and the endpoint energy 1.2 MeV of the positron decay of ${}^{13}_{7}$N.

[Ans: (a) 4.94 MeV; (b) and (c) 4.95 MeV]

5.20 The effect of a (d, p) reaction is to add a neutron to the target nucleus. Show that the binding energy of the last neutron in the product nucleus is given by the sum of the Q-value for the (d, p) reaction and the binding energy of the deuteron. Q-values of 4.48 MeV, 5.14 MeV and 1.64 MeV have been obtained for the reactions ${}^{206}_{82}\text{Pb}(d, p)\,{}^{207}_{82}\text{Pb}$, ${}^{207}_{82}\text{Pb}(d, p)\,{}^{208}_{82}\text{Pb}$ and ${}^{208}_{82}\text{Pb}(d, p)\,{}^{209}_{82}\text{Pb}$, respectively. What are the binding energies of the last neutron in ${}^{207}_{82}$Pb, ${}^{208}_{82}$Pb and ${}^{209}_{82}$Pb?

[Ans: Product nucleus: (${}^{207}_{82}$Pb, 6.71 MeV), (${}^{208}_{82}$Pb, 7.37 MeV), (${}^{209}_{82}$Pb, 3.87 MeV)]

5.21 The reaction ${}^{45}_{21}\text{Sc}(d, p)\,{}^{46}_{21}\text{Sc}$ has a Q-value of 6.54 MeV. It has a resonance when the incident deuteron kinetic energy (laboratory frame) is 2.76 MeV. Can we expect the same resonance to be excited in the reaction ${}^{43}_{20}\text{Ca}(\alpha, n)\,{}^{46}_{22}\text{Ti}$ and if so at what value of the laboratory kinetic energy of the α-particle? Given that the following β-decay

$${}^{46}_{21}\text{Sc} \rightarrow {}^{46}_{22}\text{Ti} + e^- + \bar{\nu}_e$$

has a Q-value of 2.37 MeV and the mass difference between the neutron and a hydrogen atom is 0.78 MeV/c^2. **[Ans:** 11.7 MeV]

5.22 For the following reaction, the Q-value is –7.83 MeV.

$${}^{48}_{20}\text{Ca} + {}^{16}_{8}\text{O} \rightarrow {}^{49}_{21}\text{Sc} + {}^{15}_{7}\text{N}$$

At minimum kinetic energy of bombarding ${}^{16}_{8}$O ions to initiate the reaction, estimate the orbital angular momentum in units of $\hbar$ of the ions for grazing collision.

[Ans: 17]

BIBLIOGRAPHY

Cohen, B.L., *Concepts of Nuclear Physics*, Tata McGraw-Hill, New Delhi, 1971.

Cottingham, W.N. and Greenwood, D.A., *An Introduction to Nuclear Physics*, Cambridge University Press, Cambridge, 1986.

Devanathan, V., *Nuclear Physics*, Narosa, New Delhi, 2006.

Garg, J.B., *Nuclear Physics*: Basic Concepts, Macmillan, New Delhi, 2011.

Harvey, B.G., *Introduction to Nuclear Physics and Chemistry*, Prentice-Hall, New Jersery, 1970.

Kaplan, I., *Nuclear Physics*, Addison-Wesley, Boston, 1962.

Krane, K.S., *Introductory Nuclear Physics*, Wiley, New York, 2008.

Library of Experimental Nuclear Reaction Data, 1999. IAEA-NDS-CD-05 (Austria).

Lilley, J., *Nuclear Physics: Principles and Applications*, Wiley, New York, 2002.

Mittal, V.K., Verma, R.C., Gupta, S.C., *Introduction to Nuclear and Particle Physics*, PHI Learning, Delhi, 2013.

Sood, D.D., Ramamoorthy, A.V.R., *Fundamentals of Radiochemistry*, IANCAS, Mumbai, 2000.

6 Interaction of Radiations with Matter

"Concerning matter, we have been all wrong! What we have called matter is energy, whose vibration has been so lowered as to be perceptible to the senses. There is no matter."

—Albert Einstein

There are two broad categories in which the nuclear radiations can be classified:

(a) Charged particulate radiations, e.g., p, e^-, e^+, α
(b) Uncharged radiations, e.g., n, X-rays, γ-rays

It is important to understand how various radiations interact with matter, because the operation of a radiation detector itself depends on the manner in which the radiation to be detected interacts with the active volume of the detector material. Charged particulate radiations, because of the electric charge carried by the particle, interact through the Coulomb forces with the electrons of the medium through which they pass and lose their energy. On the other hand, uncharged radiations undergo interaction in two steps. An X-ray or γ-radiation can transfer all or part of its energy to the electrons within the medium through various processes described in the subsequent sections. The resulting secondary electrons are very similar to the fast electrons (such as β-particles) and their energies serve as the basis of the detector signal. In contrast, neutrons collide with nuclei of the atoms of medium and may get captured resulting in a nuclear reaction, which may in turn result in the emission of secondary charged particles which would interact with matter through the Coulomb force.

6.1 INTERACTION OF CHARGED PARTICLES

Upon entering any medium, the charged particle immediately interacts either with the orbital electrons or directly with the nucleus. The difference in the size, the mass and the binding energy of the nucleus and the electrons determines the type of interaction that the incoming radiation will undergo. All the interaction processes of charged particles with matter are mainly due to Coulomb forces. The net outcome of such interactions is the reduction in the kinetic energy of the incoming radiation during the interaction, resulting in either complete absorption of the radiation or stopping of the particle along with its charge neutralization. These interaction processes include excitation, ionization, scattering and nuclear transmutation. Table 6.1 illustrates the various absorption processes.

Table 6.1 Various Absorption Processes for Interaction of Charged Radiation with Matter

Type of reactions	*Reacting fields*	*Cross-section* σ (b)#	*Process*
Particle energy loss through atomic excitation and ionization	Orbital electrons	$> 10^5$	Ionization, (atomic) excitation
Particle elastically scattered	Atomic nucleus	< 10	Nuclear scattering
Particle inelastically scattered	Atomic nucleus	< 1	Nuclear (Coulomb) excitation
Particle captured, formation of compound nucleus	Atomic nucleus	< 0.1	Nuclear transmutation

#: The reaction cross-section gives only the order of magnitude at about 1 MeV in $Z \approx 20$.

Inelastic collisions with the orbital electrons, is usually the predominant mechanism by which a charged particle loses its kinetic energy in an absorber. Due to each collision, one or more orbital electrons experience a transition to an excited state, known as excitation or to an unbound state, known as ionization. Elastic collision with the orbital electrons, on the other hand, where the energy and the momentum are conserved and the energy transfer is less than the lowest excitation potential of the electrons, is significant only for the case of very low energy (<100 eV) incident β-particle. This leads to large-angle deflections of a fraction of the incident electrons, known as backscattering.

In the elastic collision with a nucleus, also known as *Rutherford scattering*, the incoming charged particle may be deflected but does not radiate or excite the nucleus. The incident particle loses only the kinetic energy required for the conservation of energy and momentum between two particles. Usually, the incoming particles are lighter than the target nuclei, because of which the collision leads to a change in their direction, while the change in their energies can almost be neglected. It is also sometimes referred to as *Coulomb scattering*, since the interaction is mainly due to Coulomb forces. If the collision with the nucleus is inelastic, and if the energy of the particle is near the relativistic range, a quantum of radiation is emitted and the corresponding amount of kinetic energy is lost by the incident particle. This type of electromagnetic radiation is called the bremsstrahlung radiation or continuous X-radiation.

For low energy incident particle, the formation of compound nucleus is also a possibility.

We will now study the nature of interactions in terms of relatively heavy charged particles (e.g., α, p), light charged particles (e.g., e^-, e^+), electromagnetic radiations (e.g., γ-rays, X-rays), and neutrons.

6.1.1 Interaction of Heavy Charged Particles

The interaction of heavy charged particles with matter is less complicated as compared to the interaction of light charged particles. Heavy particles lose most of their energy through the ionization and excitation of the atoms in the absorber. Rutherford scattering and bremsstrahlung are generally not significant in the response of the radiation detectors. The incident particle must lose its kinetic energy in many such interactions during its passage through the detector, before it is completely stopped, since only a finite amount of energy is used to ionize or excite an atom.

Stopping Power

The mechanism by which charged particles transfer their energy in inelastic collisions with matter are expressed in the form of stopping power, relative stopping power or specific ionization.

Specific ionization (SI) is defined as the total number of ion-pairs formed per unit length, both primary and secondary. At high energies, the average energy required to create a single ion pair is nearly constant and has similar value for all incoming charged particles. Table 6.2 lists ion-pair generation energy for different materials.

Table 6.2 Ion-pair Generation Energy for Different Materials

Material	*Ion-pair generation energy* (eV)
Air	33.9
Silicon	3.6
Germanium	2.8
Silicon oxide	17
Hydrogen	37
Helium	46
Nitrogen	36
Oxygen	32
Neon	37

EXERCISE 6.1: If a 5 MeV α-particle is passing through an air-filled detector, on an average how many primary ion-pairs have to be formed before the incident particle is stopped?

Solution: The number of primary ion-pair formed is

$$= \frac{5 \times 10^6 \text{ eV}}{33.9 \dfrac{\text{eV}}{\text{ion pair}}} = 1.47 \times 10^5 \text{ ion pairs}$$

In a given absorber, the stopping power for the charged particles is defined as the ratio of the differential energy loss for that particle within the material and the corresponding differential path length.

$$S = -\frac{dE}{dx} \tag{6.1}$$

where E denotes the kinetic energy of the incident particle. The value of $-\frac{dE}{dx}$ along a particle track is also known as its specific energy loss or linear energy transfer. A quantum-mechanical expression including relativistic effects that describes the specific energy loss, first carried out in 1930, is known as the Bethe–Bloch formula. For heavy charged particles, it is written as follows:

$$-\frac{dE}{dx} = \frac{4\pi z^2 e^4}{m_e v^2} NB \tag{6.2}$$

$$B = Z\left[\ln\left(\frac{2m_e v^2}{I}\right) - \ln\left(1 - \frac{v^2}{c^2}\right) - \frac{v^2}{c^2}\right]$$

where v is the velocity of the charged particle, z is the charge of the charged particle, e is the electronic charge, N is the number density and Z is the atomic number of the absorber atoms, m_e is the rest mass of the electron, B is stopping number and the parameter I is the experimentally evaluated average excitation and ionization potential of the absorber. Typical values of I are given in Table 6.3.

Table 6.3 Typical Values of Average Excitation and Ionization Potential

Material	H	He	C	N	*Air*	O	Al	Fe	Cu	Pb
Z	1	2	6	7	7.2	8	13	26	29	82
I (eV)	19	44	77	88	94	100	166	300	371	1070

An empirical relation used for the average ionization (excitation) potential is given as follows:

$$I \cong 9.1Z(1 + 1.9Z^{-2/3}) \tag{6.3}$$

The expression for B varies slowly with the particle energy. Note that $-\frac{dE}{dx}$ is independent of the mass of the incident particle.

Equation (6.2) is valid for charged particles with velocities that are large compared to the velocity of the orbital electrons. For charged particles with $v << c$ (non-relativistic particles), only the first term in the stopping number equation for B is needed, and equation (6.2) reduces to

$$-\frac{dE}{dx} = \frac{4\pi z^2 e^4}{m_e v^2} NZ\left[\ln\left(\frac{2m_e v^2}{I}\right)\right] \tag{6.4}$$

For a given non-relativistic particle, the stopping power, therefore, is inversely proportional to the energy of the charged particle. For different charged particles with the same velocity, the particles carrying more charge will have the largest specific energy loss. For example, alpha particles will lose energy at a greater rate than protons of the same velocity. For different absorbing materials, the materials with the higher density and atomic number will have greater stopping power. Figure 6.1 shows a plot of the specific energy loss as a function of energy for protons and alpha particles travelling in aluminium and nitrogen.

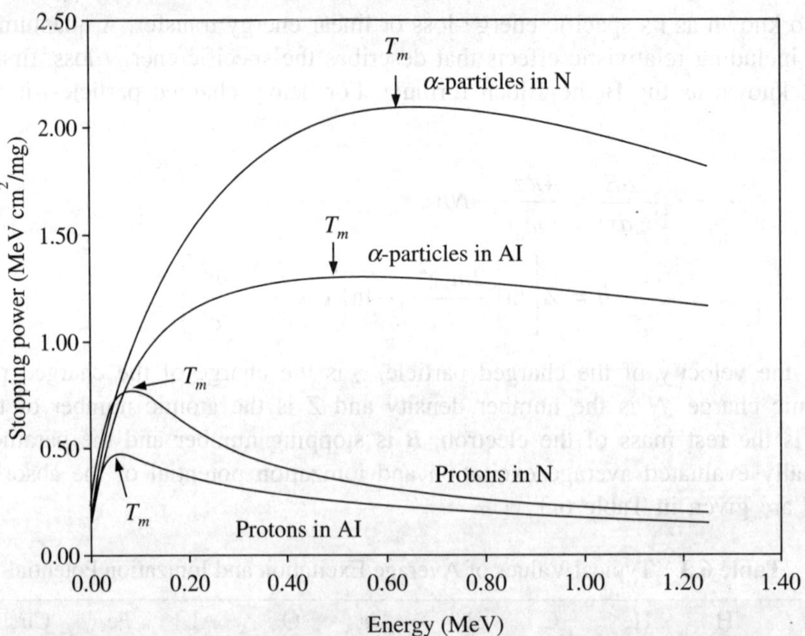

FIGURE 6.1 Specific energy loss vs. energy for protons and alpha particles. T_m indicates the energy at which the specific energy loss is maximized. [Data from National Institute of Standards and Technology: PSTAR and ASTAR Database]

The energy deposited by the incident charged particle in thin absorbers (or detectors) can be calculated as follows:

$$\Delta E = \left[-\frac{dE}{dx} \right]_{\text{avg}} t \tag{6.5}$$

where t is the absorber thickness and $\left(-\frac{dE}{dx} \right)_{\text{avg}}$ is the linear stopping power averaged over the energy of the particle whilst it is in the absorber. If the energy loss is small, the stopping power does not vary much and hence it can be approximated by its value at the incident particle energy.

EXERCISE 6.2:

(a) Show that the specific ionization of 480 MeV α-particle is approximately equal to that of a 30 MeV proton.

(b) Show that the rate of change of ionization over a given distance is different for the two particles. Indicate how this can be used to identify one particle, assuming we know the identity of the other particle.

Solution:

(a)
$$-\frac{dE}{dx} \propto \frac{z^2}{v^2}$$

or
$$-\frac{dE}{dx} \propto \frac{z^2}{E/m}$$

$$\therefore \quad \frac{\left(-\dfrac{dE_\alpha}{dx}\right)}{\left(-\dfrac{dE_p}{dx}\right)} = \frac{m_\alpha z_\alpha^2 E_p}{m_p z_p^2 E_\alpha}$$

$$= \frac{(4)(2)^2(30\text{ MeV})}{(1)(1)^2(480\text{ MeV})} = 1$$

(b) The change in ionization over a given distance is different for different particles in a medium and is proportional to $\frac{z^2}{v^2}$. Calibration curves can be drawn for different particles which may be later used to identify the unknown particle.

The plot between $-\frac{dE}{dx}$ and the penetration distance (x) is known as the *Bragg curve*. Bragg curve for the α-particles emitted by different radionuclides is shown in Figure 6.2. The position of the Bragg peaks is correlated with the initial energy of the α-particles. The rate of energy loss is low in the initial stages because the charged particle is moving so fast that it has only a short time for interaction with the electrons. The time that a charged particle spends within a given distance is proportional to $\frac{1}{v}$, which would be low in the initial stages. The particle energy reduces as it passes through the matter. This results in increased specific ionization along its path, which reaches a maximum when the charged particle velocity becomes comparable to the velocity of the orbital electrons of the medium. The charged particle starts picking up the electrons and the rate of energy loss starts diminishing. Further, during the atomic collisions, $-\frac{dE}{dx}$ is greatly reduced as the ion will pick up and then lose electrons many times near the end of the path.

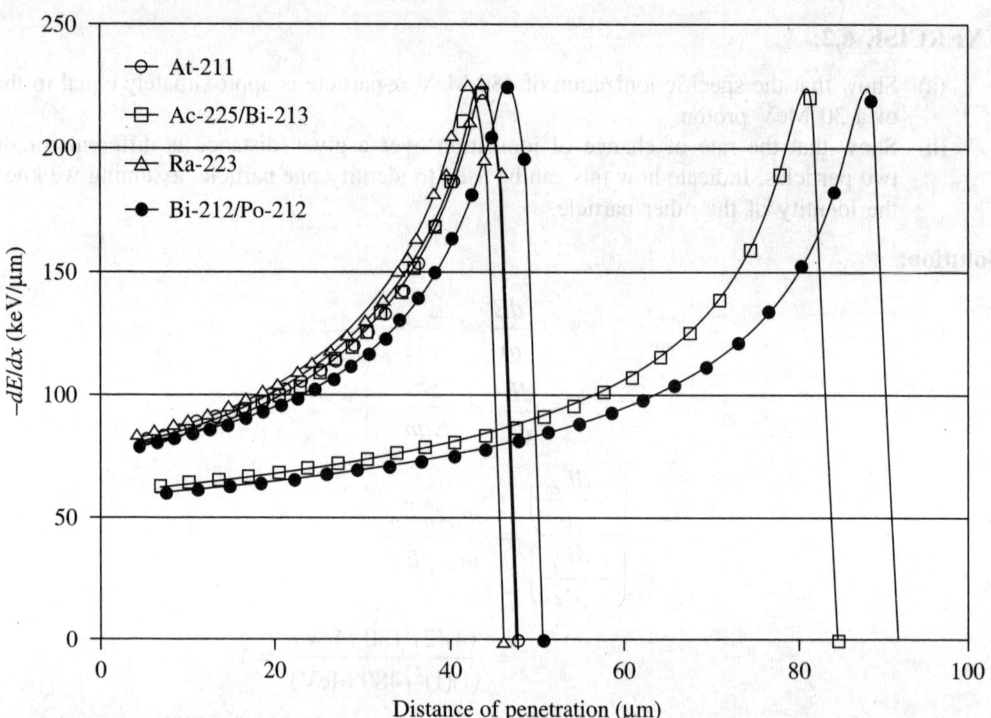

FIGURE 6.2 The specific energy loss vs. distance in water travelled by typical alpha particles emitted by radionuclides, ^{225}Ac (5.829 MeV)/^{213}Bi (8.375 MeV), ^{211}At (5.867 MeV), ^{212}Bi (6.08 MeV)/^{212}Po (8.78 MeV), ^{223}Ra (5.716 MeV). [Song H, Senthamizhchelvan S, Hobbs RF, Sgouros G., Antibodies, 2012; 1(2):124-148.]

EXERCISE 6.3: Assuming the Bethe–Bloch formula is valid for low energies, show that the rate of ionization has a maximum (the Bragg peak) and find the kinetic energy of the protons in silver ($Z = 47$) for which the maximum would occur.

Solution: For small velocity v, Bethe–Bloch formula can be written as follows:

$$S = -\frac{dE}{dx} = \frac{4\pi z^2 e^4}{m_e v^2} NZ \left[\ln\left(\frac{2m_e v^2}{I} \right) \right]$$

Differentiating with respect to v, we get

$$\frac{dS}{dv} = \frac{8\pi z^2 e^4}{m_e v^3} NZ \left[1 - \ln\left(\frac{2m_e v^2}{I} \right) \right]$$

For maxima, $\frac{dS}{dv}$ should be zero, which is when

$$v^2 = \frac{eI}{2m_e}$$

where e is the Euler's number, the base of the natural logarithm. Using equation (6.3) for I, the kinetic energy of a proton in silver at maxima is,

$$E_p = \frac{1}{2}m_p v^2 = \frac{m_p e I}{4m_e}$$

$$= \frac{m_p e}{4m_e} 9.1Z(1 + 1.9Z^{-2/3})$$

$$= \frac{1836(2.718)(9.1)(47)(1 + 1.9 \times 47^{-2/3})}{4} \approx 611 \text{ keV}$$

The relative stopping power for a charged particle interacting with a given material is defined as the ratio of the loss in particle energy per atom of the given material to the energy loss per atom in the standard material, which is aluminium for the α-particle and air for the β-particle.

$$S_m = \frac{N_{st}\left(\dfrac{dE}{dx}\right)_{\text{material}}}{N_{\text{material}}\left(\dfrac{dE}{dx}\right)_{st}}$$

Energy Straggling

When a mono-energetic beam of charged particulate radiation travels a fixed absorber thickness Δx, there are fluctuations in the energy of the emerging beam about the mean value due to finite number of collisions with the atoms of the medium along the path. Therefore, the mono-energetic beam will show a distribution in the energy with the peak shifted down by the mean energy loss. This phenomenon is known as energy straggling as depicted in Figure 6.3.

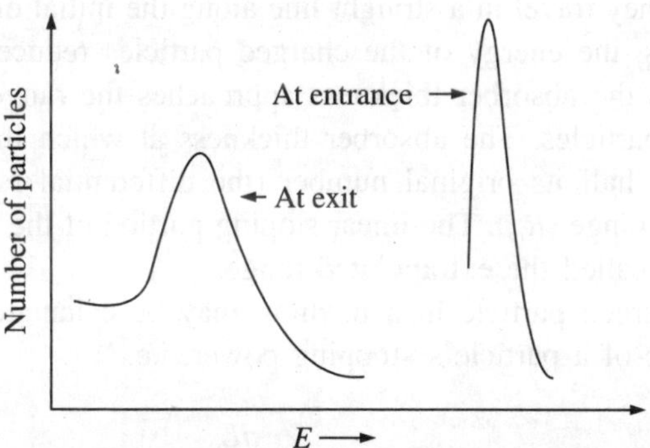

FIGURE 6.3 Energy straggling.

Particle Range

The distance that a charged particle traverses before coming to rest is known as its range (R). It is inversely related to the stopping power of the absorber, which means that if a particle has a high or a short range, the absorber has a low or high stopping power, respectively. The range depends on the absorber material and the type as well as the energy of the charged particle. The range is longer for the particles of higher energy, and shorter for the heavier particles. Ranges of positively charged particles are determined by the absorption method either with thin metallic absorbers or using a gas in a container whose pressure can be varied. A typical plot of distance travelled in the absorber and the intensity of the charged particles is shown in Figure 6.4.

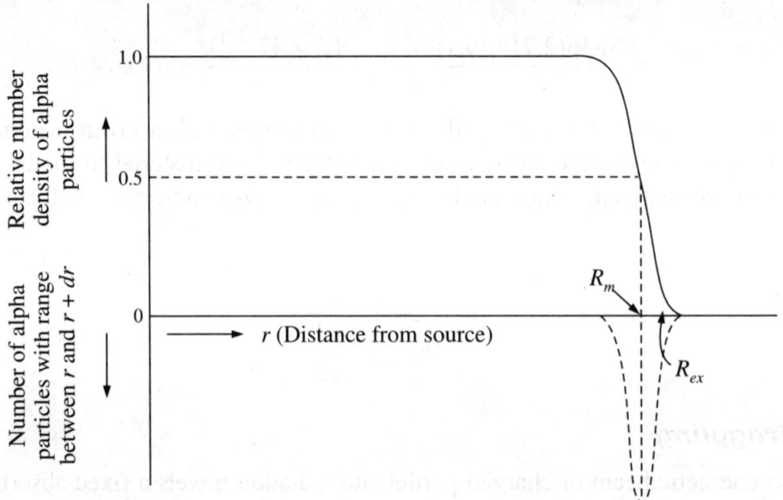

FIGURE 6.4 Number of alpha particles from a point source as a function of distance from the source and the derivative (dashed curve).

The heavy charged particles suffer only small deflection due to collisions with the orbital electrons and therefore, they travel in a straight line along the initial direction. For small values of the absorber thickness, the energy of the charged particles reduces with no attenuation of the charged particles. As the absorber thickness approaches the range, there is a rapid fall in the number of charged particles. The absorber thickness at which the number of transmitted particles becomes nearly half its original number (the differential range is maximum at this point) is called the mean range (R_m). The linear sloping portion of the curve when extrapolated, cuts the x-axis at R_{ex}, is called the extrapolated range.

The range of a charged particle in a medium may be calculated theoretically from the integration of the inverse of a particle's stopping power, i.e.,

$$R(E_{in}) = \int_0^{E_{in}} \frac{dE}{dE/dx} \tag{6.6}$$

where E_{in} is the initial kinetic energy of the incident particle. The final velocity is taken as zero. The evaluation of the above integral is a bit complicated, although it can be shown that the range of particle of mass m and charge ze can be represented by:

$$R(v) = \frac{m}{z^2}F(v) \tag{6.7}$$

where $F(v)$ represents a unique function of the particle with initial velocity v, which means for particles of the same initial velocity, the factor is identical.

EXERCISE 6.4: Calculate the mean range of 20 MeV deuterons and 40 MeV α-particles, if the mean range of 10 MeV protons in lead is 0.316 mm.

Solution: The velocity v for non-relativistic particles of mass m and kinetic energy E is given as $v = \sqrt{\frac{2E}{m}}$. It can be seen that for all the three particles, the velocity is coming out to be same, i.e.,

$$\sqrt{\frac{2\times 10}{1}} = \sqrt{\frac{2\times 20}{2}} = \sqrt{\frac{2\times 40}{4}}$$

Hence, $F(v)$ is same for all the three particles. Therefore, we can write the mean range for deuteron and α-particle as follows:

$$R_d = \frac{\left(\frac{m_d}{z_d^2}\right)}{\left(\frac{m_p}{z_p^2}\right)}R_p = \frac{\left(\frac{2}{1^2}\right)}{\left(\frac{1}{1^2}\right)}(0.316 \text{ mm}) = 0.632 \text{ mm}$$

$$R_\alpha = \frac{\left(\frac{m_\alpha}{z_\alpha^2}\right)}{\left(\frac{m_p}{z_p^2}\right)}R_p = \frac{\left(\frac{4}{2^2}\right)}{\left(\frac{1}{1^2}\right)}(0.316 \text{ mm}) = 0.316 \text{ mm}$$

There can be a case when the data for the range or energy loss of precisely the same particle-absorber combination is not available. For example, the linear stopping power of α-particles in a metallic oxide may be estimated from the separate data on the stopping power in both the pure metal and in oxygen. By applying the following *Bragg–Kleeman rule*, the stopping power in a compound medium may be known.

$$\frac{1}{N_c}\left(\frac{dE}{dx}\right)_c = \sum_i a_i \frac{1}{N_i}\left(\frac{dE}{dx}\right)_i \tag{6.8}$$

where N is the atom density and a_i is the atom fraction of the component.

The range of a charged particle in a compound medium can also be estimated if the range data in all the constituent elements is known. The underlying assumption involved in this derivation is that the shape of the specific energy loss curve is independent of the stopping medium.

$$\frac{1}{R_c} = \sum_i \frac{w_i}{R_i} \tag{6.9}$$

where R is the range and w_i is the weight fractions of the individual components.

If the range data is not available for all the constituent elements, estimates can be made based on a semi-empirical formula known as *Bragg–Kleeman rule* as well. If R_1, ρ_1 and A_1 are the range, density and the atomic weight in medium 1, and the corresponding quantities R_2, ρ_2 and A_2 in medium 2, they are related to each other by the following relation:

$$\frac{R_2}{R_1} = \frac{\rho_1\sqrt{A_2}}{\rho_2\sqrt{A_1}} \tag{6.10}$$

The accuracy of the above estimate diminishes when the difference of the atomic weights of the two materials is large.

In air, the range of α-particles with energy of 4–10 MeV can be empirically determined using Geiger's rule.

$$R = KE^{3/2} \tag{6.11}$$

where $K \cong 0.32$, R is the range in cm at 1 atmosphere and 15 °C and E is the energy in MeV.

EXERCISE 6.5: What is the minimum energy of an α-particle that can be counted with a GM counter whose counter window is 2.5 mg cm^{-2} thick and made up of stainless steel ($A \approx 56$). The density of air can be taken as 1.226×10^{-3} g cm^{-3} and its atomic weight as 14.6.

Solution:

We are given $R_s\rho_s = 2.5 \times 10^{-3}$ g cm^{-2}. Applying the Bragg–Kleeman rule, we get

$$R_a = \frac{R_s\rho_s\sqrt{A_a}}{\rho_a\sqrt{A_s}}$$

$$= \frac{(2.5\times10^{-3}\text{ g cm}^{-2})\sqrt{14.6}}{(1.226\times10^{-3}\text{ g cm}^{-3})\sqrt{56}} = 1.04$$

Applying Geiger's rule, we get

$$E = \left(\frac{R}{0.32}\right)^{2/3} = \left(\frac{1.04}{0.32}\right)^{2/3} = 2.19\text{ MeV}$$

Therefore, α-particles of energy greater than 2.19 MeV will be registered.

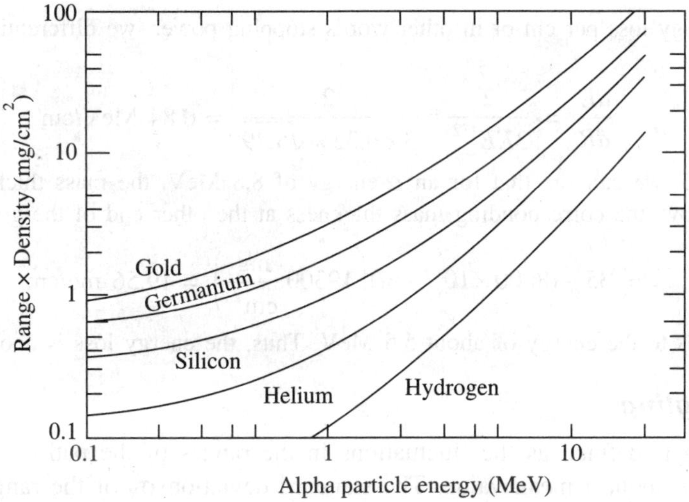

FIGURE 6.5 Range-energy curves calculated for alpha particles in different materials. Units of the range are given in mass thickness to minimize the differences in these curves. [National Institute of Standards and Technology: ASTAR Database]

When the absorber (detector) thickness is not small, the energy loss may not be obtained directly as a properly weighted $\left(-\frac{dE}{dx}\right)_{avg}$ is not known. In such cases, the deposited energy is obtained using the range-energy data of the type plotted in Figure 6.5. If R_0 represents the full range of the incident particle with energy E_0 in the absorber material, then the energy E_t of the particles that emerge from the opposite surface can be found out using the corresponding range R_t at the other end, which is calculated by subtracting the physical thickness of the absorber t from R_0. The deposited energy is simply the difference of these two energies, $\Delta E = E_0 - E_t$. The underlying assumption in this procedure is that the charged particle does not suffer significant deflection and follows a linear path in the absorber.

EXERCISE 6.6: 8.8 MeV α-particles have a range of 8.6 cm in standard air. Find their energy loss per cm in standard air at a distance 4 cm from a thin source. What will be the energy loss for the 8.8 MeV α-particles, when they pass through a gold foil of 8000 nm thickness? [ρ_{Au} = 19.3 g/cm^3]

Solution: Applying Geiger's rule, we get

$$8.6 = K(8.8)^{3/2} \quad \text{or} \quad K = 0.33$$

At a distance of 4 cm from the source, the residual range is (8.6 – 4) = 4.6 cm. Applying Geiger's rule to find the energy at this point, we get

$$E = \left(\frac{4.6}{0.33}\right)^{2/3} = 5.79 \text{ MeV}$$

To find the energy loss per cm or in other words stopping power, we differentiate the Geiger's relation to get

$$\frac{dE}{dR} = \frac{2}{3KE^{1/2}} = \frac{2}{3 \times 0.33 \times \sqrt{5.79}} = 0.84 \text{ MeV/cm}$$

From Figure 6.5, we can see that for an α-energy of 8.8 MeV, the mass thickness of gold is ~35 mg/cm^2. Now, the corresponding mass thickness at the other end of the gold plate

$$= 35 - (8000 \times 10^{-7} \text{ cm})\left(19300 \frac{\text{mg}}{\text{cm}^2}\right) = 19.56 \text{ mg/cm}^2$$

This corresponds to the energy of about 5.5 MeV. Thus, the energy loss is about 3.3 MeV.

Range Straggling

Range straggling is defined as the fluctuations in the ranges of the paths of mono-energetic beam of particles about a mean value. The standard deviation σ_R of the range distribution is related to the mean range R, and to the mass number A of the particle by:

$$\sigma_R \propto \frac{R}{\sqrt{A}} \tag{6.12}$$

6.1.2 Interaction of Light Charged Particles

The interaction of light charged particles (relativistic particles) like β-particles in matter and the mechanism of energy loss are more complex than for heavy charged particles due to the smaller mass and higher speed. Large deviation in the particle path becomes possible as it interacts with the orbital electrons of comparable mass, and in a single encounter, it can lose much of its energy. Even back scattering from the incident surface of the medium is a possibility. Because of these large deviations, the actual path length of a light charged particle (β) in a medium is always greater than its range in the medium. Thus, the ranges for beta particles are poorly defined due to enormous range straggling (Figure 6.6).

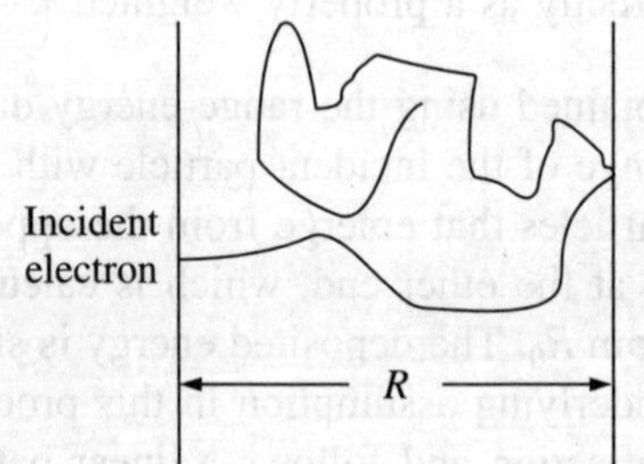

FIGURE 6.6 Range and path length of light charged particles like beta particles.

The energy loss for a β-particle in a medium has the following two components:

(i) The collision term, $\left(\frac{dE}{dx}\right)_c$, represents the energy loss due to Coulomb interactions. This includes the ionization and the excitation.

(ii) When a fast moving electron is accelerated or decelerated as it passes through the absorber, a photon is emitted and such photons are called bremsstrahlung (German for *braking* radiation). If the change in the acceleration of the electrons is due to the presence of magnetic field, then the electromagnetic radiation emitted is called *synchrotron* radiation. Another mode of interaction can be the annihilation of electron and positron, particularly in the case of positrons. The resulting γ-rays will then interact with the matter. The radiative term, $\left(\frac{dE}{dx}\right)_r$, represents the energy loss due to bremsstrahlung or electromagnetic radiation.

Therefore, the total stopping power for the β-particles as illustrated in Figure 6.7, is the sum of the collisional and radiative losses,

$$\frac{dE}{dx} = \left(\frac{dE}{dx}\right)_c + \left(\frac{dE}{dx}\right)_r \tag{6.13}$$

where the expression for the collision term for β-particle, is similar to equation (6.2), the Bethe–Bloch formula, to describe the specific energy loss due to ionization and excitation for the heavy charged particles, and is given as

$$-\left(\frac{dE}{dx}\right)_c = \frac{4\pi e^4}{m_e v^2} NB \tag{6.14}$$

where $B = Z\left[\ln\frac{m_e v^2 E}{2I^2(1-\beta^2)} - \ln 2\left(2\sqrt{1-\beta^2} - 1 + \beta^2\right) + (1-\beta^2) + \frac{1}{8}\left(1-\sqrt{1-\beta^2}\right)^2\right]$

where $\beta \equiv \frac{v}{c}$ is used for relativistic particles (like high energy electrons). The expression for the radiative term is given as follows:

$$-\left(\frac{dE}{dx}\right)_r = \frac{NEZ(Z+1)e^4}{137 m_e^2 c^4}\left(4\ln\frac{2E}{m_e c^2} - \frac{4}{3}\right) \tag{6.15}$$

The factors of E and Z^2 in the numerator of the radiative term indicate that the radiative losses are higher for high energy electrons and for absorber materials of large atomic number.

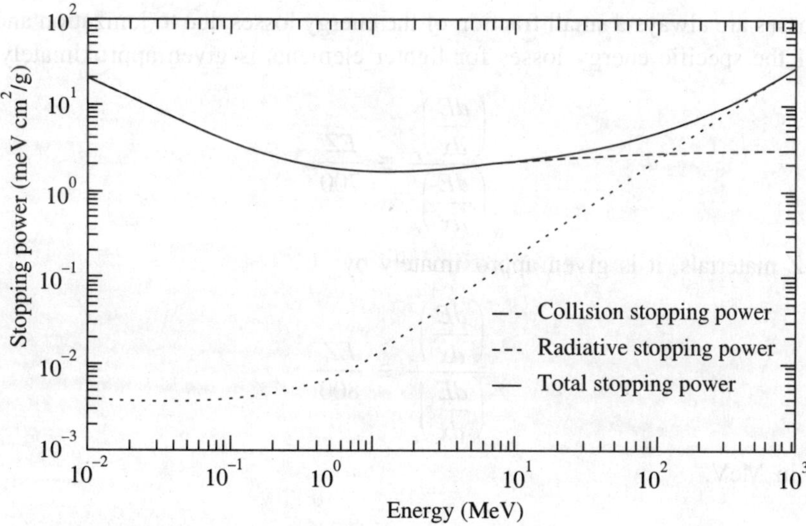

FIGURE 6.7 Components of the total stopping power for electrons in nitrogen. [National Institute of Standards and Technology: ESTAR Database]

In terms of the radiation length L_R, which is defined as the average thickness of the absorber that reduces the mean energy of an electron or positron by a factor of e (Euler's number), the average rate of radiative energy loss for the relativistic electrons is given by:

$$-\left(\frac{dE}{dx}\right)_r = \frac{E}{L_R} \tag{6.16}$$

Integrating the above equation, we get

$$E = E_0 \exp\left(-\frac{x}{L_R}\right) \tag{6.17}$$

where E_0 is the initial energy of the electron and x is the length traversed by the electron.

EXERCISE 6.7: An electron with an initial energy of 2 GeV traverses 10 cm of water with a radiation length of 36.1 cm. Calculate its final energy. How would the energy loss change if the particle were a muon rather than an electron? Muon is an elementary particle similar to the electron (mass 0.511 MeV/c^2), with unitary negative electric charge of −1 and a spin of $\frac{1}{2}$, but with a much greater mass (105.7 MeV/c^2).

Solution:

$$E = E_0 \exp\left(-\frac{x}{L_R}\right) = 2 \exp\left(-\frac{10}{36.1}\right) = 1.51 \text{ GeV}$$

From equation (6.15), we can see that the radiation losses at fixed E are proportional to m^{-2}, where m is the mass of the incident particle. Thus, for muons, which are much heavier than electrons, the radiation losses are negligible at this energy.

Radiative losses are always a small fraction of the energy losses due to ionization and excitation. The ratio of the specific energy losses for lighter elements is given approximately by

$$\frac{\left(\frac{dE}{dx}\right)_r}{\left(\frac{dE}{dx}\right)_c} \cong \frac{EZ}{700} \tag{6.18}$$

For higher-Z materials, it is given approximately by

$$\frac{\left(\frac{dE}{dx}\right)_r}{\left(\frac{dE}{dx}\right)_c} \cong \frac{EZ}{800} \tag{6.19}$$

where E is in MeV.

EXERCISE 6.8: At what energy would an electron start losing its energy equally in both bremsstrahlung and ionization/excitation while moving through water and lead (Z = 82)?

Solution:

$$\frac{\left(\frac{dE}{dx}\right)_r}{\left(\frac{dE}{dx}\right)_c} \cong \frac{EZ}{700} = 1$$

An effective atomic number for a compound or a mixture can be estimated by using the following empirical relation:

$$Z_{\text{eff}} = (f_1 Z_1^{2.94} + f_2 Z_2^{2.94} + ...)^{1/2.94} \tag{6.20}$$

where f_n is the fraction of the total number of electrons associated with each atom and Z_n is the atomic number of each element.

For water (H_2O), which is made up of two hydrogen atoms (Z = 1) and one oxygen atom (Z = 8), the total number of electrons is 10. Hence, the fraction of electrons for two hydrogen atoms is 2/10 and for the oxygen atom is 8/10. Therefore, the effective atomic number of water

$$Z_{\text{eff}} = (0.2 \times 1^{2.94} + 0.8 \times 8^{2.94})^{1/2.94} = 7.42$$

$$\therefore \qquad E = \frac{700}{7.42} = 94.3 \text{ MeV}$$

For lead,

$$\frac{\left(\frac{dE}{dx}\right)_r}{\left(\frac{dE}{dx}\right)_c} \cong \frac{EZ}{800} = 1$$

$$\therefore \qquad E = \frac{800}{82} = 9.76 \text{ MeV}$$

Range–Energy Relations for Mono-energetic Electrons

It is difficult to define the range of fast electrons as the total path length of an electron is considerably greater as compared to the distance of penetration along its initial direction. Collisions with the orbital electrons of the medium may change the direction of the electrons. As a result, they may never be able to reach the detector kept in line with the source for the measurement of the transmitted electrons. Thus, even a thin absorber can reduce the flux of electrons in an attenuation set up. A plot of the number of transmitted electrons versus the absorber thickness is shown in Figure 6.8.

It can be noticed that there is a continuous decrease in the electron flux, which gradually reduces to zero with the increase in thickness. Extrapolation of the linear portion of the curve to zero intensity represents the absorber thickness so that almost no electrons can penetrate the entire thickness. This extrapolated range (R_{ex}) is also known as the practical range. Range of electrons is much larger as compared to the α-particles of the same energy, e.g., the range of 1 MeV electrons would be around 2 mm in low density materials and about 1 mm in moderately dense materials, whereas a thin paper can stop 4 MeV α-particles. Roughly, for electrons of equal initial energy, the product of the range and the density of the absorber is a constant for different materials.

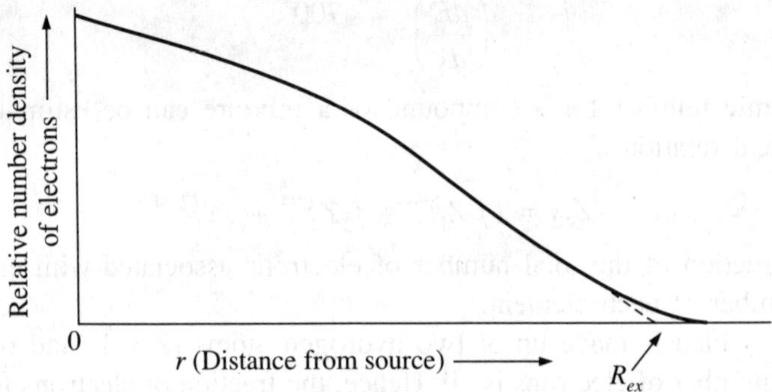

FIGURE 6.8 Transmission or attenuation curve for mono-energetic electrons.

Absorption of Continuous Beta Particle Spectrum

The transmission curve for the β-particles emitted by a radioisotope source differs significantly from Figure 6.8, which is for mono-energetic electrons, because of the continuous distribution of their energy. In the radioactive decay of nuclides emitting the β-particles, all the β-particle energies are possible from zero to a maximum, called the endpoint energy. The low energy β-particles are rapidly absorbed even for a thin absorber, and hence the initial slope on the transmission curve is greater. Beta particle absorption spectrum shows a very close exponential behaviour (Figure 6.9).

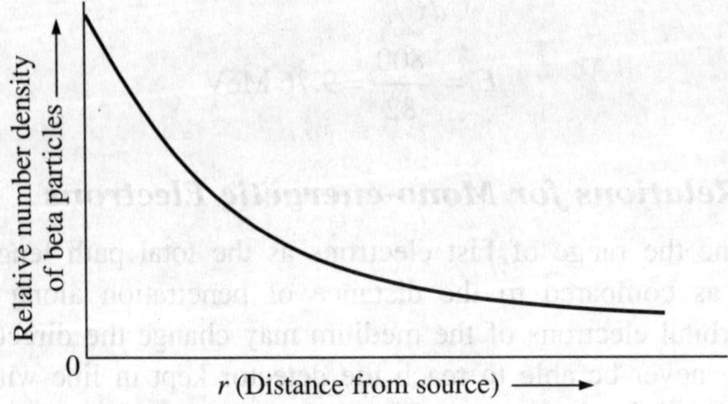

FIGURE 6.9 Transmission curve for continuous beta particle spectrum.

On a semi-log plot, therefore, the transmission curve for the β-particles is linear as shown in Figure 6.10. If n is defined as the absorption coefficient and t is the absorber thickness, the counting rate with the absorber is given as

$$I = I_0 \exp(-nt) \tag{6.21}$$

where I_0 is the counting rate without the absorber. The absorption coefficient n correlates well with the endpoint energy of the β-emitter for a specific absorbing material.

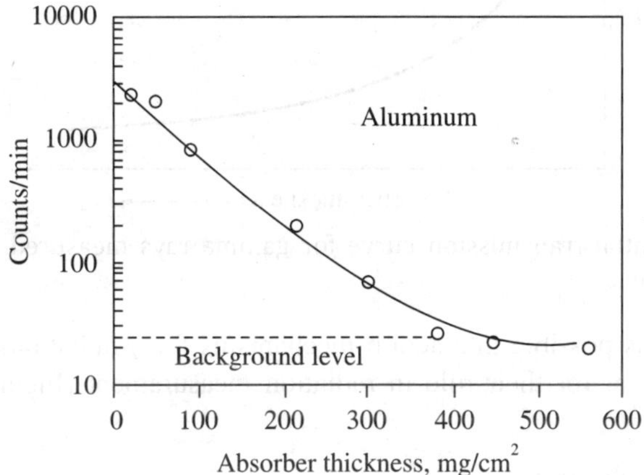

FIGURE 6.10 Transmission curve for beta particles emitted from ^{210}Bi (end point energy of 1.17 MeV) in aluminum.

6.2 INTERACTION OF ELECTROMAGNETIC RADIATIONS WITH MATTER

Many particle interactions, radioactive decays and nuclear reactions result in the emission of X-rays and γ-rays. Gamma rays are the highest energy electromagnetic radiations or photons, such that unlike α and β-particles, they are not completely stopped by any material. However, their intensity can be reduced or attenuated. The attenuation is governed by the exponential absorption law:

$$I = I_0 \exp(-\mu t) \tag{6.22}$$

where I is the γ-ray intensity transmitted through an absorber of thickness t, I_0 is the γ-ray intensity without the absorber and μ is the linear absorption or attenuation coefficient. The transmission curve for mono-energetic γ-rays, therefore, is exponential in nature for *good geometry* conditions as shown in Figure 6.11.

In good geometry conditions, the detector measures mostly the γ-rays that are directly emitted from the source. In bad geometry measurement, however, the γ-rays after having scattered in the absorber also reach the detector, and thereby an additional contribution of the scattered γ-rays that has to be taken into account. A multiplicative correction term known as the buildup factor is introduced to take care of the situation. The attenuation is given as follows:

$$\frac{I}{I_0} = B(t, E_\gamma) \exp(-\mu t) \tag{6.23}$$

where $B(t, E_\gamma)$ is the buildup factor. The buildup factor depends on the specific geometry of the experiment and the type of the detector used.

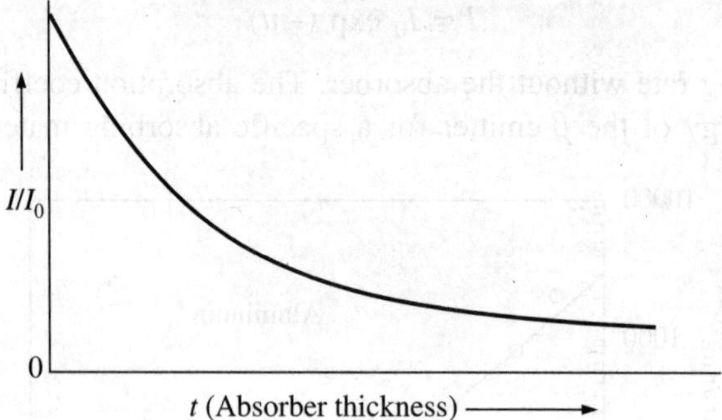

FIGURE 6.11 Exponential transmission curve for gamma rays measured under good geometric conditions.

Out of the various possible interaction mechanisms for γ-radiations in matter; only three are particularly important for their role in radiation measurement (Figure 6.12). These are as follows:

(i) Photoelectric absorption
(ii) Compton scattering
(iii) Pair production

All these processes lead to the partial or total transfer of the γ-radiation energy to the electron energy, which leads to the photon either disappearing completely or scattering through a large average angle. Unlike the charged particles discussed earlier which lose energy gradually through continuous interactions with many absorber atoms, changes in the γ-ray photon energy is sudden and abrupt.

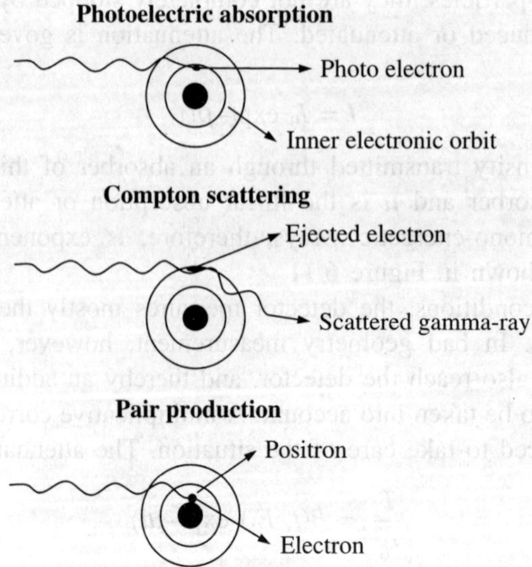

FIGURE 6.12 Schematic description of the three main processes for gamma-ray interaction with matter.

6.2.1 Photoelectric Absorption

In around 1880, Hertz and Lenard observed that when a clean metallic surface is irradiated by monochromatic light of proper frequency, electrons are emitted from it. This phenomenon of ejection of the electrons from the metal surface is called the photoelectric effect and the electrons thus ejected are called the photoelectrons.

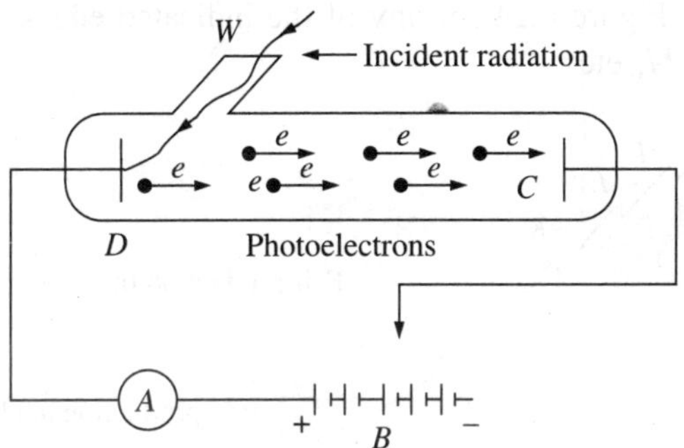

FIGURE 6.13 Schematic diagram of the experimental setup for photoelectric effect.

The schematic diagram of the experimental setup for photoelectric effect is shown in Figure 6.13. In an evacuated glass tube, two metal plates *C*, acting as the collecting anode and *D* which acts as the photosensitive plate are enclosed. These two plates are connected to a battery *B* and ammeter *A*. When the incident radiation falls on plate *D* through the window *W*, electrons are ejected out of the plate and a current flows in the circuit. With the help of this apparatus, the following dependencies of the photoelectric effect were found.

(i) The number of electrons emitted by the metal is found to be directly proportional to the intensity of the light.
(ii) The emitted electrons move faster if the light has a higher frequency.
(iii) There is a cut-off frequency for the incident photons, below which no electrons are emitted. The retarding potential for which the photoelectric current becomes zero is called cut-off or stopping potential V_c.

In the photoelectric absorption process, the incident γ-ray photon undergoes an interaction with an absorber atom in which the photon disappears entirely. This absorption results in the excitation of the atom above the binding energy of some of its orbital electrons, with the result that a pair consisting of an electron and an ion formed. The energy E_e of the emitted photoelectron is the difference between the energy of the γ-ray ($E_\gamma = h\nu$) and the binding energy for that electron E_b in the atom, i.e.,

$$E_e = h\nu - E_b \tag{6.24}$$

Nearly all the γ-ray energy is carried off by the ejected photoelectron since the electron is lighter than the recoil atom.

The ion-pair formed as a result of photoelectric absorption creates a vacancy in one of the bound shells of the absorber atom. This vacancy is quickly filled through the capture of a free electron from the medium or rearrangement of the electrons from the outer shells of the atom. This produces either the fluorescence X-rays or the Auger electrons.

The probability for the photoelectric absorption for a given orbital electron is the maximum if the energy of the incident γ-ray is equal to the binding energy for that orbital electron. It is zero when the energy of the γ-ray is less than the binding energy. As the incident photon energy increases above the binding energy, the probability for the photoelectric absorption decreases. This trend can be seen in Figure 6.14 for any of the indicated edges. The edges correspond to the electron shells, *K*, *L*, *M*, etc.

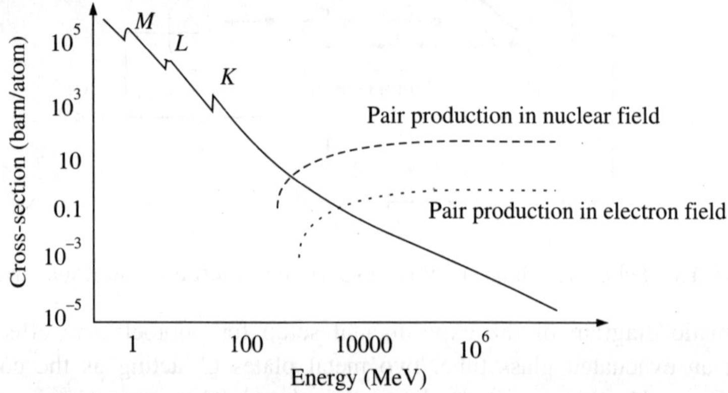

FIGURE 6.14 Photoelectric effect and pair production cross-sections of photon interactions with lead.

The atomic cross-section for the photoelectric absorption varies over different ranges of the incident photon energy E_γ and the atomic number of the absorber material *Z*. A rough approximation for the cross-section is as follows:

$$\sigma_\gamma \propto \begin{cases} \dfrac{Z^5}{E_\gamma^{3.5}}, & \text{for low photon energies} \\ \dfrac{Z^{4.5}}{E_\gamma}, & \text{for } 0.1\text{ MeV} < E_\gamma < 5\text{ MeV} \end{cases} \tag{6.25}$$

The dependence of the photoelectric absorption probability on the atomic number of the absorber material is one of the reasons for the prevalence of high-*Z* materials, such as lead, in γ-ray shields.

EXERCISE 6.9: The γ-ray photon from ^{137}Cs when incident upon a piece of uranium ejects photoelectrons from its *K*-shell. The momentum measured with a magnetic β-ray spectrometer, yields a value of $B.r = 3.083 \times 10^{-3}$ Wb/m. The binding energy of a *K*-electron in uranium is 115.59 keV. Determine the energy of the incident γ-ray photons.

Solution: The centripetal force is equal to the magnetic force for the moving charged particle in a perpendicular magnetic field, i.e.,

$$\frac{mv^2}{r} = evB$$

The momentum p is given by:

$$p = eBr$$
$$= (1.602 \times 10^{-19}\text{C})(3.083 \times 10^{-3} \text{ Wb/m})$$
$$= (1.602 \times 10^{-19}\text{C})(3.083 \times 10^{-3} \text{ Wb/m}\left(\frac{3 \times 10^8 \text{ m/s}}{1.602 \times 10^{-13} \text{ J/MeV}}\right)\text{MeV/c}$$
$$= 0.925 \text{ MeV/c}$$

Therefore, the total energy of the emitted photoelectron

$$E = \sqrt{(pc)^2 + m^2c^4} = \sqrt{0.925^2 + 0.511^2} \text{ MeV}$$

The total energy is equal to the kinetic energy plus rest mass energy. Therefore, the kinetic energy of the photoelectron

$$K = E - mc^2 = \sqrt{0.925^2 + 0.511^2} - 0.511 = 0.546 \text{ MeV}$$

From equation (6.24), the energy of the γ-ray photon

$$E_g = E_e + E_b = 0.546 + 0.116 = 0.662 \text{ MeV}$$

6.2.2 Compton Scattering

In 1923, A.H. Compton, directing a monochromatic beam of X-rays at a thin slab of carbon, observed that the X-rays that were scattered from the carbon at various angles had a longer wavelength λ' as compared to the incident wavelength λ_0. The amount of wavelength shift, $\Delta\lambda = \lambda' - \lambda_0$ was the same regardless of the target material, implying that it is an effect involving electrons rather than the atom as a whole.

In the Compton scattering, the incident photon is deflected through an angle θ with respect to its original direction, and part of its energy is transferred to the electron, which is assumed to be initially at rest (Figure 6.15). The electron is known as the recoil electron. Because all the scattering angles are possible, the energy transferred to the recoil electron can vary from

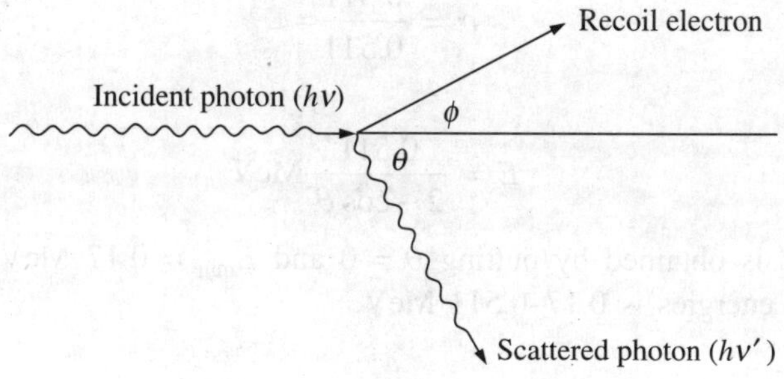

FIGURE 6.15 Compton scattering.

a large fraction of the incident photon energy to zero. From conservation of momentum and energy, one can obtain the following expression, which relates the energy transfer and the scattering angle:

$$h\nu' = \frac{h\nu}{1 + \dfrac{h\nu}{m_e c^2}(1 - \cos\theta)} \tag{6.26}$$

where $m_e c^2$ is the rest mass energy of electron. The difference between $h\nu$ and $h\nu'$ is the kinetic energy of the recoil electron ejected in the Compton scattering. For small scattering angles, the energy of the scattered photon is slightly less than the incident photon energy and very little energy is transferred to the electron. The energy of the scattered photon is the minimum, when the photon is back scattered ($\theta = \pi$).

The kinetic energy of the recoil electron is therefore given by:

$$E_e = h\nu - h\nu' = h\nu\left[\frac{\dfrac{h\nu}{m_e c^2}(1 - \cos\theta)}{1 + \dfrac{h\nu}{m_e c^2}(1 - \cos\theta)}\right] \tag{6.27}$$

The probability of Compton scattering per atom of the absorber is governed by the number of electrons available as scattering targets. Therefore, the probability increases with increasing Z.

EXERCISE 6.10: What is the range of the energies of γ-rays from the annihilation radiation which are Compton scattered? Plot the variation of scattered γ-ray energy with the scattering angle.

Solution: From equation (6.26), the energy of the scattered γ-rays equation

$$E = \frac{E_0}{1 + \alpha(1 - \cos\theta)}$$

where $E_0 = h\nu$, $\alpha = \dfrac{E_0}{m_e c^2}$.

In e^+–e^- annihilation, each γ-ray has energy $E_0 = 0.511$ MeV. Therefore,

$$\alpha = \frac{0.511}{0.511} = 1$$

and,

$$E = \frac{0.511}{2 - \cos\theta}\ \text{MeV}$$

$E_{max} = 0.511$ MeV is obtained by putting $\theta = 0$ and $E_{min} = 0.17$ MeV by putting $\theta = \pi$. Thus, the range of energies is 0.17–0.511 MeV.

We use the following Matlab code to get the plot (Figure 6.16).

```
clear all

syms x real positive;
f=0.511/(2-cos(x));
ezplot(f,[0,pi])
xlabel('Scattering angle (rad)')
ylabel('Photon energy (MeV)')
title('')
```

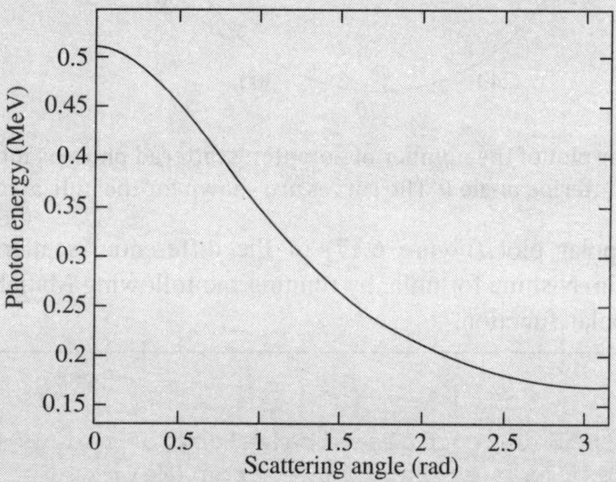

FIGURE 6.16 Scattered gamma ray energy vs. scattering angle.

As can be verified from the above plot, the range of energies is 0.17–0.511 MeV.

The angular distribution of the scattered γ-rays is predicted by Klein–Nishina formula for the differential scattering cross-section, i.e.,

$$\frac{d\sigma}{d\Omega} = r_e^2\left[\frac{1}{1+\alpha(1-\cos\theta)}\right]^3\left[\frac{1+\cos^2\theta}{2}\right]\left[1+\frac{\alpha^2(1-\cos\theta)^2}{(1+\cos^2\theta)[1+\alpha(1-\cos\theta)]}\right] \tag{6.28}$$

where $\alpha = \dfrac{h\nu}{m_e c^2}$ and r_e is the classical electron radius.

EXERCISE 6.11: Use polar coordinates to plot the number of photons that are Compton scattered as a function of the scattering angle for the initial photon energies of 1 keV, 100 keV and 10 MeV. See how there is a strong tendency for forward scattering at higher values of the γ-ray energy.

Solution: To simplify equation (6.28), we write the expression using two sub-expressions, $f(\theta) = (1 - \cos\theta)$ and $g(\theta) = (1 + \cos^2\theta)$.

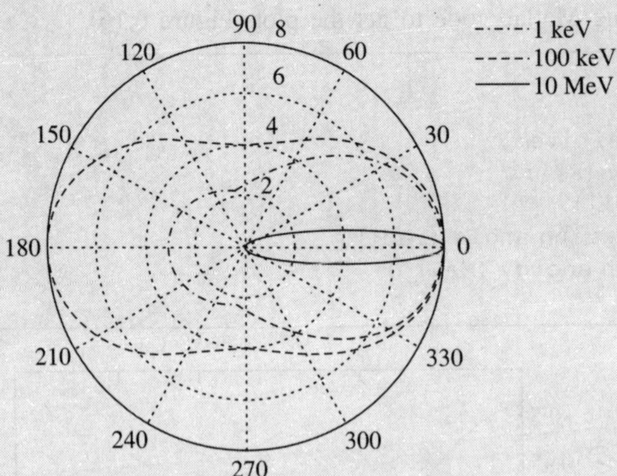

FIGURE 6.17 A polar plot of the number of Compton scattered photons into a unit solid angle at the scattering angle θ. The curves are shown for the indicated initial energies.

We get the above polar plot (Figure 6.17) of the differential scattering cross-section as predicted by the Klein–Nishina formula, by running the following Matlab code, which makes use of the built-in polar function.

```
clear all

re2=2.81794^2; %Square of classical electron radius (fm^2)
me = 0.511; %Rest mass energy of electron (MeV)
E=[0.001 0.1 10]; %Incident photon energy (MeV)
a=E./me; %alpha
t = 0:0.01:2*pi; %Scattering angle (radians)
f=1-cos(t);
g=1+(cos(t)).^2;

rho1=re2.*((1./(1+a(1,1).*f)).^3).*(g/2).*(1+((a(1,1).
*f).^2)./...
  (g.*(1+(a(1,1).*f))));
rho2=re2.*((1./(1+a(1,2).*f)).^3).*(g/2).*(1+((a(1,2).
*f).^2)./...
  (g.*(1+(a(1,2).*f))));
rho3=re2.*((1./(1+a(1,3).*f)).^3).*(g/2).*(1+((a(1,3).
*f).^2)./...
  (g.*(1+(a(1,3).*f))));

polar(t,rho1,'--b')
hold on
polar(t,rho2,'-.b')
polar(t,rho3,'-b')
legend('1 keV','100 keV','10 MeV','Location','BestOutside')
hold off
```

6.2.3 Pair Production

If an incident γ-ray photon has energy that exceeds twice the rest mass energy of an electron ($2m_ec^2 = 1.022$ MeV), the pair production process is energetically possible, which involves the disappearance of the photon to create an electron–positron pair,

$$\gamma \rightarrow e^+ + e^-$$

Any additional photon energy appears as kinetic energy is shared by the electron-positron pair.

$$E_{e^-} + E_{e^+} = h\nu - 2m_ec^2 \tag{6.29}$$

Electric charge is conserved in the reaction because of the equal and opposite charges constituting a pair. Momentum is conserved by the presence of the nucleus, which absorbs usually a negligible amount of the kinetic energy. The same process can also occur in the Coulomb field of an electron, with a threshold energy of $4m_ec^2$. However, in the electric field of a nucleus, the probability of pair production is significantly higher.

The pair production process is complicated because the positron is not a stable particle. Positron slows down in the medium, and once its kinetic energy is comparable to the thermal energy of electrons in the absorber material, the positron annihilates, i.e., it combines with an electron in the absorber material. In this process, both the particles vanish simultaneously to form two annihilation γ-ray photons of energy $m_ec^2 = 0.511$ MeV each, which are emitted in the opposite directions in order to conserve the linear momentum. The annihilation radiation appears almost simultaneously with the original pair production process because of the very small time required for the positron to slow down and annihilate.

The probability of pair production per atom varies approximately as the square of the absorber atomic number. It also increases with incident photon energy. The cross-section for pair production per atom is given as follows:

$$\sigma_{\text{pair}} \propto Z^2 \ln h\nu \tag{6.30}$$

EXERCISE 6.12: A positron and an electron having negligible kinetic energy annihilate one another, producing two γ-rays of equal energy. What is the wavelength of these γ-rays? Why cannot a single γ-ray photon of twice the energy be emitted instead of two photons?

Solution: In the annihilation process, the energy released is equal to the sum of the rest mass energy of the electron–positron pair. Because of momentum and energy conservation, the two photons must carry equal energy. Therefore, each γ-ray photon carries energy equal to

$$E_\gamma = m_ec^2 = 0.511 \text{ MeV}$$

The wavelength of each photon is

$$\lambda = \frac{hc}{E_\gamma} = \frac{197.329 \times 2\pi \text{ MeV fm}}{0.511 \text{ MeV}} = 2425 \text{ fm} = 2.425 \times 10^{-12} \text{ m}$$

If a single photon is emitted, the linear momentum cannot be conserved.

The energy dependence of the three γ-ray interaction processes is shown in Figure 6.18 for sodium iodide.

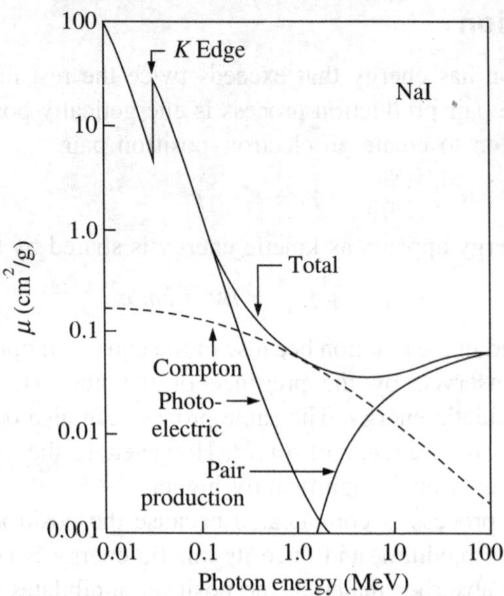

FIGURE 6.18 Energy dependence of the various gamma-ray interactions in sodium iodide. [G.R. White, National Bureau of Standards. (US), Report 1003, 1952]

Gamma-ray Attenuation

We have already been mentioned that the γ-ray attenuation is governed by the following exponential absorption law

$$I = I_0 \exp(-\mu t)$$

Each of the three γ-ray interaction processes removes the photon either by scattering away from the detector direction or by absorption, and can be characterized by a fixed probability of occurrence per unit path length of the absorber. The sum of these three probabilities gives the total probability per unit path length that the γ-ray photon is removed from the beam. This is how the linear attenuation coefficient is defined. Thus,

$$\mu = \mu_{ph} + \mu_C + \mu_{\text{pair}} \tag{6.31}$$

where μ_{ph}, μ_C and μ_{pair} are the attenuation coefficients for photoelectric absorption, Compton scattering and pair production, respectively.

EXERCISE 6.13: A 4 cm diameter and 1 cm thick NaI crystal is used to detect the 660 keV γ-rays emitted by a 100 μCi point source of ^{137}Cs placed on its axis at a distance of 1 m from its surface. The linear absorption coefficients for the photoelectric absorption and the Compton processes are 0.03 and 0.24 per cm, respectively. Calculate separately the number of photoelectrons and Compton electrons released in the crystal. What is the number of 660 keV γ-photons that pass through the crystal without interacting?

Solution: The solid angle subtended at the point source by the circular face of the crystal is essentially conical, where the apex angle of the cone is 2θ, as shown in Figure 6.19.

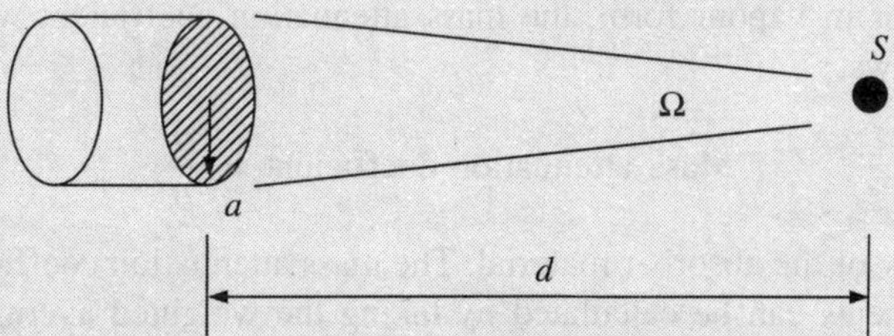

FIGURE 6.19 Solid angle subtended by a circular crystal face at the point source S.

For a cone, the solid angle

$$\Omega = 2\pi(1 - \cos\theta)$$

or

$$\Omega = 2\pi\left(1 - \frac{d}{\sqrt{d^2 + a^2}}\right)$$

For $d >> a$, using binomial expansion, the above expression reduces to

$$\Omega = \frac{\pi a^2}{d^2}$$

Assuming γ-rays are emitted isotropically, the fraction of γ-rays entering the crystal

$$F = \frac{\Omega}{4\pi} = \frac{a^2}{4d^2}$$

$$= \frac{2^2}{4 \times 100} = 10^{-4}$$

The number of photons entering the crystal per second from the source of strength S is given by

$$I_0 = SF$$

$$= \left[(100 \times 10^{-6}\,\text{Ci})\left(3.7 \times 10^{10}\,\frac{\text{dps}}{\text{Ci}}\right)\right](10^{-4}) = 370\ \text{s}^{-1}$$

The number of photons that come out at the end of the crystal of thickness 1 cm, without being scattered or absorbed will be

$$I = I_0 \exp\,[-(\mu_{\text{ph}} + \mu_C)t]$$
$$= 370 \exp\,[-(0.03 + 0.24\ \text{cm}^{-1})\,(1\ \text{cm})]$$
$$= 280$$

The attenuation coefficient varies with the density of the absorber. For a given material and γ-ray energy, the mass attenuation coefficient does not vary with the physical state of the absorber. It is therefore used more frequently than μ. For example, in case of water whether it is in liquid form or in vapour form, the mass attenuation coefficient will remain the same. It is defined as follows:

$$\text{Mass attenuation coefficient} = \frac{\mu}{\rho} \tag{6.32}$$

where ρ is the density of the absorber material. The mass attenuation coefficient for a compound or a mixture of elements can be calculated by taking the weighted average,

$$\left(\frac{\mu}{\rho}\right)_c = \sum_i w_i \left(\frac{\mu}{\rho}\right)_i \tag{6.33}$$

where w_i is the weight fraction of the ith element in the mixture.

In terms of the mass attenuation coefficient, the exponential attenuation law takes the following form:

$$I = I_0 \exp\left[-\left(\frac{\mu}{\rho}\right)\rho t\right] \tag{6.34}$$

where the product ρt, is known as the absorber's mass thickness. The thickness of the absorbers used in the radiation measurements is often measured in mass thickness rather than physical thickness. For different absorber materials, a particle should encounter about the same number of electrons while travelling through the absorbers of equal mass thickness. Therefore, the stopping power and the range, when expressed as ρt, are roughly the same for materials that do not differ in Z.

The γ-ray photons can also be characterized by their mean free path λ, which is the average distance traversed by the photon in the absorber before an interaction takes place. It is simply the reciprocal of the linear attenuation coefficient.

$$\lambda = \frac{\int_0^\infty t\, e^{-\mu t}\, dt}{\int_0^\infty e^{-\mu t}\, dt} = \frac{1}{\mu} \tag{6.35}$$

EXERCISE 6.14: The soft tissue consists of large percentage of water, therefore can be approximated to have a density of 1 g/cm^3. It is further approximated to be composed of 76.1826% of oxygen, 10.1174% of hydrogen, 11.1% of carbon and 2.6% of nitrogen. Determine the linear attenuation coefficient for 0.6 MeV γ-rays for which the mass attenuation coefficients in these four elements are 0.08070 cm^2/g in oxygen, 0.1599 cm^2/g in hydrogen, 0.08058 cm^2/g in carbon, and 0.08063 cm^2/g in nitrogen.

Solution: The mass attenuation coefficient for the soft tissue

$$\left(\frac{\mu}{\rho}\right)_c = \sum_i w_i \left(\frac{\mu}{\rho}\right)_i$$

$$= 0.101174 \times 0.1599 + 0.111 \times 0.08058 + 0.026 \times 0.08063 + 0.761826 \times 0.08070$$

$$= 0.088698 \text{ cm}^2/\text{g}$$

Therefore,

$$\mu = (1 \text{ g/cm}^3)\,(0.088698 \text{ cm}^2/\text{g}) = 0.088698 \text{ cm}^{-1}$$

6.3 INTERACTION OF NEUTRONS

Neutrons, because of being neutral, do not have any Coulomb interactions with the orbital electrons in passing through the matter, and are not continuously slowed down. They interact with the atomic nuclei through nuclear force, which has an extremely short range. Since, atomic nuclei are much smaller than the atoms, the probability of an energetic neutron hitting a nucleus is quite low. For this reason, the neutrons can traverse large thickness of the absorber material before being stopped.

Neutrons interact in matter by means of nuclear reactions, which depend to a great extent on their energies, which can range from hundreds of MeV to fractions of an eV. Neutrons may be scattered, in which case, the energy is transferred to the recoiling nuclei. On the other hand, they may be absorbed in a variety of different processes. Most neutron detectors utilize some sort of conversion of the incident neutron into secondary charged particles, which can then be detected directly.

The most common neutron interactions are as follows:

Elastic scattering: The elastic interaction of the neutrons with atomic nuclei is most significant at neutron energies below the threshold for the nuclear reactions. In collisions with heavy nuclei, the neutrons lose little energy. In collisions with light nuclei, such as hydrogen, deuteron, helium and carbon, the neutrons lose most of their energies. A neutron can lose all of its kinetic energy in a single collision with a proton, as they have similar masses.

Inelastic scattering: A neutron may strike a nucleus and rather than deflecting off, as in the case of elastic scattering, a compound nucleus may be formed. This nucleus is unstable and emits a neutron of lower energy together with a γ-ray photon, which takes up the remaining energy. This process is called inelastic scattering, and is most effective at high neutron energies in heavy materials.

Radiative capture: The neutron is captured by a nucleus, which then emits a γ-photon. This reaction is most important for the neutrons with very low energy. The product nuclei of (n, γ) reactions are typically radioactive and are β or γ emitters.

Transmutation: When the neutrons form a compound nucleus which in turn ejects a different particle, a transmutation is said to have occurred. In addition to the (n, p) reactions, many other reactions like $(n, 2n)$ and (n, α) are possible.

All these interaction have already been outlined in Chapter 5.

The number of neutrons that collide with the absorber material nuclei is proportional to the neutron beam intensity and the total number of nuclei in the absorber. The probability for each type of interaction is generally expressed in terms of the effective cross-sectional area, frequently called the microscopic cross-section σ expressed in barn. Examples of cross-section plots for neutron are shown in Chapter 7 (Figure 7.5). When multiplied by the number of nuclei N per unit volume, also known as the number density, the microscopic cross-section σ is converted into the macroscopic cross-section Σ, which is the probability per unit length for the specific interaction, i.e.,

$$\Sigma = N\sigma \tag{6.36}$$

By adding together the cross-sections for individual interaction, we can get the total interaction cross-section.

$$\Sigma_{\text{total}} = \Sigma_{\text{scatter}} + \Sigma_{\text{radiative}} + \ldots \tag{6.37}$$

This is equivalent to the linear absorption coefficient for the γ-rays as defined earlier.

EXERCISE 6.15: Calculate the microscopic and macroscopic absorption cross-section for natural uranium if σ_{235} = 681 b, and σ_{238} = 2.7 b. The density of uranium is 19 g/cm^3. The abundances of ^{238}U and ^{235}U isotopes in natural uranium are 99.28% and 0.72%, respectively.

Solution: The number densities for the respective isotopes

$$N_{238} = 99.28\%\frac{N_a\rho}{A} = 99.28\%\frac{(6.023\times10^{23})\times19}{238}$$

$$= 4.77 \times 10^{22} \text{ nuclei/cm}^3$$

$$N_{235} = 0.72\%\frac{N_a\rho}{A} = 0.72\%\frac{(6.023\times10^{23})\times19}{235}$$

$$= 3.50 \times 10^{20} \text{ nuclei/cm}^3$$

Therefore, the total microscopic cross-section of natural uranium

$$\sigma = \frac{N_{235}\sigma_{235} + N_{238}\sigma_{238}}{N_{235} + N_{238}} = 7.64 \text{ b}$$

and, the total macroscopic cross-section of natural uranium

$$\Sigma = N_{235}\sigma_{235} + N_{238}\sigma_{238} = \text{s}(N_{235} + N_{238}) = 0.367 \text{ cm}^{-1}$$

The neutron beam attenuation also follows the exponential law for good geometry conditions, like that for the γ-rays. The number of detected neutrons falls off exponentially with the absorber thickness.

$$I = I_0 \exp(-\Sigma t) \tag{6.38}$$

where I_0 is the initial intensity of the neutron beam and t is the thickness of the absorber material. The ratio $\frac{I_0}{I}$ is called the attenuation factor.

Like the γ-ray photons, neutrons too can also be characterized by their mean free path λ, which is the average distance traversed by the neutron in the absorber before an interaction takes place. It can be calculated from the probability that a neutron will interact in the distance interval between t and $t + dt$, and is simply the reciprocal of the macroscopic cross-section for the medium.

$$\lambda = \frac{\int_0^\infty t e^{-\Sigma t} dt}{\int_0^\infty e^{-\Sigma t} dt} = \frac{1}{\Sigma} \tag{6.39}$$

EXERCISE 6.16: What is the mean free path and the time required by a neutron with energy 100 eV to have its first interaction in natural uranium?

Solution: The neutron free path is the reciprocal of the macroscopic cross-section (value from the previous example)

$$\lambda = \frac{1}{\Sigma} = \frac{1}{0.367 \text{ cm}^{-1}} = 2.72 \text{ cm}$$

Using non-relativistic relation, the neutron velocity can be obtained from its kinetic energy

$$v = \sqrt{\frac{2E}{m_n}} = \sqrt{\frac{2(100 \times 10^{-6} \text{ MeV})}{939.57 \text{ MeV/c}^2}}$$

$$= 4.614 \times 10^{-4}\ c = 1.38 \times 10^5 \text{ m/s}$$

Therefore, the time for the first interaction

$$t = \frac{\lambda}{v} = \frac{0.0272 \text{ m}}{1.38 \times 10^5 \text{ m/s}} = 0.197\ \mu\text{s}$$

Reaction Rate and Neutron Flux

In most of the situations, rather than analyzing an individual neutron, it is more convenient to analyze the neutron population as a whole. The total distance that the neutron travels in unit time per unit volume of a given material is called the neutron flux, and has units of number of neutrons per unit time and unit area. If all the neutrons have a single energy or fixed velocity v and the neutron number density n, then the neutron flux, Φ may be written as

$$\Phi = nv \tag{6.40}$$

For a majority of applications, determining the neutron reaction rates is important. A neutron while interacting with the nuclei of a medium scatters from one nucleus to another, until it is absorbed or it escapes the boundary of the material. The neutron travels a distance equal to the mean free path before interacting with the medium; therefore, the reaction rate density (reaction rate per unit volume of the medium and unit time) is given as the neutron flux divided by the mean free path, i.e.,

$$R = \frac{\Phi}{\lambda} = \Phi\Sigma \tag{6.41}$$

If the neutrons have different velocities or in other words, different energies, a more generalized equation for the reaction rate density may be written to include an energy-dependent neutron flux and cross-section.

$$R = \int_0^\infty \Phi(r, E)\Sigma(E)\, dE \tag{6.42}$$

EXERCISE 6.17: In ^{235}U isotope, a neutron flux is sustained at 5×10^{14} neutrons/cm^{2}s. If the reaction rate density is 1.5×10^{13} reactions/cm^{3}s, calculate the macroscopic and microscopic cross-sections for the medium.

Solution: From equation (6.41), the macroscopic cross-section is given as follows:

$$\Sigma = \frac{R}{\Phi} = \frac{1.5 \times 10^{13}}{5 \times 10^{14}} = 0.03 \text{ cm}^{-1}$$

Therefore, the microscopic cross-section is (using the number density N_{235} from Exercise 6.15)

$$\sigma = \frac{\Sigma}{N_{235}} = \frac{0.03 \text{ cm}^{-1}}{3.50 \times 10^{20} \text{ nuclei/cm}^3} = 85.7 \text{ b}$$

6.4 RADIATION EXPOSURE AND DOSE

If the human tissue is exposed to nuclear radiations, the primary physical interactions are similar to those in any other material, but there are important secondary effects involving chemical alteration of the biological molecules which are influenced by the ionization or excitation caused by the radiation. Radiation exposure, thus, may result in adverse effects in living organisms. For the study of its biological effects, radiation protection and the safe handling of radiation sources, it is important to have a way to measure the radiations. The concepts of radiation exposure and dose are useful for such radiation measurement.

6.4.1 Gamma Ray Exposure

The concept of γ-ray exposure is roughly analogous to the strength of an electric field created by a point charge. The basic unit of γ-ray exposure is defined in terms of the charge dQ due to ionization created by the secondary electrons formed within a volume element of air of mass dm, when these secondary electrons are completely stopped in air. $\frac{dQ}{dm}$ gives the exposure value X. The historical unit of γ-ray exposure has been the Roentgen (R), defined as the amount of γ or X-radiation which will produce one electrostatic unit (esu) of charge of either sign $\left(\frac{1}{3}\times 10^{-9}\text{ C}\right)$ in one cm^3 of air at NTP (0 °C, 1 atm).

$$1\text{ R} = \frac{1\text{ esu}}{0.001293\text{ g}} = \frac{\frac{1}{3}\times 10^{-9}\text{ C}}{1.293\times 10^{-6}\text{ kg}} = 2.58\times 10^{-4}\text{ C/kg}$$

6.4.2 Absorbed Dose

Rather than exposure, to study the effects of radiation on a person, we are more interested in the dose absorbed by the body. For radiation protection purposes, it is useful to define an average absorbed dose for a tissue or an organ:

$$D_T = \frac{\varepsilon_T}{m_T} \tag{6.43}$$

where m_T is the mass of the tissue in which ε_T amount of energy is deposited.

The historical unit of absorbed dose has been the rad defined as 100 ergs/g. The SI unit is the gray (Gy), which is defined as 1 J/kg. The two units are related as follows:

$$1\text{ Gy} = 100\text{ rad}$$

6.4.3 Dose Equivalent

Types of ionizing radiations differ in the way in which they interact with biological materials, so that equal doses (i.e., equal amounts of energy deposited) do not necessarily have equal biological effects. In order to put all the different types of ionizing radiations on an equal basis with respect to their potential for causing harm, we need to define another quantity. The severity of the biological damage is directly related to the local rate of energy deposition along the path taken by the charged particle, known as the linear energy transfer (LET). LET is very similar to the stopping power $\left(-\frac{dE}{dx}\right)$ that was defined earlier. The radiations with higher values of LET, such as heavy charged particles, tend to result in greater biological damage as compared to those with lower LET, such as electrons, even though the total energy deposited per unit mass is the same.

A quantity, called *dose equivalent*, has therefore been defined to more adequately quantify the probable biological effect of a given radiation exposure. The dose equivalent H_T in a tissue or organ is given by:

$$H_T = w_R D_{T,R} \tag{6.44}$$

where $D_{T,R}$ is the average absorbed dose in tissue T from a given type of radiation R, and w_R is the radiation weighing factor, which gives the average biological response to a given radiation dose relative to that induced by 250 keV X-rays or γ-rays. Table 6.4 lists the values of w_R for different radiations.

Table 6.4 Weighing Factors for Different Radiations

Type of radiation	*Energy range*	*Weighing factor* (w_R)
Photons, electrons	All energies	1
Neutrons	<10 keV	5
	10–100 keV	10
	100 keV–2 MeV	20
	2–20 MeV	10
	>20 MeV	5
Protons	<20 MeV	5
Alpha particles, fission fragments, heavy nuclei		20

Thus, we can see that a 1-Gy dose given by 1 MeV neutrons (w_R = 20) is biologically equivalent to a 20-Gy dose of γ-radiation (w_R = 1). If there are several types of radiations, the dose equivalent is given by the weighted average of all the contributions.

$$H_T = \sum_R w_R D_{T,R} \tag{6.45}$$

The units for dose equivalent depend on the corresponding units of absorbed dose. If the dose is expressed in the historical unit of rad, dose equivalent is in units of rem. The SI unit for dose equivalent is called the sievert (Sv), when the absorbed dose is measured in gray (Gy). Because, 1 Gy = 100 rad, the units for the dose equivalent are interrelated, i.e.,

$$1 \text{ Sv} = 100 \text{ rem}$$

Guidelines for radiation exposure limits are generally quoted in units of dose equivalent, in order to put different types of exposures and energies on a common footing.

EXERCISE 6.18: During a diagnostic X-ray, a broken leg with mass of 5 kg receives a dose equivalent of 0.5 mSv. If the X-ray energy is 50 keV, how many X-ray photons were absorbed?

Solution:

$$D_{T,R} = \frac{H_T}{w_R} = \frac{0.5 \times 10^{-3}}{1} = 0.5 \times 10^{-3} \text{ Gy}$$

From equation (6.43), the total energy deposited by the X-ray photons

$$\varepsilon_T = D_T m_T = (0.5 \times 10^{-3} \text{ Gy})\ (5 \text{ kg}) = 2.5 \times 10^{-3} \text{ J}$$

Therefore, the number of photons

$$N = \frac{\varepsilon_T}{E_{\text{photon}}} = \frac{2.5 \times 10^{-3} \text{ J}}{(50 \times 10^{-3} \text{ MeV})(1.602 \times 10^{-13} \text{ J/MeV})} = 3.12 \times 10^{11}$$

PROBLEMS

6.1 For each ion-pair formed, an α-particle loses about 35.5 eV. Calculate the number of ion-pairs produced by an α-particle with a kinetic energy of 5.5 MeV.

[**Ans:** 154,930 ion-pairs]

6.2 Assuming the Bethe–Bloch formula is valid for low energies; find the kinetic energy of protons in iron for which the Bragg peak would occur. [**Ans:** 374 keV]

6.3 Show that the range of a deuteron of energy E is twice that of a proton of energy $E/2$.

6.4 Compare the stopping power of a 2 MeV proton and a 4 MeV α-particle in the same medium.

6.5 Show that the range of α-particles and protons of energy 1–10 MeV in aluminum is $\frac{1}{1600}$ of the range in air at 15°C and 1 atm. Take the air density as 1.226×10^{-3} g cm^{-3} and the density of aluminum as 2.7 g cm^{-3}.

6.6 The range of 5 MeV α-particles in air at NTP is 3.8 cm. Calculate the range of 10 MeV α-particles using Geiger's rule. [**Ans:** 10.75 cm]

6.7 Protons and deuterons lose the same amount of energy in passing through a thin sheet of material. How are their energies related?

6.8 The range of a 15 MeV proton is 1100 μm in nuclear emulsions. A second particle whose initial ionization is the same as that of proton, has a range of 165 μm. What is the mass of the particle in terms of the mass of an electron? The rate at which a single ionized particle loses energy E, by ionization along its range is given by

$$\frac{dE}{dR} = \frac{K}{(\beta c)^2} \text{ MeV μm}$$

where βc is the velocity of the particle and K is a constant depending on the emulsion.

[**Ans:** $275m_e$]

6.9 Show that except for small ranges, the straggling of a beam of ^{3}He particles is greater than that of a beam of ^{4}He particles of equal range.

6.10 Compute the maximum range of a proton in aluminum, air, silicon and water for the range of energies from 0.001 eV to 1 GeV.

6.11 Use Bethe–Bloch formula to calculate the energy loss of an α-particle and a proton as a function of the particle velocity, varying it from zero to the speed of light.

6.12 The range of a β-particle is measured to be 0.111 mm in aluminum. Calculate its range if the β-emitter is placed in air 1 cm away from the aluminum sheet and with a 1.7 mg/cm^2 mica absorber between the counter and the aluminum sheet. The density of air can be taken as 1.293 mg cm^{-3}. [**Ans:** 32.96 mg cm^{-3}]

6.13 Ultraviolet light of wavelength 2537 Å from a mercury arc falls upon a silver photocathode. If the photoelectric threshold wavelength for silver is 3250 Å, calculate the least potential difference which must be applied between the anode and the photocathode to stop the electrons from the photocathode. **[Ans:** 1.08 V]

6.14 Show that the photoelectric effect cannot take place with a free electron.

6.15 A 30 keV X-ray photon strikes the electron, which is initially at rest. As a result, the photon is scattered through an angle of 30°. What is the recoil velocity of the electron? **[Ans:** 0.03*c*]

6.16 Plot the Compton scattering energy of the scattered beam for the initial photon energies of 0.05, 0.1, 1, 5 and 10 MeV as a function of the photon scattering angle.

6.17 Calculate the maximum wavelength of γ-rays which in passing through matter can lead to the creation of electrons? **[Ans:** 1.214×10^{-12} m]

6.18 Show that the electron–positron pair cannot be created by an isolated photon.

6.19 Calculate the attenuation factor for thermal neutrons passing through a layer of water 2.5 cm thick. The macroscopic cross-section for the thermal neutrons is 0.02 cm^{-1}. **[Ans:** 1.05]

6.20 Hazardous levels of ionizing radiation are signified by the trefoil sign (Figure 6.20) on a yellow background, which are usually posted at the boundary of a radiation controlled area or at any location where the radiation levels are significantly higher.

FIGURE 6.20 Ionizing radiation hazard symbol.

Within a given year, a radiation worker received a low-energy (<10 keV) neutron dose of 2 mGy. How many kBq of ^{131}I could he be allowed to ingest as treatment (ingestion dose = 22 μSv/kBq) for a thyroid problem if he is to avoid exceeding the recommended dose limit of 20 mSv/y? [Ans: 0.45 kBq]

BIBLIOGRAPHY

Jevremovic, T., *Nuclear Principles in Engineering*, Springer, New York, 2009.

Knoll, G.F., *Radiation Detection and Measurement*, Wiley, New York, 2010.

Leo, W.R., *Techniques for Nuclear and Particle Physics Experiments*, Narosa, New Delhi, 1995.

Radiation Detection and Measurement, IANCAS Bulletin, Vol. 3, No. 3, Mumbai, 2004.

7 Detectors and Instrumentation

"What we observe is not nature itself, but nature exposed to our method of questioning."

—Werner Heisenberg

Experimental studies in the various areas of nuclear science and engineering involve detection of nuclear radiations, which refer to all those particles and electromagnetic radiations emitted as a result of nuclear reactions including radioactive decay. A radiation detector, also known as a particle detector, is a device used for this purpose. The term counter is often used rather than detector, when the detector counts the particles but does not resolve its energy.

Detection does not simply mean localization of the radiation/particle. To be useful, detection has to have a resolution sufficient enough to enable particles to be separated in both space and time in order to understand their association with a particular event. Measurement of its energy, momentum and spin is also required. No single detector is ideal and is optimal with respect to all these requirements. Nevertheless, detectors are such an integral part of an experiment to probe atomic and nuclear phenomena that they are also designed and fabricated for different specific needs. Detector development is a rapidly moving major area of research. Over the years, improved detector systems have been continuously evolving, making use of the developments in material processing, detector design and fabrication, electronic pulse processing techniques and data recording and analysis methods, which have resulted in marked improvements in the detection efficiency and resolution.

Although, there are different types of radiation detectors, their basic properties like efficiency, energy resolution, etc. remain common to all them.

7.1 GENERAL PROPERTIES OF RADIATION DETECTORS

Some of the important general characteristics of radiation detectors are as follows:

(i) Energy proportionality
(ii) Pulse shape
(iii) Energy resolution
(iv) Detection efficiency
(v) Time resolution

We discuss each of these briefly.

7.1.1 Energy Proportionality

The pulse size is usually nearly proportional to the energy deposited by the incident charged quantum of radiation within the detector. If V_{max} represents the maximum amplitude of the signal pulse, it should be directly proportional to the total charge Q created within the detector during one radiation interaction, since the capacitance C of the equivalent load circuit is generally fixed. It is given by the following relation:

$$V_{max} = \frac{Q}{C} \tag{7.1}$$

For scintillating detectors, the ionization density of the radiation influences the transformation of energy into light, e.g., the light pulse of an α-particle is only about 1/10 of that of an electron depositing the same amount of energy in liquid scintillators.

EXERCISE 7.1: What is the amplitude of the voltage pulse produced when a charge equal to that carried by 10^6 electrons is collected on a capacitance of 100 pF?

Solution:

$$V_{max} = \frac{Q}{C} = \frac{(10^6)(1.602 \times 10^{-19}\ \text{C})}{100 \times 10^{-12}\ \text{F}} = 1.602\ \text{mV}$$

7.1.2 Pulse Shape

If we examine a large number of signal pulses, their amplitudes will not all be the same. Variations may be due to the differences in the radiation energy or the fluctuation within the detector itself. The pulse amplitude distribution can therefore, be used to deduce information about the incident radiation by discriminating true signal pulses from interfering pulses.

Most commonly, the pulse amplitude information is displayed through the *differential pulse height distribution*, which has the pulse height H on the x-axis and dN/dH on the y-axis, where N represents the number of pulses. Therefore,

$$\text{Number of pulses with amplitude between } H_1 \text{ and } H_2 = \int_{H_1}^{H_2} \frac{dN}{dH}\, dH \tag{7.2}$$

Hence, one can infer that the peaks in the spectrum indicate the amplitudes where more number of pulses may be found, and the valleys in the spectrum are indicative of the pulse amplitude near which relatively few pulses occur.

When the detectors are operated in the pulse mode, the pulses are commonly fed to a counting device. In order to cut off the noise and the background, the signal amplitudes exceeding the fixed discrimination level H_d of the counting device only get registered. The gain or the amplification of the counting system is adjusted to find the optimum value for carrying out the counting. In some types of radiation detectors like Geiger–Mueller counters or scintillation counters, the gain can be conveniently adjusted by varying the applied voltage to the detector.

EXERCISE 7.2: A hypothetical differential pulse height distribution in shown in Figure 7.1. In practical, the spectrum is a smooth curve.

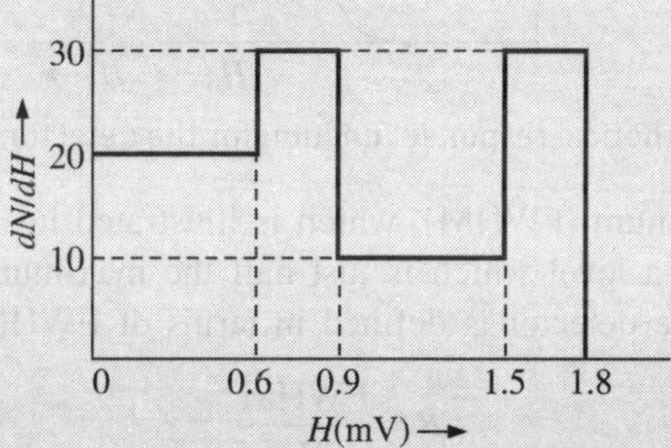

FIGURE 7.1 A hypothetical differential pulse height distribution.

Find the total number of pulses under the entire spectrum. If the discrimination level of the counting device is set at 0.3 mV, what will be the total number of pulses?

Solution: The total number of pulses N_0 is obtained by integrating the area under the entire spectrum, i.e.,

$$N_0 = \int_0^{1.8} \frac{dN}{dH} dH$$

$$= 20 \times 0.6 + 30 \times (0.9 - 0.6) + 10 \times (1.5 - 0.9) + 30 \times (1.8 - 1.5) = 36$$

If H_d is set at 0.3 mV, the total number of pulses will be

$$N = \int_{0.3}^{1.8} \frac{dN}{dH} dH$$

$$= 20 \times (0.6 - 0.3) + 30 \times (0.9 - 0.6) + 10 \times (1.5 - 0.9) + 30 \times (1.8 - 1.5) = 30$$

7.1.3 Energy Resolution

The ability of a detector to distinguish between two incident radiations of nearby energies is called the energy resolution of the detector. It can be determined from the response function of the detector, which is the differential pulse height distribution for a detector in response to a mono-energetic source of radiation. A hypothetical response function is shown in Figure 7.2.

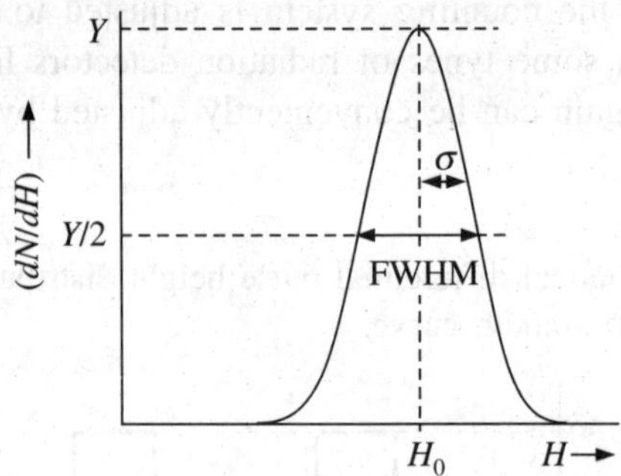

FIGURE 7.2 Hypothetical response function for the detector, illustrating FWHM.

The full width at half maximum (FWHM), which is illustrated in Figure 7.2, is defined as the width of the distribution at a level which is just half the maximum *y*-coordinate of the peak. The energy resolution of the detector is defined in terms of FWHM as follows:

$$R = \frac{\text{FWHM}}{H_0} \tag{7.3}$$

where H_0 is the pulse amplitude (where the peak is located). The energy resolution R is expressed in percentage. Semiconductor diode detectors used in α-spectroscopy can have an energy resolution of less than 1 percent, whereas scintillation detectors used in γ-spectroscopy generally have an energy resolution in the range of 7–10 percent. Needless to say, smaller the value of the energy resolution, better the detector will be able to distinguish between two radiations of similar energy.

The count N is typically a large number and therefore, the response function can be assumed to follow a Gaussian[1] or Normal distribution with a standard deviation of σ. The response of many detectors is found to be approximately linear, so the pulse amplitude H_0 is given by

$$H_0 = KN$$

where K is the proportionality constant. If one only considers statistical fluctuations, for the total count of charge carriers N, assuming that the formation of each charge carrier is a Poisson[2] process, one would expect the standard deviation to be proportional to $\sqrt{N}$. Hence, the standard deviation of the peak is given by:

$$\sigma = K\sqrt{N}$$

1. See Appendix C.
2. See Appendix C.

The FWHM for a Gaussian-shaped peak is 2.35σ, therefore the limiting value of the energy resolution, taking into account only the statistical fluctuations, is given by

$$R = \frac{2.35K\sqrt{N}}{KN} = \frac{2.35}{\sqrt{N}} \tag{7.4}$$

Repeated measurements of the energy resolution for radiation detectors have shown that the achievable values of resolution can be lower than the minimum predicted value by only considering the statistical fluctuations. The *Fano factor* is introduced to quantify this observation, which is defined as follows:

$$F = \frac{\text{Observed variance}}{\text{Poisson predicted variance } (N)} \tag{7.5}$$

Thus, the observed energy resolution is given as follows:

$$R = 2.35\sqrt{\frac{F}{N}} \tag{7.6}$$

The Fano factor has different values for different detectors. For semiconductor diode detectors and proportional counters, it is substantially less than unity, whereas for scintillation detectors, it is close to unity.

EXERCISE 7.3: In a semiconductor detector with a Fano factor of 0.1, what should be the minimum number of charge carriers per pulse so that a statistical energy resolution limit of 0.5% is achieved?

Solution:

$$R = 2.35\sqrt{\frac{F}{N}}$$

$$\therefore \quad N = \frac{(2.35)^2}{R^2}F = \frac{(2.35)^2}{(0.005)^2}(0.1) = 22090$$

7.1.4 Detection Efficiency

All radiation detectors, in principle, give rise to an output pulse for each quantum of radiation which interacts with the detector. If all the radiation quanta incident on the detector deposit their energy in the detector, then the intrinsic detector efficiency is said to be 100%. The intrinsic efficiency of a detector is defined as follows:

$$\varepsilon_{\text{int}} = \frac{\text{Number of pulses recorded}}{\text{Number of quanta incident on the detector}} \tag{7.7}$$

In practice, it is also important to know the absolute efficiency of the detector, which is given by

$$\varepsilon_{\text{abs}} = \frac{\text{Number of pulses recorded}}{\text{Number of radiation quanta emitted by the source}} \tag{7.8}$$

The intrinsic efficiency of a detector depends mainly on the detector material, the radiation energy and the physical thickness of the detector. In case of the absolute efficiency, the geometric dependence also plays a major role. Thus, it is much more convenient to tabulate values of intrinsic rather than absolute efficiencies.

For isotropic sources, i.e., for sources emitting radiations in all directions equally, the two efficiencies are related as follows:

$$\varepsilon_{\text{int}} = \varepsilon_{\text{abs}} \frac{4\pi}{\Omega} \tag{7.9}$$

where Ω is the solid angle of the detector seen from the actual source position (in steradian).

Counting efficiencies are also classified according to the nature of the event recorded. If we accept all the pulses from the detector, i.e., all the interactions no matter how low in energy are counted, we use total efficiencies. Here, the entire area under the differential pulse height spectrum is measured taking into account all the pulses regardless of their amplitudes. On the other hand, if we are just interested in recording the interactions which deposit the full energy of the incident radiation, then we use peak efficiency. The number of full energy events can simply be obtained by integrating the total area under the peak. The total and peak efficiencies are related by the 'peak-to-total' ratio, which is sometimes tabulated separately.

$$r = \frac{\varepsilon_{\text{peak}}}{\varepsilon_{\text{total}}} \tag{7.10}$$

It is often preferable to use peak efficiencies for an experiment, as the number of full energy events is not sensitive to perturbing effects such as scattering from the surrounding objects or the background noise.

EXERCISE 7.4: A point source is located along the axis of a right circular cylindrical detector (Figure 7.3), where the source-detector distance is 20 cm and detector radius is 5 cm. The detector has intrinsic peak efficiency of 12% at 1 MeV. The point source emits a 1 MeV γ-ray in 80% of its decays, and has an activity of 20 kBq. Neglecting attenuation between the source and the detector, calculate the number of counts that will appear under the 1 MeV full-energy peak in the pulse height spectrum from the detector over a 100 s count.

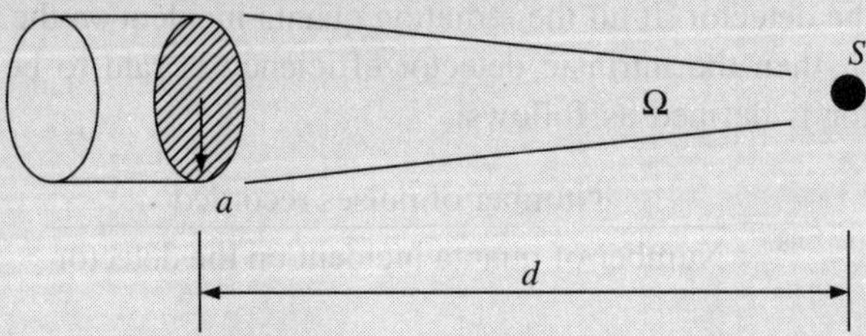

FIGURE 7.3 A point source S located at a distance d from the right cylindrical detector (with radius a) and subtending a solid angle Ω.

Solution: If we join the point source with the edge of the cross-section of the detector, we basically get a conical shape. For a cone, the solid angle

$$\Omega = 2\pi(1 - \cos\theta)$$

where the apex angle of the cone is 2θ. Therefore, we have

$$\Omega = 2\pi\left(1 - \frac{d}{\sqrt{d^2 + a^2}}\right)$$

$$= 2\pi\left(1 - \frac{20}{\sqrt{20^2 + 5^2}}\right) = 0.1875$$

To calculate the radiation quanta emitted by the source, we have

$$S = (80\%)(20 \times 10^3 \text{ dps})(100 \text{ s}) = 1.6 \times 10^6 \text{ disintegrations}$$

From equation (7.9), we have

$$\varepsilon_{\text{abs}} = \frac{\varepsilon_{\text{int}}}{\left(\dfrac{4\pi}{\Omega}\right)} = \frac{(12\%)(0.1875)}{4\pi}$$

Therefore, the number of full energy pulses recorded

$$S\varepsilon_{\text{abs}} = \frac{(1.6 \times 10^6)(12\%)(0.1875)}{4\pi} = 2866 \text{ counts}$$

7.1.5 Time Resolution

When the radiation quantum interacts with the detector, it produces a pulse. While this pulse is processed, the detector is unavailable for processing the next pulse that may be generated during this interval. Since radioactive decay is a random process, there is always some probability that a true event will be lost because it occurs too quickly following a preceding event. In nearly all detector systems, there is a minimum amount of time which should separate two events so that they are recorded as two separate events. This minimum time is called the *dead time* of the counting system. As the counting rates become higher, the dead time losses increase and any measurements made when the counting rates are relatively high must include some correction factor.

Models for Dead Time Behaviour

There are two commonly used models.

Non-paralysable detector: The dead time period is of fixed duration. Any pulse appearing in this duration is lost.

Paralysable detector: If another pulse occurs during the dead time period, the dead period is extended by the same duration of dead time. Therefore, this detector will experience greater counting losses with increased activity.

These models represent the idealized behaviour. The dead time behaviour of a detector, however, is somewhere in between these two extremes. Their behaviour is illustrated in Figure 7.4. In this example, there are six true events which occur at certain intervals. In the case of non-paralysable detector, two events are lost with only four counts being recorded, while in the case of paralysable detector, only three counts are recorded.

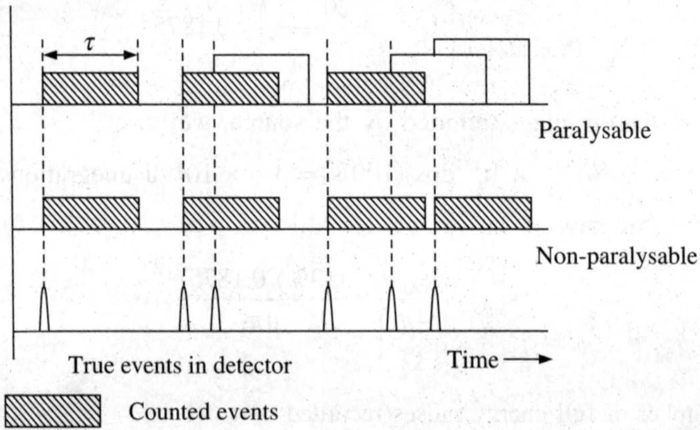

FIGURE 7.4 Illustration of paralysable and non-paralysable models of dead time behaviour for radiation detectors.

If n represents the true interaction rate, m is the recorded rate and τ is the system dead time, then the non-paralysable detector is dead for $m\tau$ fraction of time. Therefore, the rate at which the true events are lost is given by $nm\tau$. Hence,

$$n - m = nm\tau$$

or $$n = \frac{m}{1 - m\tau} \quad \text{(for non-paralysable model)} \qquad (7.11)$$

In the paralysable case, dead periods are not always of fixed length. In this case, we will have to first derive a probability distribution function to describe the time intervals between adjacent random events.

If t is the time interval between adjacent random events, then we say that if an event occurred at $t = 0$, there should not be any event occurring between 0 and t and the next event should occur between t and $t + dt$.

$$P_1(t)dt = P(0) \times (ndt)$$

where $P_1(t)$ is the probability distribution function for intervals between random events and $p_1(t)dt$ represents the probability of the next event occurring in dt interval after a delay of time

t, $P(0)$ is the Poisson distribution value and represents the probability of no events occurring between time 0 and t, and ndt is the probability of an event occurring during dt interval.

The average number of recorded events, $\bar{x} = nt$. Therefore,

$$P_1(t)dt = \left[\frac{(nt)^0 e^{-nt}}{0!}\right] ndt$$

$$\therefore \qquad P_1(t)dt = ne^{-nt}\,dt \tag{7.12}$$

The probability of intervals larger than the system dead time τ can be obtained by integrating this distribution between τ and ∞, i.e.,

$$P_2(\tau) = \int_{\tau}^{\infty} ne^{-nt}\,dt = e^{-n\tau} \tag{7.13}$$

The rate of occurrence of such intervals is obtained by multiplying the above probability with n, the true interaction rate. Thus,

$$m = ne^{-n\tau} \tag{7.14}$$

For paralysable model, we cannot directly solve for the true rate n. Equation (7.14) must be solved iteratively if n is to be calculated for given values of m and τ.

EXERCISE 7.5: A source of ^{116m}In ($T_{1/2}$ = 54 m) is counted using a G–M tube. Successive 1 minute observations gave 131340 counts at 12:00 noon, and 93,384 counts at 12:40 pm. Neglecting the background, calculate the true interaction rate in the G–M tube at 12:00 using both the models for dead time losses. Also calculate the respective dead time of the detector.

Solution:

$$\lambda = \frac{0.693}{T_{1/2}} = \frac{0.693}{54} = 1.28 \times 10^{-2}\ \text{m}^{-1}$$

Since, the half-life of the radioactive source is small, we will have to take into account that its activity will change with time. Hence, the true interaction rate n will change with time. If n_0 is the true interaction rate at noon, then

$$n = n_0 e^{-\lambda t}$$

For non-paralysable model: From equation (7.11), we have

$$n_0 e^{-\lambda t} = \frac{m}{1 - m\tau}$$

or,

$$\tau = \frac{1}{m} - \frac{e^{\lambda t}}{n_0}$$

Now the system dead time should remain constant, therefore equating the τ at $t = 0$ and $t = 40$ m, we have

$$\frac{1}{131340} - \frac{1}{n_0} = \frac{1}{93384} - \frac{e^{(1.28\times10^{-2})(40)}}{n_0}$$

Solving for n_0, we get $n_0 = 216054$. The detector dead time at $t = 0$ is given by

$$\tau = \frac{1}{131340} - \frac{1}{216054} = 2.98\ \mu\text{m}$$

For paralysable model: From equation (7.14), we have

$$m = (n_0 e^{-\lambda t}) e^{-(n_0 e^{-\lambda t})\tau}$$

or
$$\lambda t + \ln\left(\frac{m}{n_0}\right) = -n_0 e^{-\lambda t}\tau$$

∴
$$\tau = \frac{\lambda t + \ln\left(\frac{m}{n_0}\right)}{-n_0 e^{-\lambda t}}$$

Now the system dead time should remain constant, therefore equating τ at $t = 0$ and $t = 40$ m, we have

$$\frac{\ln\left(\frac{131340}{n_0}\right)}{-n_0} = \frac{(1.28\times10^{-2})(40) + \ln\left(\frac{93384}{n_0}\right)}{-n_0 e^{-(1.28\times10^{-2})(40)}}$$

or
$$0.6 \times (11.786 - \ln n_0) = 11.956 - \ln n_0$$

∴
$$n_0 = \exp\left(\frac{11.956 - 0.6\times11.786}{0.4}\right) = 200988$$

The detector dead time

$$\tau = \frac{\ln\left(\frac{131340}{200988}\right)}{-200988} = 2.1\ \mu\text{m}$$

The variation of the observed rate m versus the true rate n for both the models is shown in Figure 7.5. For lower rates, the two models give the same result, but as the rates become higher, they diverge. A non-paralysable detector approaches an asymptotic value when $m = 1/\tau$. This represents a situation in which the counter barely has time to finish one dead period before starting another. In case of paralysable detector, the observed rate goes through a maximum.

At very high rates, the observed rates again start decreasing. Therefore, one must be careful when using a counting system which may be paralysable, and one may make a mistake of interpreting the low observed rates correspond to low interaction rates rather than very high rates on the other side of the maximum. The ambiguity can be resolved by changing the true rate in the known direction and observing whether the observed rate increases or decreases.

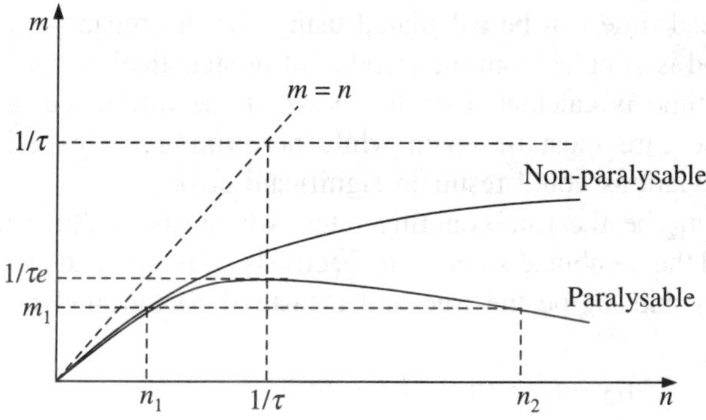

FIGURE 7.5 Variation of the observed rate m vs. the true rate n for two models of dead time losses.

At low rates ($n << 1/\tau$) the following approximations could be used.

Non-paralysable: $$m = \frac{n}{1 + n\tau} \approx n(1 - n\tau) \qquad (7.15a)$$

Paralysable: $$m = ne^{-n\tau} \approx n(1 - n\tau) \qquad (7.15b)$$

EXERCISE 7.6: The measured counting rate rises to a maximum and then decreases, as a source is brought progressively closer to a paralysable detector. If a maximum counting rate of 50000 per second is reached, what is the dead time of the detector?

Solution: From equation (7.14), we know that

$$m = ne^{-n\tau}$$

At maxima, the slope *dm*/*dn* should be zero. Thus,

$$e^{-n\tau}(1 - n\tau) = 0$$

$$\therefore \qquad n = \frac{1}{\tau}$$

Therefore, the maximum observed rate

$$50000 = \frac{1}{\tau}e^{-1}$$

or $$\tau = 7.36\ \mu s$$

Methods of Dead Time Measurement

In order to use both the models discussed before, one should know the system dead time τ. Often the dead time may not be known or may vary depending on the operating conditions. In such situations, it has to be measured directly.

By assuming that one of the specific models is applicable and by measuring the observed rate for at least two different true rates which differ by a known ratio, the dead time can be calculated.

One of the common methods is the two-source method, which is based on observing the counting rate from two sources individually and in combination. The observed rate due to the combined sources is going to be less than the sum of the true rates, since the counting losses are nonlinear. The dead time can be calculated using this discrepancy. Although conceptually the two-source method is simple, from the standpoint of statistical errors, this is not a preferred method as the dead time is calculated on the basis of the difference of two large numbers, which means extreme care must be taken while performing the experiment, as even small approximations or deviations could result in significant errors.

Let n_1, n_2 and n_{12} be the true counting rates, which also include the background, with source 1, source 2 and the combined sources, respectively. If m_1, m_2 and m_{12} be the corresponding observed rates and n_b and m_b be the true and measured background rates with both sources removed, then

$$n_{12} - n_b = (n_1 - n_b) + (n_2 - n_b)$$

or

$$n_{12} + n_b = n_1 + n_2 \tag{7.16}$$

EXERCISE 7.7: A counter gives exactly 10,000 counts in 1 s when a standard source is in place. An identical source is placed next to the first source, and the counter now records 19000 counts in 1 s. What is the counter dead time? The background can be neglected.

Solution: Assuming the non-paralysable model, from equation (7.16), we have

$$\frac{m_{12}}{1-m_{12}\tau} + \frac{m_b}{1-m_b\tau} = \frac{m_1}{1-m_1\tau} + \frac{m_2}{1-m_2\tau}$$

Solving the quadratic equation in τ, we get

$$\tau = \frac{A(1 \pm \sqrt{1-C})}{B}$$

where $A \equiv m_1 m_2 - m_b m_{12}$, $B \equiv m_1 m_2(m_{12} + m_b) - m_b m_{12}(m_1 + m_2)$, and

$$C \equiv \frac{B}{A^2}(m_1 + m_2 - m_{12} - m_b)$$

If $m_b = 0$, and $m_{12} = m_1 + m_2$, the dead time should vanish as there is nò discrepancy in counting. This happens only when the following solution is considered out of the above two possibilities.

$$\tau = \frac{A(1 - \sqrt{1-C})}{B}$$

Now, neglecting the background ($m_b = 0$), we get

$$\tau = \frac{(10000)^2 - \sqrt{(10000)^2(19000 - 10000)^2}}{(10000)^2 \times 19000} = 5.26\ \mu\text{s}$$

The other method is carried out when a short-lived radioisotope is available, similar to what we did in Exercise 7.5. The advantage of this method is that if more than two values are given, then we can also test the validity of the assumed models and decide which model best describes the counting system.

EXERCISE 7.8: A source of ^{116m}In ($T_{1/2}$ = 54 m) is counted using a G–M tube. Successive 1 minute observations give the following data. Calculate the true interaction rate in the G–M tube at t = 0 using both the models for dead time losses. Also, calculate the respective dead time of the detector. Which of the two models for dead time losses best describes the counting system? [Neglect the background]

Time (min)	0	10	20	30	40
Counts	131340	121352	111620	102298	93384

Solution:

$$\lambda = \frac{0.693}{T_{1/2}} = \frac{0.693}{54} = 1.28 \times 10^{-2}\ \text{m}^{-1}$$

For non-paralysable model

From equation (7.11), we have

$$n_0 e^{-\lambda t} = \frac{m}{1 - m\tau}$$

or

$$m e^{\lambda t} = -n_0 \tau m + n_0$$

Now, when we plot $me^{\lambda t}$ vs. m, if the detector follows the non-paralysable model, we should expect a straight line, which has a slope of $(-n_0\tau)$ and y-intercept at n_0. We have five data points available in our problem, so we draw a best fit straight line using Excel and note down the value of variance.

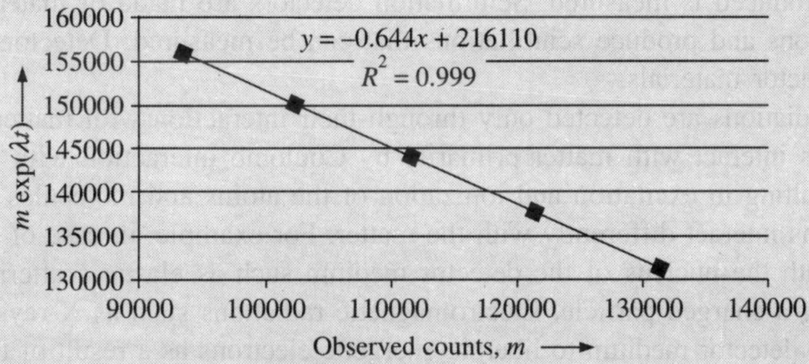

FIGURE 7.6 Application of the decaying source method to determine dead time for non-paralysable model.

From Figure 7.6, we get n_0 = 216,110 and $\tau = \dfrac{0.644}{216110} = 2.98\ \mu\text{m}$.

For paralysable model

From equation (7.14), we have

$$m = (n_0 e^{-\lambda t}) e^{-(n_0 e^{-\lambda t})\tau}$$

or $$\lambda t + \ln m = -n_0 \tau e^{-\lambda t} + \ln n_0$$

Again, we plot $(\lambda t + \ln m)$ vs. $e^{-\lambda t}$, and if the detector follows the paralysable model, we expect a straight line, which has a slope of $(-n_0\tau)$ and y-intercept at $\ln n_0$. Using the five data points available, we draw the best fit straight line using Excel and note down the value of variance.

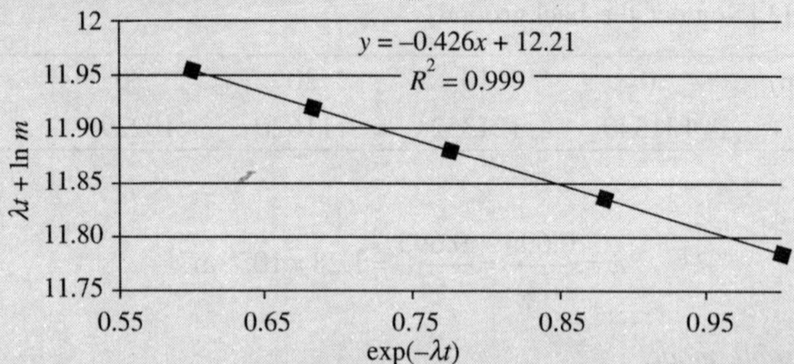

FIGURE 7.7 Application of the decay source method to determine dead time for paralysable model.

From Figure 7.7, we get $n_0 = \exp 12.21 = 200787$ and $\tau = \dfrac{0.426}{200787} = 2.12$ µm.

Both the models for dead time behaviour seem to describe the counting system fairly well, since the counting rate is lower than $1/\tau$.

There are a variety of radiation detectors. They can be classified based on the detector material or on the type of measurement they carry out. For example, in gas-filled detectors, the ionization produced is measured. Scintillation detectors are made of materials that absorb energetic radiations and produce scintillations that can be measured. Detectors are also made up of semiconductor materials.

Nuclear radiations are detected only through their interaction with matter. The energetic charged particles interact with matter primarily by Coulomb interaction with the electrons of the medium resulting in excitation and ionization of the atoms and molecules of the medium. Neutral radiations interact differently with the matter. For example, in case of the neutron, the interaction is with the nucleus of the detector medium such as elastic scattering or a nuclear reaction emitting a charged particle. Electromagnetic radiations such as X-rays and γ-rays can interact with the detector medium to liberate energetic electrons as a result of the photoelectric absorption. Some of the commonly used detectors are discussed in the following sections:

7.2 GAS-FILLED DETECTORS

Many of the oldest and most widely used radiation detectors are based on measuring the effects produced by the ionizing radiation as it passes through a gas filled chamber. As the radiation passes through the gaseous medium, the radiation excites and ionizes the gas molecules along its track. The gas-filled detectors include ionization chambers, proportional counters, Geiger tubes, etc., and in all these devices, the ionization produced by the incident radiation generates an electrical signal. They can be constructed with metal and ceramic components and are usually filled with inert gas. Therefore, these devices can be made relatively robust and tolerant to high radiation and temperature levels. They can be installed in the inaccessible areas in the nuclear facilities and their signals can be monitored remotely to obtain information about the radiation.

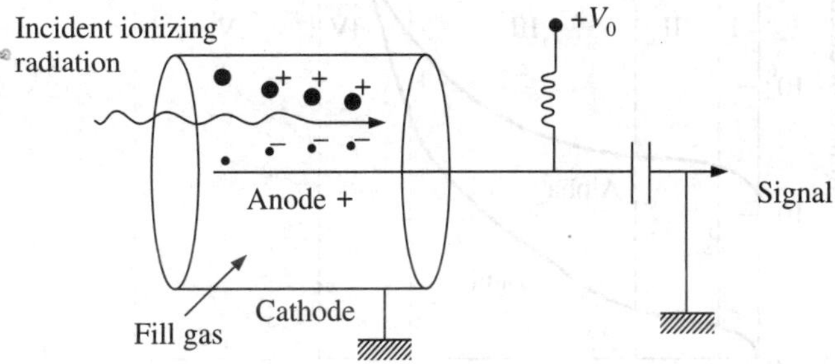

FIGURE 7.8 Schematic diagram of a gas-filled detector.

As the name suggests, the gas-filled detectors consist of a gas-filled chamber with a central electrode well insulated from the chamber walls as shown in Figure 7.8. A voltage V is applied between the wall and the central electrode. The neutral gas molecules are ionized by the radiation and the resulting positive ion and the free electron called an ion-pair is generated. The average energy W needed to create an ion-pair in a medium is almost independent of the particle type and energy. This ensures a unique linear relationship between the energy lost by a charged particle and the average quantity of the resulting ionization and forms the basis of energy measurement for a gas-filled detector. The average energy deposited per ion-pair for some gases commonly used in gas-filled detectors is presented in Table 7.1.

Table 7.1 Average Energy Deposited per Ion-pair for Some Gases Commonly Used in Gas-filled Detectors

Gas	He	Ne	Ar	Kr	Xe	CH_4	CO_2	Air
W (eV)	31.7	36.2	26.4	24	21.7	29.1	20	35

If the electric field that is applied is high enough, it prevents recombination of the ion-pair, and the positive ions start migrating to the cathode and the electrons towards the anode. Electrons are collected more rapidly since they are lighter. Measurement of this charge gives an indication of the presence of the ionizing radiations, and information regarding its strength may be estimated. However, at lower voltages, the ion-pairs may recombine. The free electrons may be taken up by the positively charged ions to form neutral atom. This reduces the current, which is then no longer proportional to the incident radiation. As the applied voltage is increased, recombination losses reduce and a saturation current (constant current) is obtained. The ionization chambers operate in this voltage region. The voltage required to obtain the saturation current is called the saturation voltage and its value depends mainly on the type of gas used and its pressure.

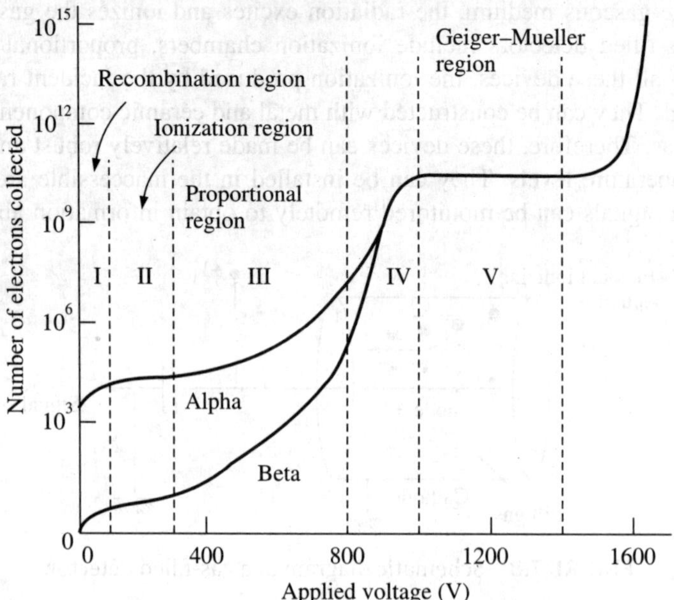

FIGURE 7.9 Characteristic curve of a gas-filled detector signal as a function of applied voltage.

At higher voltages, because of the secondary ionization, charge ionization takes place and the pulse height becomes proportional to the energy of the radiation. Proportional counters operate in this region. At still higher voltages, the pulse height becomes independent of the energy. Geiger tubes operate in this region. A characteristic curve of a gas-filled detector signal as a function of the applied voltage is shown in Figure 7.9.

The transport of charges inside the chamber is mainly governed by two kinds of motion—diffusion and drift. When there is no applied electric field, the electrons and the ions generated due to radiation diffuse uniformly away from the point of creation. They suffer multiple collisions with the gas molecules and eventually recombine. From Maxwell–Boltzmann distribution, which describes particle speeds in idealized gases, we can write:

$$v_{\text{diff}} = \sqrt{\frac{8k_BT}{\pi m}} \tag{7.17}$$

where k_B is the Boltzmann constant, T is the temperature and m is the mass of the particle. At room temperature, the speed at which an electron diffuses is of the order of 10^4 m/s.

If an external electric field is applied, the electrons and the positive ions acquire mobility and drift towards the anode and the cathode, respectively. For ions in a gas, the drift velocity is given by the following relation:

$$v_{\text{drift}} = \mu\frac{E}{p} \tag{7.18}$$

where μ is the mobility, E is the electric field strength and p is the gas pressure. The mobility tends to remain fairly constant over wide ranges of electric field and gas pressure, and does not greatly differ for either positive or negative ions in the same gas. Free electrons behave differently though, because of their much lighter mass, the mobility is typically 1000 times greater than that for ions. Therefore, the collection times for electrons are generally of the order of microseconds rather than milliseconds (for ions). The linearity with electric field as predicted by equation (7.18) begins to fail at higher values of the electric field, and a saturation velocity is reached.

7.2.1 Ionization Chambers

Ionization chambers are the simplest of all the gas-filled detectors. As the applied voltage is increased, more and more ion-pairs reach the electrodes, and at certain field strength, all the ion-pairs produced are collected. A further increase in the field strength does not increase the pulse height, resulting in saturation current. The pulse height in this region is directly proportional to the ionization produced by the incident radiation. Thus, this region is useful for radiation detection and the gas-filled detectors operating in this region are called ionization chambers. One of the most important applications of the ion chambers is in the measurement of the γ-ray exposure. Under the proper conditions, an estimation of the ionization charge in an air-filled ionization chamber can provide a correct measure of the exposure. The exposure rate will be reflected by the ionization current. Other applications of the ion chamber include radiation survey, radiation source calibration and remote sensing ionization.

EXERCISE 7.9: Calculate the net charge created when a 5.5 MeV α-particle is stopped in helium. Find the corresponding saturation current if 300 α-particles per second enter a helium filled ion chamber.

Solution: Using the value of W for helium from Table 7.1, we get the number of ion-pairs produced by one 5.5 MeV α-particle as follows:

$$n_0 = \frac{E'}{W} = \frac{5.5 \times 10^6 \text{ eV}}{31.7 \text{ eV/ion-pair}} = 1.735 \times 10^5$$

The net charge created

$$q = \text{(Number of charge carriers) (Elementary charge)}$$
$$= (1.735 \times 10^5)\ (1.602 \times 10^{-19} \text{ C}) = 2.78 \times 10^{-14} \text{ C}$$

Corresponding saturation current produced by 300 α-particles per second

$$i = \frac{dq}{dt} = (300)(2.78 \times 10^{-14}) = 8.34 \times 10^{-12} \text{ A}$$

Most applications of the ionization chambers involve their use in the current mode in which the average rate of ion formation within the chamber is measured. They can also be operated in the pulse mode. The equivalent circuit of an ion chamber operated in pulse mode is shown in Figure 7.10.

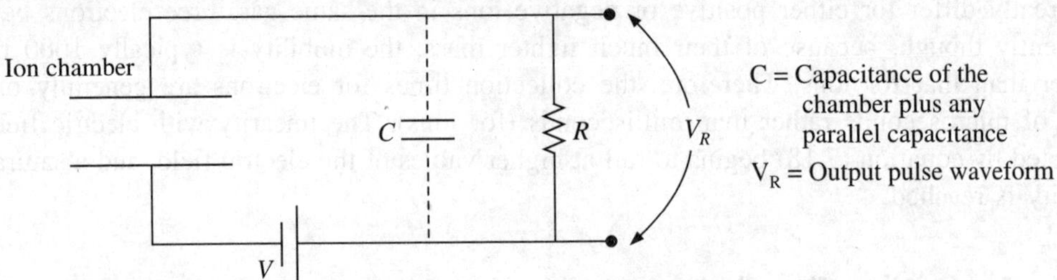

FIGURE 7.10 Equivalent circuit of an ion chamber operated in pulse mode.

The basic electrical signal is the voltage across the load resistance *R*. When there is no ionization charge within the ion chamber, this signal voltage is zero. The typical amplitude of an ion chamber pulse is relatively small. The maximum pulse amplitude of the signal pulse is given by:

$$V_{\max} = \frac{n_0 e}{C} \tag{7.19}$$

where n_0 is the number of ion-pairs and e is the electronic charge.

EXERCISE 7.10: For typical ion chambers and associated wiring, the capacitance C is of the order of 10^{-10} farad. Calculate the pulse amplitude, if a 1 MeV charged particle loses all of its energy within an air-filled chamber.

Solution:

$$n_0 = \frac{E}{W} = \frac{1 \times 10^6 \text{ eV}}{35 \text{ eV/ion-pair}} = 2.86 \times 10^4$$

$$V_{\max} = \frac{n_0 e}{C} = \frac{(2.86 \times 10^4)(1.602 \times 10^{-19} \text{ C})}{10^{-10} \text{ F}} = 4.58 \times 10^{-5} \text{ V}$$

7.2.2 Proportional Counters

Proportional counters operate in region III—the proportional region as depicted in Figure 7.9, at a higher voltage gradient than the ion chambers. Unlike an ion chamber, proportional counters are almost always operated in the pulse mode and rely on the phenomena of *gas multiplication to* amplify the charge represented by the ion-pairs created within the gas. Pulses are therefore significantly larger than those from the ion chambers used under the same conditions, and for this reason, they can be used in situations in which the number of ion-pairs generated by the radiation is considerably small. Therefore, one of the important applications of the proportional counters has been in the detection and measurement of the low energy X-rays. They are also widely used in the detection of the neutrons. Since, the pulse heights for α- and β-particles are significantly different, both α- and β-particles can be effectively counted using the proportional counters. A typical spectrum of ^{57}Co radioactive source for the low-energy X-ray measurement in the proportional counter is shown in Figure 7.11.

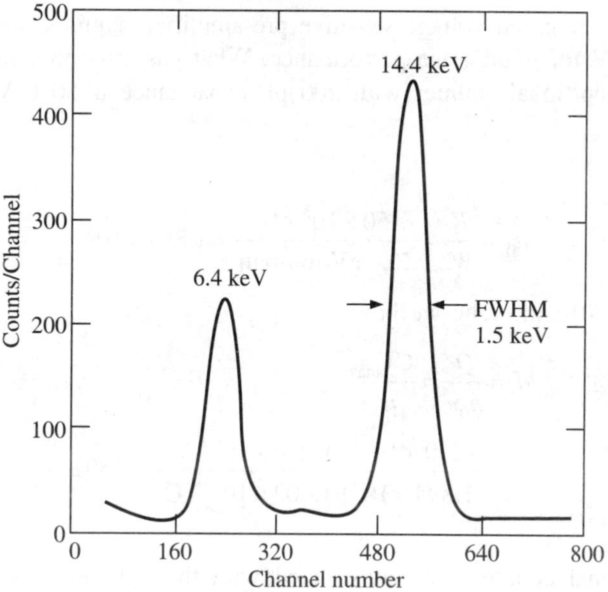

FIGURE 7.11 A typical Co-57 spectrum (proportional counter).

Gas multiplication is a result of the higher field strength within the gas chamber. At lower values of the field, the free electrons and the ions that are created by the incident radiation just drift to their respective collecting electrodes. During this migration, the charges collide with the neutral gas molecules. But because of the low mobility of the ions, they attain very small average energy between collisions. On the other hand, free electrons are easily accelerated and have significant kinetic energy. If this energy exceeds the ionization energy of the neutral gas molecule, an additional ion-pair may be created in the collision. There is a threshold field above which the secondary ionization occurs. In typical gases, at atmospheric pressure, it is of the order of 10^6 V/m. The electrons liberated by this secondary ionization can also in turn create additional ionization and this can lead to a cascade effect. The gas multiplication process, therefore, takes the form of a cascade, known as *avalanche*, in which each free electron created in such a collision can create more free electrons by the same process. In the proportional counter, the avalanche terminates when all the free electrons are collected at the anode. The number of secondary ionization events can be maintained under proper conditions to be proportional to the number of primary ion-pairs formed and the total number of ions can be multiplied by a multiplication factor. This charge amplification within the detector itself results in significantly improved signal-to-noise ratio compared with the ion chambers operated in the pulse mode.

If M is the average multiplication factor, which characterizes the counter operation, then the total charge Q created by n_0 original ion-pairs

$$Q = Mn_0e \tag{7.20}$$

EXERCISE 7.11: A given voltage-sensitive pre-amplifier requires minimum input pulse amplitude of 10 mV for good/noise performance. What gas multiplier factor is required in an argon-filled proportional counter with 200 pF capacitance if 50 keV X-rays are to be measured?

Solution:

$$n_0 = \frac{E}{W} = \frac{50 \times 10^3 \text{ eV}}{26.4 \text{ eV/ion-pair}} = 1.894 \times 10^3$$

From equations (7.20) and (7.1), we have

$$M = \frac{Q}{n_0 e} = \frac{CV_{\text{max}}}{n_0 e}$$

$$= \frac{(200 \times 10^{-12} \text{ F})(10 \times 10^{-3} \text{ V})}{(1.894 \times 10^3)(1.602 \times 10^{-19} \text{ C})} = 6591$$

Most proportional counters are constructed with the cylindrical geometry with a thin wire acting as anode along the axis of a large hollow tube, which serves as the cathode (see Figure 7.8). The electric field inside a cylinder at a radius r is given by:

$$E(r) = \frac{V}{r \ln\left(\frac{b}{a}\right)} \tag{7.21}$$

where V is the voltage applied between the anode and the cathode, a is the radius of the anode wire and b is the cathode inner radius. Based on the above formula, Diethorn derived a widely used expression for the average gas multiplication factor

$$\ln M = \frac{V}{\ln\left(\frac{b}{a}\right)} \frac{\ln 2}{\Delta V}\left[\ln \frac{V}{pa \ln\left(\frac{b}{a}\right)} - \ln K\right] \tag{7.22}$$

where p is the gas pressure. ΔV is the potential difference through which an electron moves between successive ionizing events and K represents the minimum value of E/p below which the multiplication cannot occur. Both K and ΔV are constants for a given fill gas. Diethorn parameters for some commonly used proportional gases are listed in Table 7.2.

Table 7.2 Diethorn Parameters for Proportional Gases

Gas Mixture	90% Ar, 10% CH_4 [P-10]	95% Ar, 5% CH_4 [P-5]	100% CH_4	100% C_3H_8	75% Ar, 15% Xe, 10% CO_2
ΔV (eV)	23.6	21.8	36.5	29.5	20.2
K × 10^{-4} (V/cm.atm)	4.8	4.5	6.9	10	5.1

EXERCISE 7.12: A proportional counter with an anode wire radius of 0.003 cm and cathode radius of 1 cm is filled with P-10 gas at 1 atmospheric pressure. Calculate the voltage required to achieve a gas multiplication factor of 1000. At the same voltage, by what factor would the multiplication factor change if the anode were twice as large? If instead, the cathode radius is doubled, by what factor will the multiplication factor change?

Solution: From the Diethorn expression, we get

$$\ln 1000 = \frac{V}{\ln\left(\dfrac{1\text{ cm}}{0.003\text{ cm}}\right)} \frac{\ln 2}{(23.6)} \left[\ln \frac{V}{(1\text{ atm})(0.003\text{ cm})\ln\left(\dfrac{1\text{ cm}}{0.003\text{ cm}}\right)} - \ln 4.8 \times 10^4\right]$$

On simplifying, we get

$$f(V) = 1366.27 - V(\ln V - 6.728) = 0$$

From Figure 7.12 of $f(V)$, we can see that $V \approx 1790$ V for $f(V) = 0$.

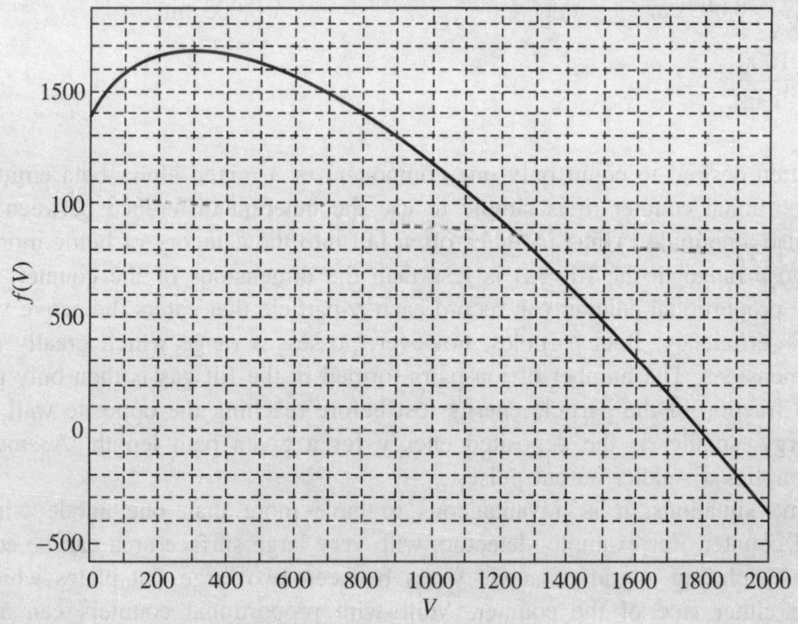

FIGURE 7.12 $f(V)$ vs. V plot.

The following Matlab code is used to get the above plot.

```
clear all

syms x real positive;
f=1366.27-x*(log(x)-6.728);
ezplot(f,[0,2000])
hold on
plot([0 2000],[0 0],'r') %x-axis
grid minor
```

```
xlabel('V')
ylabel('f(V)')
title('')
```

If the anode radius is doubled, i.e., $a = 0.006$ cm, we get

$$M_1 = \exp\left\{\frac{1790\text{ V}}{\ln\left(\dfrac{1\text{ cm}}{0.006\text{ cm}}\right)}\frac{\ln 2}{(23.6)}\left[\ln\frac{1790\text{ V}}{(1\text{ atm})(0.006\text{ cm})\ln\left(\dfrac{1\text{ cm}}{0.006\text{ cm}}\right)} - \ln 4.8\times10^4\right]\right\} = 7.5$$

$$X = \frac{M}{M_1} = \frac{1000}{7.5} \approx 133$$

If instead, the cathode radius is doubled, i.e., $b = 2$ cm, we get

$$M_2 = \exp\left\{\frac{1790\text{ V}}{\ln\left(\dfrac{2\text{ cm}}{0.003\text{ cm}}\right)}\frac{\ln 2}{(23.6)}\left[\ln\frac{1790\text{ V}}{(1\text{ atm})(0.003\text{ cm})\ln\left(\dfrac{2\text{ cm}}{0.003\text{ cm}}\right)} - \ln 4.8\times10^4\right]\right\} = 192$$

$$X = \frac{M}{M_2} = \frac{1000}{192} \approx 5.2$$

It is often desired to count only one component of a mixed alpha–beta emitting source. With a proportional counter, it is simple to use the inherent difference between the typical α- and β-pulse amplitude. Alpha particles often fall into the category of being mono-energetic particles whose range in the fill gas is less than the dimensions of the counter. Because of this fact, the proportional counter can record each α-particle that enters the active volume with almost 100% efficiency. Beta particles, however, possess a range which greatly exceeds the chamber dimensions. The number of ion-pairs formed in the fill gas is then only proportional to the small fraction of the particle energy lost before reaching the opposite wall. Higher the particle energy, smaller is the deposited energy for a given path length. A smaller energy deposition implies a smaller output pulse.

In some situations, it is advantageous to have more than one anode wire within a proportional counter. For example, detectors with very large surface area can be economically constructed by placing a grid of anode wires between two large flat plates which serve as cathodes on either side of the counter. Multi-wire proportional counters can also help in designing position-sensing or fast response proportional counters, because the electrons drift to the nearest anode wire and the signal appears only on one anode wire. Because the speed of response is largely related to the drift time of the initial electron to the multiplying region, which is close to the anode wire where large values of electric field occur, the response time can be minimized by providing many multiplying regions throughout the volume of the detector.

Another variant of the proportional counter design is the 4 *pi proportional counter*. The term 4 pi refers to the solid angle by which the detector views the source. In 4 pi geometry, the detector completely surrounds the source. This is done by placing one counting chamber above the source, and another below it. In whichever direction the radiation leaves the source, it can produce a pulse. Thus, if the source is isotropic, the intrinsic efficiency will be equal to the absolute efficiency of the detector.

7.2.3 Geiger–Mueller Counters

Geiger–Mueller counter, which is commonly referred to as GM counter or simply Geiger tube, is one of the oldest radiation detectors. It was invented by Geiger and Mueller in 1928. Due to its simplicity, low cost and ease of operation, it is still being used for radiation measurements in various areas of research. They operate in region V—the Geiger–Mueller region as shown in Figure 7.9. Just like proportional counters, Geiger tubes are almost always operated in the pulse mode. In Geiger tube, the gas multiplication is very high as compared to the ion chambers or the proportional counters. In Geiger tubes, using proper conditions, a situation is created wherein an avalanche can trigger a second avalanche at a different location within the tube, resulting in a self-propagating chain reaction. Once this Geiger discharge reaches a certain size, the collective effects of all the individual avalanches come into play and ultimately the chain reaction terminates, and a plateau is reached. Because the limiting point is always reached after about the same number of avalanches have been formed, all the pulses from a Geiger tube are of the same amplitude, regardless of the number of original ion-pairs that initiated the process. Thus, a Geiger tube can only function as a simple counter to measure radioactivity and cannot be applied to the radiation spectroscopy since all the information on the energy deposited by the incident radiation is lost. For example, Geiger tubes do not have the ability to differentiate the heavy charged particle and the electrons, and the pulse amplitude remains the same regardless of the form of ionization. In Geiger discharge, the gas multiplication factor M is much higher (10^5–10^8).

EXERCISE 7.13: Estimate the voltage at which the counter in the previous exercise enters the Geiger region of operation. Assume that the average multiplication factor, $M \geq 3.33 \times 10^4$, in the Geiger region.

Solution: From equation (7.22) and the data from the previous exercise, we have

$$\ln(3.33\times10^4) = \frac{V}{\ln\left(\dfrac{1\text{ cm}}{0.003\text{ cm}}\right)}\frac{\ln 2}{(23.6)}\left[\ln\frac{V}{(1\text{ atm})(0.003\text{ cm})\ln\left(\dfrac{1\text{ cm}}{0.003\text{ cm}}\right)} - \ln 4.8\times10^4\right]$$

On simplifying, we get

$$f(V) = 2060 - V(\ln V - 6.728) = 0$$

Just like before, we can plot $f(V)$ to get $V \approx 2164$ V. Another way is to use iteration to get the answer by using the Matlab function fzero.

```
X = fzero(@(x) 2060-x*(log(x)-6.728),2000)
```

Besides the lack of energy information, a major disadvantage of Geiger tubes is their large dead time in comparison to other radiation detectors. These detectors are therefore, limited to relatively low counting rates, and a dead time correction must be applied to situations which involve even moderate counting rates. A full Geiger discharge occurs when the positive ion concentration reduces the electric field to the point where no gas multiplication takes place. Any event that takes place during this period will not cause an avalanche. Because of this, the dead time will tend to be constant and follow non-paralysable model.

In Geiger tubes, as the value of the multiplication factor is high, even trace amounts of gases which form negative ions, e.g., oxygen, are avoided. The noble gases like helium and argon are widely used as the principal fill gas. In addition, a fraction of quench gas, chosen with a higher ionization potential and more complex molecular structure than the primary gas, is also added to avoid the problem of multiple pulsing. The quench gas absorbs the radiated photons, which are emitted when the excited atoms of the fill gas de-excite. These high energy photons can ionize the cathode and cause additional avalanches.

EXERCISE 7.14: Continuing with the Geiger counter in Exercise 7.13, estimate the critical radius for the avalanche formation if the threshold electric field for the onset of avalanche formation is 2 × 10^6 V/m. Thus, estimate the average drift time of an electron from the cathode to the multiplying region. The mobility of a free electron is 1.5 × 10^{-4} m^{2}atm/sV.

Solution: Using equation (7.21), the critical radius for the avalanche formation

$$r_c = \frac{V}{E\ln\left(\dfrac{b}{a}\right)} = \frac{2160\text{ V}}{(2\times 10^6\text{ V/m})\ln\left(\dfrac{1\text{ cm}}{0.003\text{ cm}}\right)} = 9.86\times 10^{-4}\text{ m}$$

The average electric field between $b = 1 \times 10^{-2}$ m and $r_c = 9.86 \times 10^{-4}$ m in the multiplying region

$$E_{avg} = \frac{1}{b-r_c}\int_{r_c}^{b}\frac{V}{r\ln\left(\dfrac{b}{a}\right)}dr = \frac{1}{b-r_c}\frac{V}{\ln\left(\dfrac{b}{a}\right)}\int_{r_c}^{b}\frac{1}{r}dr$$

$$= \left(\frac{1}{b-r_c}\right)\frac{V}{\ln\left(\dfrac{b}{a}\right)}\ln\left(\frac{b}{r_c}\right)$$

$$\therefore \quad E_{avg} = \frac{(2160)\ln\left(\dfrac{1\times 10^{-2}}{9.86\times 10^{-4}}\right)}{(1\times 10^{-2} - 9.86\times 10^{-4})\ln\left(\dfrac{1\times 10^{-2}}{3\times 10^{-5}}\right)} = 9.56\times 10^4\text{ V/m}$$

From equation (7.18), the drift velocity in the multiplying region is given by

$$v_{drift} = \mu\frac{E_{avg}}{p} = (1.5\times 10^{-4}\text{ m}^2\text{atm/sV})\frac{(9.56\times 10^4\text{ V/m})}{(1\text{ atm})}$$

The average drift time from the cathode to the multiplying region is given as follows:

$$t_{\text{drift}} = \frac{b - r_c}{v_{\text{drift}}} = \frac{(1\times 10^{-2} - 9.86\times 10^{-4})}{(1.5\times 10^{-4})(9.56\times 10^{4})} = 0.63 \text{ ms}$$

7.3 SCINTILLATION DETECTORS

Darkening of a photographic plate and the scintillation of fluorescent materials are one of the oldest techniques known. These detectors work on the principle that when a charged particle interacts with the detector material, the energy lost via ionization and excitation processes is partly converted into the characteristic scintillation photons. These photons are detected in devices such as photomultiplier tubes or photodiodes to produce electronic pulses.

Some of the desired characteristics of the scintillation detector are as follows:

(i) Large light output should be there for a given energy deposited by the incident radiation.

(ii) The response should be linear, i.e., the light yield should be proportional to the deposited energy over a wide range.

(iii) The scintillation material should be transparent to the emitted light.

(iv) The decay time of scintillation should be short ($\sim 10^{-9}$ s) for fast signal generation.

(v) There should be little or no afterglow.

(vi) There should be physical and chemical stability against radiation damage, in normal environment conditions.

(vii) It should be easy to manufacture and cost effective.

(viii) The material should be of good quality to make detectors in large size and desired shape.

(ix) Its refractive index should be near to that of glass (~1.5) so that there is an efficient coupling of the scintillation light with a photomultiplier tube (PMT).

Inorganics like thallium activated NaI, CsI and bismuth germanate oxide (BGO), and organic-based liquid mixtures, anthracene and plastics are often used as scintillation materials. No material meets all the criteria listed above. The inorganic scintillators tend to have the best light output and energy proportionality, but are relatively slow in response time. Organic scintillators are much faster, but yield less light. The choice of the scintillator also depends on the intended application. The high *Z*-value of the constituents and high density of the inorganic crystals make them suitable for the γ-ray spectroscopy, whereas the organic scintillators are often preferred for the β-spectroscopy and fast neutron detection. One of the most common detectors used for the γ-ray detection are based on thallium-activated (Tl) sodium iodide crystal (NaI). The thallium ions serve as activation centres in the inorganic crystals where de-excitation of the absorbed γ-ray energy and the subsequent emission of the visible light or scintillation occur. Properties of some inorganic and organic scintillators are listed in Table 7.3.

Table 7.3 Comparison of Properties of Some Inorganic and Organic Scintillators

Scintillator	*Density* (g/cc)	*Refractive index*	*Decay time* (ns)	λ_{max} (nm)	Photons/ MeV	*Remarks*
			Inorganic			
NaI (Tl)	3.67	1.85	230	410	34000	Cheap, large sizes, good pulse shape discrimination
CsI (Tl)	4.51	1.80	1000	540	52000	Excellent pulse shape discrimination
CsI (pure)	4.51	1.80	1000	480	17000	Needs UV glass photomultiplier tube
BGO	7.13	2.15	300	460	8200	High *Z*, good electron detector
BaF_2	4.88	1.50	600	325	6500	Ultra fast, needs quartz window photomultiplier
CaF_2 (Eu)	3.19	1.44	940	435	24000	Low *Z*, good electron detector
LiI (Eu)	4.08	1.96	1400	470	11000	For thermal neutron
$PbWO_4$	8.28	2.16	5-15	440–500	4000	Good for high E_γ calorimetry
			Organic			
Anthracene	1.25	1.62	30	447	16300	High vapour pressure
BC400	1.032	1.58	2.4	423	10000	Cheap, large sizes
BC422Q	1.032	1.58	0.7	370	1600	Ultrafast, low light output
BC501A	0.874	1.58	3/32/270	425	12700	Liquid, good pulse shape discrimination, in fast neutron detectors

EXERCISE 7.15: Estimate the scintillation efficiency of anthracene if 1 MeV energy loss of the particle creates 20300 photons with an average wavelength of 447 nm.

Solution: The energy of one photon is given as follows:

$$E_p = \frac{hC}{\lambda} = \frac{(4.135 \times 10^{-15}\ \text{eVs})(3 \times 10^8\ \text{m/s})}{(447 \times 10^{-9}\ \text{m})} = 2.775\ \text{eV}$$

The scintillation efficiency

$$\varepsilon_{\text{scint}} = \frac{\text{Energy converted to light}}{\text{Total energy}} = \frac{(20300)(2.775\ \text{eV})}{(1 \times 10^6\ \text{eV})} = 5.63\%$$

The process of *fluorescence* is the prompt emission of the visible radiation from a substance following its excitation. *Phosphorescence* corresponds to the emission of light of longer wavelength compared to fluorescence and with longer characteristic time. Delayed fluorescence results in the same emission spectrum as the prompt fluorescence but longer emission time after excitation. If τ represents the fluorescence decay time, then the intensity of the principal scintillation light, i.e., prompt fluorescence, at time t following excitation is given as follows:

$$I = I_0 e^{-t/\tau} \tag{7.23}$$

In most organic scintillators, τ is of the order of a few nanoseconds, and therefore, the prompt scintillation component is relatively fast.

EXERCISE 7.16: Assuming an inverse decay constant of 230 ns, how much time is required for a NaI (Tl) scintillation event to emit 99% of the total light yield?

Solution:

$$I = I_0 e^{-t/\tau}$$

or

$$0.99 = e^{-t/\tau}$$

$$t = -\tau \ln (0.99)$$

$$= -(230 \text{ ns}) (-0.01005) = 2.31 \text{ ns}$$

Since the scintillation light is emitted in all the directions, only a limited fraction can travel directly to the surface at which the photomultiplier tube, which converts the weak light output of a scintillation pulse into a corresponding electric signal, is located. The remainder, if it has to be collected, must be reflected one or more times at the scintillator surfaces. If the incidence angle θ made by the photon when it reaches the reflector surface is greater than the critical angle θ_c, there will be total internal reflection. If, however, $\theta < \theta_c$, partial reflection and partial transmission through the surface will occur. The critical angle θ_c, is determined from the refraction indices for the scintillation medium n_0 and the surrounding medium n_1, which is air in most of the cases,

$$\theta_c = \sin^{-1} \frac{n_1}{n_0} \tag{7.24}$$

In order to recapture the light, the scintillator is normally surrounded by a reflector at all surfaces except that at which the photomultiplier tube is mounted.

Various events occur in the vicinity of a typical source-photomultiplier tube and shield configuration, as shown in Figure 7.13. The photocathode depicted in the figure is a photosensitive layer, which serves to convert as many of the incident light photons as possible into low-energy electrons. The photocathode is coupled to an electron multiplier structure. If the light is a pulse from a scintillation crystal, the photoelectrons that are produced will also be a similar time duration pulse. But their number is too small to serve as a detectable electrical signal. The photomultiplier section in a photomultiplier tube provides an efficient geometry for the collection of the photoelectrons as well as it amplifies their number. The multiplier section of a photomultiplier tube is based on the phenomenon of secondary electron emission. The photoelectrons are accelerated and made to strike an electrode surface, called a dynode, as depicted in the figure. With the right kind of the dynode material, the energy deposited by the incident electron can again emit more than one electron from the same surface.

A characteristic X-ray is emitted by the absorber atom in the photoelectric absorption process. In most of the cases, this X-ray energy is reabsorbed near the original interaction site. However, if the photoelectric absorption takes place near a surface of the detector, the X-ray photon may escape, in which event the energy deposited in the detector is decreased by an amount equal to the photon energy. The secondary radiations created near the source include annihilation radiation and bremsstrahlung. The surrounding material or the shielding is also a potential source of secondary radiations, which can be produced by the interaction of the primary γ-rays emitted by the source.

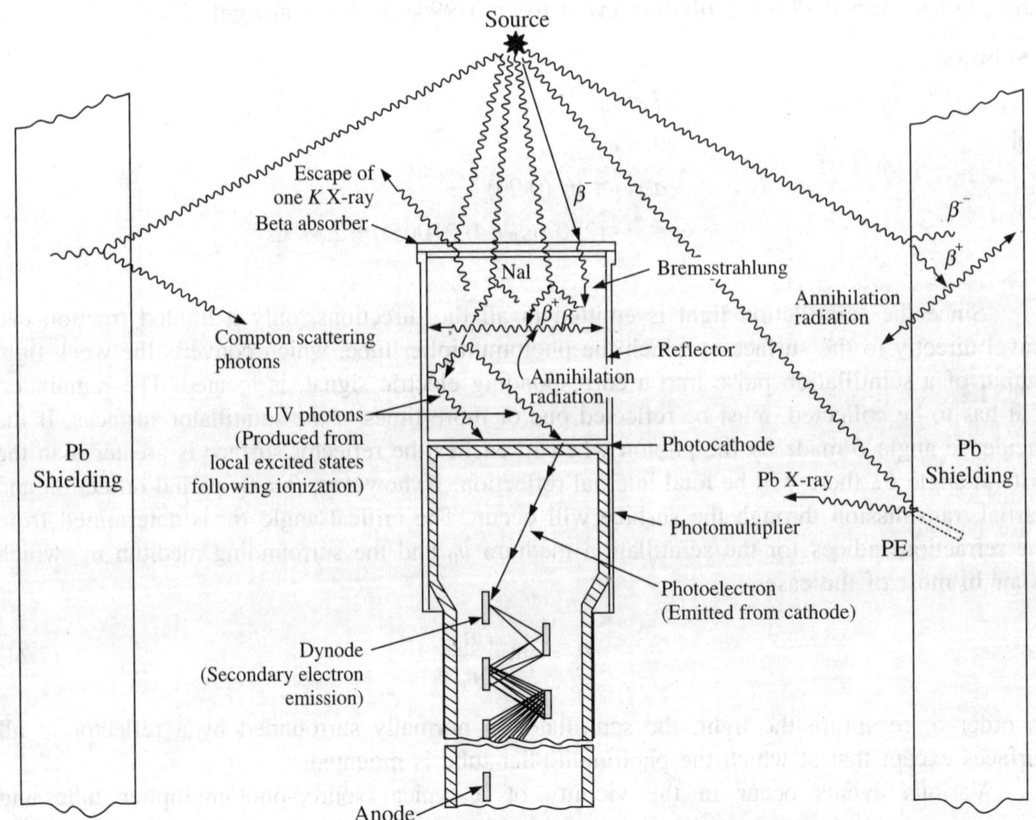

FIGURE 7.13 Various events in the vicinity of a typical source—crystal detector—shield configuration. [*Experiments in Nuclear Science*, ORTEC, 1976 © AMETEK, Inc.].

A typical spectrum of ^{137}Cs radioactive source with NaI (Tl) is shown in Figure 7.14.

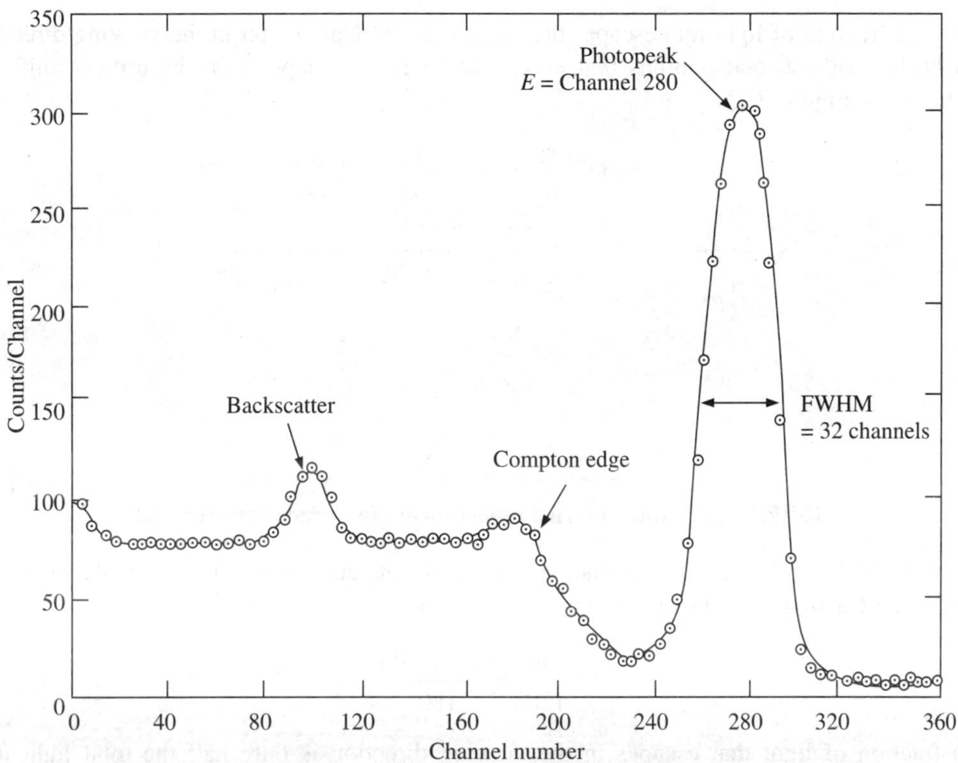

FIGURE 7.14 A typical gamma ray spectrum of Cs-137, using a NaI (Tl) detector.

EXERCISE 7.17: The human eye can detect as few as ten visible photons as a single flash. Would an observer with the pupil diameter of 3 mm be able to see individual scintillation events that are caused by a 1 MeV β-particle in NaI (Tl), if he is viewing the scintillator surface from a 10 cm distance? The refractive index of air is 1. Assume 13% efficiency for NaI crystal.

Solution: From Table 7.3, λ_{max} for NaI (Tl) is 410 nm. Therefore, the energy of one photon

$$E_p = \frac{hc}{\lambda} = \frac{(4.135 \times 10^{-15} \text{ eVs})(3 \times 10^8 \text{ m/s})}{(410 \times 10^{-9} \text{ m})}$$

$$= 3.026 \text{ eV}$$

Total number of photons emitted is given as follows:

$$\text{Number of photons} = (13\%)\frac{1 \times 10^6 \text{ eV}}{3.026 \text{ eV/photon}}$$

$$= 4.3 \times 10^4 \text{ photons}$$

Now, the fraction of light that escapes the surface of the NaI crystal in the viewing direction will be half of the total photons that escape and will also depend on the critical angle θ_c as shown in Figure 7.15.

$$\theta_c = \sin^{-1}\frac{n_1}{n_0} = \sin^{-1}\frac{1}{1.85} = 32.72°$$

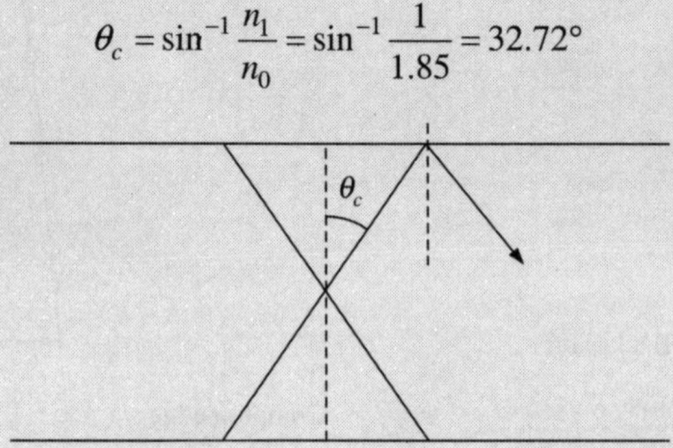

FIGURE 7.15 Total internal reflection at the surface of the crystal.

All the light that subtends greater than θ_c angle with the surface will be internally reflected. Hence, the fraction of light that escapes the surface of the NaI crystal

$$= \frac{2\theta_c}{180°} = \frac{2(32.72°)}{180°}$$

The fraction of light that escapes in the viewing direction is only half the total light that escapes. Therefore, the number of photons that escape in the viewing direction

$$= \frac{2(32.72°)(50\%)(4.3 \times 10^4 \text{ photons})}{180°} = 7810 \text{ photons}$$

If Ω is the solid angle subtended by the light on the pupil of the eye, then the number of photons that enter the pupils

$$= \left(\frac{\Omega}{4\pi}\right)(7810 \text{ photons})$$

$$= \frac{2\pi\left(1 - \dfrac{10}{\sqrt{10^2 + 0.15^2}}\right)}{4\pi}(7810) = 0.44$$

$$\approx 1 \text{ photon}$$

It is less than 10 photons. Therefore, the observer will not be able to see the scintillation events.

7.3.1 Variants of the Scintillator Design

Phoswich Detector

It is a hybrid scintillation detector known as phoswich, developed for specialized applications such as in low background counting of X-rays and β-particles. A phoswich is a sandwich of two phosphors (luminescent substance that emits light when excited by radiation), having different physical and optical properties viewed by the same photomultiplier tube. Events occurring in the two phosphors are separately identified by their noticeably dissimilar scintillation decay times, because the shape of the output pulse from the photomultiplier tube depends on the relative contribution of the light from the two scintillators.

Figure 7.16 shows a typical low background phoswich detector. In this arrangement, a thin sheet of CaF_2(Eu) is coupled together with a NaI (Tl) crystal to a single photomultiplier tube. A dead layer of quartz is placed between the two phosphors. Alpha and low energy β-particles are stopped fully in the CaF_2(Eu) detector. More energetic β-particles are stopped in quartz. On the other hand, Compton scattered γ-rays from CaF_2 pass easily through the low *Z* quartz and are stopped in NaI (Tl). Proper pulse shape analysis can distinguish such coincident events.

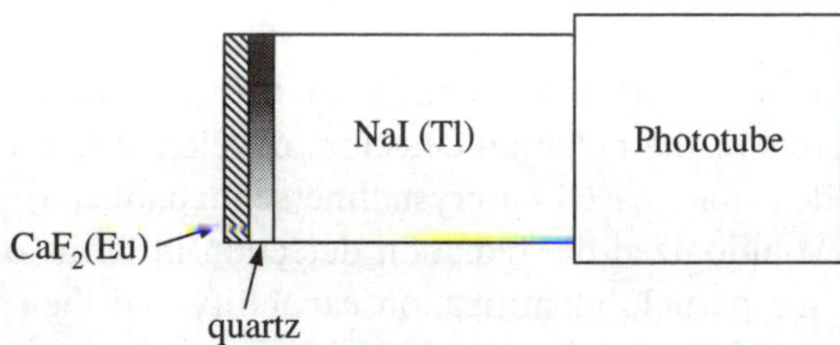

FIGURE 7.16 Schematic diagram of a phoswich detector designed for alpha, beta and gamma counting.

Liquid Scintillation Detectors

Liquid scintillator falls in the category of the organic scintillators and consists of one or two solvents and one or two solutes along with optimal materials like solubilizers and chemiluminiscent (light generated by chemical reactions within the sample-scintillator solution) inhibitors. In liquid scintillators, the sample to be counted is directly dissolved in the liquid scintillator. By doing this, problems relating to sample self–absorption, attenuation of the particles by the detector windows, and β-backscattering from the detector are completely avoided. These advantages make them the preferred choice for the low-energy β-particle detection. They are also used for the detection of the α-particles.

Gas Proportional Scintillation (GPS) Counters

It is a hybrid detector, which combines some of the properties of a proportional counter with those of scintillation detector. In a conventional gas scintillator, the excited gas molecules are created by direct interaction of the incident radiation as it passes through the gas. In

GPS counter, however, the visible and the ultraviolet photons emitted by the excited gas atoms or molecules are detected by a photomultiplier tube, which generates an electrical pulse proportional to the number of photons incident on the tube. The energy resolution for the low-energy radiations is better for GPS counters as compared to that for the proportional counters.

EXERCISE 7.18: If the energy resolution of a particular liquid scintillation detector is 5% for ^{14}C β-rays (160 keV), estimate its energy resolution for the 18 keV β-rays from tritium.

Solution: From equation (7.4), the energy resolution of a detector is inversely proportional to $\sqrt{E}$. Therefore,

$$R_t = (R_C)\sqrt{\frac{E_C}{E_t}} = (5\%)\sqrt{\frac{160}{18}} = 14.9\%$$

7.4 SEMICONDUCTOR DETECTORS

Before discussing the semiconductor detector, let us first look at Figure 7.17, to compare a γ-ray spectrum as seen by a Ge(Li) semiconductor detector and by a NaI(Tl) scintillation detector. One of the marvels of the radiation detectors developed in the last seven decades has been the semiconductor detectors, based on crystalline semiconductor materials such as silicon and germanium. They revolutionized the radiation detection in terms of the energy resolution, the rapid response time, the particle identification capability and their compact size, and they were quickly adopted in nuclear physics research for charged particle energy measurements and the γ-ray spectroscopy (precise determination of the photon energy). More recently, semiconductor devices are also being used in high energy physics for precise determination of the particle tracks. Because the solid densities are about 1000 times greater than that for a gas, semiconductor detectors have compact size as compared to the gas-filled counterparts. Also, the solids have relatively small stopping length for nuclear radiations and hence fast response time. The effective thickness may also be varied depending on the application. When it comes to the energy resolution, the small size of the band gap between the valence band and the conduction band, which is of the order of 1 eV in a semiconductor means that the energy required to create one electron–hole pair analogous to the ion-pair in the gas-filled detectors, is approximately 3 eV which is significantly smaller than that for a gas detector where the energy required to create an ion-pair is around 20–30 eV. The corresponding value for the scintillator detectors is 100 eV. The superior energy resolution along with fast response time has made possible the identification of wide range of particles. Drawbacks of semiconductor detectors include their susceptibility to radiation, light and temperature damage, their high cost and large power consumption for cooling.

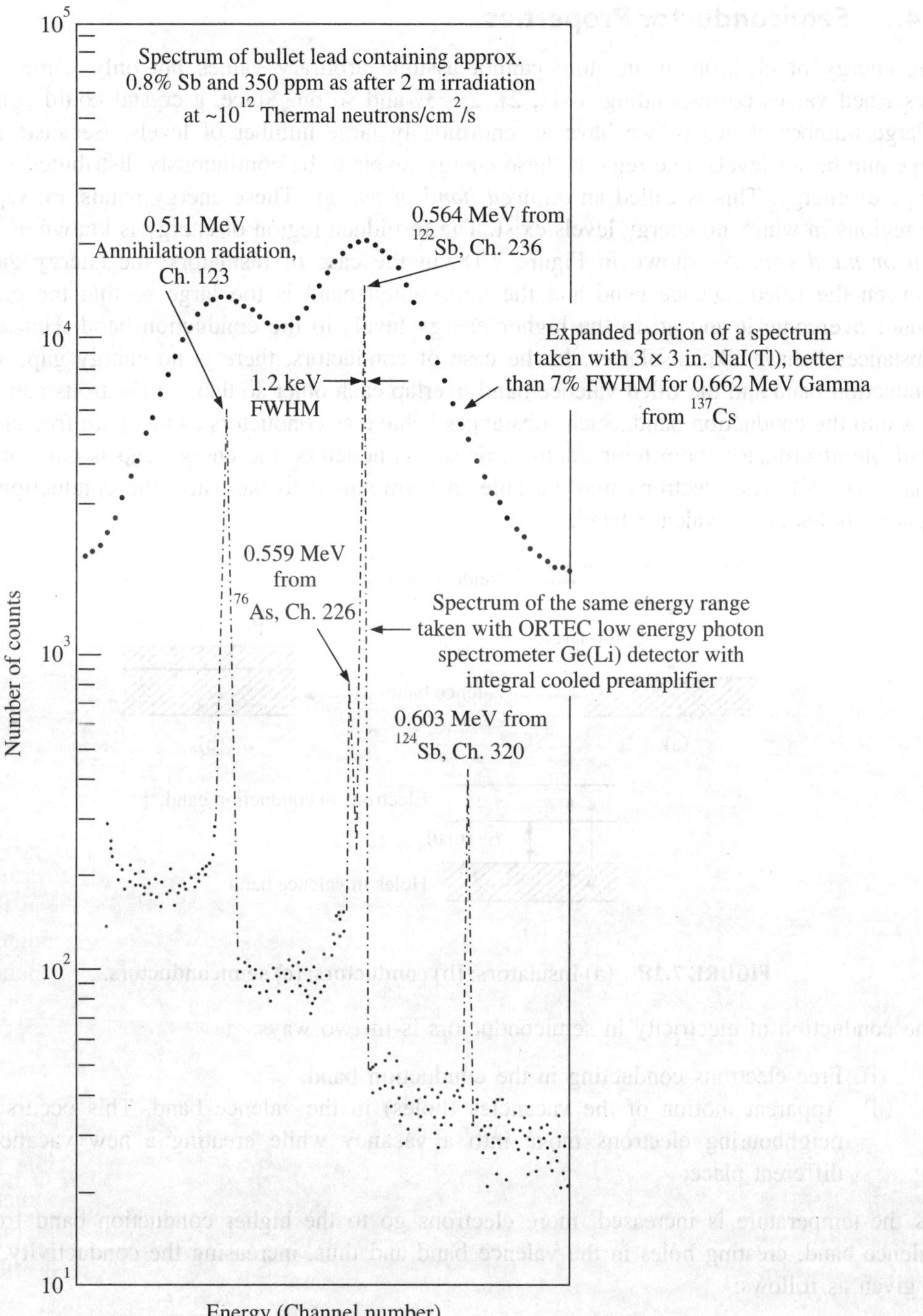

FIGURE 7.17 Comparative spectra taken with Ge(Li) and NaI(Tl) detectors. [*Experiments in Nuclear Science*, ORTEC 1976 © AMETEK, Inc.]

7.4.1 Semiconductor Properties

The energy of electron in an atom cannot assume arbitrary values but only some definite prescribed values corresponding to 1*s*, 2*s*, 2*p*, 3*s* and so on. Since, a crystal could consist of a large number of atoms, we have an enormously large number of levels. Because of such large number of levels, one regards these energy levels to be continuously distributed within a range of energy. This is called an *allowed band of energy*. These energy bands are separated by regions in which no energy levels exist. The forbidden region of energy is known as *energy gap or band gap*. As shown in Figure 7.18, in the case of insulators, the energy gap (E_g) between the filled valence band and the unoccupied band is too large so that the electrons cannot overcome it and go to the higher energy levels in the conduction band. Hence, such substances behave like insulators. In the case of conductors, there is no energy gap, and the conduction band and the filled valence band overlap each other so that the electrons can readily pass into the conduction band. Such substances behave as conductors as there are free electrons available at ordinary room temperature. For semiconductors, the energy gap is comparatively small (~1 eV). The electrons may be able to surmount it to pass into the conduction band, creating holes in the valence band.

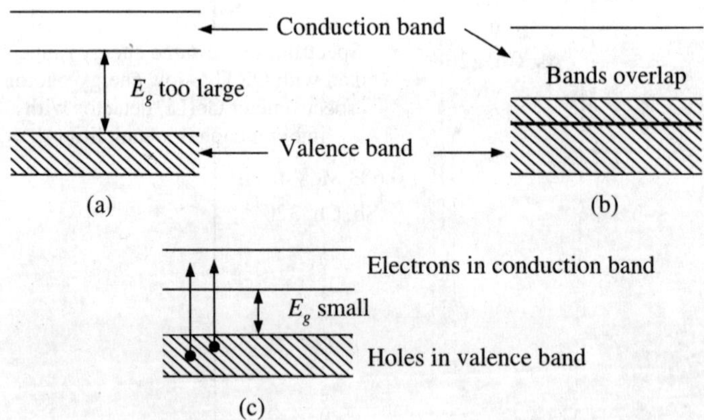

FIGURE 7.18 (a) Insulators; (b) conductors; (c) semiconductors.

The conduction of electricity in semiconductors is in two ways:

(i) Free electrons conducting in the conduction band.
(ii) Apparent motion of the vacancies (holes) in the valence band. This occurs as the neighbouring electrons move into a vacancy while creating a new vacancy at a different place.

As the temperature is increased, more electrons go to the higher conduction band from the valence band, creating holes in the valence band and thus, increasing the conductivity, which is given as follows:

$$\sigma = e(n_e\mu_e + n_h\mu_h) \tag{7.25}$$

where n_e and n_h are the number densities of the electrons and the holes, respectively and μ_e and μ_h are their respective mobility. The mobility is defined as the drift velocity per unit electric field, i.e.,

$$\mu_e = \frac{v_e}{E} \quad \text{and} \quad \mu_h = \frac{v_h}{E} \tag{7.26}$$

where v_e and v_h are the drift velocities of the electrons and the holes, respectively. At higher electric field values, the drift velocity increases slowly with the electric field and ultimately reaches a saturation velocity, after which it becomes independent of further increase in the electric field.

Intrinsic semiconductors are pure semiconductors having no impurities. The electrical conduction in these is completely due to the electrons excited from the valence band into the conduction band. In an intrinsic semiconductor, at 0 K the valence band is totally filled and the conduction band is completely empty having no electrons available for conduction. However, as the band gap is very small, at room temperature, some electrons move into the conduction band due to the thermal energy. Taking the example of silicon (or germanium, see Table 7.4 for comparison of some properties of Si and Ge), which has four valence electrons; each valence electron is shared by four neighbouring Si atoms. At room temperature (or above), the thermal energy causes the valence electrons to be free for conduction. Each valence electron creates one vacancy when it breaks from the bond and becomes available for conduction, as shown in Figure 7.19.

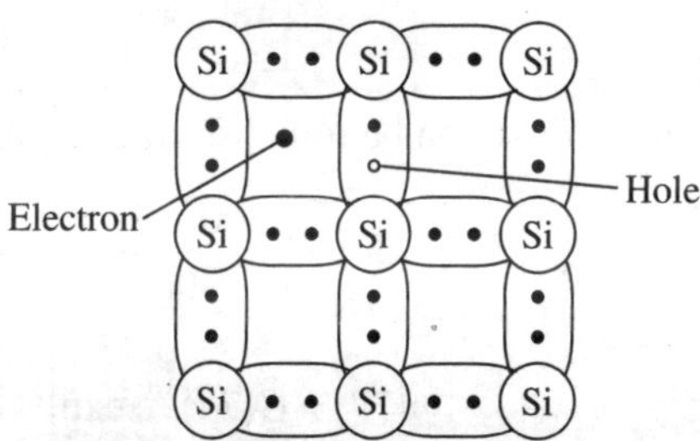

Figure 7.19 Crystal structure of Si.

For an intrinsic semiconductor, the number of electrons in the conduction band is equal to the number of holes, i.e.,

$$n_e = n_h = n_i \tag{7.27}$$

where n_e and n_h are the number densities of the electrons in the conduction band and of the holes in the valence band, respectively. n_i is the intrinsic carrier density.

Table 7.4 Properties of Intrinsic Silicon and Germanium

	Si	Ge
Atomic number	14	32
Density (g/cm^2)	2.33	5.33
Forbidden energy gap (E_g) at 300 K (eV)	1.115	0.665
Forbidden energy gap (E_g) at 0 K (eV)	1.165	0.746
Intrinsic carrier density at 300 K (m^{-3})	1.5×10^{16}	2.4×10^{19}
Electron mobility at 300 K (cm^2/Vs)	1350	3900
Hole mobility at 300 K (cm^2/Vs)	480	1900
Energy per hole-electron pair at 77 K (eV)	3.76	2.96

Probability per unit time of an electron-hole pair being thermally generated is given by

$$p(T) = CT^{3/2} \exp\left(-\frac{E_g}{2k_BT}\right) \tag{7.28}$$

where T is the absolute temperature, E_g is the band-gap energy, k_B is the Boltzmann constant and C is the proportionality constant characteristic of the material.

EXERCISE 7.19: By what factor is the rate of thermal generation of the electron–hole pairs in germanium decreased by cooling from room temperature to the liquid nitrogen temperature (77 K)?

Solution: From equation (7.28), we have

$$p(300) = C(300)^{3/2} \exp\left[-\frac{0.665 \text{ eV}}{2\left(8.62\times10^{-5}\dfrac{\text{eV}}{\text{K}}\right)(300 \text{ K})}\right] = 0.0135\, C$$

E_g at 77 K can be found from interpolation as we know its value at 330 K and 0 K. Therefore,

$$E_g(77 \text{ K}) \approx \left(\frac{300-77}{300}\right)(0.746-0.665) = 0.725 \text{ eV}$$

$$p(77) = C(77)^{3/2} \exp\left[-\frac{725 \text{ eV}}{2\left(8.62\times10^{-5}\dfrac{\text{eV}}{\text{K}}\right)(77 \text{ K})}\right] = 1.29\times10^{-21}\, C$$

Thus, the rate is reduced by a factor,

$$\frac{0.0135\, C}{1.29\times10^{-21}\, C} = 1.05\times10^{19}$$

In *extrinsic semiconductors*, donor (electron giving) or acceptor (hole-creating) impurities are added. The impurities create electrons or holes having energy levels just below the conduction band (for electrons) or just above the valance band (for holes). The process of addition of impurities into a pure semiconductor is called *doping*. Doped semiconductors are of two types:

n-type semiconductor: This type is produced when pentavalent or 'donor-type' impurities are added to a pure semiconductor. Consider the case when some arsenic (As) atoms (valency 5) are added to a silicon (Si) crystal. Four electrons of arsenic take part in forming covalent bonds with the silicon atoms. The one extra electron for each arsenic atom remains available as free electron helping in the conduction. These extra electrons occupy the energy levels just below the conduction band (Figure 7.18). Hence, they can easily be excited into the conduction band. In *n*-type semiconductors, the electrons are the majority carrier and the holes are the minority carriers.

p-type semiconductor: This type is produced when trivalent or 'acceptor-type' impurities are added to a pure semiconductor. Consider the case when some boron (B) atom (valency 3) is added to a silicon (Si) crystal. The three valence electrons of boron atom take part in forming covalent bonds with three electrons of the neighbouring silicon atoms. The fourth neighbouring silicon atom remains unbounded and a vacancy is created. This vacancy is equivalent to the presence of a hole in the valence band, which can help in the conduction. The impurities produce the allowed energy levels just above the valence band (Figure 7.20). The electrons from the valence band can move to occupy these energy levels creating new holes for the conduction. In *p*-type semiconductors, the majority carriers are the holes and the minority carriers are the electrons.

For extrinsic semiconductors,

$$n_e n_h = (n_i)^2 \tag{7.29}$$

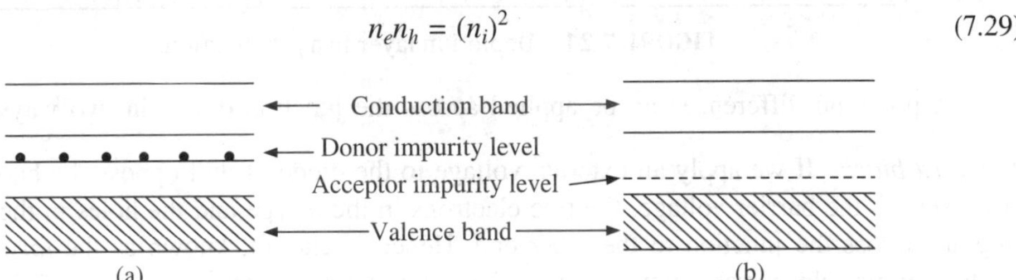

FIGURE 7.20 (a) Donor level created in the silicon band gap; (b) acceptor level created in the silicon band gap.

EXERCISE 7.20: A semiconductor is known to have an electron concentration of 8×10^{13} cm^{-3} and a hole concentration of 5×10^{12} cm^{-3}. Is the semiconductor *n*-type or *p*-type? Calculate the resistivity of the sample, if the electron mobility is 23000 cm^2/Vs and the hole mobility is 100 cm^2/Vs.

Solution: Since, $n_e = 8 \times 10^{13}$ cm^{-3} > $n_h = 5 \times 10^{12}$ cm^{-3}, the semiconductor is *n*-type.

From equation (7.25), the conductivity of the sample is given by:

$$\sigma = e(n_e\mu_e + n_h\mu_h) = 1.6 \times 10^{-19}[8 \times 10^{13}(23000) + 5 \times 10^{12}(100)]$$

$$= 0.29448 \text{ Scm}^{-1}$$

Thus, the resistivity of the sample

$$\rho = \frac{1}{\sigma} = \frac{1}{0.29448 \text{ Scm}^{-1}} = 3.396\ \Omega\text{cm}$$

7.4.2 Semiconductor Junction

The arrangement obtained by joining a *p*-type and an *n*-type semiconductor end to end is called a *p-n junction* or a *junction diode*.

Due to their high concentration, some free electrons in the *n*-type diffuse through the junction to the *p*-type and likewise some holes in the *p*-type diffuse into the *n*-region. Transfer of these charges to either region uncovers positive immobile charges in the *n*-region and negative immobile charges in the *p*-region. This results in the establishment of an electric field at the junction. Potential difference thus produced is called the *potential barrier* and is of the order of 0.1 to 0.3 V (for Ge) depending upon the extent of impurity. This barrier voltage opposes any further movement of the charges through the junction. This barrier voltage must be overcome if some current is to flow across the junction. As the region around the junction does not contain any mobile charge carrier, it is known as the *depletion layer* (Figure 7.21).

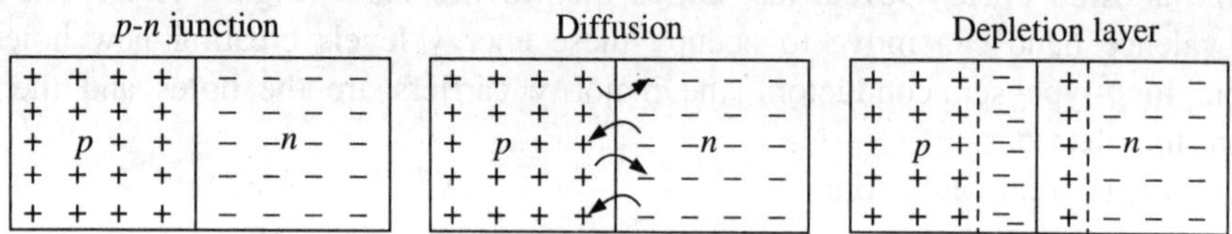

FIGURE 7.21 Depletion layer in a *p-n* junction.

A potential difference can be applied across the junction diode, in two ways:

Forward bias: If we apply an external voltage to the diode, it will oppose the barrier voltage. If it exceeds the barrier voltage, the free electrons in the *n*-type and the holes in the *p*-type will migrate across the junction to the other end. Hence, a current, known as the *forward current* starts flowing. No current flows until the barrier voltage is overcome (Figure 7.22).

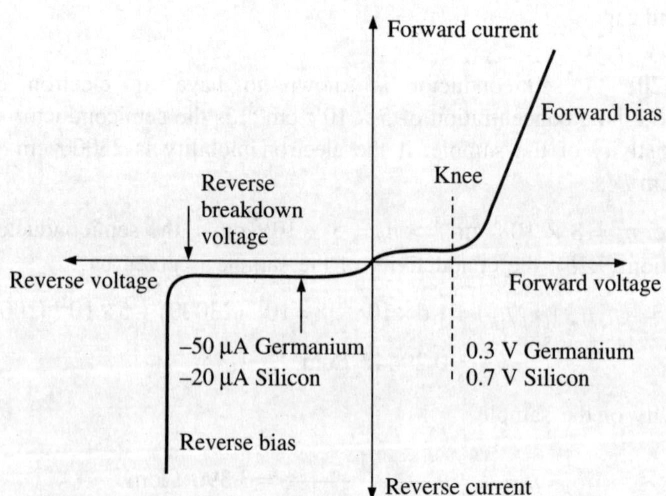

FIGURE 7.22 Characteristic curve for *p-n* junction diode.

Reverse bias: The voltage applied to the junction with opposite polarity will aid the barrier voltage in preventing the flow of charge carrier across the junction. Such voltage is called the reverse-biased voltage. This results in high resistance across the junction. In reverse-bias, no electrons flow from *n*-type to *p*-type and no holes can flow from *p*-type to *n*-type. However, a few charge carriers accelerated by the reverse-bias voltage do cross the junction and constitute a current in reverse direction known as the *leakage current* (Figure 7.22). Above the maximum operating voltage known as the breakdown voltage, the diode properties deteriorate drastically.

The reverse *p-n* junction provides the necessary conditions for the detection of an ionizing radiation. Semiconductor radiation detectors have a *PIN diode structure*, where *I* stand for the intrinsic carrier free region created by the depletion of the charge carriers when a reverse bias is applied to the diode. The width of the depletion layer represents the active volume of the detector, and can be changed by varying the reverse bias voltage. The variable active volume of the semiconductor junctions is unique among radiation detectors, and can be used to good advantage. When an energetic particle passes through the depletion layer of a semiconductor diode, it ionizes the atoms along its track, creating equal number of free electrons and holes. The electrons are collected by the anode and the holes by the cathode, leading to a short electric pulse. The detection process is similar to that occurring in the ionization chambers, where the incident radiation created ion-pairs. Therefore, the semiconductor diode detectors are also referred to as solid state ionization chambers. A typical alpha spectrum from ^{234}U showing α-particles leading to the two states in ^{230}Th is shown in Figure 7.23.

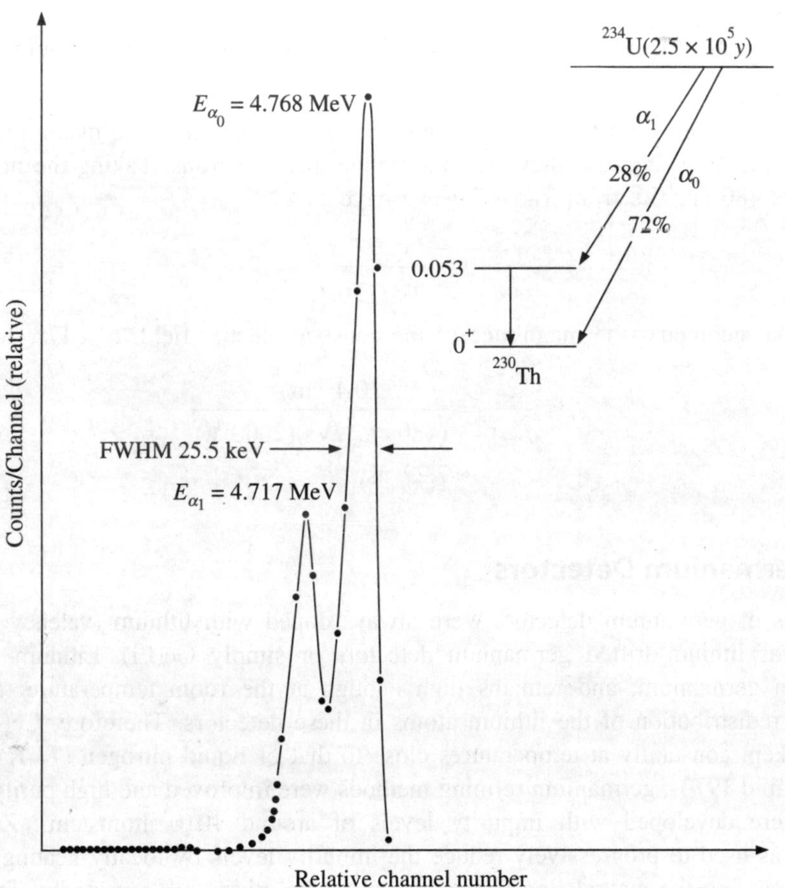

FIGURE 7.23 A typical alpha spectrum from U-234.

7.4.3 Lithium-drifted Silicon Detectors—Si(Li)

It is not yet possible to produce an intrinsic silicon crystal with net impurity concentration less than about 10^{12} atoms/cm^3. Therefore, silicon detectors are doped with lithium, which hardly migrates at room temperature. Silicon has low Z and therefore, it is only used for the detection of charged particle, mainly α-particles. When the surface of an n-type silicon crystal doped with lithium is oxidized, a thin p-type surface layer rich in holes is created. The depletion layer depth can reach as much as 100–1000 nm.

The Si(Li) detectors are usually operated at the room temperature if they are employed for the spectroscopy of the long range charged particles but must be cooled for the high resolution X-ray spectroscopy. The low temperature operation considerably reduces the thermally generated leakage current and the associated noise; it also decreases the charge collection time due to an increase in the charge carrier mobility. The larger band gap assures that the thermally generated leakage current at any given temperature is smaller in silicon than in germanium. With other factors being roughly equivalent, the limiting energy resolution using equivalent electronic components should be superior with the silicon detectors.

EXERCISE 7.21: Estimate the maximum charge collection time for a 4 mm thick Si(Li) detector operated at 2000 V.

Solution: The collection time will depend on the drift velocity. The maximum collection time is for the holes because they are less mobile than electrons. Taking the hole mobility at 300 K is 480 cm^2/Vs from Table 7.4, we have

$$t_c = \frac{x}{v_d} = \frac{x}{\mu_H E}$$

In the planar geometry, the magnitude of the constant electric field, $E = V/x$. Thus

$$t_c = \frac{x^2}{\mu_H V} = \frac{(0.4 \text{ cm})^2}{(480 \text{ cm}^2/\text{Vs})(2000 \text{ V})}$$
$$\approx 7.27 \text{ ns}$$

7.4.4 Germanium Detectors

Earlier types of germanium detectors were always doped with lithium (valency 3) and were referred to as lithium-drifted germanium detectors or simply Ge(Li). Lithium ion mobility is greater in germanium, and remains high enough at the room temperature to permit the undesirable redistribution of the lithium atoms in these detectors. Therefore, Ge(Li) detectors have to be kept constantly at temperatures close to that of liquid nitrogen (77 K).

In the mid 1970s, germanium refining methods were improved and high purity germanium detectors were developed with impurity levels of around 1010 atoms/cm^3. Zone refining technique was used to progressively reduce the impurity levels by locally heating the material and slowly passing the melted ring from one end to the other end, repeatedly. The impurities get transferred to the molten ring and are swept away from the sample. Since 1980s, all new germanium diode detectors are made of high purity germanium, generally referred to as HPGe. These new crystals can be kept at room temperature when not in use. The conductivity of pure germanium is so high at the room temperature that the weak current pulses produced by the ionizing particles, are drowned in small current fluctuations. In order to suppress the current fluctuations to an acceptable level, the diode is cooled to 77 K with liquid nitrogen.

Since, germanium has high Z, these detectors are useful for γ-ray spectroscopic measurements, because unlike in α-measurements, thick detectors are required for the γ-ray measurements. When a γ or an X-ray photon enters a detector, it should create a recoil electron by one of the three processes described in the last chapter, before it is recorded as an event: the photoelectric absorption, the Compton scattering, or the pair production. The relative probability of each of the three types of interactions is shown in Figure 7.24.

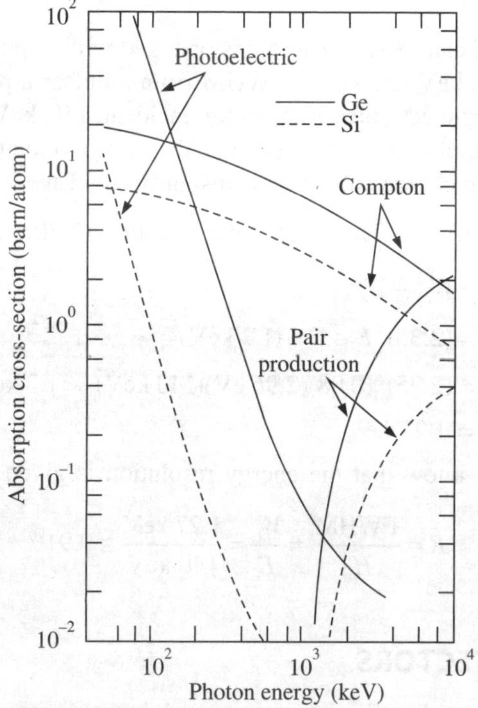

FIGURE 7.24 Relative probability for different interactions as a function of energy in Ge and Si. [*Experiments in Nuclear Science*, ORTEC 1976 © AMETEK, Inc.]

Both HPGe and Ge(Li) of the same size have nearly identical performance characteristics with respect to the energy resolution and the detection efficiency. The overall energy resolution achieved in a germanium detector system is generally determined by a combination of three factors: variations in the charge collection efficiency, the inherent statistical spread in the number of charge carriers, and the noise. The detector size and the quality along with the energy of the radiation determine which of these factors is dominant. The FWHM of a typical peak (W_t) in the spectrum for a mono-energetic γ-ray is given as follows:

$$W_t^2 = W_D^2 + W_X^2 + W_e^2 \tag{7.30}$$

where W_D represents the inherent statistical fluctuations in the number of charge carriers created and is given by:

$$W_D^2 = (2.35)^2 F\varepsilon E \tag{7.31}$$

where F is the Fano factor, ε is the energy necessary to create one electron–hole pair, and E is the γ-ray energy. W_X is due to the incomplete charge collection. For detectors of large size and low average electric field, its contribution increases. Its magnitude is often experimentally estimated by carrying out a series of FWHM measurements by varying the applied voltage. The assumption involved is that if the electric filed is made infinitely large, the charge collection would be complete. W_e represents the electronic noise. Its magnitude can be experimentally measured by using a precision pulser with highly stable amplitude, as input and recording the corresponding peak in the pulse height spectrum.

EXERCISE 7.22: A planar Ge(Li) detector is operated with a pulse processing system that produces a peak with 1.2 keV equivalent FWHM from a pulser input. Determine the energy resolution of the detector-electronics system for incident 140 keV γ-rays. The detector is operated with sufficient applied voltage to saturate the carrier velocities. Assume that the term accounting for incomplete charge collection is insignificant. Take Fano factor for Ge as 0.08.

Solution: W_X^2 is insignificant. Therefore, from equation (7.30), we have

$$\begin{aligned} W_t^2 &\approx W_D^2 + W_e^2 \\ &= (2.35)^2 F \in E + (1.2\ \text{keV})^2 \\ &= (2.35)^2 (0.08)(2.96\ \text{eV})(140\ \text{keV}) + (1.2\ \text{keV})^2 \end{aligned}$$

$$\therefore \quad W_t = 1.27\ \text{keV}$$

From equation (7.3), we know that the energy resolution is given as follows:

$$R = \frac{\text{FWHM}}{H_0} = \frac{W_t}{E} = \frac{1.27\ \text{keV}}{140\ \text{keV}} \cong 0.91\%$$

7.5 NEUTRON DETECTORS

A neutron detector does not directly detect a neutron but responds to the secondary radiation (generally fast charged particles) that is emitted when the neutron undergoes a nuclear reaction in the detector medium. For slow and thermal neutrons, the (n, p), (n, α) or (n, fission) reactions on light nuclei are among the most commonly used in detectors. For fast neutrons of several MeV energy, scattering off a light target can give enough energy to a recoiling nucleus for detection.

7.5.1 Slow Neutron Detection

The isotope ${}^{10}_{5}\text{B}$ is most commonly used in the form of BF_3 gas inside a proportional counter. The ${}^{10}_{5}\text{B}(n, \alpha){}^{7}_{3}\text{Li}$ reaction may be written as follows:

$$ {}^{10}_{5}\mathrm{B} + {}^{1}_{0}n \rightarrow \begin{cases} {}^{7}_{3}\mathrm{Li} + {}^{4}_{2}\alpha & Q = 2.792 \text{ MeV} \\ {}^{7}_{3}\mathrm{Li}^{*} + {}^{4}_{2}\alpha & Q = 2.310 \text{ MeV} \end{cases} $$

The branching indicates that the reaction product ${}^{7}_{3}\mathrm{Li}$ may be left either in the ground state or the first excited state. When thermal neutrons are used to induce the reaction, about 94% of all reactions lead to the excited state with just 6% directly leading to the ground state. The Q-value is shared between the outgoing α-particle and the ${}^{7}_{3}\mathrm{Li}$ ion. The charged particles that are emitted in the reaction cause ionization in the detector gas, which gives rise to an output signal.

In the ${}^{3}_{2}\mathrm{He}$ proportional counter, which is also widely used, the reaction used is ${}^{3}_{2}\mathrm{He}(n, p){}^{3}_{1}\mathrm{H}$, which may be written as follows:

$$ {}^{3}_{2}\mathrm{He} + {}^{1}_{0}n \rightarrow {}^{3}_{1}\mathrm{H} + {}^{1}_{1}p \qquad Q = 0.765 \text{ MeV} $$

The energy released in the reaction is shared between the proton and the triton.

Another popular reaction for the detection of slow neutrons is ${}^{6}_{3}\mathrm{Li}(n, \alpha){}^{3}_{1}\mathrm{H}$, which may be written as follows:

$$ {}^{6}_{3}\mathrm{Li} + {}^{1}_{0}n \rightarrow {}^{3}_{1}\mathrm{H} + {}^{4}_{2}\alpha \qquad Q = 4.78 \text{ MeV} $$

The energy released is shared between the α-particle and the triton. Because a stable lithium-containing proportional gas does not exist, lithium proportional counter is not available. Though, lithium-containing scintillators are quite common as slow neutron detectors. The scintillation mechanism is similar to that of NaI. Because of their fast response, lithium-containing glass scintillators have also become quite popular in neutron physics research.

Figure 7.25 shows the energy dependence of some of these reactions. The $1/v$ dependence of these reaction cross-sections offers the advantage that the count rate is proportional to the neutron density at the detector and is independent of the neutron velocity distribution.

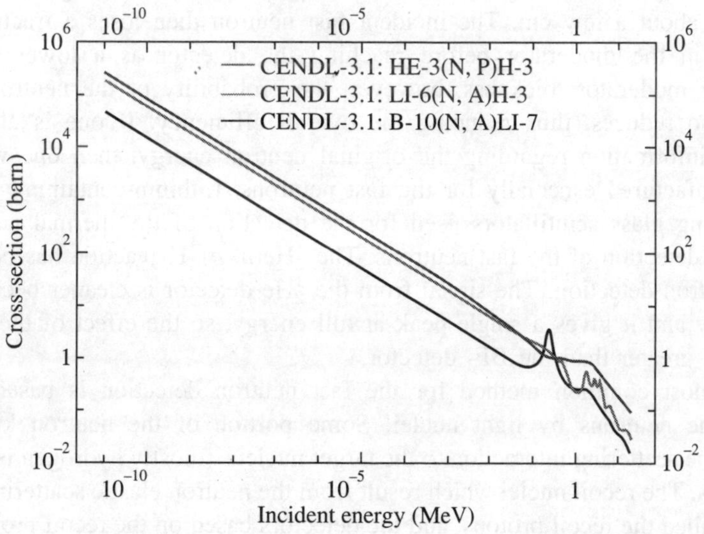

FIGURE 7.25 Energy dependence of cross-sections for neutron reactions on ^{3}He, ^{6}Li and ^{10}B. [National Nuclear Data Center: CENDL-3.1, China, 2009]

EXERCISE 7.23: When operated at a gas multiplication factor of 1000, estimate the pulse amplitude produced by the interaction of a thermal neutron (0.025 eV) in a ^{3}He proportional counter of 100 pF capacitance.

Solution: The following reaction takes place in the tube:

$$^3_2\text{He} + ^1_0n \rightarrow ^3_1\text{H} + ^1_1p \qquad Q = 0.765 \text{ MeV}$$

Number of ion-pairs produced (from Table 7.1, for helium, W = 31.7 eV/ion-pair)

$$n_0 = \frac{E}{W} = \frac{0.765 \times 10^6 \text{ eV}}{31.7 \text{ eV/ion pair}} = 2.413 \times 10^4$$

The pulse amplitude produced

$$V = \frac{n_0 eM}{C} = \frac{(2.413 \times 10^4)(1.602 \times 10^{-19} \text{ C})(1000)}{100 \times 10^{12} \text{ F}} = 38.66 \text{ mV}$$

7.5.2 Fast Neutron Detection

At higher energies, the neutron's kinetic energy adds to the Q-value of the reaction and increases the energies of the reaction products. In principle, the reactions that can be used to detect slow neutrons could also be applied to detect fast neutrons. However, with increasing neutron energy, as can be seen in Figure 7.25, the probability that a neutron will interact by one of these reactions decreases rapidly, and the detector efficiency becomes very low for fast neutrons.

The inherently low detection efficiency for fast neutron of any slow neutron detector can be improved by surrounding the detector with hydrogen-containing moderating material of thickness of about a few cm. The incident fast neutron then loses a fraction of its initial kinetic energy in the moderator before reaching the detector as a lower energy neutron. With increasing moderator thickness, however, the probability of the neutron to ever reach the detector also reduces, thus reducing the overall efficiency. If one is also interested in preserving the information regarding the original neutron energy, then one will have to use a detector manufactured especially for the fast neutrons. Lithium-containing scintillators and lithium-containing glass scintillators used for the detection of the thermal neutrons can also be used for the detection of the fast neutrons. The $^3_2\text{He}(n, p)^3_1\text{H}$ reaction has been widely used for the fast neutron detection. The signal from the $^3_2$He detector is cleaner because its Q-value is relatively low and it gives a single peak at full energy, so the effect of the neutron energy on the signal is greater than the BF_3 detector.

But the most common method for the fast neutron detection is based on the elastic scattering of the neutrons by light nuclei. Some portion of the neutron kinetic energy is transferred by the scattering interaction to the target nucleus (mostly hydrogen is used), resulting in recoil nucleus. The recoil nuclei which result from the neutron elastic scattering from ordinary hydrogen are called the recoil protons, and the detectors based on the recoil protons are referred to as the *proton recoil detectors*. Scintillator materials containing hydrogen are most commonly used for the fast neutron detection, because the signal comes from the energy of the recoil protons scattered by the neutrons within the scintillator itself.

Based on the assumption of elastic scattering, we use the conservation of energy (non-relativistic kinetic energy) and the conservation of momentum to arrive at the following relation for the recoil nucleus energy in terms of its own angle of recoil (Figure 7.26),

$$E_R = \frac{4A}{(1+A)^2}(\cos^2\theta)E_n \tag{7.32}$$

where A is the ratio of the mass of the target nucleus to the neutron mass, θ is the scattering angle of the recoil nucleus, E_R is the recoil nucleus kinetic energy and E_n is the incoming neutron kinetic energy. If the target nucleus is a proton, it must be noted that the angle between a recoil proton and the corresponding scattered neutron is always 90°.

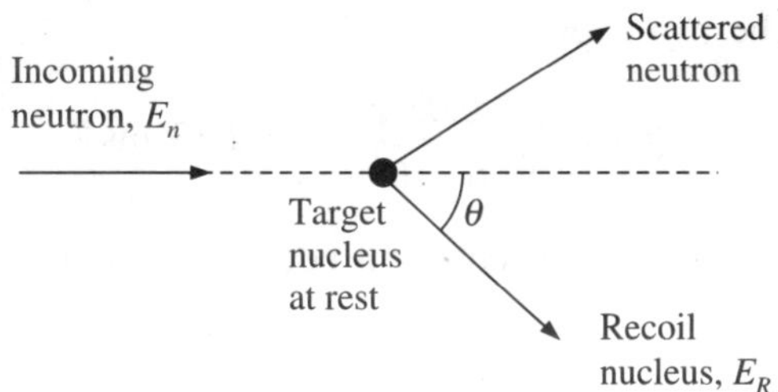

FIGURE 7.26 Neutron elastic scattering diagram in the laboratory frame.

The detection efficiency of a device based on the recoil protons or other recoil nuclei can be calculated from the scattering cross-section σ_s. If nuclei of only one species are present in the detector, the intrinsic efficiency is given by

$$\varepsilon = 1 - \exp(-N\sigma_s d) \tag{7.33}$$

where N is the number density of target nuclei and d is the incident neutron path length through the detector. If there are two species present in the detector, for example, carbon often appears in combination with hydrogen in the proton recoil detectors, then the competing effects of scattering have to be taken into account and equation (7.33) takes the following form:

$$\varepsilon = \frac{N_H\sigma_H}{N_H\sigma_H + N_C\sigma_C}\{1 - \exp(-(N_H\sigma_H + N_C\sigma_C)d)\} \tag{7.34}$$

where the subscripts H and C refer to the separate hydrogen and carbon values.

EXERCISE 7.24: A 1 MeV neutron enters a plastic scintillator and undergoes two sequential scatterings from hydrogen nuclei before escaping. If the first scattering deflects the neutron at an angle of 40° with respect to its original direction and the scattering sites are 3 cm apart, calculate the time that separates the two events. If the photomultiplier anode time constant is 20 ns, will the two events be resolved?

Solution: Since, the mass of a proton is almost equal to the neutron's mass,. we get $A \approx 1$. Therefore, from equation (7.32), we have

$$E_R \approx (\cos^2 \theta)\, E_n$$

or
$$E_R = \{\cos^2(90° - 40°)\}(1 \text{ MeV}) = 0.413 \text{ MeV}$$

Therefore, the neutron energy after the first scattering

$$E'_n = E - E_R = 1 - 0.413 = 0.587 \text{ MeV}$$

Now,
$$v'_n = \sqrt{\frac{2E'_n}{m}} = \sqrt{\frac{2(0.587 \times 10^6 \text{ eV})(1.602 \times 10^{-19} \text{ J/eV})}{1.675 \times 10^{-27} \text{ kg}}} = 1.06 \times 10^7 \text{ m/s}$$

The time that separates the two events

$$t = \frac{x}{v'_n} = \frac{0.03 \text{ m}}{1.06 \times 10^7 \text{ m/s}} = 2.83 \text{ ms}$$

It is less than 20 ns, hence, the two events will not be resolved.

7.6 SOME OTHER DETECTORS

7.6.1 Cerenkov Detectors

Electromagnetic shockwave (light) emitted by a high-speed charged particle when the particle passes through a transparent, non-conducting, solid material at a speed greater than the speed of light in the material is known as Cerenkov radiation. First predicted by P.A. Cerenkov in 1934, this radiation is used as a signal for the indication of high-speed particles. Particle detectors that utilize Cerenkov radiation are called Cerenkov detectors. Cerenkov detectors play an important role in the detection of particles with speeds approaching that of light such as those generated in large accelerators and in cosmic rays. Dielectrics such as glass, water, or clear plastic may be used in the Cerenkov detectors. The choice of the material depends on the range of particle velocities to be measured. Gas, notably CO_2, may also be used as the dielectric in the Cerenkov detectors. In such detectors, the intensity of light emitted is smaller than in solid or the liquid dielectric detectors, but the velocity required to produce a count is higher because of the low refractive index of the gas. When two Cerenkov detectors are used in coincidence, one after the other, with proper choice of dielectric, the combination will be sensitive to a given velocity range of particles.

In contrast to the scintillation light, which is emitted isotropically, the Cerenkov photons are emitted preferentially along the direction of the particle velocity as shown in Figure 7.27. The light is confined to a cone with vertex angle θ, where

$$\cos\theta = \frac{1}{\beta n} \tag{7.35}$$

where n is the refractive index of the medium and the particle velocity $v = \beta c$.

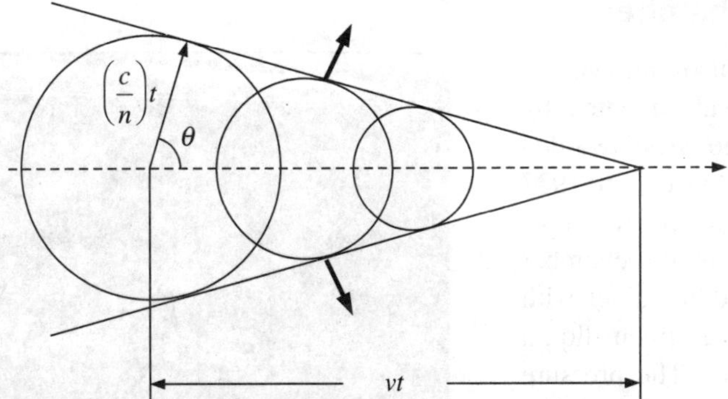

FIGURE 7.27 Cerenkov radiation: electromagnetic shockwave emitted when the particle travels faster than the speed of light in the same medium.

EXERCISE 7.25: An electron incident on a glass block of refractive index 1.5 emits Cerenkov radiation at an angle of 45° to its direction of motion. At what speed is the electron travelling?

Solution: From equation (7.35), we have

$$\beta = \frac{1}{n\cos\theta} = \frac{1}{1.5\cos 45^\circ} = 0.943$$

Therefore, the particle velocity

$$v = 0.943c = 2.828 \times 10^8 \text{ m/s}$$

7.6.2 Track-etch Detectors

When an ionizing charged particle travels through a dielectric material, a trail of damaged molecules along the particle track is generated with the transfer of energy to the electrons. In certain dielectric materials, this damaged track can be exposed through chemical etching of the material surface using a strong acid or base solution. If the charged particles irradiated the surface at some point of time, then each of these particles leaves a trail of damaged material that begins at the surface, extending to a depth equal to the range of the particle. In the materials of choice, the rate of chemical etching along this track is higher than the etching rate of the undamaged surface. Therefore, as the etching progresses, a depression (pit) is formed at the location of each track. Within few hours, these pits become large enough to be seen directly under a microscope. A measurement of the number of these pits per unit area gives a measure of the particle flux to which the surface has been exposed.

The minimum density of damage along the track is related to the stopping power or the specific energy loss ($-dE/dx$), that is required in order for the etching rate to create a pit. For heavier charged particles, the density of the damage is more. This threshold behaviour makes such detectors completely insensitive to β-particles and γ-rays, and can be exploited for one's advantage. Automated methods can be used to measure the density of the pits by using microscope stages coupled to computers.

7.6.3 Cloud Chamber

FIGURE 7.28 Tracks of ionizing radiation in cloud chamber.

Cloud chamber, also known as Wilson chamber, was invented by C.T.R. Wilson, who was awarded the Nobel Prize (Physics) in 1927 for its development. In its most basic form, a cloud chamber involves an enclosed container with air saturated with a given liquid like water or alcohol. The pressure in the chamber is then lowered by moving a piston attached to it and the vapours become supersaturated. At this point, when an ionizing charged particle passes through the chamber, the supersaturated vapours condense on the ions and a trail of droplets along the path of the charged particles is seen. These tracks known as cloud tracks have distinctive shapes, like an α-particle track is broad and short while an electron's track is light and thin (see Figure 7.28). Use of magnetic field allows the sign and the momentum of the particles to be measured by observing the curvature of their tracks. The particle tracks are photographed for later study.

EXERCISE 7.26: A certain electron–positron pair generated cloud chamber tracks of radius of curvature 3 cm lying in a plane perpendicular to the applied magnetic field of magnitude 0.11 tesla as shown in Figure 7.29. What was the energy of the γ-ray which produced the pair?

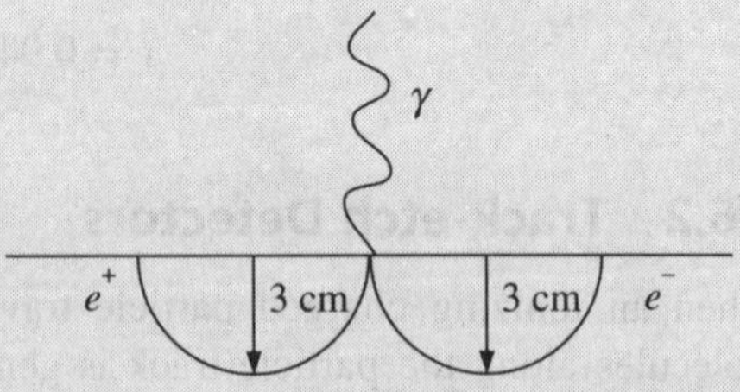

γ e^+ 3 cm 3 cm e^-

FIGURE 7.29 Cloud chamber tracks of an electron-positron pair in presence of 0.11 tesla magnetic field.

Solution: Centripetal force will be provided by the magnetic field. Thus,

$$evB = \frac{m_\gamma v^2}{r} = \frac{pv}{r}$$

or

$$pc = ceBr$$

$$= \frac{(3\times10^8 \text{ ms}^{-1})(1.6\times10^{-19}\text{ C})(0.11\text{ T})(0.03\text{ m})}{1.6\times10^{-13}\text{ MeV/J}} = 0.99 \text{ MeV}$$

Therefore, the energy of the electron/positron

$$E = \sqrt{(pc)^2 + m_e^2c^4} = \sqrt{0.99^2 + 0.511^2} = 1.1 \text{ MeV}$$

Hence, the energy of the γ-ray that produced the electron–positron pair is approximately

$$E_\gamma = 2E = 2.2 \text{ MeV}$$

7.6.4 Bubble Chamber

The basic drawback of the cloud chamber is that it is not possible to observe high energy particles, because of the low density of the supersaturated vapour. In 1952, D.A. Glaser, who was later awarded the Nobel Prize (Physics) in 1960, conceived the idea of using superheated liquid in place of superheated vapour to display the tracks of ionizing particles, and the bubble chamber was born. Each track is a string of small bubbles marking the path of the elementary particle. The boiling bubbles grow rapidly, and so a photograph is quickly taken before the resolution of the individual track is destroyed. In order to make the bubble chamber sensitive to the arrival of another charged particle, the boiling must be stopped, and the liquid must be superheated again. Due to the high density of the liquid, even high energy cosmic rays can be recorded using a bubble chamber. Bubble chambers vary in size, in the liquid that is used inside them and in how fast the photographs of the tracks can be taken.

The first bubble chamber had only a few centimetres dimension. Bubble chambers have steadily become larger and are now built as large as 15 ft across. The main reason for making such large chamber is to be able to study neutrino interactions. Since, neutrinos interact only weakly with other particles, one uses large volume of liquid to increase the chance of an interaction occurring. In Figure 7.30, the schematic diagram of Fermi Lab's 15-ft bubble

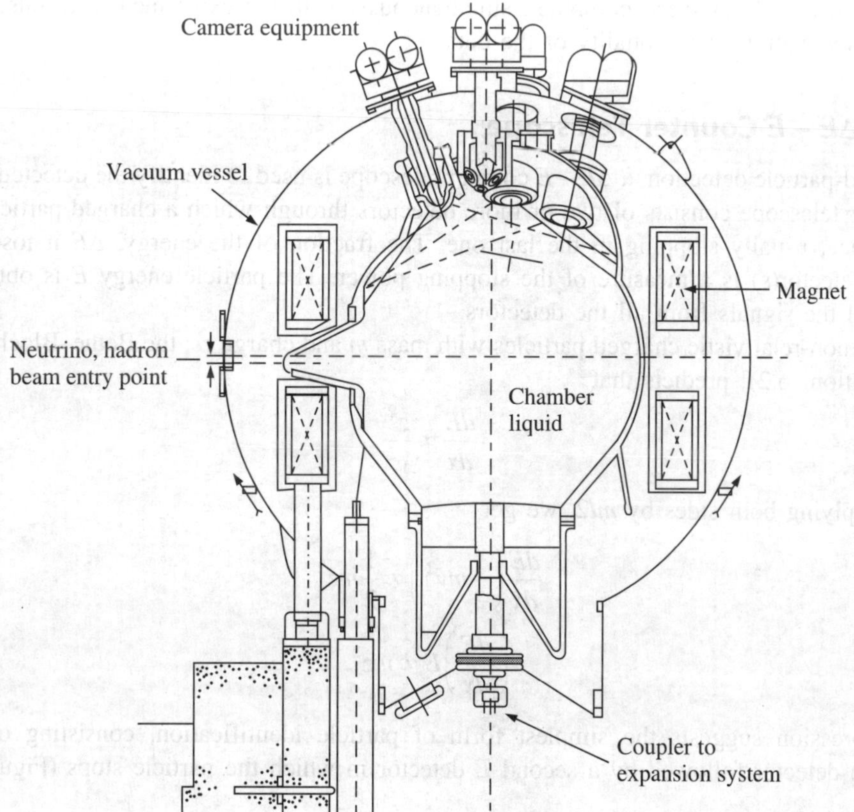

FIGURE 7.30 Line drawing showing major elements of Fermi Lab's 15 ft bubble chamber.

chamber is shown. Bubble chambers have been combined with other particle detectors such as spark chambers, Cerenkov counters, etc. to build hybrid systems to be able to further study the particles before and after they leave the chamber.

7.6.5 Spark Chamber

Spark chamber consists of a stack of metal plates or wire grids placed in a sealed box filled with a high pressure gas such as helium, neon or a mixture of the two. The size of the plates and their number vary depending upon the purpose and the nature of the experiment. A high voltage is maintained between the alternate plates. When the ionizing particle travels through the box, sparks jump between adjacent, oppositely charged plates and the trail of sparks left by the particle is seen as a series of dashes.

Although spark chamber is characterized by somewhat lower resolution than bubble chambers, the spark chamber offers the advantage that it operates much more rapidly and instead of indiscriminately recording all the events that occur, like in a bubble chamber, it has a unique triggering capability that selects only those events that have passed the logical selection tests. Spark chambers can be highly automated, with electronic data collection and storage in contrast to the bubble chamber, where it is done photographically. The data can be analyzed by a high-speed computer, simultaneously with the experiment and thus, helps in real-time evaluation of the quality of the data.

7.6.6 $\Delta E - E$ Counter Telescope

In charged-particle detection, a $\Delta E - E$ counter telescope is used to identify the detected particle. A counter telescope consists of two or more detectors through which a charged particle passes in sequence, usually stopping in the last one. The fraction of the energy, ΔE it loses in the passing detector(s) is a measure of the stopping power. The particle energy E is obtained by adding all the signals from all the detectors.

For non-relativistic charged particles with mass m and charge ze, the Bethe–Bloch formula [see equation (6.2)] predicts that

$$\frac{dE}{dx} \propto \frac{z^2}{v^2}$$

On multiplying both sides by $m/2$, we get

$$\frac{dE}{dx}\left(\frac{1}{2}mv^2\right) \propto \frac{1}{2}mz^2$$

or

$$\left(\frac{dE}{dx}\right)E \propto mz^2 \tag{7.36}$$

This expression suggests the simplest form of particle identification, consisting of a very thin (ΔE) detector followed by a second E detector in which the particle stops (Figure 7.31).

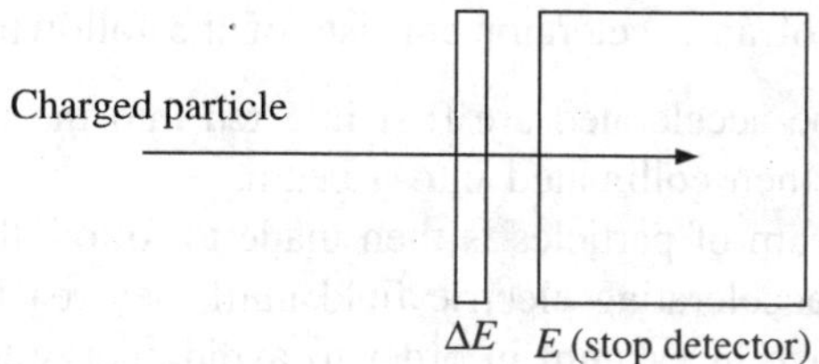

FIGURE 7.31 Simplest form of $\Delta E - E$ counter telescope.

As an example, in one of the studies, two silicon surface barrier (SSB) detector telescopes each having ΔE detectors of thickness 150 and 100 μm, respectively and an E detector of thickness 1 mm were used to detect projectile like fragments (PLFs). The proton, deuteron, triton and α-particles are uniquely identified by plotting ΔE against the residual energy (E_{res}) in the E detector. This plot was converted to effective particle identification (PI) versus the total energy plot. Figure 7.32 shows a typical PI vs. total energy plot, where all the PLFs are clearly identified as different hyperbola as per their unique mz^2 values.

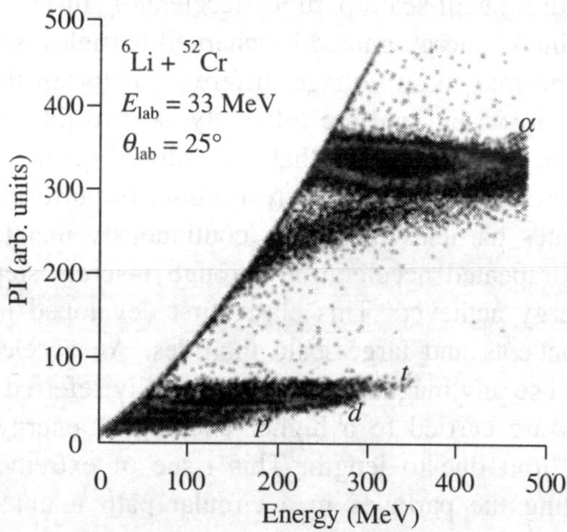

FIGURE 7.32 Particle Identification plot of PLF produced in $^{6}Li + {}^{52}Cr$ reaction at E_{lab} = 33 MeV. [B. Pandey, H.M. Agrawal et al., "Measurement of Fe55(n,p) cross sections by the surrogate-reaction method for fusion technology applications," *Phys. Rev. C 93*, 021602(R) (2016)]

7.7 PARTICLE ACCELERATORS

After Rutherford's discovery of the atomic nucleus in 1919, scientists turned their effort into building machines that could probe matter at short wavelength scales. Most of our understanding about the sub-atomic particles and the structure of the atomic nucleus, has been gained through the study of nuclear reactions caused by the energetic charged particles such as protons, deuterons, alpha particles, etc. Hence, over the years, several types of particle accelerators have been developed to accelerate the charged particles to kinetic energies large enough to be able to overcome the height of the Coulomb barrier due to the repulsive force between the positively charged nucleus in the target material and the impinging beam of charged particle, and penetrate the nucleus in order to investigate the matter at shorter distances.

In principle, the operation of an accelerator consists of the following steps:

(i) The particles to be accelerated are first injected into the machine from source.
(ii) The particles are then collimated into a beam.
(iii) The collimated beam of particles is then made to follow the desired trajectory, under the action of an accelerating electric field until they reach the required energy. The particles must move in vacuum in order to avoid energy loss and being deflected by collisions with the gas molecules.
(iv) At this point, the accelerated beam is made to hit a target material producing nuclear reactions.
(v) The secondary particles produced in the reaction can be used to study further reactions.

There exist two basic classes of accelerators: electrostatic (DC) accelerators that use a steady DC potential and oscillating field (RF) accelerators that employ various combinations of time-varying electric and magnetic fields. Examples of DC accelerators include the Cockcroft–Walton accelerator and the Van de Graff accelerator. The cathode ray tube in an ordinary old television set is nothing but a small-scale particle accelerator. In DC accelerators, because the field is electrostatic, the kinetic energy gained by charged particles is restricted to the potential energy corresponding to the maximum voltage difference between the terminals. These types of accelerators are simpler in design and are relatively easy to fabricate.

The second type is the RF accelerators that use radio frequency electromagnetic fields to accelerate particles, and circumvent the above restriction, because unlike a high-voltage type accelerator which accelerates particles through a continuously maintained potential, the high energies are achieved by repeated acceleration through discrete steps in energy, which are smaller than the final energy achieved. This class, first developed in 1920s, is the basis for all modern accelerator concepts and large-scale facilities. An accelerator that varies only in electric field and does not use any magnetic field is generally referred to as a linear accelerator (LINAC). As accelerators are carried to a higher and higher energy, LINACs are restricted by practical construction limit due to length. This issue of extreme length is circumvented conveniently by accelerating the particles in a circular path maintained by time-varying or static magnetic fields.

FIGURE 7.33 A section of the underground accelerator tunnel of the CERN Large Hadron Collider. [Courtesy CERN]

At present, the largest particle accelerator facility with the record of highest particle energies up to 7 TeV, is the Large Hadron Collider (LHC) at CERN[3]. A section of the underground accelerator tunnel of LHC is shown in Figure 7.33. The accelerator complex at CERN is a succession of machines with progressively more energies. Each one of these machines injects the beam into the next one, which takes over to bring the beam to an even higher energy, and so on.

Another unique international accelerator facility for the research with anti-protons and ions (FAIR) is under construction near Darmstadt in Germany. India is also one of the forty countries participating in this collaborative project. Unprecedented high intensity beams will be available at this facility.

7.7.1 Cockcroft–Walton Accelerator

In 1932, Cockcroft and Walton[4] developed an electrostatic accelerator based on the voltage multiplier principle. This was the first accelerator used to obtain accelerated protons to produce a nuclear reaction. This machine is still quite useful for experimental work at moderate particle energies.

The primary source of energy for the Cockcroft–Walton accelerator is a high voltage transformer which provides alternating voltage $V(t) = V_0 \sin \omega t$, where V_0 is of the order of 100 kV. This energy source is connected to a cascade voltage multiplier having two columns of capacitors connected in series and the two columns are connected by a chain of diodes which conduct only in the direction of the arrows as shown in Figure 7.34. If no current is drawn from the transformer, after some time a stationary state is reached where all the capacitors, except for C_1, are charged to a voltage of $2V_0$, and the diodes do not conduct any current. The potentials are fixed on the plates of the capacitors that are on the right, whereas those on the left oscillate between the limits as indicated in the figure. When the system is not in a stationary state, the diodes conduct current until the system is back in the stationary state. Each stage of the voltage multiplier is connected to a metallic tubular electrode. The ion beam, which is confined in an evacuated accelerator tube to avoid any scattering from the air molecules, thus is accelerated across the gap between the two electrodes. In modern Cockcroft–Walton accelerators, solid state rectifiers[5] are used instead of diodes, as the rectifiers do not require a separate power supply and they also have the advantage of operating without difficulty inside a pressure chamber.

The earliest Cockcroft–Walton accelerator used a 4-stage stack to obtain 710 keV protons which were used to bombard the lithium target to produce two alpha particles.

$$p + {}^7_3\text{Li} \rightarrow 2\alpha$$

3. CERN stands for Conseil Européen pour la Recherche Nucléaire. It is a European organization for nuclear research, the world's largest particle physics centre located in Geneva. India has an observer status at CERN.
4. In 1951, they were awarded Nobel Prize (Physics) for their work.
5. A diode refers to a discrete device, while a rectifier or rectifier assembly has multiple diodes. Rectifiers are designed to handle higher voltage and current. They are typically found in power supplies.

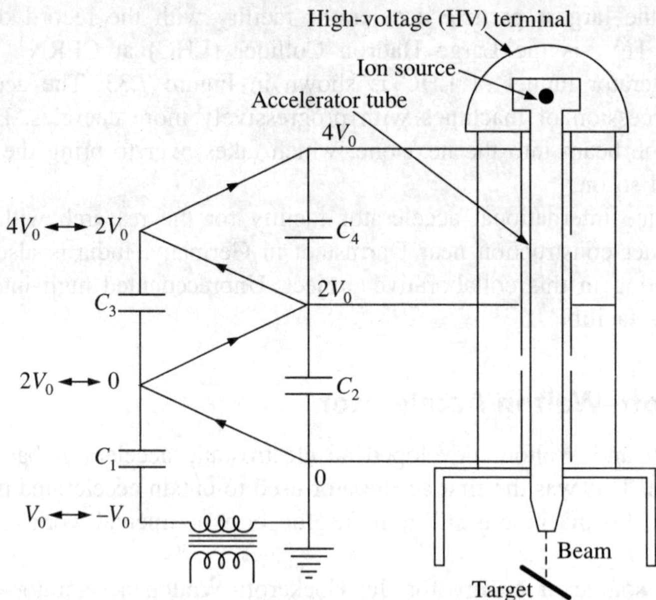

FIGURE 7.34 Schematic diagram of Cockcroft–Walton accelerator.

A million-volt Cockcroft–Walton accelerator was installed at Tata Institute of Fundamental Research (TIFR), Mumbai in 1953 which could generate beams of protons and deuterons of energies up to 1 MeV and beam currents up to 200 micro-amperes.

7.7.2 Van de Graff Accelerator

Van de Graff accelerator's design is based on the principle that if a charged conductor is brought into internal contact with a second hollow conductor, all of its charge transfers to the hollow conductor irrespective of how high the potential of the latter is.

Schematic representation of the Van de Graff accelerator is shown in Figure 7.35. It consists of a hollow conducting sphere which serves as the high voltage (HV) terminal supported by an insulating column structure. The entire structure resides inside a pressure vessel filled with an insulating gas, which is typically SF_6 or a mixture containing 80% nitrogen and 20% carbon dioxide. The charge is supplied to the hollow conductor by means of a charge conveyor belt made up from an insulating material like neoprene, on which positive charge is sprayed at the grounded end and is carried up towards the top terminal, where it is removed by the collector comb attached to the top terminal. An intense electrostatic field is set up near the sharp points of the comb and the gas near the spray comb is ionized. The resulting corona discharge (an electrical discharge accompanied by the ionization of the surrounding air) transfers the charge to the belt which is carried up. If the comb has positive potential, then the electrons produced in the discharge are attracted to it and the positive charge is carried by the belt. The Van de Graff machine can be placed in both vertical as well as horizontal position.

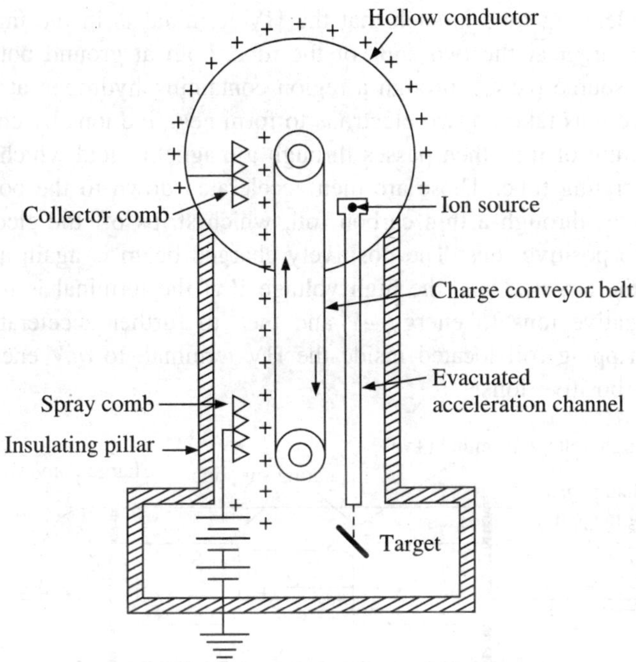

FIGURE 7.35 Schematic diagram of Van de Graff accelerator.

When the electric potential on the sphere produces a field equal to the dielectric strength of the gas, then the gas starts ionizing and potential leakage takes place. This sets the maximum limit for the potential above which it cannot be increased. If Q is the charge supplied to the HV terminal, the maximum potential V to which it is raised is given by the following relation:

$$V = \frac{Q}{C} \tag{7.37}$$

where C is the capacitance of the capacitor formed by the high voltage terminal and the pressure vessel.

Van de Graff accelerators have been used to accelerate protons and α-particles up to energies of 10 – 20 MeV. Compared to the Cockcroft–Walton accelerator, Van de Graff has the advantage that it reaches high voltages more readily. It also gives greater stability of particle energy (up to 0.01%), since its terminal voltage is highly stable. Cockcroft–Walton accelerator gives higher beam intensities more readily but its voltage has fluctuations of the order of 1%.

Tandem Accelerator

Unlike Van de Graff accelerator where the ion source is located inside the HV terminal, thus restricting the type of ion sources that can be mounted due to the problems of power and space, in tandem accelerator which is a two-stage accelerator, the ion source is located outside the HV terminal and is at ground potential, where unlimited power can be available. Here the positive ions are accelerated to twice the energy corresponding to the voltage of the HV terminal. The schematic diagram of the tandem accelerator is shown in Figure 7.36.

Here the accelerating tube is such that the HV terminal is in the middle of it, with the ion source and the target at the two ends of the tube, both at ground potential. The positive beam from the ion source passes through a region containing hydrogen at low pressure where some of the positive ions take up two electrons to form negative ions by charge exchange with hydrogen. The mixture of ions then passes through a magnetic field which filters the negative ions into the accelerating tube. These are then accelerated down to the positive HV terminal. Here the beam passes through a thin carbon foil, which strips off the electrons, transforming the negative ions to positive ions. The positively charged beam is again accelerated from the terminal down to the target. Thus, the high voltage V at the terminal is used for accelerating singly charged negative ions to energy eV and then to further accelerate the positive ions produced in the stripping foil located inside the HV terminal, to qeV energy, where q is the charge state of the positive ions.

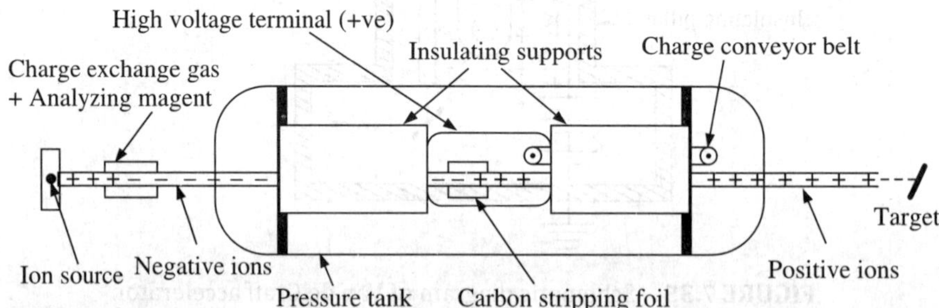

FIGURE 7.36 Schematic diagram of a two-stage tandem accelerator.

On the flip side, the beam intensity of the accelerated particles is low because of the lower efficiency of the charge exchange and stripping processes. Adding further stages can increase the energy even more, although needless to mention that the beam intensity will reduce further. Since, the ion source is outside the HV terminal and is at ground potential, this type of accelerator gives high energies and even heavy ions can be accelerated using tandem accelerator. Also, unlike Van de Graff accelerator, in tandem accelerator the restrictions of power and space are not significant, since both the ion source and the target are outside the HV terminal.

Pelletron Accelerator

Instead of the insulating belt and corona discharge in the standard tandem accelerator, nowadays, metallic pellet chains are used. The pellets are charged inductively by using high voltage. Negative potential is used for inducing positive charge on the chain. The chain is charged more uniformly as the charging current is ripple free, and the terminal voltage is more stable. Also the pellet chain can operate at higher velocities as compared to the usual conveyer belt, which means higher particle energies can be attained. Tandem accelerators with pellet chain charging system are called Pelletrons.

In India, a medium energy heavy ion accelerator was setup in the eighties under a collaborative project of BARC and TIFR, at TIFR campus in Mumbai. The facility is based on a 14 million volt Pelletron accelerator capable of providing accelerated beams of particles such as protons, α-particles and heavy ions. This was the first accelerator facility of its kind in India, which enabled scientists to undertake studies of nucleus–nucleus collisions with heavy ions at medium energies. Another identical machine was installed at Nuclear Science Center, now known as Inter University Accelerator Center (IUAC), New Delhi, to facilitate the university users to do frontline research in nuclear physics, material science and atomic physics.

EXERCISE 7.27:

(a) A 5 MV Van de Graff accelerator is equipped to accelerate protons, deuterons and α-particles. What is the kinetic energy of the different beams as they hit the target? Is it possible to prepare ${}^{15}_{8}O$ isotope using this machine in each case?

(b) In a tandem accelerator, the HV terminal is at 17 MV. Ag^- ions are injected into the accelerating tube, which after passing through the carbon stripping foil reach the target with 15 electrons removed. What is the kinetic energy of the beam as it hits the target?

Solution:

(a) The energy gained by an ion having q charge as it passes through a potential of V volt is qV eV. Therefore, for protons (${}^{1}_{1}H^{+}$), deuterons (${}^{2}_{1}H^{+}$) and α-particles (${}^{4}_{2}He^{++}$), the kinetic energy gained in the accelerating tube is 5 MeV, 5 MeV and 10 MeV, respectively.

The possible reactions to produce ${}^{15}_{8}O$ isotope at these energies are as follows:

$$p + {}^{14}_{7}N \to {}^{15}_{8}O + \gamma$$

$$d + {}^{14}_{7}N \to {}^{15}_{8}O + n$$

$$\alpha + {}^{12}_{6}C \to {}^{15}_{8}O + n$$

(b) The beam reaching the target is made up of Ag^{15+} ions. The kinetic energy of the beam should be equal to the energy gained by the negative ions until they reach the stripping foil inside the HV terminal plus the energy gained by the positive ions, i.e.,

$$\begin{aligned} E_f &= (1 + q) \text{ eV} \\ &= (e + 15e)(17 \text{ MV}) = 272 \text{ MeV} \end{aligned}$$

7.7.3 Linear Accelerator (LINAC)

Unlike electrostatic machines like Cockcroft–Walton and Van de Graff accelerators which use static electric fields to accelerate particles, in linear accelerator, also called 'linac,' radio frequency electromagnetic field is used to accelerate particles. Because of using oscillating fields, the achievable kinetic energy for particles in these devices is not limited by electrical breakdown as in the case of electrostatic accelerators. The electrodes can either be arranged to accelerate particles in a linear or a circular path, depending on whether the particles are subjected to a magnetic field causing their trajectories to arc. In linear accelerators, the acceleration occurs along approximately straight trajectories inside the accelerator tube, made up of several sections of narrow metallic cylinders or drift tubes of increasing lengths. Inside the cylinders, the electric field is zero and in the gaps between two neighbouring cylinders, the electric field alternates with the frequency of the RF source. As the particles approach a cylinder, they are accelerated towards it because of the opposite polarity charge applied to the cylinder. As they pass through the cylinder, the polarity is switched so that the particular cylinder now repels them and they are accelerated towards the next drift tube. The length of the cylinders is chosen to be equal to the distance travelled by the ion beam during one half cycle so that whenever the accelerating particle crosses the gap, it encounters the field with the same phase as in the previous gap, and so it gains energy at each transit.

The schematic representation of the drift-tube type linear accelerator is shown in Figure 7.37.

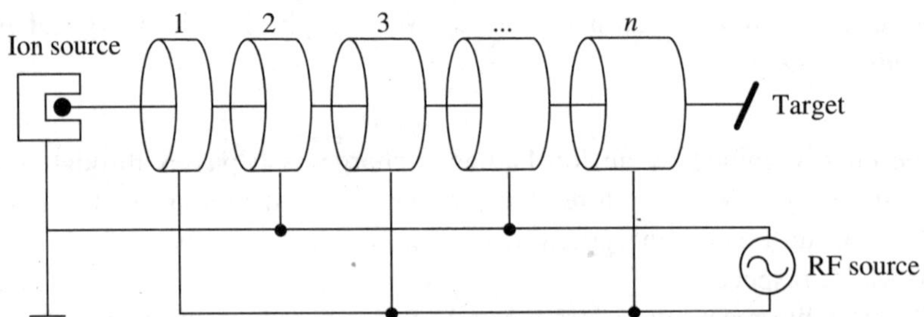

FIGURE 7.37 Schematic diagram of the drift-tube type linear accelerator.

The length of each cylinder L, therefore, is given by

$$L = \frac{vT}{2} = \frac{v}{2f} \tag{7.38}$$

where v is the velocity of the accelerating particle and T and f are the period and frequency of the oscillating field. As the particle travels, it gains energy and v increases, and therefore L must continuously increase. In order for the velocity of the particle to be large, either the accelerator has to be really long or the frequency of the RF source must be high. If, however, very high frequencies are used in order to avoid enormous machine lengths, the transmission losses in the conductors become really high. Linear accelerators are also based on another acceleration mechanism, where electromagnetic waves propagate inside the waveguide, which is a hollow cylindrical metallic conducting tube, inside which the particle travels.

Linacs are often used to provide an initial low-energy kick to the particles before they are injected into the circular accelerators. The longest linear accelerator in operation is the Stanford Linear Accelerator, SLAC, which is 3 km long.

EXERCISE 7.28: A section of linear accelerator has 5 drift tubes and is driven by a 50 MHz oscillator. Assuming that the protons are injected into the first drift tube at 100 kV and gain 100 kV in every gap crossing, what is the total length of the particular section of the accelerator tube? The gap lengths can be ignored.

Solution: After the ions pass though n drift tubes, i.e., $(n - 1)$ gaps, their energy

$$E = E_0 + (n - 1)qV$$

where q is the charge carried by the ion, V is the voltage drop across the gap and E_0 is the initial ion energy.

We can do non-relativistic calculations for the velocity as the ion energy is not large. Therefore, if L_n is the length of the nth drift tube and v_n is the velocity of the ion beam in the nth drift tube, from equation (7.38), we have

$$L_n = \frac{1}{2f}\sqrt{\frac{2E}{m}} = \frac{1}{2f}\sqrt{\frac{2[E_0 + (n-1)qV]}{m}}$$

Therefore, the total length of the section

$$\begin{aligned} L &= \sum_{n=1}^{5} L_n \\ &= \frac{1}{2f}\sqrt{\frac{2}{m}}\left(\sqrt{E_0} + \sqrt{E_0 + qV} + \sqrt{E_0 + 2qV} + \sqrt{E_0 + 3qV} + \sqrt{E_0 + 4qV}\right) \\ &= \frac{1}{2(50\times 10^6 \text{ Hz})}\sqrt{\frac{2(1.6\times 10^{-19}\text{ J/eV})(10^5\text{ eV})}{1.67\times 10^{-27}\text{ kg}}}\left(\sqrt{1} + \sqrt{2} + \sqrt{3} + \sqrt{4} + \sqrt{5}\right) \\ &= 0.367\text{ m} \end{aligned}$$

7.7.4 Cyclotron

Cyclotron was first developed by Lawrence and Livingston at the University of California, Berkeley in 1931. It is important historically because it was the first accelerator to make use of magnetic field to guide the particles in nearly circular path so that the same electrode is used over and over again to give repeated acceleration to the particles in this cyclic process. This way, the downside of the linear accelerator, its enormous length and high cost vis-à-vis a ring accelerator of equal energy, can be overcome. Lawrence was awarded the Nobel Prize (Physics) in 1939 for his invention of the cyclotron.

The basic operating principle of a cyclotron is shown in Figure 7.38. Particles are accelerated in spiral orbits inside two semicircular flat metallic electrodes, called Dees, separated along their straight edges by a small gap. The Dees are connected to an RF source and a constant and uniform magnetic field is applied perpendicular to the Dees, causing the particles to orbit in circular paths in a plane called median plane. An ion source at the centre of the gap injects the particles into one of the Dees. The particles are accelerated, each time they cross the gap, by the RF electric field applied across the gap, and therefore, they follow a spiral path as they gain energy. The important feature of a cyclotron is that the time it takes for the particle to travel one semicircle is independent of the radius of the path because as the particles spiral to larger radii, they also gain energy and hence move at a greater speed. If the half-period of the RF voltage on the Dees is set equal to the semicircular orbit time, then the electric field alternates in exact synchronization as the particles cross the gap, and the particles are accelerated each time as they pass the gap. At the edge of the magnet, a full energy particle beam is extracted out as an external beam by an electrostatic deflector. The whole setup is enclosed in a vacuum chamber to avoid the beam loss by scattering.

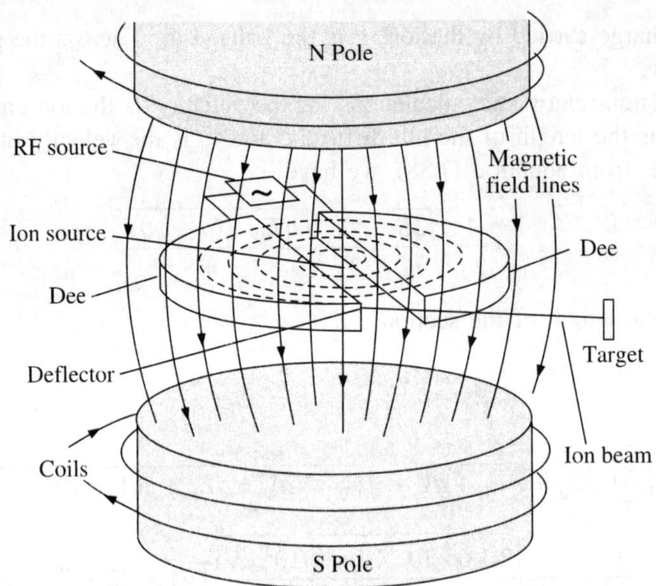

FIGURE 7.38 Schematic layout of a cyclotron showing the two Dees driven by an RF source. The charged particles move in a spiral orbit from their point of injection until they spiral outward to the maximum radius when they emerge out of the accelerator and hit the target.

If r is the instantaneous radius of the circular path of a particle of mass m and charge q moving with velocity v in a uniform magnetic field B, then the Lorentz force qvB provides the necessary centripetal force required to maintain the circular path, i.e.,

$$qvB = \frac{mv^2}{r} \tag{7.39}$$

and the time required by a particle to cover a semicircular path is given by

$$t = \frac{\pi r}{v} = \frac{m\pi}{qB} \tag{7.40}$$

Therefore, the orbiting frequency, called the *cyclotron frequency* or *cyclotron resonance frequency*, is given by

$$f = \frac{1}{2t} = \frac{qB}{2\pi m} \tag{7.41}$$

The particle velocity increases gradually as it spirals outward and the maximum velocity at radius R, when it is extracted out, according to equation (7.39), is as follows:

$$v_{\max} = \frac{qBR}{m} \tag{7.42}$$

giving a final energy of

$$T = \frac{1}{2}mv_{\text{max}}^2 = \frac{(qBR)^2}{2m} \tag{7.43}$$

Although conceptually, the conventional cyclotron is simple, it has several drawbacks. The resonance condition for the cyclotron is in conflict with the vertical focusing requirement (B should decrease radially), need to avoid loss of ions that have a velocity component in the vertical direction. Another severe difficulty comes from the relativistic behaviour of the accelerated particles. As the energy of the particle increases so does its mass, which it f does not remain constant at larger radii. As a result, the resonance condition requiring the orbiting frequency to be equal to the frequency of the electric field is broken and because of the phase shift, the particle will reach the accelerating gap progressively later to finally fall out of resonance with the RF field and be no longer accelerated. This provides the upper limit in the energy attainable in a conventional cyclotron, which is about ~2% of the rest mass energy of the particle.

EXERCISE 7.29: In a cyclotron driven at a frequency of 10 MHz, α-particles are accelerated up to a maximum radius of curvature of 50 cm. The effective voltage applied to the Dees is 50 kV. Neglecting the gap between the Dees, determine the total acceleration time of the particles and the total distance covered by the particles during the complete cycle of acceleration.

Solution: The maximum kinetic energy acquired by the particles can also be calculated as follows:

$$T = NqV \tag{7.44}$$

where q is the charge carried by the ion, V is voltage drop as the particle moves across the gap and N is the total number of times that the particle passes the accelerating gap before being extracted.
From equations (7.40), (7.41), (7.43) and (7.44), time needed to cover N circular orbits

$$t = N\left(\frac{2m\pi}{qB}\right) = \left(\frac{T}{qV}\right)\left(\frac{2m\pi}{qB}\right)$$

$$= \left\{\frac{\frac{(qBR)^2}{2m}}{qV}\right\}\left(\frac{2m\pi}{qB}\right) = \frac{\pi^2 mR^2 f}{qV}$$

$$= \frac{\pi^2(4\times1.66\times10^{-27}\text{ kg})(0.5\text{ m})^2(10\times10^6\text{ Hz})}{(2\times1.6\times10^{-19}C)(50\times10^3\text{ V})} = 10\ \mu\text{s}$$

The total distance covered by the particles inside the cyclotron

$$L = \pi\sum_{1}^{N} r_n$$

where r_n is the radius of the semicircular orbit after the nth passing of the accelerating gap.

From equations (7.41) and (7.44), we get

$$r_n = \frac{v_n}{2\pi f} = \frac{\sqrt{\frac{2T}{m}}}{2\pi f} = \frac{\sqrt{\frac{2nqV}{m}}}{2\pi f}$$

Therefore,

$$L = \frac{\sqrt{\frac{2qV}{m}}}{2f} \sum_{1}^{N} \sqrt{n}$$

Since, N is large, we can write the above equation as follows:

$$L \approx \frac{\sqrt{\frac{2qV}{m}}}{2f} \int \sqrt{n}\, dn$$

$$= \frac{\sqrt{\frac{2qV}{m}}}{2f} \left(\frac{N^{3/2}}{3/2}\right) = \frac{4\pi^3 m f^2 R^3}{3qV}$$

$$= \frac{4\pi^3 (4 \times 1.66 \times 10^{-27}\ \text{kg})(10 \times 10^6\ \text{Hz})^2 (0.5\ \text{m})^3}{3(2 \times 1.6 \times 10^{-19}\ \text{C})(50 \times 10^3\ \text{V})} = 200\ \text{m}$$

It is clear that in order to maintain the resonance condition, either the frequency of the RF field or the magnetic field or both must be modified to compensate for the relativistic effect. The first solution results in frequency-modulated cyclotron, called a *synchrocyclotron*. These machines are not built nowadays due to their complexity, cost and limitation on the beam intensity.

Cyclotron based on the alternative solution of increasing the magnetic field with increasing orbital radius to compensate for the increase in the relativistic mass of the accelerating particles, is known as *sector-focusing* or *azimuthally varying field (AVF)* cyclotron. In order to avoid the undesirable effect of defocusing the beam as the radius of the orbit beam increases, an extra source of axial force is introduced through azimuthal variation in the magnetic field. The magnetic field is modified by making sectors of higher field strength followed by sectors of lower field strength. This is achieved by planting extra soft iron pieces on both the pole pieces of the magnet as shown in Figure 7.39. When the charged particle encounters a series of alternating regions in which the magnetic field repeatedly increases and then decreases, it moves in a distorted orbit that is not circular.

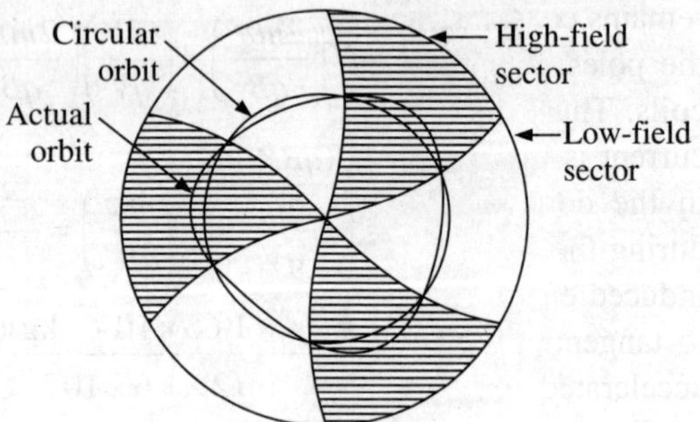

FIGURE 7.39 High-field and low-field regions in an AVF cyclotron. Particles move in a distorted orbit, performing radial oscillations about the circular orbit.

EXERCISE 7.30: The conventional cyclotron is limited in energy because of the relativistic increase in the mass of the accelerated particle. What percentage increment to the magnetic flux density at extreme radius is necessary to preserve the resonance condition for protons of energy 20 MeV?

Solution: The cyclotron resonance frequency

$$f = \frac{qB}{2\pi m}$$

Now, because of relativistic effects, the effective mass of the accelerated particle is $m_0\gamma$. Therefore, to preserve the resonance condition, the magnetic field should be $B\gamma$.

For 20 MeV protons,

$$\gamma = \left(\frac{T}{m_0c^2}\right) + 1 = \left(\frac{20 \text{ MeV}}{938.28 \text{ MeV}}\right) + 1 = 1.0213$$

Therefore, the percentage increase in the magnetic flux density

$$= \frac{B(\gamma - 1)}{B} \times 100 = (1.0213 - 1) \times 100 = 2.13\%$$

7.7.5 Betatron

A cyclotron cannot be used for accelerating electrons or the β-particles, since m_e is very small and the change of mass occurs at low energies. For instance, a 1 MeV electron has a velocity of $0.9c$ and its moving mass is 2.3 times that of its rest mass; in comparison with a proton of the same energy, the mass factor is only 1.001.

D.W. Kerst in 1940 built a new accelerator called the betatron, which could accelerate electrons up to 300 MeV. Instead of using the electrostatic method of accelerating, betatron employs time varying magnetic field, which gives rise to induced electric field on the electrons moving in a fixed circular orbit, thus accelerating the electrons. Since, the radius of the orbit remains constant, the acceleration chamber can be made in the shape of a donut, placed between the poles of a specially shaped magnet, which are energized by an alternating current in the coils. Thus, the betatron can be considered as the analog of a transformer, where the primary current is the alternating current which excites the magnet, and the electron current circulating in the donut shaped vacuum chamber is the secondary current. The electrons are injected during the first quarter of a cycle when the magnetic field is nearly zero. When the alternating induced electric field reaches its maximum, the high energy electrons leave their circular paths to tangentially eject and hit the target (which as a result emits X-rays). So, the electrons are accelerated only for about a quarter cycle of the period of the alternating current and the accelerator has a pulsed operation with the same frequency as the current.

Maxwell's equations govern that a charged particle when not moving in a straight line or accelerating, radiates energy in the form of electromagnetic waves. This also occurs for particles in accelerators, especially for electrons in circular accelerators. The betatrons are limited by large radiative losses suffered by the electrons moving at nearly the speed of light in a relatively small orbit.

EXERCISE 7.31: In the betatron, the magnetic field has a twin purpose—inducing an electric field that accelerates the electrons and exerting Lorentz force on the electrons to keep them in a circular orbit. If B is the time varying magnetic field at the circular orbit traced by the electron and $\bar{B}$ is the average field within the area enclosed by the circular orbit, derive the relation, known as the *betatron relation,* so that the magnetic field can both guide and accelerate the electrons in the orbit of fixed radius R.

Solution: From Faraday's law, the induced emf in the electron orbit is given by rate of change of the magnetic flux. The induced emf is also equal to the work done by an electron in going around the circular orbit. Therefore,

$$2\pi RE = -\frac{d\varphi}{dt} = -\pi R^2 \frac{d\bar{B}}{dt}$$

Thus, the force acting on the electron

$$F = -eE = \frac{eR}{2}\frac{d\bar{B}}{dt} \tag{7.45}$$

The Lorentz force, evB provides the necessary centripetal force pv/R required to maintain the circular path. Therefore,

$$evB = \frac{pv}{R} \quad \text{or} \quad p = eRB$$

The force acting on the electron is equal to the rate of change of its momentum, i.e.,

$$F = \frac{dp}{dt} = \frac{d}{dt}(eRB) = eR\frac{dB}{dt} \tag{7.46}$$

From equations (7.45) and (7.46), we have

$$\frac{dB}{dt} = \frac{1}{2}\frac{d\bar{B}}{dt}$$

On integration, we get the betatron condition,

$$B(t) = \frac{1}{2}d\bar{B} + \text{Constant}$$

7.7.6 Synchrotron

In order to extend the frequency-modulated cyclotron or synchrocyclotron to higher energies, magnet with a bigger radius will be required. The magnet is the principal contributing factor to the cost of a cyclotron, and the cost of the magnet becomes exorbitant for high energy range. The solution to this problem is the synchrotron accelerator, in which both the magnetic field and the resonant frequency vary with time. The essential feature that keeps the cost low in a synchrotron is that the orbit in which the particles move is kept constant by synchronous adjustment of the resonant frequency and the magnetic field strength. The guiding magnetic field, therefore, needs to be applied only at the circumference, not throughout the entire circular region, as in the case of a cyclotron. The cost of such a ring-shaped magnet is roughly proportional to the orbit radius, whereas for a magnet with circular poles, it is roughly proportional to the cube of the radius. The charged particles travel in evacuated pipes placed along the orbit as in the betatron, but unlike a betatron, in synchrotron, the acceleration is produced by an alternating electric field. Physically, the donut-shaped vacuum tube can be very large, up to hundreds of metres. Also, there need not be a single magnet to keep the particles in orbit. Most synchrotrons are built using sector bending magnets separated by field free straight sections. Dipole magnets help to keep the particles in their (almost) circular orbits; the quadrupole magnets focus on the ion beam. The RF oscillators to accelerate the particles are placed at appropriate locations along the donut in the straight sections between the dipole magnets. The particles get an energy increment each time they cross the accelerating gap. The particles are injected into the synchrotron only after an initial acceleration in an appropriate machine, usually a linear accelerator. After making several hundreds and thousands of turns, the full energy particles are extracted out from the synchrotron. Figure 7.40 shows the schematic diagram of a typical modern synchrotron.

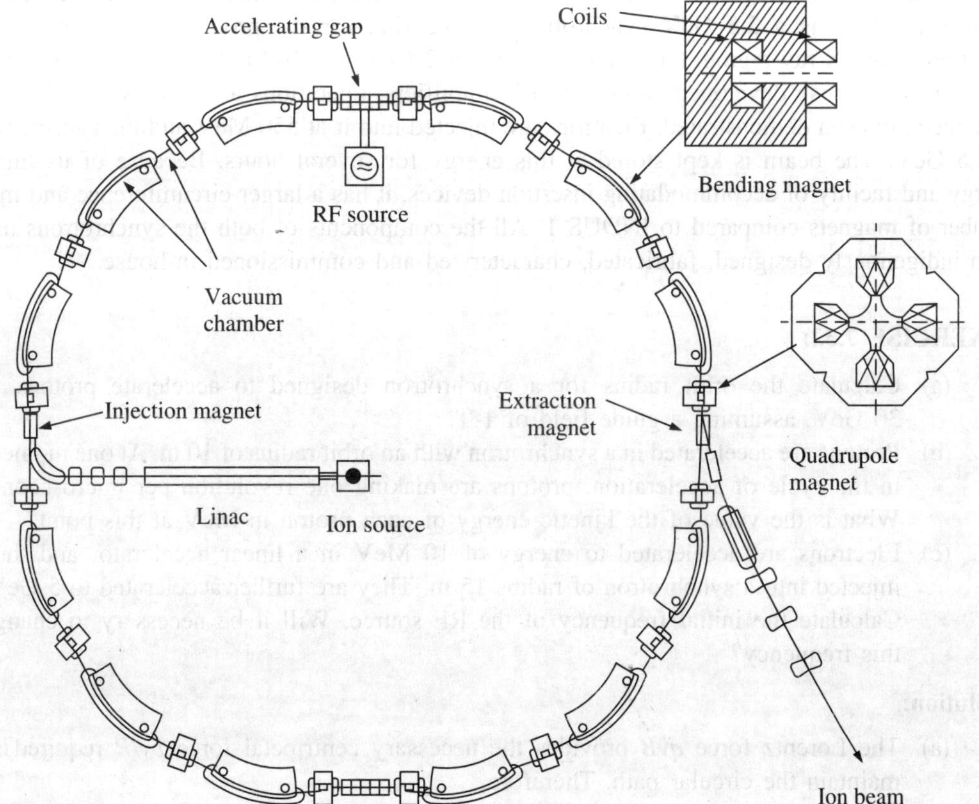

FIGURE 7.40 Schematic diagram showing the principal components of a typical synchrotron.

In an electron synchrotron, the speed of the particles is quite close to the speed of light, and the time taken by the electrons to complete one orbit is essentially constant. Therefore, the accelerating gap can have single frequency. A proton synchrotron differs from an electron synchrotron since the speed of the proton is not fixed, because of which the time taken to complete one orbit is not fixed. Therefore, for synchronization, the RF source has to be modulated.

Synchrotron radiation is an electromagnetic radiation, which is emitted by the charged particles circulating in a synchrotron when they are deflected by the magnetic field in the dipole magnets. Since, protons have lower speeds in comparison to the electrons, there is much less energy loss by radiation. When the loss of energy by synchrotron radiation equals the energy acquired from the accelerating field in the same interval of time, further acceleration becomes impossible. Storage rings, which are ring-shaped vacuum chambers, are designed to keep the particles circulating at a constant energy for several hours. These are used for colliding beam experiments and production of synchrotron radiation.

India's first synchrotron source INDUS-1, a 450 MeV electron storage ring for the production of the radiation in the UV region, became operational at the Raja Ramanna Center for Advanced Technology (RR-CAT), Indore in 1999. Another synchrotron radiation source, INDUS-2, which is a 2.5 GeV synchrotron storage ring for the production of synchrotron radiation in the X-ray region, has been operating at 2 GeV and 100 mA, since 2010. With the addition of in-house developed solid state RF amplifiers, operations at 2.5 GeV and 100 mA have recently been demonstrated. Electrons are injected into it at 550 MeV and then accelerated to 2.5 GeV. The beam is kept stored at this energy for several hours. Because of its higher energy and facility of accommodating insertion devices, it has a larger circumference and more number of magnets compared to INDUS-1. All the components of both the synchrotrons have been indigenously designed, fabricated, characterized and commissioned in-house.

EXERCISE 7.32:

(a) Calculate the orbit radius for a synchrotron designed to accelerate protons to 30 GeV, assuming a guide field of 1 T.

(b) Protons are accelerated in a synchrotron with an orbit radius of 10 m. At one moment in the cycle of acceleration, protons are making one revolution per microsecond. What is the value of the kinetic energy of each proton in MeV at this point?

(c) Electrons are accelerated to energy of 10 MeV in a linear accelerator and then injected into a synchrotron of radius 15 m. They are further accelerated to 5 GeV. Calculate the initial frequency of the RF source. Will it be necessary to change this frequency?

Solution:

(a) The Lorentz force qvB provides the necessary centripetal force pv/R required to maintain the circular path. Therefore,

$$R = \frac{p}{qB} \tag{7.47}$$

The total energy E is the sum of kinetic energy T and the rest mass energy m_0c^2. Therefore, the relativistic invariant condition ($E^2 = p^2c^2 + m_0^2c^4$) can be written as follows:

$$T^2 + 2Tm_0c^2 = p^2c^2 \tag{7.48}$$

or
$$p = \frac{\sqrt{T^2 + 2Tm_0c^2}}{c}$$

$$= \frac{\sqrt{(30 \text{ GeV})^2 + 2(30 \text{ GeV})(0.938 \text{ GeV})}}{c} = 30.924 \text{ GeV}/c$$

Therefore, the radius of the orbit

$$R = \frac{\left\{\dfrac{(30.924 \times 10^9 \text{ eV})(1.6 \times 10^{-19} \text{ J/eV})}{3 \times 10^8 \text{ m/s}}\right\}}{(1.6 \times 10^{-19} \text{C})(1 \text{ T})} = 103.08 \text{ m}$$

(b) The orbital frequency or the resonant frequency

$$f = \frac{1}{(10^{-6})} = 10^6 \text{ cycle/s}$$

Taking relativistic effects into account, the resonant frequency

$$f = \frac{qB}{2\pi(m_0\gamma)} \tag{7.49}$$

From equations (7.47) and (7.59), we get

$$f = \frac{p}{2\pi R(m_0\gamma)} = \frac{(m_0\gamma)v}{2\pi R(m_0\gamma)}$$

or
$$v = 2\pi Rf$$

or
$$\gamma = \frac{1}{\sqrt{1 - \dfrac{v^2}{c^2}}} = \frac{1}{\sqrt{1 - \dfrac{\{2\pi(10 \text{ m})(10^6 \text{ Hz})\}^2}{(3 \times 10^8 \text{ m/s})^2}}} = 1.0226$$

We know that the relativistic kinetic energy

$$T = m_0c^2(\gamma - 1) = (938 \text{ MeV})(0.0226) = 21.2 \text{ MeV}$$

(c) We know that

$$\gamma = \left(\frac{T}{m_0c^2}\right) + 1$$

At injection,

$$\gamma = \left(\frac{10\text{ MeV}}{0.511\text{ MeV}}\right) + 1 = 20.57$$

or
$$v = \left(\sqrt{1 - \frac{1}{\gamma^2}}\right)c = \left(\sqrt{1 - \frac{1}{20.57^2}}\right)c = 0.9988c$$

and
$$f = \frac{v}{2\pi R} = \frac{0.9988(3\times 10^8\text{ m/s})}{2\pi(15\text{ m})} = 3.18\times 10^6\text{ cycle/s}$$

At extraction,

$$v = \left[\sqrt{1 - \frac{1}{\left\{\left(\frac{5000\text{ MeV}}{0.511\text{ MeV}}\right)+1\right\}^2}}\right]c \approx c$$

Since, the orbiting velocity of the electron is nearly constant, there is hardly any need to change the frequency of the RF source.

PROBLEMS

7.1 In Figure 7.41, three examples of response functions for detectors are shown. Which one of them has the best resolution as compared to others?

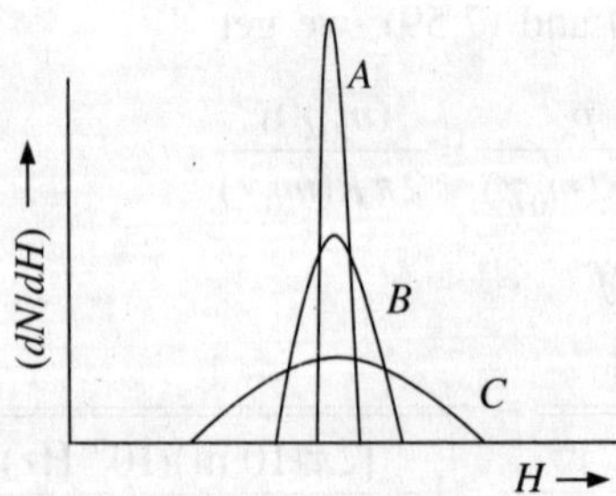

FIGURE 7.41 Examples of detector response function.

7.2 If the peak from the response function generated by a detector is Gaussian-shaped, what should be the least number of charge carriers generated per event, so that the energy resolution is better than 1%? [**Ans:** > 55000 using equation (7.4)]

7.3 Counters A and B are non-paralysable with dead time of 30 and 100 μs, respectively. At what true event rate will the dead time losses in counter B be twice as great as those for counter A? [**Ans:** n = 1.33 × 10^4 s^{-1}]

7.4 A source is counted for 1 minute and gives 561 counts. The source is removed and a 1 minute background count gives 410 counts. Estimate the net count due to the source alone and its associated standard deviation? [**Ans:** 151 ± 31.2]

7.5 A 10 minute count of a source plus the background gives a total of 846 counts. A total of 73 counts are given by the background alone for 10 minute. What is the net counting rate due to the source alone, and what is its associated standard deviation? **[Ans:** 77.3 ± 3.0 m^{-1}]

7.6 A flow counter shows an average background rate of 2.87 counts/minute. Determine the probability that a two minute count would contain:

(a) Exactly five counts

(b) At least one count

What length of counting time is required to ensure that at least one count is recorded, with 99% probability?

[Ans: (a) $P(5) = 0.167$ (b) $1 - P(0) = 0.9968$; and $t = 1.6$ m]

7.7 Find the saturation current produced by 1000 Bq of radioactive sample that emits 4 MeV α-particle and enters a neon-filled chamber. **[Ans:** 1.74×10^{-11} A]

7.8 The average β-particle energy emitted by radiocarbon is 49 keV. Calculate the saturated ion current if 150 kBq of the isotope in the form of carbon-dioxide gas is introduced into a large volume ion chamber filled with pressurized argon.

[Ans: 4.46×10^{-11} A]

7.9 Fission fragments of energy 200 MeV enter an air-filled ion chamber. If the capacitance of the detection system is 25 μF, calculate the pulse height.

[Ans: 3.65×10^{-2} V]

7.10 A gas counter operated at 1 kV has the anode wire of radius 1 mm, the inner radius of the chamber is 10 mm. What is the maximum strength of the electric field at the wire? **[Ans:** 4.3×10^5 V/m]

7.11 In a given gas-filled counter operated at half the atmospheric pressure, the mobility of a free electron is 1.5×10^{-4} m^2atm/sV. The threshold electric field for the onset of avalanche formation is 2×10^6 V/m. If this gas is used in a cylindrical tube with anode radius of 0.005 cm and cathode radius of 2 cm, calculate the drift time of an electron from the cathode to the multiplying region for an applied voltage of 1500 V. **[Ans:** 1.04 ms]

7.12 Calculate the long-wavelength limit of the sensitivity of a photo-layer with a work function of 1.5 eV. **[Ans:** 827 nm]

7.13 The energy resolution of a particular NaI (Tl) scintillation detector is 7% for ^{137}Cs γ-rays (0.662 MeV). Estimate its energy resolution for the 1.28 MeV γ-rays from ^{22}Na. **[Ans:** 5.03%]

7.14 Pure Si at 300 K has equal electron and hole concentration of 1.5×10^{16} m^{-3}. Doping by indium increases the hole concentration to 4.5×10^{22} m^{-3}. Calculate the electron concentration in the doped semiconductor. **[Ans:** $n_e = 5 \times 10^9$ m^{-3}]

7.15 Assuming that the charge collection is complete and that the electronic noise is negligible, find the expected energy resolution of a Ge(Li) detector for the 0.662 MeV γ-rays from ^{137}Cs. **[Ans:** 0.14%]

7.16 An incident fast neutron is moderated and then diffuses a total path length of 10 cm before it is captured in the BF_2 proportional counter. Determine the time lag between the time of neutron incidence and the leading edge of the output pulse. Neglect the moderation time. Assume that the neutron travels 10 cm as a thermal neutron (0.025 eV). **[Ans:** 45 μs]

7.17 Show that the angle between a recoil proton and the corresponding scattered neutron is always 90°.

7.18 Show that except for small ranges, the straggling of a beam of ^{3}He particles is greater than that of a beam of ^{4}He particles of equal range.

7.19 Calculate the detection efficiency of a methane-filled proportional counter for incident 100 keV neutrons if the gas pressure is 1 atmosphere and the neutron path length through the gas is 5 cm. The cross-sections for hydrogen and carbon can be taken as 15 b and 4.5 b, respectively. **[Ans:** 24.4%]

7.20 Consider Cerenkov radiation emitted at an angle θ relative to the direction of a charged particle in a medium of refractive index n. Show that its rest mass energy mc^2 is related to its momentum by $mc^2 = pc(n^2\cos^2\theta - 1)^{1/2}$.

7.21 High energy neutrino beams at the Fermi Lab are made by first forming a mono-energetic π^+ beam and then allowing the pions to decay as follows:

$$\pi^+ \rightarrow \mu^+ + \nu$$

Find the energy of the decay neutrino in the rest frame of the π^+. Recall that the rest mass of a pion is 140 MeV/c^2 and the rest mass of a muon is 140 MeV/c^2. **[Ans:** 30 MeV]

7.22 In a Cockcroft–Walton accelerator, α-particles are injected in the HV terminal kept at 600 kV. Calculate the kinetic energy of the beam when it hits the target, which is kept at ground potential? **[Ans:** 1.2 MeV]

7.23 Protons of 2 MeV energy, enter a linear accelerator that has 97 drift tubes connected alternately to a 200 MHz oscillator. The final energy of the protons is 50 MeV. What are the lengths of the second and the last drift tubes? **[Ans:** 0.055 m and 0.235 m]

7.24 What is the length of the longest drift tube in a linear accelerator, which is operating at a frequency of 20 MHz and is capable of accelerating ^{12}C ions to a maximum energy of 100 MeV? **[Ans:** 1 m]

7.25 Calculate the magnetic field and the Dee radius of a cyclotron which could accelerate protons to a maximum energy of 5 MeV if the available radio frequency is of 8 MHz. **[Ans:** $B = 0.52T$, $R = 0.616$ m]

7.26 Alpha particles are accelerated in a cyclotron operating with a magnetic field of magnitude 0.8 T. If the extracted beam has an energy of 12 MeV, calculate the extraction radius and the orbital frequency of the beam. **[Ans:** $R = 0.623$ m, $f = 6.15$ MHz]

7.27 What is the energy acquired by an electron in one rotation, if the magnetic field inside the betatron is changing at 50 Wm^{-2}s^{-1} and the radius of the electronic orbit is 1 m? **[Ans:** 50 eV]

7.28 A proton synchrotron has a magnetic field of 1.1 T. When the accelerated protons move in a circular path at a speed of $0.800c$, what is the radius of the path they follow? **[Ans:** 3.79 m]

7.29 A modern accelerator produces two counter-rotating 30 GeV proton beams which collide head-on. What is the total energy of collision in the CMS? Determine the energy that is required for a conventional proton accelerator, in which the proton beam strikes a stationary hydrogen target to give the same center-of-mass energy? **[Ans:** 60 GeV, 1918 GeV]

7.30 For non-relativistic particle, the total radiant flux emitted at any instant is given by

$$P = \frac{q^2 a^2}{6\pi \varepsilon_0 c^3}$$

where q is the charge on the particle and a is its instantaneous acceleration. In SI units, P is in watt. A proton of kinetic energy of 50 MeV is travelling in a circular orbit of radius 1 m, through the uniform magnetic field of a cyclotron. What is the energy that is lost to radiation in one orbiting? Can the effect be ignored in designing the cyclotron?

BIBLIOGRAPHY

Accelerators for Frontiers in Science and Technology, IANCAS Bulletin, Vol. 5, No. 1, Mumbai, 2006.

Kapoor, S.S. and Ramamurthy, V.S., *Nuclear Radiation Detectors*, Wiley Eastern, New Delhi, 1986.

Knoll, G.F., *Radiation Detection and Measurement*, Wiley, New York, 2010.

Leo, W.R., *Techniques for Nuclear and Particle Physics Experiments*, Narosa, New Delhi, 1995.

Littlefield, T.A. and Thorley, N., *Atomic and Nuclear Physics,* Van Nostrand Reinhold Company, London, 1970.

Parker, S.P., *Encyclopedia of Physics*, McGraw-Hill, New York, 1993.

Persico, E., Ferrari, E. and Segre, S.E., *Principles of Particle Accelerators,* W.A. Benjamin, Inc., New York, 1968.

Radiation Detection and Measurement, IANCAS Bulletin, Vol. 3, No. 3, Mumbai, 2004.

Theodorsson, P., *Measurement of Weak Radioactivity*, World Scientific, Singapore, 1996.

8 Particle Physics

"We have no right to assume that any physical laws exist, or if they have existed up to now, that they will continue to exist in a similar manner in the future."

—Max Planck

The past three decades have seen a remarkable impact in the understanding of elementary particles. With the advent of high energy particle accelerators and detectors, scientists are engaged in the pursuit of the most elementary constituents of the matter. A systematic treatment of particle interactions is beyond the level of this text because particle physics is a separate discipline at the graduate level.

8.1 HISTORICAL DEVELOPMENT

The desire to identify the smallest constituents of matter is indeed very old. Ancient Indian schools of philosophy proposed that space, water, fire, air and earth are the elementary constituents of nature. An almost similar view of the basic constituents of nature is also present in ancient Greek civilization who regarded water, fire, air and earth as four unchangeable elements of nature. The philosophers at that time also coined the word 'atom.' The modern view of atom may be traced back to the discovery of the atomic model by Dalton in 1803 and later the Mendeleev's periodic table of chemical elements. The modern genesis of the quest for the origin of the ultimate matter and kind of space-time we live in is over a hundred years old. With the discovery of the electron in 1897 and of the proton in 1900, the story of elementary particles beyond the physics of atoms started. The probing beam of α-particles from radioactive sources were limited in energy and to get a closer look at the nucleus, physicists needed a better source of higher energy particle beams. It was in 1930 when E. O. Lawrence built the first successful particle accelerator, that it was possible to make a detailed exploration of the inner structure of the atom. Since then, one accelerator followed another and with increasing sophistication and energy, it became the most powerful tool in the hands of particle physicists to explore and unravel the inner structure of the atom and to discover one elementary particle after another.

The problem of continuous β-spectrum from radioactive nuclei and the discovery of neutronic constituents of the atomic nucleus by Chadwick led Yukawa to propose the strong nuclear force based on quanta of finite mass (mesons). Further, with the Fermi theory of β-decay and the theoretical prediction of neutrino by Pauli, the weak nuclear forces were identified by the slow process of β-decay. The weak and strong nuclear forces along with the familiar long range gravitational and electromagnetic interactions thus complete the list of four fundamental interactions found in nature. The last century saw the birth of two major conceptual frameworks: quantized mechanics and relativity theory. The concepts were merged together by P.A.M. Dirac, resulting in the formulation of relativistic quantum mechanics. His famous equation, known as Dirac equation, predicted the existence of antimatter. Subsequent developments further provided firm basis to our understanding of the origin of matter in the atomic and subatomic regime. The current view of the basic constituents and their interactions will be presented in the following sections.

8.2 INTERACTIONS

There are four fundamental interactions which account for all the known forces in the universe (Figure 8.1):

(i) Gravitational interaction
(ii) Electromagnetic interaction
(iii) Strong interaction
(iv) Weak interaction

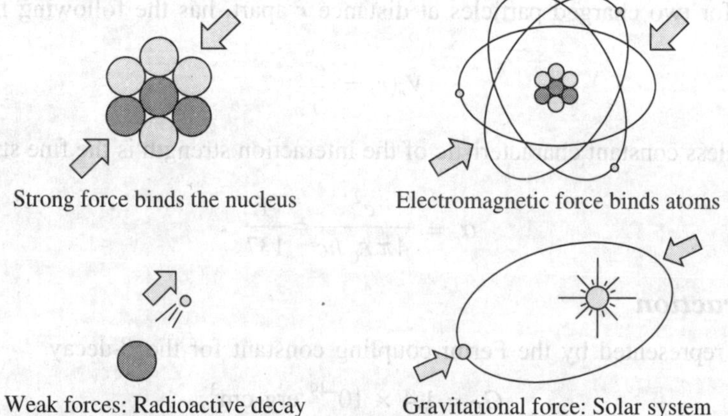

FIGURE 8.1 Four fundamental interactions of nature.

The gravitation interaction, which is the weakest of all, is the force that holds the earth together, binds the sun and the planets into the solar system, and binds stars into galaxies. Between individual particles, however, it is extremely feeble that it can usually be ignored. The electromagnetic force acts between particles which carry an electronic charge. It holds the cloud of negatively charged electrons around the positively charged protons in the nucleus, binds atoms together to form molecules and crystals. It is the significant interaction for all of chemistry and biology. The strong interaction glues the nucleons together; it binds neutrons and protons to form the nuclei of all the elements. The strongest force known in nature, it is also of quite short range. It is the dominating interaction in high-energy physics. The weak interaction causes the light elementary particles (the leptons: e, ν, μ and τ) to interact with each other and with heavier particles. It is responsible for the radioactivity of a nucleus which can emit an electron when a neutron breaks up.

The interactions can be classified according to the value of a characteristic dimensionless constant related to the interaction strength.

Strong Interaction

Range of interaction is ~10^{-15} m. The interaction potential between two nucleons r distance apart has the form

$$V_s(r) = \frac{g_h}{r} \exp\left(-\frac{r}{R}\right) \tag{8.1}$$

where $R = \dfrac{\hbar}{m_\pi c}$ is the Compton wavelength of pion. The exponential function indicates a short interaction length. The dimensionless constant α_s gives the interaction strength.

$$\alpha_s = \frac{g_h^2}{\hbar c} \approx 1$$

Electromagnetic Interaction

The potential for two charged particles at distance r apart, has the following form:

$$V_e(r) = \frac{e^2}{r} \tag{8.2}$$

The dimensionless constant characteristic of the interaction strength is the fine structure constant

$$\alpha_e = \frac{e^2}{4\pi\,\varepsilon_0\,\hbar c} = \frac{1}{137}$$

Weak Interaction

Its strength is represented by the Fermi coupling constant for the β-decay

$$G_F = 1.4 \times 10^{-49} \text{ erg cm}^3$$

The potential of weak interaction has the form

$$V_w(r) = \frac{g_W}{r} \exp\left(-\frac{r}{R_w}\right) \tag{8.3}$$

where $R_w = \dfrac{\hbar}{2m_w c}$ (as we did in the Yukawa theory). The dimensionless constant characteristic of the interaction strength is

$$\alpha_w = \frac{g_h^2}{\hbar c} = G_F \frac{m_p^2 c}{\hbar^3} \approx 10^{-5}$$

EXERCISE 8.1: Use the mass of the W^- particle to determine the range of the weak interaction responsible for the neutron β-decay. Show that the lifetime of the neutron is larger than the decay time through weak interaction. [Given: $m_w = 80.4$ GeV/c^2]

Solution:

$$R_w = \frac{\hbar c}{2m_w c^2} = \frac{197.3 \text{ MeV fm}}{2\left(80.4 \dfrac{\text{GeV}}{c^2}\right)(c^2)} = 1.2 \times 10^{-18} \text{ m}$$

We have assumed that the W^- exchange particle travels with c in the above expression for range. This represents an upper limit because W^- particle is too massive to travel with c.

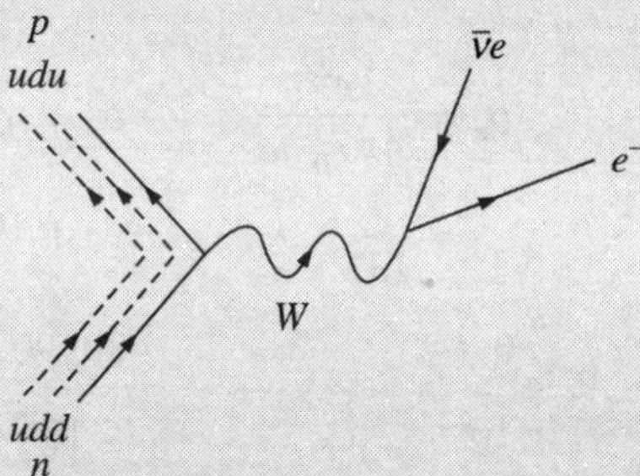

FIGURE 8.2 Feynman diagram for beta-minus decay of a neutron into a proton, electron and antineutrino associated with electron, via an intermediate heavy W boson.

W^- boson is created virtually (Figure 8.2) and the energy required is equal to the rest mass energy ($m_w c^2$) is available for very short duration according to the Heisenberg uncertainty principle

$$\Delta E \cdot \Delta t \sim \frac{\hbar}{2}$$

In this case, the violation of the conservation of energy is massive, so the violation must occur over a very short period of time.

$$\Delta t = \frac{\hbar}{2m_w c^2} = \frac{1.2 \times 10^{-18} \text{ m}}{3 \times 10^8 \text{ m/s}} = 4 \times 10^{-27} \text{ s}$$

The lifetime of the neutron is longer than the time it takes for the decay process itself.

Gravitational Interaction

The interaction potential between two protons has the following form:

$$V_G(r) = G\frac{m_p^2}{r} \tag{8.4}$$

The dimensionless constant characteristic of interaction strength is

$$G\frac{m_p^2}{\hbar c} \approx 6 \times 10^{-39}$$

Table 8.1 summarizes the characteristics of the interactions.

Table 8.1 Characteristics of the Fundamental Interactions

	Strong	*Electromagnetic*	*Weak*	*Gravitational*
#Carrier of the force [Gauge boson's spin-parity J^π]	Gluon 1^-	Photon 1^-	$W^\pm$, Z^0 $1^-, 1^+$	Graviton $?2^+$
†Coupling constant	$\alpha_s \le 1$	$\alpha_e = \frac{e^2}{4\pi\varepsilon_0 \hbar c} = \frac{1}{137}$	$\alpha_w = G_F\frac{m_p^2 c}{\hbar^3} \approx 10^{-5}$	$G\frac{m_p^2}{\hbar c} \approx 6 \times 10^{-39}$
Mass	0	0	80, 90 GeV	0
Relative strength	1	10^{-2}	$\le 10^{-5}$	10^{-38}
Cross-section	10 mb	10^{-3} mb	10^{-12} mb	
Time scale	10^{-23} s	$10^{-18} - 10^{-20}$ s	$\ge 10^{-13}$ s	
Range	$\le 10^{-15}$ m	∞	10^{-18} m	∞
Source	Colour charge	Electric charge	Weak charge	Mass

\# Gluons, photons, $W^\pm$ and Z^0 are called gauge bosons and are responsible for the strong and electroweak interactions. Gravitons are also bosons. Fermions exert attractive or repulsive forces on each other by exchanging gauge bosons, which are the force carriers.

† The square of the charge is known as the coupling constant in strong, electromagnetic and weak interactions. Strictly speaking, the coupling constants are not constant but vary gradually with the particle energy.

8.3 CONSTITUENT PARTICLES OF MATTER

Not long ago, protons, neutrons and electrons were regarded as the fundamental particles of nature when we learned in the 1900s through the experiments of Rutherford and others that atoms consist mostly of empty space with clouds of electrons surrounding a dense central nucleus constituting of the protons and the neutrons. The science of particle physics surged ahead with the arrival of particle accelerators that could accelerate protons and electrons to high energies, which are then made to collide with nuclei to produce many new particles. Based on these experiments and the theoretical studies, it became apparent that leptons and quarks are the basic building blocks of matter, i.e., they are now understood to be the 'elementary particles.' Quarks and leptons, both have half-integral spins and are therefore fermions [see Section 1.8.3 in Chapter 1].

The developments of the past few years have revealed that quarks and leptons can themselves be grouped in families of four (quartet of quarks and leptons). Astonishingly, all the components of our normal everyday world can be explained in terms of just one such family—the first generation constituents, containing 'up' and 'down' quarks, the electron and the electron-type neutrino. At higher energies (that can be observed through some phenomena in outer space, or can be created artificially using particle accelerators) two further generations of quarks and leptons come into play. These quartets appear to be heavier copies of the normal quartet. Table 8.2 illustrates the basic constituents of matter. An open question is "Are there any further generations also?"

In addition to these component particles, there are others (photons, bosons and gluons) which communicate the forces between the components.

Table 8.2 Basic Constituents of Matter

	Quarks	*Charge*	*Leptons*
In everyday matter	Up	2/3	Electron
	Down	–1/3	Electron-type neutrino
In matter existing in higher energy environments	Strange	–1/3	Muon
	Charm	2/3	Muon-type neutrino
	Bottom	–1/3	Tau
	Top	2/3	Tau-type neutrino

8.3.1 Leptons

The electron was the first of the leptons[1], actually, the first of the elementary particles observed by J.J. Thomson in 1897, for which an acceptable theory was developed by Paul A.M. Dirac in 1928. The intrinsic angular momentum (spin) of the electron was found to be $\hbar/2$ and therefore, constrained by the Pauli exclusion principle, a fact that has key implications for the creation of the periodic table of the elements. The existence of the antiparticles for the electrons was predicted by Dirac's relativistic quantum theory of the electron. Carl D. Anderson, in 1932, during the study of the cosmic rays discovered the predicted positron, the first known antiparticle.

The lepton family has twelve members. The present standard model assumes that there are no more than three generations of having four members in each. The properties of the leptons are listed in Table 8.3. All the leptons are believed to have $J^{\pi} = 1/2^{+}$. The neutrinos are uncharged; their masses are unknown but unlikely to exceed a few eV/c^2.

The muon, the member of the second family of leptons, was discovered in a cloud chamber by Anderson and Neddermayer in 1937. The muon decays through the following reaction:

$$\mu^{\pm} \rightarrow e^{\pm} + \nu_{\mu} + \overline{\nu}_e$$

1. Lepton in Greek means light particles.

where $m_\mu = 207m_e$. Neutrinos associated with muon were seen in 1962. The reaction

$$\nu_\mu + n \rightarrow \mu^- + p$$

is seen, whereas

$$\nu_\mu + n \rightarrow e^- + p$$

is not seen.

Table 8.3 Properties of Leptons

Particle	*Antiparticle*	*Rest mass* (MeV/c^2)	*L* (*e*)	*L* (muon)	*L* (tau)	*Lifetimes* (s)
Electron e^-	e^+	0.511	+1	0	0	Stable
Electron-type neutrino ν_e	$\bar{\nu}_e$	$< 7 \times 10^{-6}$	+1	0	0	Stable*
Muon μ^-	μ^+	105.7	0	+1	0	2.2×10^{-6}
Muon-type neutrino ν_μ	$\bar{\nu}_\mu$	< 0.27	0	+1	0	Stable*
Tau τ^-	τ^+	1777	0	0	+1	2.96×10^{-13}
† Tau-type neutrino ν_τ	$\bar{\nu}_\tau$	< 31	0	0	+1	Stable*

† Tau-type neutrino and antineutrino are predicted theoretically, but yet to be observed directly.
* The neutrinos are shown as stable; neutrino labs do talk of 'neutrino oscillations.'

A yet heavier lepton, tau lepton ($m_\tau = 3490m_e$), was discovered in 1975 in the reaction

$$e^- + e^+ \rightarrow \tau^- + \tau^+$$

followed by the decays

$$\tau^- \rightarrow e^- + \bar{\nu}_e + \nu_\tau$$

and

$$\tau^+ \rightarrow \mu^+ + \nu_\mu + \bar{\nu}_\tau$$

Leptons participate in the weak interaction and if charged, they also interact in the electromagnetic interaction. Besides the leptons of Table 8.3, there also exist six antiparticles. These antiparticles have the opposite electric charge, but they have exactly the same mass and spin as the corresponding particles.

8.3.2 Quarks

In the present standard model, there are six 'flavours' of quarks. Quarks are observed to be only in combinations of two quarks (*mesons*) and three quarks (*baryons*). Mesons and baryons are collectively designated as *hadrons*, or strongly interacting particles, because they interact via the strong force.

The properties of six quarks are listed in Table 8.4. Besides the quarks, there also exist six antiquarks.

Table 8.4 Properties of Six Flavours of Quarks

Quark	*Spin, parity*	*Charge Q/e*	*Baryon number*	*Isospin I*	*S*	*C*	*B*	*T*	*Mass** (GeV/c^2)
Up (u)	$1/2^+$	+2/3	1/3	1/2	0	0	0	0	~0.3
Down (d)	$1/2^+$	−1/3	1/3	1/2	0	0	0	0	~0.3
Charm (c)	$1/2^+$	+2/3	1/3	0	0	+1	0	0	1.6
Strange (s)	$1/2^+$	−1/3	1/3	0	−1	0	0	0	0.5
Top (t)	$1/2^+$	+2/3	1/3	0	0	0	0	+1	174
Bottom (b)	$1/2^+$	−1/3	1/3	0	0	0	+1	0	4.5

Antiquarks $\bar{u}, \bar{d}, \bar{c}, \bar{s}, \bar{t}$ and $\bar{b}$ have opposite signs for charge, baryon number, S, C, B and T.

* The masses should not be taken too seriously, because the non-confinement of quarks implies that we cannot isolate them to measure their masses in a direct manner.

Each of these six flavours of quarks can have three different 'colours'—red, green and blue. This colour designation has absolutely nothing to do with the visual colours that we see. 'Colour' is the 'charge' of the strong nuclear force, analogous to the electric charge for electromagnetism. The force between quarks is attractive only in the 'colourless' combinations of three quarks (baryons) and quark–antiquark pairs (mesons)[2]. The interaction potential, $V(r)$ between two quarks inside a nucleon is given by:

2. Gluons are the particles, which hold the quarks together. Figure 8.3 shows a Feynman diagram depicting the exchange of a gluon ($B\bar{R}$) between the interaction of a red quark coming from left with a blue quark coming in from the right. The colours of the quarks are changed in the interaction.
Based on the three colours red, blue and green, nine colour-anticolour combinations are possible for a gluon. They are $B\bar{B}, B\bar{R}, B\bar{G}, R\bar{B}, R\bar{R}, R\bar{G}, G\bar{B}, G\bar{R}$ and $G\bar{G}$. The three net colourless combinations, $B\bar{B}$, $R\bar{R}$ and $G\bar{G}$, must be dealt with carefully in view of the symmetry properties of the colour fields. This requires that they should be coupled in three ways:

$$\frac{1}{\sqrt{2}}(R\bar{R} - G\bar{G}),\quad \frac{1}{\sqrt{6}}(R\bar{R} + G\bar{G} - 2B\bar{B}),\quad \frac{1}{\sqrt{3}}(R\bar{R} + G\bar{G} + B\bar{B})$$

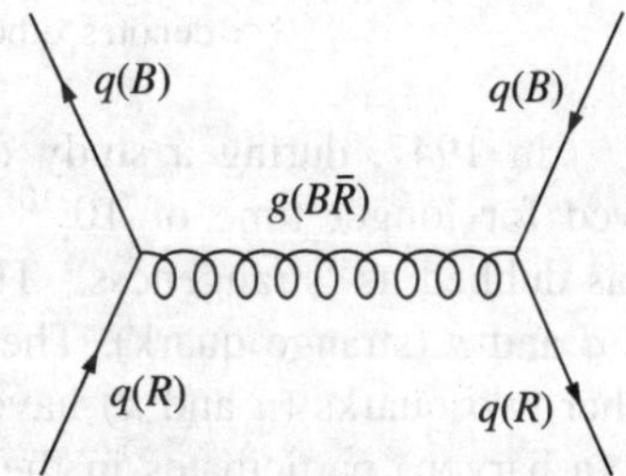

FIGURE 8.3 Feynman diagram depicting the exchange of a gluon ($B\bar{R}$).

The first and the second of these colourless combinations can transmit colour, but the third term cannot. Hence, there are only eight independent gluons, as also predicted by QCD (quantum chromodynamics). Gluons may interact with each other since each gluon carries a colour charge. This situation differs for the photons, which are mediators of the electromagnetic force.

$$V(r) = -\frac{a}{r} + br \tag{8.5}$$

where r is the distance between the quarks and a and b are positive constants of suitable dimensions

The up and down quarks are stable quarks. The other quarks, however, are unstable and decay. Their life-times and major decays are shown in Table 8.5. Quarks get transformed by the exchange of *W* bosons and the rate and nature of the decay of hadrons by the weak interactions are determined by these transformations. Up and down quarks are the most common and least massive quarks, being the constituents of the protons and the neutrons and therefore, of most ordinary matter. In the β^- decay, a neutron transforms to a proton via weak interaction

$$n(udd) \rightarrow p(duu) + e^- + \bar{\nu}_e$$

is considered to be the result of a more fundamental quark process where a down quark becomes an up quark:

$$d \rightarrow u + e^- + \bar{\nu}_e$$

Table 8.5 Life-times and Major Decay Paths of Unstable Quarks

Quark	*Life-times* (s)	*Major decays*
Strange (s)	10^{-8}–10^{-10}	$s \rightarrow u + x$†
Charm (c)	10^{-12}–10^{-13}	$c \rightarrow \begin{cases} s + x \\ d + x \end{cases}$
Bottom (b)	10^{-12}–10^{-13}	$b \rightarrow c + x$
Top (t)	~10^{-25}	$t \rightarrow b + x$

† x denotes other particles allowed by the appropriate conservation laws.

In 1947, during a study of cosmic ray interactions, a new particle Λ (lambda), which lived for longer time of 10^{-10} s than the expected 10^{-23} s, was discovered. This behaviour was dubbed as 'strangeness.' The lambda is a baryon, which was proposed to be made up of u, d and s (strange quark). The strange quark has the strangeness value of (–1), whereas the other two quarks (u and d) have $S = 0$. The shorter life-time of 10^{-23} s was expected because Λ (a baryon) participates in the strong interaction. The longer observed lifetime led to a new conservation law for such decays, i.e., 'conservation of strangeness.' Particles decaying by the strong or the electromagnetic interactions preserve the strangeness quantum number.

The decay process for the Λ particle, however, must violate this rule, since there is no lighter particle containing a strange quark, to which it could decay. The following decay processes show that strangeness is not conserved:

$$\Lambda^0(uds) \rightarrow p(uud) + \pi^-(\bar{u}d) \qquad S = -1 \neq 0 + 0$$

$$\Lambda^0(uds) \rightarrow n(udd) + \pi^0\left(\frac{\bar{u}u + \bar{d}d}{\sqrt{2}}\right) \qquad S = -1 \neq 0 + 0$$

Strange quark gets transformed into another quark in the process. This can only occur by the weak interaction.

All the known hadrons could be specified by some combination of these three quarks and their antiquarks. But there was still a problem related to the discrepancy in the life-times of some of the known particles, and a fourth quark called the charmed quark (c) was proposed in 1970. In 1974, a meson called the J/ψ particle was discovered, which required the existence of the charmed quark and its antiquark. In 1977, an experimental group at Fermi lab found a new resonance at 9.4 GeV/c^2. This new resonance was interpreted as a bottom–antibottom quark pair and called the upsilon meson.

Convincing evidence of top quark (t) was reported by Fermi lab's Tevatron facility in April 1995, in the collision products of 0.9 TeV protons with equal energetic antiprotons in the $p - \bar{p}$ collider.

The interaction that was envisioned is as follows:

$$q\bar{q} \rightarrow \bar{t}t$$

From the proton–antiproton collision, a quark and an antiquark interact to form a top–antitop pair. The top quark decays to form a W boson and a bottom quark.

$$t \rightarrow W + b$$

which in turn decays as follows:

$$b\begin{cases}\bar{\nu} \text{ lepton} + \text{hadrons} \\ \text{hadrons only}\end{cases}$$

$$W^+\begin{cases}\text{lepton } \nu \\ u\bar{d}, c\bar{s}, \ldots\end{cases}$$

Confinement

Till date, no one has ever seen an isolated quark. There are good reasons for the lack of direct observation. It seems that the colour force does not decrease with distance like the other observed forces. There is now a belief among the physicists that perhaps free quarks cannot be observed and that they can only exist within hadrons. Basically, isolated (free) quark cannot be seen because the energy required to separate them produces $q\bar{q}$ pairs long before they are far enough apart to be observed separately.

Figure 8.4 shows an event where a high energy γ-ray is scattered from a neutron. A free quark cannot escape because of the confinement. For high energies, a $q\bar{q}$ pair is created and a pion and a proton are the final particles.

Another way to visualize the quark confinement is called the 'bag model.' Here, one can imagine the quarks as enclosed in an elastic bag, which allows the quarks to move freely around. But, when one tries to pull a quark out, the bag stretches and resists.

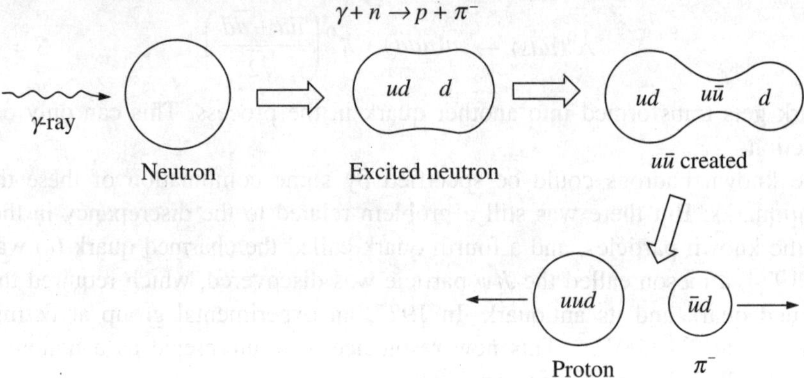

FIGURE 8.4 Event of high energy gamma-ray scattering from a neutron.

8.4 CLASSIFICATION OF PARTICLES

Table 8.6 gives the mass, mean-lifetimes (τ) and common decay modes of the elementary particles excluding resonances and graviton. Their classification into hadrons, photon and leptons is also indicated. The subdivision of hadrons into baryons (nucleons and hyperons) and mesons (pions and kaons) is also shown. Bubble chamber tracks for the positive kaon decay are shown in Figure 8.5.

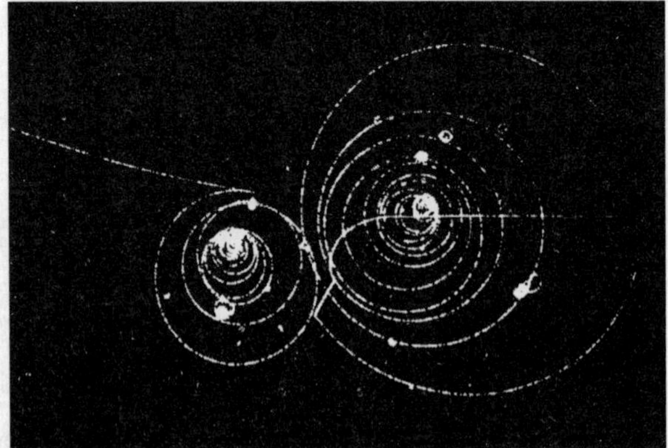

FIGURE 8.5 This image is taken from one of CERN's bubble chambers, and shows the decay of a positive kaon in flight. The decay products of this kaon can be seen spiraling in the magnetic field of the chamber. [Courtesy CERN]

Electron (e^-), muon (μ^-), tau (τ^-), and the three neutrinos (ν_e, ν_μ, ν_τ) constitute the class of leptons. A hadron stands for a strongly interacting particle distinguished from lepton which has only weak or electromagnetic interactions. Photon is the massless carrier of the electromagnetic field. Mesons and photon are bosons (a particle of integral spin). Bosons obey the Bose–Einstein statistics; the wave function of two identical bosons is symmetric under particle exchange. The baryons and leptons are fermions (a particle with half integral spin: $1/2\hbar$, $3/2\hbar$, ...). Fermions obey Fermi–Dirac statistics; the wave function that describes two identical particles is anti-symmetric, i.e., changes sign under particle exchange.

Table 8.6 Mass, Mean-lifetimes and Common Decay Modes of Elementary Particles

			Particle	*Mass* (MeV/c^2)	τ (s)	*Common decay mode*
Hadrons	Mesons	Pions	π^-, π^+	139	2×10^{-8}	$\mu\nu$
			π^0	135	1.8×10^{-16}	$\gamma\gamma$
		Kaons	K^-, K^+	494	1.2×10^{-8}	$\mu\nu$
			K^0	498		$\pi^\pm\pi^0$
			Mixture of K_1, K_2			
			K_1		0.89×10^{-10}	$\pi^+\pi^-$ $\pi^0\pi^0$
			K_2		5.18×10^{-8}	$\pi^0\pi^0\pi^0$ $\pi^+\pi^-\pi^0$
		Eta meson	η	550	10^{-18}	$\gamma + \gamma$ $\pi^+ + \pi^- + \pi^0$
			η'	958	10^{-21}	$\pi^0 + \pi^0 + \gamma$
	Baryons	Nucleons	p	938.2	$>10^{38}$	Stable
			n	939.5	10^3	$pe^- \nu$
		Hyperons †	Λ	1115	2.6×10^{-10}	$p\pi^-$, $n\pi^0$
			Σ^+	1189	0.8×10^{-10}	$p\pi^0$, $n\pi^+$
			Σ^0	1192	10^{-20}	$\Lambda\gamma$
			Σ^-	1197	1.6×10^{-10}	$n\pi^-$
			Ξ^0	1314	3×10^{-10}	$\Lambda\pi^0$
			Ξ^-	1321	1.8×10^{-10}	$\Lambda\pi^-$
			Ω^-	1675	1.3×10^{-10}	$\Xi\pi$ ΛK^-
		Photon	γ	0		Stable
		Leptons	τ^-	1784	3.4×10^{-13}	$e^- \nu$
			μ^-	105	2×10^{-6}	$e\nu\bar{\nu}$
			e^-	0.51		Stable
			ν_e	0		Stable
			ν_μ	< 0.5		Stable
			ν_τ	< 164		Stable

† Λ = lambda, Σ = sigma, Ξ = xi, Ω = omega

Every particle has an associated antiparticle, which has exactly the same mass and lifetime but opposite values of the electric charge, magnetic moment, baryon number, lepton number, and flavour. Thus, positron (e^+) is the antiparticle of electron, antiproton (p^-) is that of proton, antineutrino ($\overline{\nu}_e$) is that of neutrino, etc. Photon is the antiparticle of itself, so also π^0. These antiparticles have not been shown explicitly in Table 8.6.

Properties and quark content of some baryons and mesons are given in Tables 8.7 and 8.8.

Table 8.7 Properties and Quark Content of Some Baryons ($J^\pi = 1/2^+$, $B = 1$)

Particle	*I*	I_Z	$Y = S + B$	*Quark content*
p	1/2	1/2	1	*uud*
n	1/2	–1/2	1	*udd*
Λ	0	0	0	*uds*
Σ^+	1	+1	0	*uus*
Σ^0	1	0	0	*uds*
Σ^-	1	–1	0	*dds*
Ξ^0	1/2	1/2	–1	*uss*
Ξ^-	1/2	–1/2	–1	*dss*

Table 8.8 Properties and Quark Content of Some Mesons

Particle	*I*	I_Z	$Y = S$	*Quark content*
π^-	1	–1	0	$\overline{u}d$
π^+	1	1	0	$u\overline{d}$
π^0	1	0	0	$\dfrac{u\overline{u} - d\overline{d}}{\sqrt{2}}$
K^-	1/2	–1/2	–1	$\overline{u}s$
K^+	1/2	+1/2	1	$u\overline{s}$
K^0	1/2	–1/2	1	$d\overline{s}$
$\overline{K^0}$	1/2	+1/2	–1	$\overline{d}s$
η	0	0	0	$\dfrac{u\overline{u} + d\overline{d} + 2s\overline{s}}{\sqrt{6}}$
η'	0	0	0	$\dfrac{u\overline{u} + d\overline{d} + s\overline{s}}{\sqrt{3}}$

EXERCISE 8.2: On the basis of the additive quark model, the total interaction cross-section is assumed to come from the sum of the cross-sections of various pairs. Assuming that $\sigma(qq) = \sigma(q\bar{q})$, prove that

$$\sigma(\Lambda p) = \sigma(pp) + \sigma(K^- n) - \sigma(\pi^+ p)$$

Solution: In the quark model, (Tables 8.7 and 8.8),

$$\Lambda = uds,\ p = uud,\ n = udd,\ K^- = s\bar{u},\ \pi^+ = u\bar{d}$$

$$\therefore \qquad \sigma(\Lambda p) = \sigma(uds)(uud)$$

$$= \sigma(uu + uu + ud + du + du + dd + su + su + sd) = 9\sigma(qq)$$

Similarly,

$$\sigma(pp) = 9\sigma(qq)$$

$$\sigma(K^- n) = 3\sigma(qq) + 3\sigma(q\bar{q})$$

$$\sigma(\pi^+ p) = 3\sigma(qq) + 3\sigma(q\bar{q})$$

Assuming $\sigma(qq) = \sigma(q\bar{q})$, we get

$$\sigma(\Lambda p) = \sigma(pp) + \sigma(K^- n) - \sigma(\pi^+ p)$$

8.4.1 Baryon Number

The baryons are the most abundant group of particles. For nucleons and hyperons, $B = +1$, for antibaryon, $B = -1$, for pions, kaons and other particles $B = 0$. The baryon number is an additive quantum number and is conserved in all the three (i.e., strong, weak and electromagnetic) interactions.

EXERCISE 8.3: Which of the following processes are allowed by the conservation laws?

(i) $n \to p + \gamma$
(ii) $p \to e^+ + \gamma$
(iii) $p \to \pi^+ + \gamma$
(iv) $n + \bar{p} \to \pi^- + \pi^0$

Solution:

(i) Does not conserve electric charge. Forbidden.
(ii) Does not conserve baryon number. Forbidden.
(iii) Does not conserve baryon number. Forbidden.
(iv) Allowed.

8.4.2 Isospin

Isospin is a quantum number applicable to hadrons and is conserved in the strong interactions. The concept of isospin is rooted in the charge independence of the strong interaction,

$nn \approx np \approx pp$. This doublet of neutrons and protons is said to have isospin $\frac{1}{2}$ with the projection $I_Z = +\frac{1}{2}$ for the proton and $I_Z = -\frac{1}{2}$ for the neutron. The three pions (π^+, π^-, π^0) compose a triplet, suggesting isospin $I = 1$. The projections $I_Z = +1$ (for π^+), 0 (for π^0) and -1 (for π^-). At the quark level, an isospin doublet $\left(I = \frac{1}{2}\right)$ is formed by the up and down quarks with the projection $I_Z = +\frac{1}{2}$ assigned to up quark and $I_Z = -\frac{1}{2}$ to down quark. The strange quark is in a class by itself and has isospin $I = 0$.

I is the additive quantum number and Clebsch–Gordon coefficients are used for the addition of isospins. The charge multiplicity is given by $2I + 1$. The antiparticle has the same I as the particle but the opposite I_Z. Total isospin (I) is conserved in the strong interaction, but breaks down in the electromagnetic and the weak interactions. The third component I_Z of the system of hadrons is conserved in the strong and the EM interactions, but is violated in the weak interactions $\left(\Delta I_Z = \pm\frac{1}{2}\right)$.

For example, the process

$$\Sigma^+ \rightarrow p + \eta^0$$

has not been observed even though it conserves the charge, angular momentum and the baryon number. It is forbidden by the fact that it does not conserve the isospin, $I = 1 \neq \frac{1}{2} + 0$, as required by the strong interaction.

Isospin is related to other quantum numbers for the particles by Gell-Mann formula,

$$\frac{Q}{e} = I_Z + \frac{S + B}{2} \tag{8.6}$$

where S is the strangeness and B is the baryon number.

Baryons and mesons can be grouped into isospin multiplets as shown in Figure 8.6.

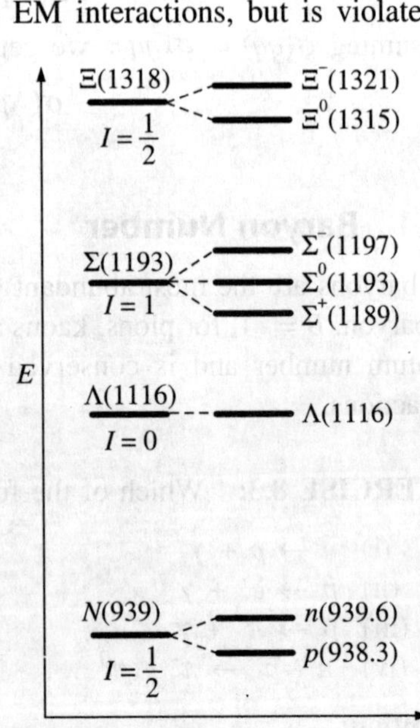

FIGURE 8.6 Isospin multiplets: nucleon doublet, Λ singlet, the Σ triplet, and Ξ doublet.

8.4.3 Resonance Particles

Most of the particles in Table 8.6, exist long enough to travel as distinct entities and their modes of decay can be observed experimentally. Ultra-short lived particles ($\tau \sim 10^{-23}$ s) cannot be detected by recording their creation and subsequent decay because distance at the most they cover is $\approx 3 \times 10^8 \times 10^{-23} \approx 3 \times 10^{-15}$ m. However, such particles appear as resonant state in the interactions of longer-lived particles.

Let us consider the bombardment of protons by energetic π^0 meson and a certain reaction is studied, for instance:

$$\pi^+ + p \rightarrow \pi^+ + p + \pi^+ + \pi^- + \pi^0$$

In this reaction, the new mesons have a certain total energy that consists of their rest mass energies plus their kinetic energies in the CMS. If we plot the number of events observed vs. the total energy of the new mesons in each event, a graph (Figure 8.7) is obtained. Evidently, there is a high tendency for the total meson energy to be equal to 783 MeV and a somewhat low tendency for it to be 549 MeV. In other words, the reaction exhibits resonances at 548 and 783 MeV or equivalently, we can say that this reaction proceeds via the creation of an intermediate particle of 549 or 783 MeV/c^2.

We can estimate the mean-lifetime (τ) of these uncharged intermediate particles (η and ω mesons) using the following formula:

$$\tau = \frac{\hbar}{\Gamma} \tag{8.7}$$

where Γ is the width of the resonance peak as shown in Figure 8.7. The values come out to be $\tau_\eta \sim 5 \times 10^{-19}$ s and $\tau_\omega \sim 7 \times 10^{-25}$ s.

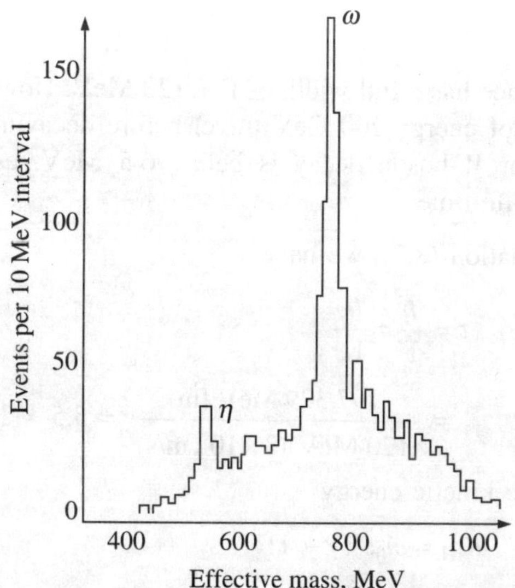

FIGURE 8.7 Resonant states in the reaction $\pi^+ + p \rightarrow \pi^+ + p + \pi^+ + \pi^- + \pi^0$ occur at effective masses of 549 and 783 MeV/c^2. By effective mass it is meant the total energy, including the rest mass energy, of the three new mesons relative to their centre of mass.

Properties and quark content of some baryon resonances are given in Table 8.9. A more extensive listing is available online at <http://pdg.lbl.gov/>.

Table 8.9 Properties and Quark Content of Some Baryon Resonances ($J^\pi = 3/2^+$, $B = 1$)

Particle	*Mass* (GeV/c^2)	*I*	I_Z	$Y = S + B$	*Quark content*
Δ^{++}	1.232	3/2	+3/2	1	*uuu*
Δ^{+}	1.232	3/2	+1/2	1	*uud*
Δ^{0}	1.232	3/2	−1/2	1	*udd*
Δ^{-}	1.232	3/2	−3/2	1	*ddd*
Σ^{*+}	1.383	1	+1	0	*uus*
Σ^{*0}	1.385	1	0	0	*uds*
Σ^{*-}	1.387	1	−1	0	*dds*
Ξ^{*0}	1.532	1/2	+1/2	−1	*uss*
Ξ^{*-}	1.535	1/2	−1/2	−1	*dss*
Ω^{-}	1.672	0	0	−2	*sss*

EXERCISE 8.4:

(a) The Δ^{++} resonance has a full width of $\Gamma = 120$ MeV. How far on an average would such a particle of energy 200 GeV travel before decaying?

(b) If the width for *W*-boson decay is below 6.5 MeV, estimate the limit for the corresponding life-time.

Solution: (a) From equation (8.7), we have

$$\tau = \frac{\hbar}{\Gamma} = \frac{\hbar c}{\Gamma c}$$

$$= \frac{197.329 \text{ MeV fm}}{(120 \text{ MeV})(3 \times 10^8 \text{ m/s})} = 5.5 \times 10^{-24} \text{ s}$$

We know that relativistic kinetic energy

$$K_{\text{rel}} = m_0 c^2(\gamma - 1)$$

or

$$\gamma = 1 + \frac{K_{\text{rel}}}{m_0 c^2} = 1 + \frac{200 \text{ GeV}}{1.236 \text{ GeV}} \cong 163$$

Therefore, the average distance travelled

$$d = v(\gamma\tau) = c(\gamma\tau)$$

$$= (3 \times 10^8 \text{ ms}^{-1})(163)(5.5 \times 10^{-24} \text{ s}) \cong 2.7 \times 10^{-13} \text{ m}$$

The distance is smaller than the best resolution obtainable in photographic emulsions, which is in micron range.

(b) Using the uncertainty principle, we have

$$\Gamma\tau \geq \hbar$$

$$\therefore \qquad \tau \geq \frac{\hbar c}{\Gamma c} = \frac{197.329 \text{ MeV fm}}{(6.5 \text{ MeV})(3\times10^{8} \text{ m/s})} = 10^{22} \text{ s}$$

EXERCISE 8.5: The decay of Ξ^{*-} initiates the decay sequence as shown in Figure 8.8. The quark content of the hadrons involved is as follows:

$$\Xi^{*-}\,(ssd),\ \Sigma^0\,(sud),\ \Lambda^0\,(sud),\ p\,(uud),\ K^-\,(s\bar{u}),\ \pi^-\,(d\bar{u}),\ \pi^0\,(u\bar{u}-d\bar{d})$$

Classify the decays as strong, electromagnetic or weak.

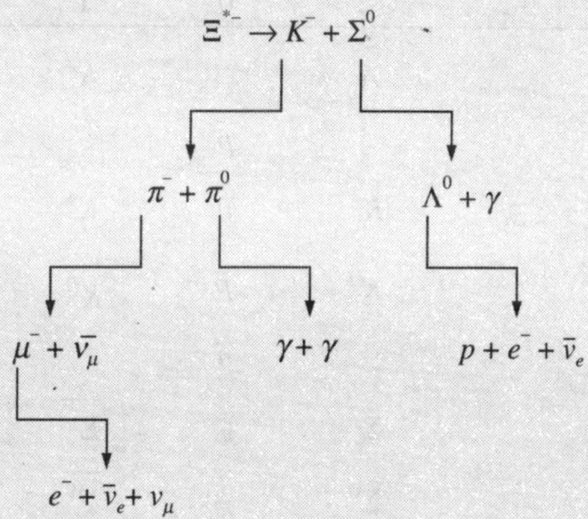

FIGURE 8.8 Decay sequence of negative *Xi*.

Solution:

(i) $\Xi^{*-} \to K^- + \Sigma^0$ Strong interaction. Does not require the weak or the electromagnetic interaction.

(ii) $\Sigma^0 \to \Lambda^0 + \gamma$ Electromagnetic interaction.

(iii) $\Lambda^0 \to p + e^- + \bar{\nu}_e$ Involves an antineutrino, therefore weak interaction.

(iv) $K^- \to \pi^- + \pi^0$ An *s* quark change to *d* quark, therefore weak interaction.

(v) $\pi^0 \to \gamma + \gamma$ Electromagnetic interaction.

(vi) $\pi^- \to \mu^- + \bar{\nu}_\mu$ Weak interaction.

(vii) $\mu^- \to e^- + \bar{\nu}_e + \nu_\mu$ Weak interaction.

8.4.4 Strangeness and Strange Particles

Heavy unstable particles such as the Λ and Σ baryons, and the *K* mesons are produced at a rapid rate in the high energy collisions but decay slowly, i.e., they have exceptionally long lifetimes. A typical production and decay sequence is

$$\pi^- + p \rightarrow K^0 + \Lambda^0$$
$$K^0 \rightarrow \pi^+ + \pi^-, \quad \Lambda^0 \rightarrow \pi^- + p$$

with lifetimes $\tau_K \sim 0.89 \times 10^{-10}$ s and $\tau_\Lambda \sim 2.63 \times 10^{-10}$ s which is large as compared to the strong interaction time scale of 10^{-23} s.

They are named strange particles. A new quantum number S, strangeness is introduced to distinguish them from other particles. Gell–Mann formula, equation (8.6) relates S with other quantum numbers. Strangeness (S) is an additive quantum number like the electric charge, and is thus, conserved in the production process. Table 8.10 summarizes the strangeness S for various hadron multiplets.

Table 8.10 *I* and *S* Assignments

I/S	–3	–2	–1	0	1	2	3
0	Ω^-		Λ		Λ^-		$\overline{\Omega}^+$
				p			
		Ξ^0	K^-	n	K^+	$\overline{\Xi}^0$	
1/2		Ξ^-	K^0	$\overline{p}$	$\overline{K}^0$	$\overline{\Xi}^+$	
				$\overline{n}$			
			Σ^+	π^+	$\overline{\Sigma}^-$		
1			Σ^0	π^0	$\overline{\Sigma}^0$		
			Σ^-	π^-	$\overline{\Sigma}^+$		

Strangeness S is conserved in the strong and the electromagnetic interactions, i.e., $\Delta S = 0$, but breaks down in the weak interactions, i.e., $\Delta S = \pm 1$.

EXERCISE 8.6: Which of the following processes are forbidden by the law of conservation of strangeness?

(i) $\pi^- + p \rightarrow \Sigma^- + K^+$
(ii) $\pi^- + p \rightarrow \Sigma^+ + K^-$
(iii) $\pi^- + p \rightarrow K^+ + K^- + n$
(iv) $n + p \rightarrow \Lambda^0 + \Sigma^+$
(v) $\pi^- + n \rightarrow \Xi^- + K^+ + K^-$
(vi) $K^- + p \rightarrow \Omega^- + K^+ + K^0$

Solution:

(i) $\Delta S = 0 + 0 - (-1) - 1 = 0$. Allowed.
(ii) $\Delta S = 0 + 0 - 1 - 1 = -2$. Forbidden.

(iii) $\Delta S = 0 + 0 - (-1) - 1 - 0 = 0$. Allowed.
(iv) $\Delta S = 0 + 0 - (-1) - (-1) = -2$. Forbidden.
(v) $\Delta S = 0 + 0 - (-2) - 1 - (-1) = -2$. Forbidden.
(vi) $\Delta S = -1 + 0 - 3 - (-1) - (-1) = 0$. Allowed.

EXERCISE 8.7: Explain if the reactions mentioned below proceed via the strong, electromagnetic or weak interactions, or if they do not occur at all.

(i) $\pi^- \rightarrow \mu^- + \bar{\nu}_\mu$
(ii) $\Sigma^0 \rightarrow \Lambda + \gamma$
(iii) $p \rightarrow n + e^+ + \nu_e$
(iv) $\pi^- + p \rightarrow \pi^0 + \Sigma^0$
(v) $\pi^- + p \rightarrow K^0 + \Sigma^0$
(vi) $e^+ + e^- \rightarrow \mu^+ + \mu^-$

Solution:

(i) Since, neutrino is involved, it is weak interaction.
(ii) Since, γ-ray is involved and $\Delta S = 0$, it is EM interaction.
(iii) Allowed as weak decay if proton is bound but forbidden when proton is free because proton is lighter than the sum of masses of the product particles.
(iv) Does not occur as a strong or EM interaction because $\Delta S \neq 0$.
(v) $\Delta S = 0$ and other quantum numbers are conserved therefore it is strong interaction.
(vi) Since, lepton-antilepton pair is involved, it is weak interaction.

EXERCISE 8.8: The products of a collision between a fast proton and a neutron are neutron, Σ^0 particle, and another particle. What is the other particle?

$$p + n \rightarrow n + \Sigma^0 + ?$$

Solution: The other particle should have positive charge, from charge conservation. The other particle should have (+1) strangeness number from strangeness conservation. Therefore, the particle is K^+ meson.

8.4.5 Hypercharge

A new quantum number hypercharge (Y) was introduced with the following relation:

$$Y = B + S$$

Y is a good quantum number for the strong interactions. All the observed particles satisfy the Gell–Mann–Nishijima formula, i.e.,

$$\frac{Q}{e} = I_Z + \frac{Y}{2} \tag{8.8}$$

and are arranged in isospin multiplets, all the members of which have the same value of spin, parity, baryon number, strangeness, and hypercharge.

EXERCISE 8.9: Is hypercharge conserved in the following reactions?

(i) $\pi^+ + \eta \rightarrow \Lambda + K^+$

(ii) $\Lambda \rightarrow p + \pi^-$

(iii) $K^0 \rightarrow \pi^+ + \pi^-$

Solution: In all these reactions hypercharge (and strangeness) is not conserved.

EXERCISE 8.10: Determine the strangeness S and the hypercharge Y of a neutral elementary particle whose isotopic spin projection is +1/2 and baryon charge is +1. What is this particle?

Solution: From the Gell–Mann–Nishijima formula, we have

$$\frac{Q}{e} = I_Z + \frac{Y}{2}$$

or

$$0 = \frac{1}{2} + \frac{Y}{2} \quad \text{or} \quad Y = -1$$

also,

$$Y = B + S \quad \text{or} \quad S = -2$$

Thus, the particle is Ξ^0.

8.5 EIGHTFOLD WAY

In the preceding section, the particles are grouped into isospin multiplets whose members are related by the isospin symmetry. This classification of the particles into isospin families helps to bring some order into the confusing diversity of more than 300 known particles. The most successful classification scheme—Eightfold Way was proposed by M. Gell-Mann. On the basis of the eightfold way, the baryons and mesons are arranged into geometrical patterns, according to their charge and strangeness.

The eight lightest baryons form a hexagonal array, with two particles at the centre as shown in Figure 8.9. This group is known as the baryon octet. One can note that particles of like charge lie along the diagonal lines and horizontal lines associate with particles of like strangeness (S). The eight lightest mesons fill a similar hexagonal pattern, forming the (pseudo-scalar) meson octet (Figure 8.10).

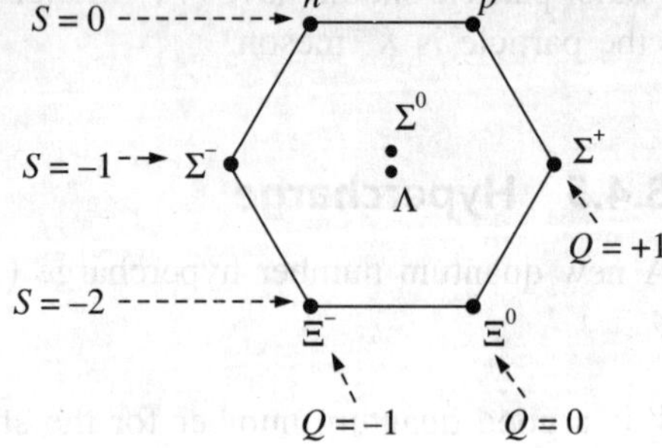

FIGURE 8.9 Baryon octet.

After 1961, a new term was introduced, called the hypercharge (Y), which was equal to S for the mesons and $S + 1$ for the baryons. Figures 8.11, 8.12 and 8.13 display supermuliplets of the eightfold way: baryon octet, baryon decuplet and meson octet, respectively. The hexagonal pattern of Figures 8.11 and 8.13 is a characteristic feature of the SU(3) symmetry. One can see that the hexagonals are not the only figures allowed by the eightfold way. Figure 8.12 displays a triangular array.

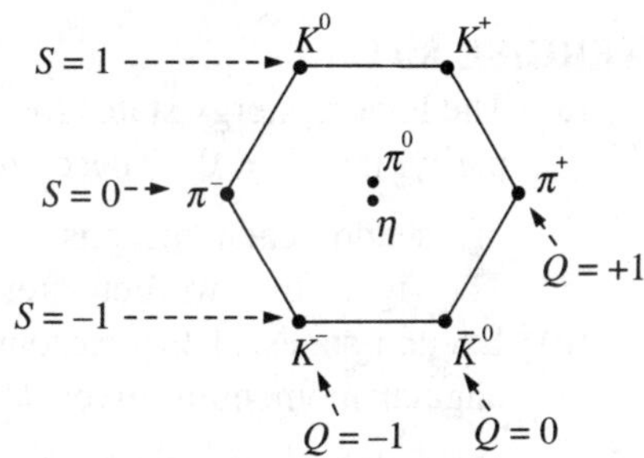

FIGURE 8.10 Meson octet

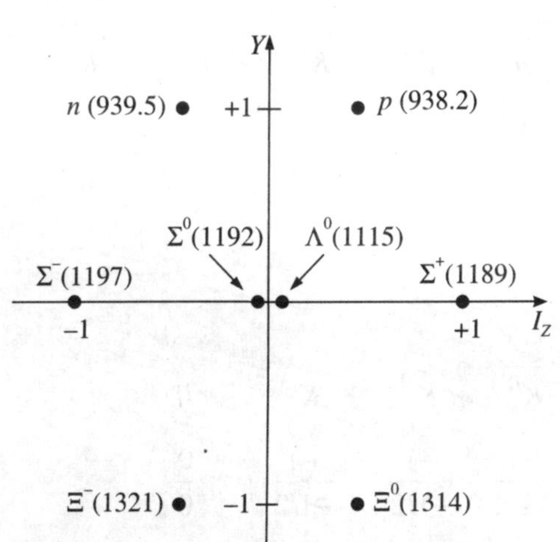

FIGURE 8.11 Hypercharge vs. isospin for the baryon octet.

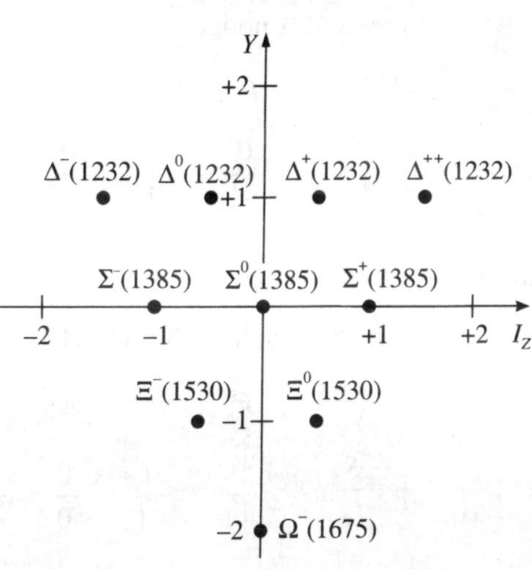

FIGURE 8.12 Hypercharge vs. isospin for the baryon decuplet.

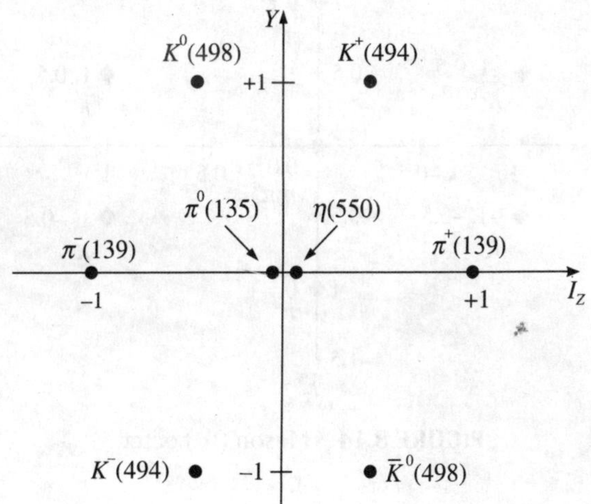

FIGURE 8.13 Hypercharge vs. isospin for the meson octet.

EXERCISE 8.11:

(a) The lowest energy states for $q\bar{q}$ pairs (mesons) correspond to zero spin and negative parity, i.e., $J^\pi = 0^-$. These are called pseudo scalar mesons. Now, show the position of pseudo scalar mesons π^+, π^-, π^0, K^0, $\overline{K^0}$, K^+, K^- and η on the S–I_Z diagram. The figure that we get after plotting the points is known as 'meson (0^-) octet'.

(b) Excited states of the mesons in which the $q\bar{q}$ pairs are aligned, with zero orbital angular momentum gives $J^\pi = 1^-$. These states are known as vector mesons. Now, show the position of vector mesons ρ^-, ρ^0, ρ^+, φ, ω, K^{*0}, K^{*+}, K^{*-} and $\overline{K^{*0}}$ on the S–I_Z diagram. The figure that we get after plotting the points is known as 'meson (1^-) nonet'.

Given:	ρ^-	ρ^0	ρ^+	φ	ω	K^{*0}	K^{*+}	K^{*-}	K^{*-}
S	0	0	0	0	0	1	1	−1	−1
I_Z	−1	0	1	0	0	−1/2	1/2	−1/2	1/2

Solution:

(a) From the Tables 8.8 and 8.10, we have

	π^+	π^-	π^0	K^0	$\overline{K^0}$	K^+	K^-	η
S	0	0	0	+1	−1	+1	−1	0
I_Z	+1	−1	0	−1/2	+1/2	+1/2	−1/2	0

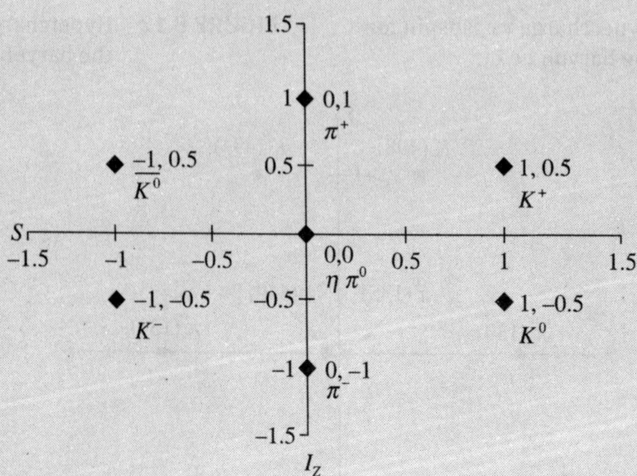

FIGURE 8.14 Meson (0^-) octet.

(b) Using Excel, we plot the given points to obtain the following $S - I_Z$ diagram

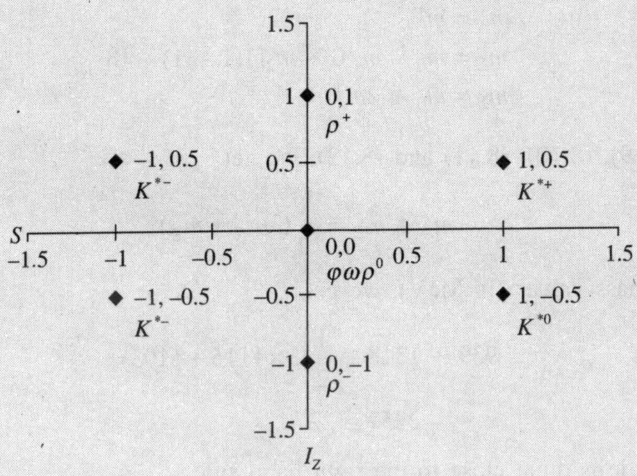

FIGURE 8.15 Meson (1^-) nonet.

EXERCISE 8.12: If we ignore the small mass differences within the isospin multiplets, then there are only four distinct masses in the baryon octet: the mass of the nucleon doublet (m_N), Σ triplet (m_Σ), Λ singlet (m_Λ), and Ξ doublet (m_Ξ). According to the Eightfold way, these masses obey the following *Gell–Mann–Okubo mass formula:*

$$m_a = m_0 + m_1 Y + m_2 \left[I(I+1) - \frac{1}{4Y^2} \right]$$

where a stands for N, Σ, Λ or Ξ; I is the corresponding isospin quantum number; Y the hypercharge and m_0, m_1 and m_2 are constants. Show that this mass formula implies

$$m_N + m_\Xi = \frac{1}{2}(3m_\Lambda + m_\Sigma)$$

Solution: According to Gell-Mann–Okubo mass formula, with approximate values of Y and T, we have

$$m_N = m_0 + m_1 + m_2 \left[\frac{1}{2}\left(\frac{1}{2} + 1 \right) - \frac{1}{4} \right]$$

or
$$m_N = m_0 + m_1 + \frac{1}{2} m_2 \tag{8.9}$$

$$m_\Xi = m_0 - m_1 + m_2 \left[\frac{1}{2}\left(\frac{1}{2} + 1 \right) - \frac{1}{4} \right]$$

or
$$m_\Xi = m_0 - m_1 + \frac{1}{2} m_2 \tag{8.10}$$

$$m_\Lambda = m_0 + m_1 \cdot 0 + m_2[0 - 0]$$

or
$$m_\Lambda = m_0 \tag{8.11}$$

$$m_\Sigma = m_0 - m_1 \cdot 0 + m_2[1(1 + 1) - 0]$$

or
$$m_\Sigma = m_0 + 2m_2 \tag{8.12}$$

From equation (8.9), (8.10), (8.11) and (8.12), we get

$$m_N + m_\Xi = \frac{1}{2}(3m_\Lambda + m_\Sigma)$$

Substituting the mass values (in MeV), we get

$$939 + 1318 \cong \frac{1}{2}(3 \times 1116 + 1193)$$

or
$$2257 \cong 2270$$

Thus, left hand side is quite close to the right hand side.

8.6 CONSERVATION LAWS

In developing the standard model for the particles, certain types of interactions and decays are observed to be common while others to be forbidden. This observation has led to a number of conservation laws, which govern them. These conservation laws are beside the classical conservation laws such as the conservation of energy, charge, etc. which are still applicable in the realm of particle interactions. There is an insightful connection between the conservation laws obeyed by a physical system and the symmetries. This connection is contained in the Noether's theorem:

'To every symmetry, there corresponds a conservation law.'

EXERCISE 8.13: What is the available energy in the collision of a moving proton with a proton at rest?

Solution: Suppose K_p (kinetic energy of proton) << m_pc^2, then the collision may be treated in a non-relativistic manner.

If the incident proton has velocity v in the laboratory frame, then

$$K_{\text{lab}} = \frac{1}{2}m_p v^2$$

In the CM frame, one proton has velocity $v/2$ and the other has velocity $-v/2$. In the CM frame, all the kinetic energy is available for the production of further particles; this kinetic energy is given by:

$$K_{CM} = \frac{1}{2}m_p\left(\frac{v}{2}\right)^2 + \frac{1}{2}m_p\left(\frac{v}{2}\right)^2 = \frac{1}{4}m_p v^2$$

From the above two equations, we get

$$\frac{K_{CM}}{K_{\text{lab}}} = \frac{1}{2}$$

Thus, one half of the energy in the lab frame is available. For example, if 200 MeV proton collides with another proton at rest, only 100 MeV of this energy may be used in a collision to create other particles.

If proton is of relativistic energy, then using the invariance property one can write the following equation:

$$\{(E_1 + E_2)^2 - (p_1 + p_2)^2c^2\}_{\text{lab}} = \{(E_1 + E_2)^2 - (p_1 + p_2)^2c^2\}_{CM}$$

By definition of the CM, $(\vec{p}_1 + \vec{p}_2)_{CM} = 0$. If the second proton is at rest, then $p_{2(\text{lab})} = 0$ and $E_{2(\text{lab})} = m_pc^2$. Using $E_{1(\text{lab})}^2 - p_{1(\text{lab})}^2c^2 = m_p^2c^4$, the aforementioned equation reduces to the following form:

$$2E_{1(\text{lab})}m_pc^2 + 2m_p^2c^4 = E_{\text{tot}\,(CM)}^2$$

$$\therefore \qquad 2E_{\text{tot(lab)}}m_pc^2 = E_{\text{tot}\,(CM)}^2 \qquad (8.13)$$

where $E_{\text{tot}(CM)} = (E_1 + E_2)_{CM}$ and $E_{\text{tot(lab)}} = E_1 + m_pc^2$.

One can write, the total energy in the laboratory frame as the sum of kinetic energies of the protons and the rest mass energies. Thus

$$E_{\text{tot(lab)}} = K + m_pc^2 + m_pc^2 \qquad (8.14)$$

From equation (8.13) and (8.14), we get

$$E_{\text{tot}(CM)} = \sqrt{2m_pc^2(K + 2m_pc^2)} \qquad (8.15)$$

Equation (8.15) gives the net available energy for the production of particles. For instance, a proton with a kinetic energy $K = 1$ TeV = 1000 GeV gives a net available energy

$$\sim \sqrt{2\text{ GeV}\,(2\text{ GeV} + 1000\text{ GeV})} \approx 45\text{ GeV}$$

By far, the largest part of the energy of motion of the projectile is tied up in the motion of the CM and cannot participate in reactions. Thus, collisions between a high-energy proton and a stationary proton are quite inefficient.

Because of this low efficiency, accelerators are designed in which two beams of particles will collide from opposite directions, so that the laboratory frame will be the centre-of-mass frame.

EXERCISE 8.14: Lambda particles and kaons are to be produced by the following reaction:

$$p + p \rightarrow p + \Lambda + K^+$$

According to the relativistic formula for the available energy in the centre of mass, what is the minimum kinetic energy required for a proton incident on a stationary proton?
Given: $m_p c^2 = 938$ MeV, $m_\Lambda c^2 = 1116$ MeV, $m_K + c^2 = 496$ MeV

Solution: In the CM frame, the minimum energy of the reaction products is the sum of their rest mass energies.

$$E_{\text{tot}(CM)} = 938 + 1116 + 496 = 2550 \text{ MeV}$$

Now, if K is the minimum kinetic energy of the proton, then we have

$$E_{\text{tot}(CM)} = \sqrt{2m_p c^2 (K + 2m_p c^2)}$$

or

$$(2250 \text{ MeV})^2 = 2 \times 938 \text{ MeV } (K + 2 \times 938 \text{ MeV})$$

$\therefore$

$$K = 1.59 \times 10^3 \text{ MeV} = 1.59 \text{ GeV}$$

8.6.1 Exact Conservation Laws

All interactions obey the following conservation laws:

(i) Conservation of the electric charge.
(ii) Conservation of the four momenta (p_x, p_y, p_z, E).
(iii) Conservation of the angular momentum.
(iv) Conservation of the baryon number.
(v) Conservation of the lepton number (the electron number, muon number and tau number are all separately conserved.)
(vi) Conservation of colour.
(vii) Invariance under CPT.

Electric Charge

The electric charge is one of the quantum numbers, which is exactly conserved. This conservation is associated with the masslessness of the photon. It is due to the conservation of charge that the electron cannot decay, for instance, via

$$e^- \rightarrow \nu_e + \nu_e + \bar{\nu}_e$$

or

$$e^- \rightarrow \nu_e + \gamma$$

Baryon Number

No known interaction or decay process in nature alters the net baryon number. The neutron along with all the heavier baryons decays directly to the proton or eventually forms proton, since the proton is the least massive baryon. This indicates that the proton cannot decay further without violating the conservation of baryon number, which means that if the conservation of baryon number holds exactly, the proton should be completely stable against any decay. One prediction of the grand unification of forces is that the proton can also decay. Until now, such a possibility has not been experimentally verified.

Conservation of baryon number forbids a decay of the type:

$$p + n \rightarrow p + \mu^+ + \mu^-$$
$$B = 1 + 1 \neq 1 + 0 + 0$$

Although, with adequate energy permits pair production in the following reaction:

$$p + n \rightarrow p + n + p + \overline{p}$$
$$B = 1 + 1 = 1 + 1 + 1 - 1$$

Lepton Number

Each of the three sets of leptons (L_e, L_μ, L_τ) are to be conserved separately. For example, observe the following decay processes

$$\pi^- \rightarrow \mu^- + \overline{\nu}_\mu$$
$$\mu^- \rightarrow e^- + \overline{\nu}_e + \nu_\mu$$

Since, a well-defined muon energy is observed from the decay, the first reaction (decay of π^-) is known to be a two-body decay. However, the decay of muon (μ^-) into electron produces a distribution of electron energies, indicating that it is at least a three-body decay. In order for both L_e and L_μ to be conserved, the other particles must be $\overline{\nu}_e$ and ν_μ.

EXERCISE 8.15: Which of the following processes are forbidden by the law of conservation of lepton number?

(i) $n \rightarrow p + e^- + \nu_e$
(ii) $\pi^+ \rightarrow \mu^+ + e^- + e^+$
(iii) $\pi^- \rightarrow \mu^- + \nu_\mu$
(iv) $p + e^- \rightarrow n + \nu_e$
(v) $\mu^+ \rightarrow e^+ + \nu_e + \overline{\nu}_\mu$
(vi) $K^- \rightarrow \mu^- + \overline{\nu}_\mu$

Solution:

(i) The process cannot occur, since there are 2 more leptons (e^-, ν_e) on the right hand side as compared to zero on the left.
(ii) The process is forbidden because this corresponds to a change of lepton number by zero on the left and –1 on the right.
(iii) The process is forbidden because μ^-, ν_μ are both leptons, which means $\Delta L = 2$.
(iv) Allowed.
(v) Allowed.
(vi) Allowed.

EXERCISE 8.16: Which of the following processes are forbidden by an absolute conservation law?

(i) $n \rightarrow e^+ + K^-$

(ii) $\mu^- \rightarrow e^- + \gamma$

(iii) $K^+ \rightarrow \pi^+ + e^+ + \mu^-$

(iv) $\tau^- \rightarrow e^- + \gamma$

(v) $n \rightarrow \bar{n}$

(vi) $\tau^- \rightarrow e^- + \pi^+ + \pi^-$

(vii) $K^+ \rightarrow \mu^+ + \nu_e$

(viii) $\pi^+ \rightarrow \mu^+ + \nu_e$

(ix) $p \rightarrow e^+ + \pi^0$

(x) $e \rightarrow \gamma + \gamma$

Solution:

(i) Forbidden by conservation of baryon number and e lepton number.

(ii) Forbidden by conservation of μ lepton and e lepton numbers.

(iii) Forbidden by conservation of e lepton and μ lepton numbers.

(iv) Forbidden by conservation of τ lepton and e lepton numbers.

(v) Forbidden by conservation of baryon number.

(vi) Forbidden by conservation of τ lepton and e lepton numbers.

(vii) Forbidden by conservation of μ lepton and e lepton numbers.

(viii) Forbidden by conservation of μ lepton and e lepton numbers.

(ix) Forbidden by conservation of baryon number and e lepton number.

(x) Forbidden by conservation of e lepton number.

8.6.2 Additional Conservation Laws

Additional conservation laws are obeyed by some but not all interactions. For example, conservation of parity charge conjugation (C), isospin (I), G-parity (G), strangeness (S) and hypercharge (Y) are obeyed only in the strong interactions.

The status of conservation laws in all the fundamental interactions is given in Table 8.11.

Table 8.11 Conservation Laws for Three Types of Interactions

Quantity	*Strong*	*Electromagnetic*	*Weak*
Q (charge)	Yes	Yes	Yes
B (baryon number)	Yes	Yes	Yes
J (angular momentum)	Yes	Yes	Yes
Mass + Energy	Yes	Yes	Yes
Linear momentum	Yes	Yes	Yes

(*Contd.*)

Table 8.11 Conservation Laws for Three Types of Interactions (*Contd.*)

Quantity	*Strong*	*Electromagnetic*	*Weak*
I (isospin)	Yes	No	No
I_Z (Third component of I)	Yes	Yes	No $\Delta I_Z = \pm 1/2$
S (strangeness)	Yes	Yes	No $\Delta S = \pm 1$
p (parity)	Yes	Yes	No
C (charge conjugation)	Yes	Yes	No
G (G-parity)	Yes	No	No
L (lepton number)	Yes	Yes	Yes
CP (charge and parity)	Yes	Yes	No †
T (time reversal)	Yes	Yes	No #
CPT	Yes	Yes	Yes

† CP violation observed only in K^0 system.

Due to CP violation and CPT conservation.

EXERCISE 8.17: Are the following particle interactions allowed by the conservation rules? If yes, indicate the force that is involved and draw a Feynman diagram for the process.

(i) $\mu^- \rightarrow e^- + \nu_\mu + \bar{\nu}_e$

(ii) $\Lambda \rightarrow \pi^+ + \pi^-$

(iii) $\nu_e + n \rightarrow p + e^-$

(iv) $\pi^0 \rightarrow \tau^+ + \tau^-$

(v) $e^+ + e^- \rightarrow \mu^+ + \mu^-$

Solution:

(i) Weak decay (Figure 8.16).

(ii) Forbidden because baryon number is violated.

(iii) Weak interaction

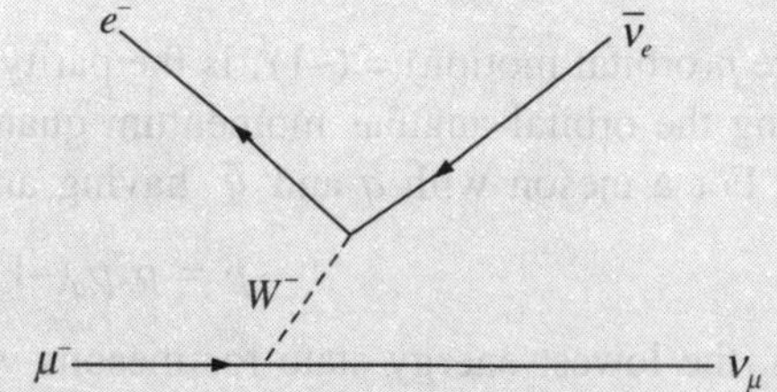

FIGURE 8.16 Feynman diagram for negative muon decay.

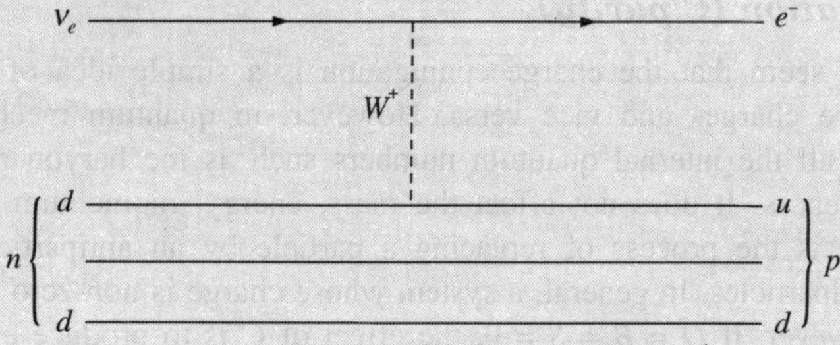

FIGURE 8.17 Feynman diagram showing the neutron decay.

(iv) Forbidden because energy is not conserved.
(v) Electromagnetic as well as weak interaction.

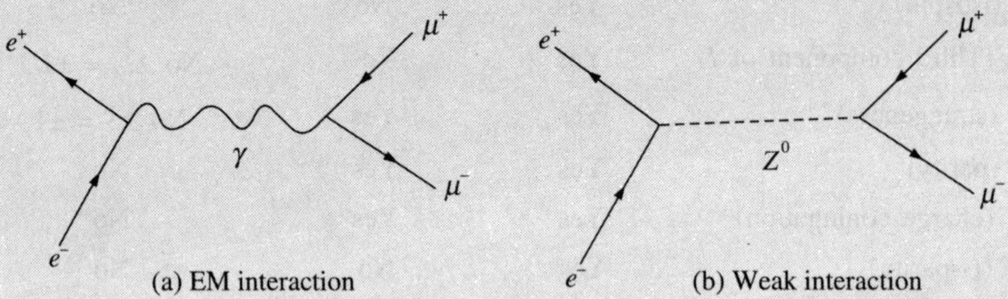

(a) EM interaction (b) Weak interaction

FIGURE 8.18 Feynman diagram showing (a) the EM interaction, and (b) the weak interaction.

Parity

Parity is associated with the reflection symmetry in ordinary space. The absolute intrinsic parity cannot be determined. The parity of a particle can only be relative to another particle. Quarks have a intrinsic parity $p_q = +1$ and an antiquark has a parity $p_q = -1$. All baryons are assigned the positive parity. All the antifermions have parity opposite to the fermions. On the contrary, bosons have the same parity for the particle and the antiparticle. Pions and kaons have odd parity.

Parity is a multiplicative number. Therefore, the parity of a composite system is equal to the parities of the parts. For a system comprising particles A and B,

$$p(AB) = p(A)p(B)p \quad \text{(Orbital motion)}$$

where p(orbital motion) = $(-1)^l$, is the parity associated with the relative motion of the particles, l being the orbital angular momentum quantum number.

For a meson with q and $\overline{q}$ having antiparallel spins ($s = 0$), the parity

$$p = p_q p_q (-1)^l = -(-1)^l = (-1)^{l+1}$$

Thus, the lowest energy state for mesons will have $J^P = 0^-$.

All the neutrinos are found to be 'left handed,' with an intrinsic parity of (–1), while the antineutrinos are right handed with parity of (+1).

Charge Conjugation (C parity)

Classically, it may seem that the charge conjugation is a simple idea of reversing positive charges by negative charges and vice versa. However, in quantum mechanical systems, it involves reversing all the internal quantum numbers such as the baryon number, the lepton number and strangeness. It does not affect the mass, energy, momentum or the spin. Thus, charge conjugation is the process of replacing a particle by an antiparticle or a system of particles by the antiparticles. In general, a system whose charge is non-zero cannot be an eigen function of C. However, if $Q = B = S = 0$, the effect of C is to produce eigen value = ±1.

For π^0, $C = +1$, for photon $C = -1$ and for n photons $C = (-1)^n$.

This implies that

$$\pi^0 \rightarrow \gamma + \gamma$$

is allowed because $C(\pi^0) = +1$ and $C(2\gamma) = (-1)^2 = +1$, while π^0 will not decay by

$$\pi^0 \rightarrow \gamma$$

which we already know because it cannot conserve momentum, but the decay

$$\pi^0 \rightarrow \gamma + \gamma + \gamma$$

can conserve momentum. This decay cannot happen because it would violate the charge conjugation symmetry.

G-Parity

The operation G consists of rotation of 180° about the y-axis or the z-axis in the isospin space followed by charge conjugation. G-parity for the pion is (–1) and for baryon, it is zero. It is a multiplicative quantum number.

EXERCISE 8.18: Show that β-decay does not obey parity or charge conjugation, separately.

Solution: Charge conjugation operation C on a neutrino (ν_e) would produce a left-handed antineutrino ($\overline{\nu}_e$), leaving spatial coordinates untouched. There is no experimental evidence for such a particle; all $\overline{\nu}_e$ are right-handed. However, the combination of the parity operation p and the charge conjugation operation C on a ν_e do produce a right-handed $\overline{\nu}_e$ in accordance with the observation. This confirms that β-decay is invariant under the combination of CP.

Time Reversal

In simple classical domain, time reversal just means replacing t by $-t$, inverting the direction of the flow of time. Reversing time reverses the time derivatives of the spatial quantities as well, so it reverses the linear momentum and the angular momentum.

Under time reversal, the strong interaction is invariant as proved by the non-existence of the electric dipole moment of the neutron consisting of three charged quarks. Verification of the predicted ratio of forward and backward reactions at the same energy in the CMS also confirms the invariance of time reversal in the strong interaction.

If $a + b \rightarrow c + d$, then

$$\frac{\sigma_{ab \rightarrow cd}}{\sigma_{cd \rightarrow ab}} = \frac{(2s_a + 1)(2s_b + 1)(p_c^*)^2}{(2s_c + 1)(2s_d + 1)(p_a^*)^2}$$

where s is the spin of particles, and p^* is the momentum. The particle beams are unpolarized.

CPT Invariance

The symmetry operations C, P, T can be combined into one operation, which is that of reversing the signs of both space and time coordinates and changing particles into antiparticles. Theorists have shown that invariance under Lorentz transformations implies invariance of the CPT operation, taken in any order.

CPT invariance itself has implications, which are crucial for our understanding of nature. These are as follows:

(i) Existence of an antiparticle for every particle.
(ii) Particles and antiparticles have identical masses and lifetimes.
(iii) All the internal quantum numbers of antiparticles are opposite to those of the particles.

EXERCISE 8.19:

(a) ρ^0 and K^0 mesons both decay mostly to π^+ and π^-. Why is the mean-lifetime of the ρ^0, 10^{-23} s, whereas that of K^0 is 0.89×10^{-10} s.

(b) Δ^0 and Λ both decay to proton and π^- meson. Why is the Δ^0 meson lifetime ~10^{-23} s, whereas that of Λ is 2.6×10^{-10} s?

(c) The particles X and Z can be produced by the following strong interaction:

$$K^- + p \rightarrow K^+ + X$$

$$K^- + p \rightarrow \pi^0 + Z$$

Identify the particles X (1321 MeV/c^2) and Z (1321 MeV/c^2) and deduce their quark content. If the decay schemes are $X \rightarrow \Lambda + \pi^-$ and $Z \rightarrow \Lambda + \gamma$, give a rough estimate of their lifetime.

Solution:

(a) In the decay $\rho^0 \rightarrow \pi^+ + \pi^-$, all the quantum numbers are conserved as required by a strong interaction. Therefore, its lifetime is characteristic of the strong interaction. On the other hand, the decay $K^0 \rightarrow \pi^+ + \pi^-$, violates strangeness. It is, therefore, a weak interaction decay, characterized by relatively long lifetime of the order of 10^{-10} s.

(b) In the decay $\Delta^0 \rightarrow p + \pi^-$, all the quantum numbers are conserved as required by the strong interaction. Therefore, its lifetime is characteristic of a strong interaction. On the other hand, the decay $\Lambda \rightarrow p + \pi^-$, I_Z is not conserved. It is, therefore, a weak interaction decay, and thus characterized by relatively long lifetime.

(c) From the conservation laws for strong interactions, the quantum numbers for X are $B = +1$, $\dfrac{Q}{e} = -1$, $S = -2$. It is Ξ^- and decays weakly with lifetime ~10^{-23} s. Its quark structure is *dss*.

The quantum numbers for Z are $B = +1$, $\dfrac{Q}{e} = 0$, $S = -1$. It is Σ^0 and decays electromagnetically with lifetime ~10^{-20} s. Its quark structure is *uds*.

8.7 BEYOND THE STANDARD MODEL

The standard model is a group of theories that embodies most of our current understanding of the fundamental particles and forces. According to this theory, which is supported by a number of experimental evidences, quarks are the building blocks of matter, and the forces act via carrier particles, exchanged between the particles of matter.

At lower energy, the basic symmetry of the electro-weak theory is broken and the forces separate in strength as the bosons gain mass. The cause for this is actually a new field, called the Higgs field. If a Higgs field suddenly permeates all the space as the universe cools, it may act as a drag on each and every particle moving in space; the drag being dependent on how well each of the particles interacts with the Higgs field. This drag appears as inertia and this is a measurable mass of the particles that were originally massless at the time of Big Bang. The particle associated with the Higgs field is called the Higgs boson. The search for the Higgs boson[3], the missing key link in the standard model, has been going on for the last several decades. Now, finally one has got the glimpse of this particle. On July 4, 2012, scientists of ATLAS and CMS experiments working at CERN's Large Hadron Collider (LHC) announced the discovery of a subatomic particle called the Higgs boson. This particle is expected to help in understanding the fact that why matter has mass, which results in imparting an object its weight due to gravity. Figure 8.19 shows the ATLAS detector at Large Hadron Collider.

FIGURE 8.19 Installing the ATLAS calorimeter. The eight torodial magnets can be seen on the huge ATLAS detector with the calorimeter before it is moved into the middle of the detector. This calorimeter measures the energies of particles produced when protons collide in the centre of the detector. ATLAS works alongside the CMS experiment to search for new physics at the 14 TeV level. [Courtesy CERN]

3. Higgs boson is named after Peter Higgs, one of the proponents for the mechanism that gives masses to particles. It also goes by the name of 'God-particle' in print and the electronic media.

While, the standard model has been successful in particle physics, it does not answer all the questions. For example, why are there only three generations – [{u, d ν_e, e}, {c, s, ν_μ, μ}, {t, b, ν_τ, t}] of the fundamental particles? Or do quarks and/or leptons in fact consist of more fundamental particles? If we look back, this may be true in future. After Chadwick's discovery of the neutron, our understanding of the nuclear particles was completed. The next level of understanding using the quark model did not arrive for another 35 years. If there indeed is another level, we may take even longer time to reach it, because experiments to unravel the mystery will require a lot of sophistication and expenses.

Also some of the recently observed phenomena, such as dark matter and the absence of antimatter in the universe remain unexplained and cannot be accounted for in this model.

8.7.1 Neutrino Oscillations

One of the most intriguing problems over the last three decades has been the solar neutrino problem. The number of solar neutrino reaching earth from the sun is a factor of 2–3, too small if our understanding of fusion process in the sun is correct. Only ν_e is produced as a result of nuclear fusion in the sun. Neutrinos come in three flavours—ν_e, ν_μ and ν_τ. In 1998, the super-Kamiokande collaboration based in Japan announced that scientists had seen neutrinos generated from cosmic rays in the earth's atmosphere, changing the flavour. This could only happen if neutrinos have mass.

Lots of data has been gathered till date and physicists have seen various oscillations (changes) between the three flavours of neutrinos. The probability that a neutrino will change flavour is dependent upon the distance the neutrino travels divided by its energy. Thus, the distance from the sun to the earth offers a measurable probability of neutrino oscillation, which does account for the low number of electron neutrinos reaching earth from the sun.

The standard model allows mixing patterns in quarks and charged leptons, but it was not expected for neutrinos. Theorists have already proposed variations of the standard model to explain the neutrino oscillation observation.

8.7.2 Matter–Antimatter

According to the existing theories of the creation of our universe in the Big Bang, it is understood that matter and antimatter should have been created in exactly the same amount. At present, it appears that matter dominates over antimatter in the universe, and physicists and cosmologists have been trying to uncover the reason behind this, if it is indeed true. Fortunately, it seems we have a window of opportunity to understand this dilemma. The tiny violation of CP symmetry in the kaon decay moves the scales towards matter over antimatter. However, the standard model estimates indicate that this CP violation is too small by at least 10 orders of magnitude to explain the predominance of matter.

8.7.3 Grand Unifying Theories

There are four different theories to account for the four forces. The electromagnetic interaction of particles is explained by the modern theory of quantum electrodynamics (QED). The weak interaction is explained by the electro-weak theory. The strong interaction between quarks and gluons has a different theory, called quantum chromodynamics (QCD), where *colour* is the equivalent of electric charge. And Einstein's general theory of relativity describes how gravity is a manifestation of the basic geometry of space–time.

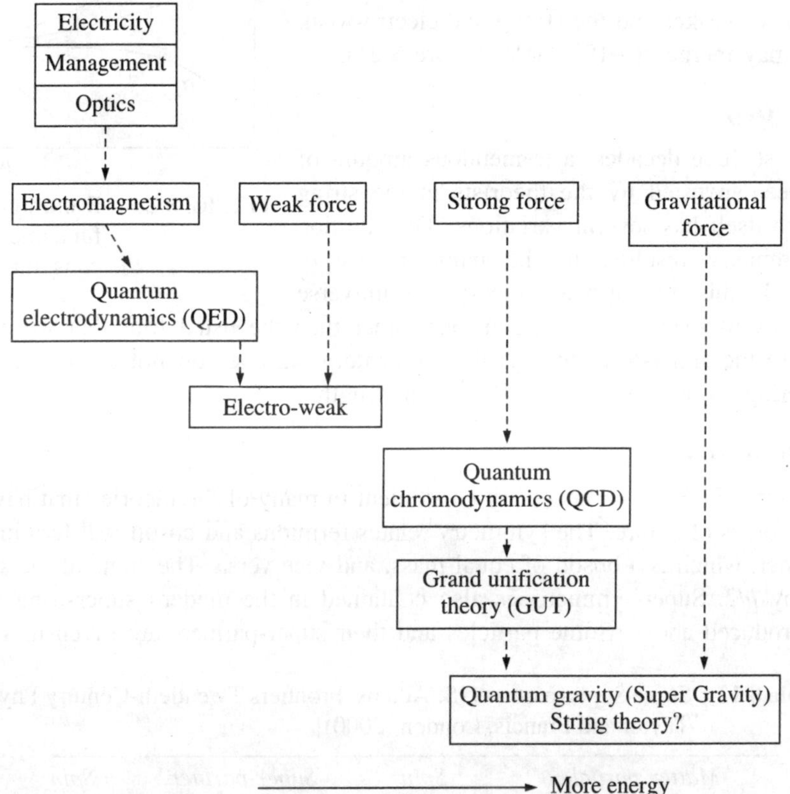

FIGURE 8.20 Unification of the fundamental forces.

One may be curious as to why there is not just one universal force. For decades, physicists have been trying to find an answer to this question and striving to unify the four forces into one universal force, which is assumed to have existed at least in the primordial universe. In such a scenario, the four forces we observe today are just the manifestations of the original single force. Nonetheless, our present science depends on having all these different forces. If the electromagnetic force were not in a delicate balance with the strong force, nuclei would disintegrate—there would be no atoms or molecules, any chemistry or biology. The weak force allows more subtle phenomena—the slow burning of stars like our sun may be impossible without the presence of weak interaction. The supernova explosions which create all the elements that are heavier than iron also depend on just the right strength of neutrino interactions, and the radioactivity inside the earth allows the earth to remain warm and hospitable.

If all the forces of nature can be explained by a single theory, such that the forces start behaving similarly only at higher energies, this is called unification. Figure 8.20 illustrates the unification of the forces. When we noted the intrinsic strengths of the four fundamental interactions in Table 8.1, we omitted the fact that the strength of the different interactions are not fixed but actually depend on the energy scales. At high energies, the strong interaction appears to grow weaker and the strong and electro-weak interactions may merge at ~10^{15} GeV (Figure 8.21).

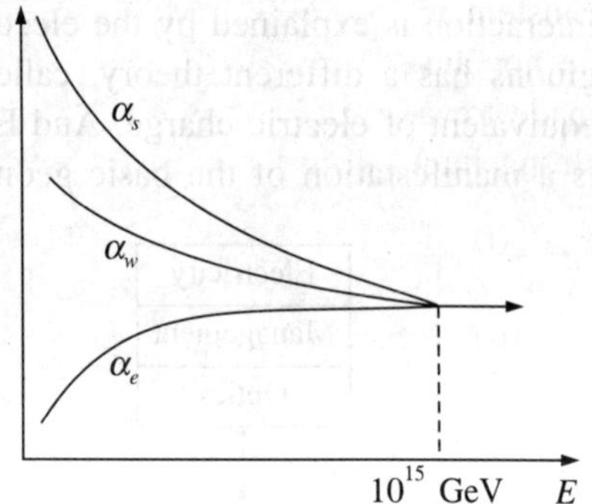

FIGURE 8.21 Evolution of the three fundamental coupling constants.

String Theory

During the last three decades, a tremendous amount of effort has been invested by the theorists in the string theory, which itself has several variations. The addition of super-symmetry resulted in the name *theory of superstrings*. In this revolutionary theory, the universe is described by as many as ten dimensions rather than the usual four that we normally use (x, y, z, t). In the superstring theory, the elementary particles do not exist as points, but as small, quivering loops that are only 10^{-35} m in length.

Super-symmetry

Super-symmetry (SUSY) is a necessary constituent in many of the theories that have attempted to unify the forces of nature. The symmetry relates fermions and boson. All fermions will have a super-partner, which is a boson of equal mass, and vice versa. The spins of the super-partner will differ by $\hbar/2$. Super-symmetry is also contained in the modern superstring theories like the ones introduced above. Some particles and their super-partners are given in Table 8.12.

Table 8.12 Some Super-partners [S. Adams, Frontiers Twentieth-Century Physics, Taylor and Francis, London (2000)]

Matter particle (fermions)	*Spin*	*Super-partner* (bosons)	*Spin*
Electron	1/2	Selectron	0
Muon	1/2	Smuon	0
Tau	1/2	Stau	0
Neutrino	1/2	Sneutrino	0
Quark	1/2	Squark	0
Force carriers (bosons)	*Spin*	*Super-partner* (fermions)	*Spin*
Graviton	2	Gravitino	3/2
Photon	1	Photino	1/2
Gluon	1	Gluino	1/2
$W^{\pm}$	1	*Wino*$^{\pm}$	1/2
Z^0	1	Zino	1/2
Higgs	0	Higgsino	1/2

Einstein is known to have worked during his last 30 years on a unified field theory. Neither he nor anyone else has completely succeeded in this attempt. Numerous attempts have been made towards the grand unified theory (GUT) to combine the weak, electromagnetic and the strong interactions. The GUT theories make many predictions, which include:

(i) The proton is unstable with a life-time of $\sim 10^{30}$ y. Current experimental measurements have shown the life-time to be greater than 10^{32} y.
(ii) Neutrinos may have a small, but finite mass. This has been confirmed.
(iii) It is possible that massive magnetic monopoles exist. Though, there is at present no confirmed experimental evidence to back this idea.

It can be expected that the work of unification will be completed by 2050[4].

PROBLEMS

8.1 Determine p_0 in the decay of K^0 meson of momentum p_0 into π^+ and π^- of momenta p_+ and p_- in the opposite direction such that $p_+ = 2p_-$. [**Ans:** 143 MeV/c^2]

8.2 Explain if the following reactions proceed via the strong, electromagnetic or the weak interactions, or if they do not occur at all.

(i) $\Xi^- \rightarrow \Sigma^- + \pi^0$ (ii) $\tau^- \rightarrow e^- + \bar{\nu}_e + \nu_e$
(iii) $\tau^+ \rightarrow \mu^+ + \gamma$ (iv) $\mu^+ + \mu^- \rightarrow \tau^+ + \tau^-$
(v) $p \rightarrow e^+ + \pi^0$ (vi) $\pi^0 \rightarrow \gamma + \gamma$
(vii) $\pi^- + p \rightarrow K^+ + \Sigma^-$ (viii) $\pi^- + p \rightarrow K^- + \Sigma^+$

8.3 Show the position of baryons with spin–parity $(1/2)^+$ on the $Y - I_Z$ diagram.

8.4 Describe the $(3/2)^+$ baryon decuplet on the $Y - I_Z$ diagram.

8.5 The baryon Ω^- has a mass of 1672 MeV/c^2 and strangeness $S = -3$. Which of the following decay modes are possible?

(i) $\Omega^- \rightarrow \Xi^- + \pi^0$
(ii) $\Omega^- \rightarrow \Sigma^0 + \pi^-$
(iii) $\Omega^- \rightarrow \Lambda + K^-$
(iv) $\Omega^- \rightarrow n + K^- + \overline{K^0}$

8.6 Which of the following reactions are allowed and which are forbidden as under weak interactions?

(i) $\nu_\mu + p \rightarrow \mu^+ + n$
(ii) $\nu_e + p \rightarrow e^- + \pi^+ + p$
(iii) $K^+ \rightarrow \pi^0 + \mu^+ + \nu_\mu$
(iv) $\Lambda \rightarrow \pi^+ + e^- + \bar{\nu}_e$

4. Weinberg, S., Scientific American, pp. 4–11 (Special edition, March 2003).

8.7 A particle X decays as rest weakly as follows:

$$X \rightarrow \pi^0 + \mu^+ + \nu_\mu$$

Determine the following properties of X

(i) Charge (ii) Baryon number
(iii) Lepton number (iv) Isospin
(v) Strangeness (vi) Spin
(vii) Boson or fermion (viii) Lower limit on its mass
(ix) Identity of X

8.8 Which of the following reaction are allowed and forbidden under the conservation of strangeness, conservation of baryon number and conservation of charge.

(i) $\pi^+ + n \rightarrow \Lambda^0 + K^+$
(ii) $\pi^+ + n \rightarrow K^0 + K^+$
(iii) $\pi^+ + n \rightarrow \overline{K^0} + \Sigma^+$
(iv) $\pi^+ + n \rightarrow \pi^- + p$
(v) $\pi^- + p \rightarrow \Lambda^0 + K^0$
(vi) $\pi^- + p \rightarrow \Lambda^0 + \pi^0$

8.9 Comment on the feasibility of the following reactions:

(i) $p + \bar{p} \rightarrow \pi^+ + \pi^-$
(ii) $p \rightarrow e^+ + \gamma$
(iii) $\Sigma^0 \rightarrow \Lambda + \gamma$
(iv) $p + p \rightarrow \Sigma^- + n + K^0 + \pi^+$
(v) $\Xi^- \rightarrow \Lambda + \pi^-$
(vi) $\Delta^+ \rightarrow p + \pi^0$

8.10 Consider the reaction:

$$K^- + p \rightarrow \Omega^- + K^+ + K^0$$

followed by the sequence of decays

$$\Omega^- \rightarrow \Xi^0 + \pi^- \qquad K^+ \rightarrow \pi^+ + \pi^0 \qquad K^0 \rightarrow \pi^+ + \pi^- + \pi^0$$

$$\downarrow \qquad\qquad \downarrow$$

$$\Xi^0 \rightarrow \pi^0 + \Lambda \qquad \pi^+ \rightarrow \mu^+ + \nu_\mu$$

$$\downarrow$$

$$\pi^0 \rightarrow \gamma + \gamma$$

Classify each process as strong, weak or electromagnetic and give your reasons.

BIBLIOGRAPHY

Cheng, D.C. and O'Neill, G.K., *Elementary Particle Physics*, Addison-Wesley, Reading (MA), 1979.

Dodd, J.E., *The Ideas of Particle Physics*, Cambridge University Press, Cambridge, 1984.

Gell-Mann, M., *What are the Building Blocks of Matter—The Nature of the Physical Universe*, Wiley, New York, 1979.

Griffiths, D., *Introduction to Elementary Particles*, Wiley, New York, 2008.

Hughes, J.S., *Elementary Physics*, Cambridge University Press, Cambridge, 1991.

Kamal, A.A., 1000 *Solved Problems in Modern Physics*, Springer, New York, 2010.

Khanna, M.P., *Introduction to Particle Physics*, Prentice-Hall, New Delhi, 1999.

Martin, B.R., *Nuclear and Particle Physics,* Wiley, New York, 2006.

Nishijima, K., *Fundamental Particles*, W.A. Benjamin, New York, 1963.

Okun, L.B., *Leptons and Quarks*, North Holland, Amsterdam, 1982.

Perkins, D.H., *Introduction to High Energy Physics*, Addison–Wesley, Reading (MA), 1982.

Scheck, F., *Leptons, Hadrons and Nuclei*, North Holland, Amsterdam, 1983.

Watkins, P., *Story of the W and Z*, Cambridge University Press, Cambridge, 1986.

9 Applications of Nuclear Physics

"No energy is costlier than no energy."
—Homi Jehangir Bhabha

Nuclear physics finds numerous applications in the fields ranging from medicine and environmental science to energy and materials. Emerging applications of nuclear technology show great potential for addressing many of our future needs. We have already touched upon a few applications of radioisotopes in Chapter 3, very briefly. It would be impossible to discuss all the possible applications of nuclear physics in a single chapter, for which a full-blown encyclopedia would be needed. But we shall strive to cover some of the important applications, which should be able to illustrate, how, in diversity of ways nuclear physics and technology have come to affect our lives and society.

9.1 ANALYTICAL APPLICATIONS

Nuclear and isotopic techniques play an essential role in many facets of our daily life. They form an integral part of our socio-economic development and enhance our understanding of nature and its processes. Nuclear analytical techniques (NATs) are based on using a radioisotope as tracer or utilizing the radioactivity produced in the sample by activation or measuring the radiations in the sample. The activating source may be a neutron or a charged particle (e.g., α, p) or a photon (e.g. γ-ray), and the radioactive products decay via radiation like α, β, γ and X-ray or delayed neutron. Using NATs, elemental concentrations are determined based on the measurement of the isotopes, in contrast to the non-nuclear techniques that utilize the properties of the atom as a whole. Most of the NATs are contamination free and often non-destructive analysis is feasible.

We will discuss some of the nuclear techniques in the next sections.

9.1.1 Neutron Activation Analysis (NAA)

Neutron activation analysis is one of the most used NATs. It is a sensitive analytical method useful for performing both quantitative and qualitative multi-elemental analysis of major (concentration exceeding 1% by mass), minor (concentration in the range of 0.1% to 1% by mass) and trace (concentration of less than 0.1% by mass) elements in samples from almost every conceivable field of scientific and technical interest. The NAA technique is based on irradiation of a sample with neutrons, usually from a nuclear reactor, and the subsequent measurement of the induced radioactivity (β, γ) and the half-life of the resulting sample to determine the elemental constituents of the sample in addition to their relative concentrations. Neutron, being a non-charged particle, interacts with the nuclei of all isotopes forming a radioisotope (activating product) in most of the cases. The sequence of events occurring during the most common nuclear reaction used for NAA, namely neutron capture and (n, γ) reaction is illustrated in Figure 9.1.

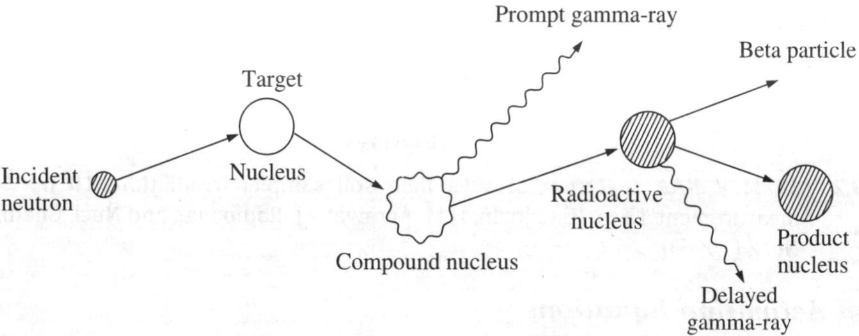

FIGURE 9.1 Interaction of neutron with target nucleus followed by emission of gamma-rays.

NAA is carried out by the measurement of the following:

(i) Prompt γ-rays known as prompt gamma-ray NAA (PGNAA), and
(ii) Delayed γ-rays known as conventional gamma-ray NAA or simply NAA.

In case of PGNAA, the measurements take place during the irradiation (on-line measurement), while in the case of conventional gamma-ray NAA, the measurements follow the radioactive decay (off-line measurement). PGNAA is generally carried out using guided neutron beam from the reactors. It is most applicable to elements (B, Cd, Sm, Gd, etc.) with extremely high neutron capture cross-sections. Conventional NAA can be used for the vast majority of elements producing radioactive nuclides.

A typical NAA spectrum is shown in Figure 9.2.

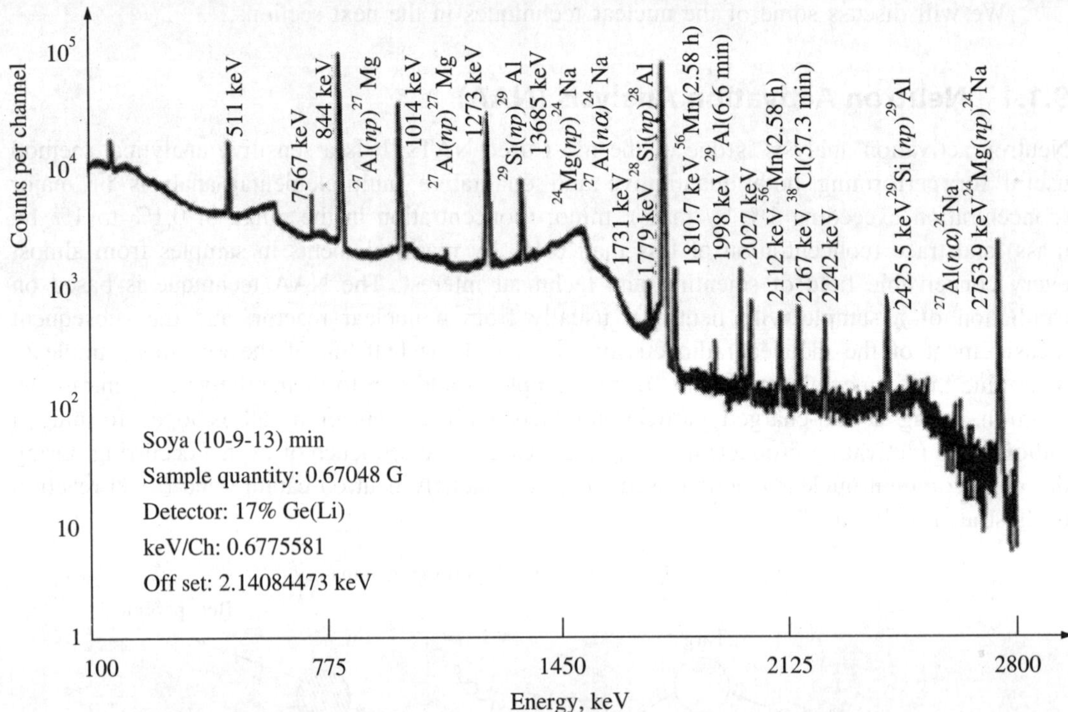

FIGURE 9.2 14 MeV NAA spectrum of soya field soil sample; irradiation: 10 m, delay 9 m, measurement 13 m. [G.C. Joshi, H.M. Agrawal – J. Radioanal. and Nucl. Chem., Vol. 198, No. 2 (1995)]

Neutron Activation Equations

The sample is irradiated for a period of time t. At the end of the irradiation or the termination of activation, the activity of the sample is given as follows:

$$A_0 = N_t \sigma \varphi S \tag{9.1}$$

where

A_0 is the number of disintegrations per second of the element in the sample when the irradiation stops,
σ is the cross-section for the (n, γ) reaction,
φ is the neutron flux,
S is the saturation factor given by $(1 - e^{-\lambda t})$,
N_t is the number of target atoms which is given by

$$N_t = \frac{N_A \alpha m}{w} \tag{9.2}$$

where

m is the mass of the target element,
N_A is the Avogadro's number,
w is the average atomic mass of the element,
α is the fraction of the target isotope in the sample.

The activity of the sample after a cooling period of T is given by:

$$A = N_t \sigma \varphi SD \tag{9.3}$$

where D is the decay factor, $e^{-\lambda T}$.

EXERCISE 9.1:

(a) A 1 mg target of natural molybdenum is irradiated for sixty-six hours in a thermal neutron flux of 10^{14} neutrons/cm^{2}s. At the end of irradiation, if the activity of $^{99}_{42}$Mo ($T_{1/2}$ = 65.94 h) formed in the sample is 9.6 × 10^6 dps, calculate the cross-section for ^{98}Mo(n, γ)^{99}Mo reaction. [Abundance of ^{98}Mo is 24.1%].

(b) Calculate the saturation activity of the sample.

(c) Calculate the activity of the sample after 24 hours of cooling.

Solution:

(a) From equation (9.1), we have

$$\sigma = \frac{A_0}{N_t \varphi (1 - e^{-\lambda t})} = \frac{A_0}{\left(\frac{N_A \alpha m}{w}\right) \varphi (1 - e^{-\lambda t})}$$

$$= \frac{9.6 \times 10^6 \text{ dps}}{\left[\frac{(6.023 \times 10^{23} \text{ atoms/g})(0.241)(10^{-3} \text{g})}{96}\right] (10^{14} \text{ cm}^{-2} \text{s}^{-1}) \left[1 - \exp\left(-\frac{0.693}{65.94 \text{ h}} \times 66 \text{ h}\right)\right]}$$

$= 1.272 \times 10^{-25}$ cm^2 = 0.127 b

(b) The saturation activity is defined at $t \to \infty$,

$$A_{\text{sat}} = N_t \sigma \varphi = \left(\frac{N_A \alpha m}{w}\right) \sigma \varphi$$

$= (1.51 \times 10^{18})(1.272 \times 10^{-25}$ cm^2) (10^{14} neutrons/cm^2 s) $= 1.92 \times 10^7$ dps

(c) From equation (9.3), we get

$$A = A_0 D$$

$$= (9.6 \times 10^6 \text{ dps}) \exp\left(-\frac{0.693}{65.94 \text{ h}} \times 24 \text{ h}\right) = 7.46 \times 10^6 \text{ dps}$$

Standardization Methods of NAA

Neutron activation analysis can be carried out by the following methods:

Absolute Method: The sample is irradiated in a neutron flux (φ) for a period of time and the radioactivity of the activation products is measured by γ-ray spectroscopy. The mass of the element can be calculated from the measured activity using the nuclear and reactor-based data. However, this method is not practiced, because it is difficult to evaluate the absolute values of φ and σ, since both the parameters vary with the neutron energy.

Relative Method: In this method, a chemical standard with a known mass of the element is co-irradiated with the sample of a known mass, and both are measured in identical conditions with respect to the detector, so that the absolute detection efficiency of the detector need not be determined. The ratio of the activities of an isotope in the sample and the standard is related to the concentration of that isotope, hence using the mass of the element in the standard (w_{std}) and activities of the standard (A_{std}) and the sample (A_{sample}), the mass of the element in the sample (w_{sample}) can be determined for the same counting time by the following relation:

$$w_{sample} = \left(\frac{A_{sample}}{A_{std}}\right) w_{std} \tag{9.4}$$

While, the relative method is accurate and simple, prior knowledge of the elements present in the sample is necessary to prepare the multi-elemental standard.

Single Comparator or k_0-NAA Method: It involves the simultaneous irradiation of a sample and a neutron flux monitor, e.g., gold, and utilizing the experimentally determined composite nuclear constant k_0. The preparation of the standards for each element and a prior knowledge of the constituents of the sample are not necessary in the single comparator method. The factor k_0 is defined as follows:

$$k_0 = \frac{M^* \theta \sigma_0 \gamma}{M \theta^* \sigma_0^* \gamma^*} \tag{9.5}$$

where M, θ, σ and γ are the atomic weight, isotopic abundance, thermal neutron capture cross-section and γ-ray abundance, respectively. The symbol '*' refers to the corresponding parameters of the comparator.

EXERCISE 9.2: A 10 mg sample of dry leaves and 10 μg of Mn standard were irradiated in a reactor with a neutron flux of 10^{12} n/cm^2 s for two hours. Sample and the standard were counted after cooling periods of 1 h and 1.5 h, respectively. Calculate the concentration of Mn in the tea leaves, if the activities of the sample and the standard are 1500 and 2000 counts per minute (cpm), respectively. Given: Half-life of the radioisotope ^{56}Mn, which is formed in the neutron activation is 2.5785 h.

Solution: Corrected activity of Mn in the sample at the time when the irradiation stops, i.e., before the cooling period of 1 h is

$$A_{sample} = (1500 \text{ cpm})\, e^{\left(\frac{0.693}{2.5785 \text{ h}}\right)(1 \text{ h})} = 1962 \text{ cpm}$$

Corrected activity of Mn in the standard at the time when the irradiation stops, i.e., before the cooling period of 1.5 h when it was measured is

$$A_{std} = (2000 \text{ cpm})\, e^{\left(\frac{0.693}{2.5785 \text{ h}}\right)(1.5 \text{ h})} = 2992 \text{ cpm}$$

From equation (9.4), the amount of Mn in tea leaves

$$w_{\text{sample}} = \left(\frac{A_{\text{sample}}}{A_{\text{std}}}\right) w_{\text{std}} = \left(\frac{1962}{2992}\right)(10\ \mu\text{g}) = 6.6\ \mu\text{g}$$

Therefore, the concentration of Mn in the tea leaves

$$= \frac{6.6\ \mu\text{g}}{10\ \text{mg}} = 6.6 \times 10^{6}\ \%$$

Applications of NAA

NAA finds extensive applications in many fields of science and technology for macro, micro and trace element analysis in the samples, like archaeology, biomedicine, animal and human tissues, environmental science, forensics, geology and geochemistry, industrial products, and nutrition. NAA is also used in quality assurance of analysis and reference materials certification.

9.1.2 Rutherford Backscattering Spectrometry (RBS)

One of the earliest experiments in nuclear physics was Rutherford's demonstration of large angle scattering of α-particles by gold nuclei. This experiment established the existence of a small nucleus within the atom. Based on Rutherford's experiment, Rutherford backscattering technique, which is a widely used multi-elemental ion beam analytical technique, involves the measurement of the energy and the yield of the ions that are backscattered (scattered at large angles) from the atoms in thin films and the near surface region of materials, using an energy sensitive detector, typically a solid state detector (Figure 9.3).

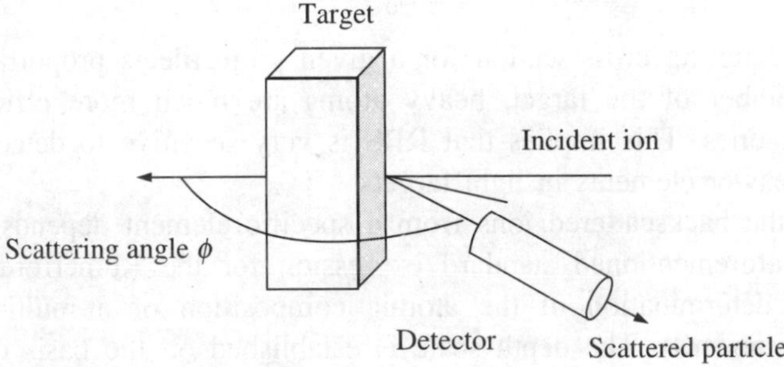

FIGURE 9.3 An example of ion beam interaction with a target showing Rutherford backscattering.

The process of backscattering is elastic in nature and is governed by Coulomb repulsion. The ratio of energy E of the backscattered ion from a target atom to the incident beam energy E_0 is known as kinematic factor K and provides identification of the elements constituting the target on the basis of their masses; its value is higher for heavier element. From conservation of non-relativistic energy and momentum, the kinematic factor when $M > m$, is given by

$$K = \left[\frac{(M^2 - m^2 \sin^2 \varphi)^{1/2} + m \cos \varphi}{M + m}\right]^2 \tag{9.6}$$

where m and M are the atomic masses of the incident ion and the target nucleus, respectively and φ is the scattering angle. For direct backscattering at 180°, the kinematic factor has its lowest value given by

$$K = \left[\frac{M - m}{M + m}\right]^2 \tag{9.7}$$

For a Rutherford scattering experiment, the most convenient way to express the results is in terms of the cross-section referred to as the differential cross-section, which is a measure of the probability per unit solid angle that the detector makes with respect to the target, for the scattering at a given angle φ. The theoretical non-relativistic expression for the Rutherford elastic scattering cross-section is given by the following formula:

$$\frac{d\sigma}{d\Omega} = 1.296\left(\frac{zZ}{E_0}\right)^2\left[\text{cosec}^4\left(\frac{\varphi}{2}\right) - 2\left(\frac{m}{N}\right)^2\right] \tag{9.8}$$

where the unit for cross-section is $\frac{mb}{sr}$, (z, m) and (Z, M) are the atomic numbers and the masses of the incident ion and the target nucleus, respectively. E_0 is the energy of the projectile in MeV. As $M >> m$, the term $2\left(\frac{m}{M}\right)^2$ is always quite small compared to $\text{cosec}^4\left(\frac{\varphi}{2}\right)$ and hence can be ignored without much error. To a good approximation, the differential cross-section is then given as follows:

$$\frac{d\sigma}{d\Omega} = 1.296\left(\frac{zZ}{E_0}\right)^2 \text{cosec}^4\left(\frac{\varphi}{2}\right) \tag{9.9}$$

Since, Rutherford scattering cross-section for a given projectile is proportional to Z^2, where Z is the atomic number of the target, heavy atoms are much more efficient scatterers as compared to lighter ones. This implies that RBS is very sensitive to detect extremely small concentrations of heavier elements in light targets.

The yield of the backscattered ions from a specific element depends on the scattering cross-section. The aforementioned standard expression for the Rutherford scattering cross-section allows the determination of the atomic composition of a multi-component target without using any standards. The depth scale is established on the basis of the energy loss of the incident ion beam and that of the backscattered ions in the material. This enables the determination of the thickness of the films and the surface profile of the elements. In a target which contains more than one type of atoms that differ in mass such that the change in the energy of the scattered particles is measurable, the masses of the different types of atoms present in the target can be determined. RBS is usually performed with 2–2.5 MeV α-particles. Heavy ions such as ^{7}Li and ^{12}C in the energy range of 6–12 MeV are also being used increasingly.

A typical RBS spectrum is shown in Figure 9.4.

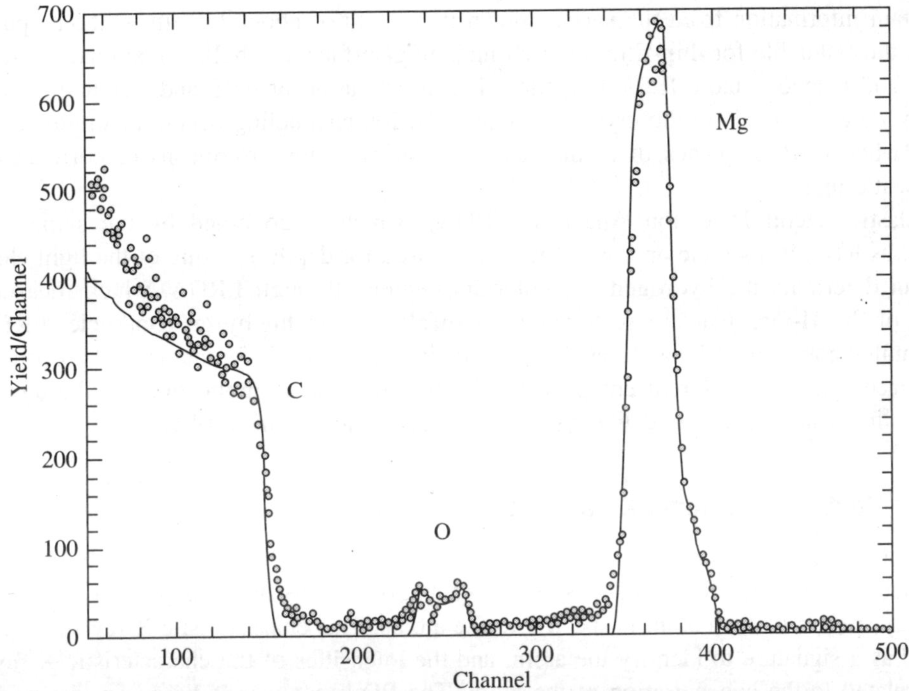

FIGURE 9.4 Typical RBS spectrum of the natural magnesium target. [B. Limata et al. 'New experimental study of low-energy (p, gamma) resonances in magnesium isotopes,' arXiv:1006.5281 [nucl-ex] (2010)].

EXERCISE 9.3: If the probability of α-particles of energy 10 MeV to be scattered through an angle greater than φ on passing through a thin foil is 10^{-3}, what is it for 5 MeV protons passing through the same foil?

Solution: From equation (9.9), we have

$$\frac{d\sigma}{d\Omega} \propto \left(\frac{z}{E_0}\right)^2$$

Hence, the probability for the scattering of 5 MeV α-particles

$$= \frac{10^{-3}}{\left(\dfrac{2}{10}\right)^2}\left(\frac{1}{5}\right)^2 = 10^{-3}$$

Applications of RBS

Backscattering spectroscopy has many applications in different disciplines. High resolution RBS, with typical depth resolution in the range of several Å, allows the analysis of ultrathin films and multi-layered structures. In thin film samples, the width of the energy spectrum of the elastically scattered incident particles gives a measure of the thickness of the film. RBS can obtain information from the surface down to a depth of about 1–2 μm without sputtering. RBS is thus, suitable for thin-film research including surface and bulk contamination, interface mixing and reaction, and diffusion profiles. The combination of RBS and channeling has been used for defect analysis in the crystalline materials. Ion channeling occurs in single crystalline targets when rows or planes of atoms can steer energetic ions by means of correlated small angle scatterings.

Elastic Recoil Detection Analysis (ERDA), which is governed by the same physical concept as RBS, is a simple and useful technique used for depth profiling of the light elements. One can determine the hydrogen or deuterium content, through ERDA, which measures the energy of the H-ions that are scattered forward after being hit by an α-particle. ERDA has also simultaneous multi-element detection capability. Atoms of different elements recoiled from the surface appear at different energies. ERDA has also found applications in the analysis of porous silicon layers, diamond like carbon films and silicon nitride films.

9.1.3 Particle Induced X-ray Emission (PIXE)

Atoms of different elements emit characteristic X-rays, when they are excited by creating the vacancies in their inner electron shells. Since, each atom has its own structure of excited levels, the X-ray spectra of each atom is specific to the atoms. This characteristic X-ray spectrum may be used as a signature to identify the atom, and the intensities of the characteristic X-rays may be correlated to the concentration of the atom. The PIXE analysis is based on this principle.

In PIXE analysis, the vacancies in the electronic shells are created by hitting the target atom with an energetic beam of protons. A Coulomb interaction takes place between the proton and the inner shell electron resulting in the removal of *K* or *L* shell electron. The vacancy thus created, is filled by an electron from a higher orbit. As a result, characteristic X-rays are emitted as schematically illustrated in Figure 9.5. The X-ray emission is subject to the following quantum mechanical rules:

$$\Delta l = 1; \quad \Delta j = 0, 1; \quad \Delta n \neq 0$$

where n is the principal quantum number, l is the orbital angular momentum and j is the total angular momentum.

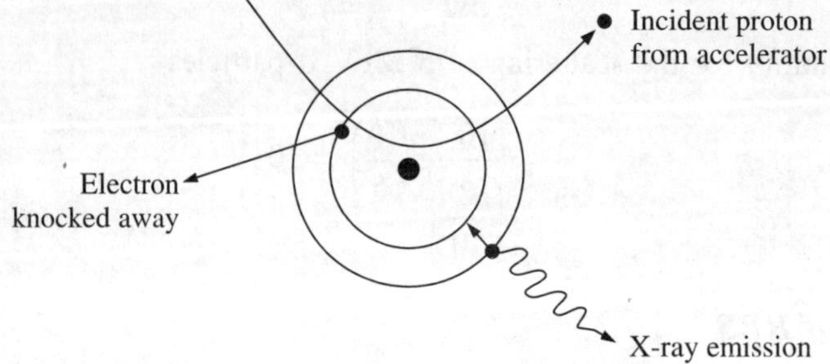

FIGURE 9.5 Schematic of inner shell vacancy creation and subsequent X-ray emission due to proton bombardment.

The shells are designated as K, L, M, N, O for n = 1, 2, 3, 4, 5, respectively. The total angular momentum j is the vector sum of the orbital angular momentum l and the spin s of the electron. Figure 9.6 shows a typical level diagram of X-ray transitions that are permitted by the quantum mechanical rules. The transitions coming to the K-shell are known as the K X-rays and that to L-shell as the L X-rays. The X-rays originating from the electronic transition from L shell and M shell to the K shell are usually referred to as K_α and K_β, respectively.

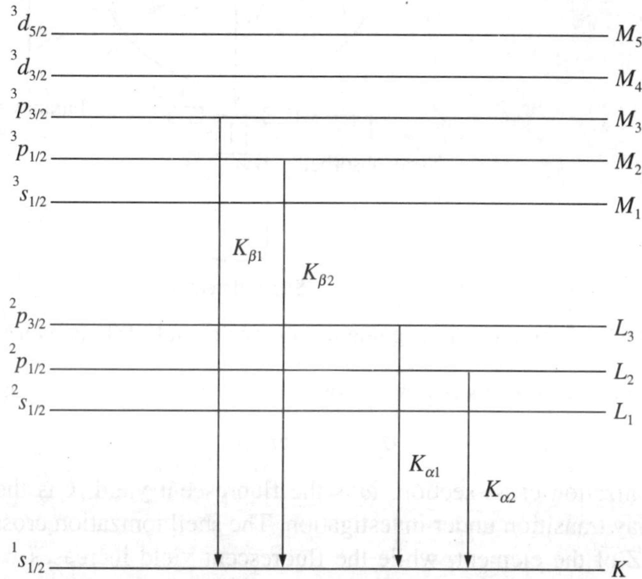

FIGURE 9.6 A typical energy level diagram of allowed X-ray transitions.

The continuous X-ray photons are also emitted in the process. The continuous X-rays, as the name suggests, have a continuous energy spectrum, which depends on the energy of the incident photon beam. An excited atom may also de-excite by other processes, such as the Auger effect and the Coster–Kronig transitions (transitions between sub-shells having same n values), in addition to the radioactive transitions leading to X-ray emission.

A PIXE setup consists of an electrostatic accelerator like the Van de Graff or a low energy Pelletron accelerator for proton acceleration, an analyzing and switching magnet, quadrupole magnet for focusing, a set of collimators, a target holder inside the scattering chamber and X-ray detecting and subsequent pulse analyzing system involving a Si(Li) detector, preamplifier, amplifier, multichannel analyzer and a computer program for analyzing the data. A well-analyzed and focused beam of protons extracted from the accelerator is injected into the beam line through an analyzing magnet and then directed horizontally over a distance of several metres to the scattering chamber. Finally, the beam falls on the target after passing through a series of collimators in the scattering chamber.

The beam charge measurements are usually carried out by charge integration from a Faraday cup, which is connected to the current integrator that provides information about the number of particles that passed through the target during a certain time interval. The characteristic X-rays produced from the targets are collected using a Si(Li) detector. A typical PIXE setup is shown in Figure 9.7.

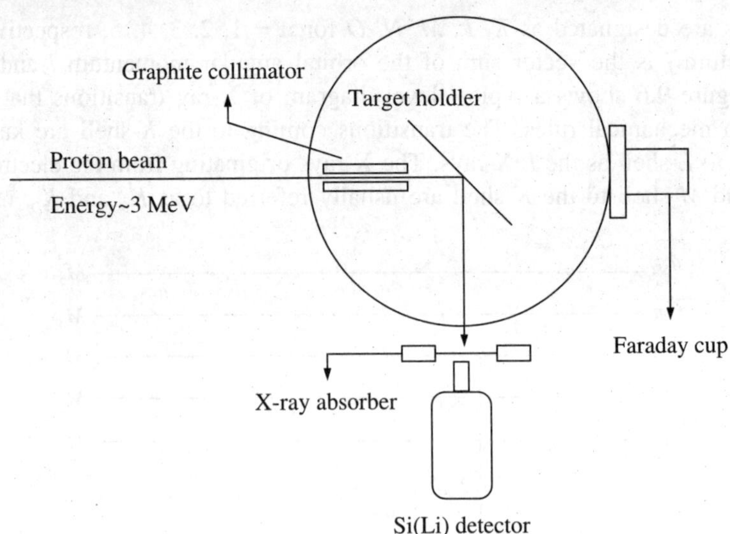

FIGURE 9.7 Schematic arrangement for carrying out PIXE measurements.

The X-ray production cross-section σ_p for a line in the X-ray spectrum is given by:

$$\sigma_p = \sigma_{\text{ion}}\,\omega k \tag{9.10}$$

where σ_{ion} is the ionization cross-section, ω is the fluorescent yield, k is the relative intensity of the particular X-ray transition under investigation. The shell ionization cross-section increases with a decrease in Z of the element, while the fluorescent yield increases with the increase in Z. The inner shell ionization also increases with incident beam energy.

If the X-ray production cross-section is known, then the rate of the production of X-rays can be calculated by the following relation:

$$R_p = I\sigma_p N dx \tag{9.11}$$

where I is the ion-beam intensity in particles per second and N is the density of the target atoms in the sample of thickness dx.

EXERCISE 9.4: In a PIXE measurement, a 0.2 mg cm^{-2} film containing 5 ppm by weight of an element of mass number 100 is bombarded with a 200 μA beam of protons for 10 m. The cross-section for exciting the L-shell of the element is 800 b and the probability of the excited atom emitting an L X-ray is 50%. Calculate the number of counts recorded if the overall detection efficiency is 0.5%.

Solution: The number N_p of X-rays counted in a time interval is equal to the X-ray production rate multiplied by the exposure time t and the overall efficiency ε for detecting X-rays. Therefore, from equation (9.11), we have

$$N_p = I\sigma_p(Ndx)\varepsilon t$$

If m is the mass of material in the sample per unit area, then

$$Ndx = \frac{mN_A}{A}$$

where N_A is the Avogadro number and A is the mass number of the target atoms. Therefore,

$$N_p = I\sigma_p\left(\frac{mN_A}{A}\right)\varepsilon t$$

$$= \left(\frac{200\times10^{-6}\,\text{A}}{1.6\times10^{-19}\,\text{C}}\right)(400\times10^{-24}\,\text{cm}^2)\left(0.2\times10^{-3}\times5\times10^{-6}\,\text{g cm}^{-2}\,\frac{6.023\times10^{-23}}{100}\right)(0.5\%)(600\,\text{s})$$

$$\therefore\ N_p = 9021$$

Applications of PIXE

In general, PIXE can be characterized as a quantitative, highly sensitive, accurate and non-destructive method for the multi-elemental analysis of materials. This technique allows the determination of most of the elements down to the ppm levels with an accuracy of about 10%. The PIXE technique offers the advantage of analysis, without the necessity for time consuming digestion, thus minimizing the likelihood for error resulting from sample preparation. Today, PIXE is well-known for trace elemental analysis of various types of samples such as, biological tissues, plant materials, aerosols, fly ash, minerals, rocks, sand, water samples, paintings, plastic, polymer, ink, semiconductor materials, etc. Art objects and archaeological specimens, which are too large to be accommodated in the scattering chamber and samples that cannot withstand vacuum, are analyzed by In-air PIXE. PIXE is routinely used in quality assurance/quality control applications for industrial processes, such as evaluating precious metal depletion of spent catalysts, monitoring adsorption by activated carbons, and determination of the contamination source. A typical PIXE spectra of standard fly ash is shown in Figure 9.8. GUFIX [Nucl. Instr. Meth. B95 (1995)] is a popular PIXE spectral data analyzing software that models a theoretical spectrum of an element using the database of K, L and M X-ray energies, relative X-ray intensities, fluorescence and Coster–Kronig probabilities, which is then fitted with the experimental spectrum with respect to the known matrix.

Particle Induced γ-ray Emission (PIGE), which involves the measurement of prompt γ-rays emanating from a nuclear reaction induced by the incident ion beam using high resolution detectors [e.g., Ge(Li) or HP(Ge)] is a complementary technique to PIXE. This technique provides sensitive determination of several low Z elements such as B, Li, F, Na, Al, Mg and Si. Therefore, a combination of PIXE and PIGE would enable the determination of several elements across the periodic table.

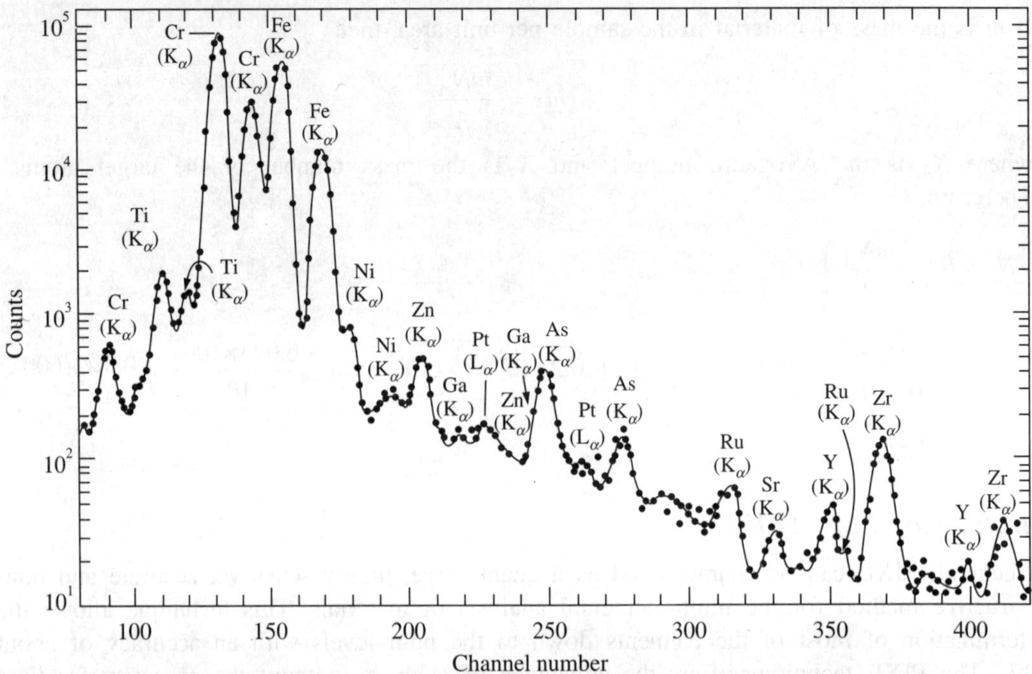

FIGURE 9.8 Typical PIXE spectrum of a geological sample. [H.M. Agrawal, IIT Kanpur – Report (1980)]

9.1.4 X-ray Fluorescence (XRF) Spectroscopy

In XRF spectrometry, high-energy primary X-ray photons emitted from a source (X-ray tube) are made to strike the sample. The photons from the X-ray tube have sufficient energy to knock the electrons out of the innermost, *K* or *L*, orbitals. When this occurs, an electron from an outer orbital, *L* or *M*, will move into the newly vacant space at the inner orbital, resulting in the emission of a secondary X-ray photon. This phenomenon is known as fluorescence. The secondary X-ray that is produced is characteristic of a specific element. The energy and the intensity of the X-rays are measured either in wavelength or the energy mode. The first mode is called the wavelength dispersive XRF mode (WDXRF), while the second one is known as the energy dispersive XRF mode (EDXRF).

All the ionizations do not result in the X-ray emission. If the energy of the X-ray is greater than the binding energy of an outer shell electron, then it is possible that the energy is spent in the ejection of the electron and this electron is known as Auger electron. The ratio of the number of emitted X-ray photons to the total number of ionizations is called the fluorescent yield ω and is related to the atomic number Z as follows:

$$\omega = \frac{Z^4}{(A_i + Z^4)} \tag{9.12}$$

where A_i is approximately 10^6 for the K shell and 10^8 for the L shell.

Mosley[1] established the basis for qualitative and quantitative X-ray spectrochemical analysis, which lies in the relationship between the atomic number Z of the element and the wavelength λ of the characteristic radiation emitted, i.e.,

$$\frac{1}{\lambda} = k(Z - s)^2 \qquad (9.13)$$

where k and s are constants that depend upon the spectral series of the emission line. Characteristic X-ray spectra are excited when a sample is irradiated with a sufficiently short wavelength X-ray.

An EDXRF system normally has three main components: an excitation source, a spectrometer/detector, and a data collection/processing unit. The incident X-ray beam is typically produced from a Rh target. W, Mo, Cr and others can also be used in place of Rh, depending on the application. Various types of detectors (proportional counter, scintillation detector, Si(Li) detector) are used to measure the emitted beam intensity. The proportional counter is commonly utilized for measuring the long wavelength (> 0.15 nm) X-rays that are typical of the K spectra from elements lighter than Zn. The scintillation detector is frequently utilized to analyze the shorter wavelengths in the X-ray spectrum (K spectra of element from Nb to I, L spectra of Th and U). Schematic diagram of an EDXRF setup is shown in Figure 9.9.

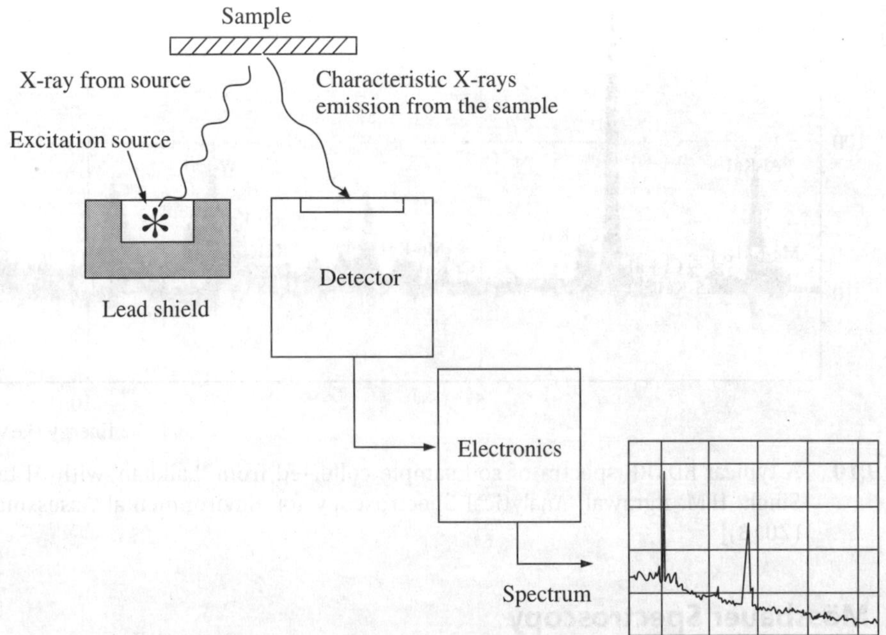

FIGURE 9.9 Schematic diagram of EDXRF setup.

1. Moseley, Henry G.J. (1913), 'The High Frequency Spectra of the Elements' Philosophical Magazine: 1024.

Applications of XRF

XRF spectrometry is widely applied in many industries and scientific fields. XRF is a non-destructive, versatile, economical, rapid technique for qualitative and quantitative analysis for chemical elements based on their spectral lines emitted, which can be used for a wide variety of samples from powders to liquids. The steel, cement, petroleum and nuclear industries regularly utilize XRF devices for material development tasks and quality control. Geologists have been using XRF for many years to determine the major and the trace components in one quick inspection without needing much sample preparation. Forensic scientists utilize XRF spectrometry to compare samples associated with the suspects (i.e., dirt or sand on the clothing or the shoes) to the samples from the crime scenes. In theory, XRF can detect X-ray emission from virtually every element, depending on the wavelength and the intensity of the incident X-rays. In practice, however, most commercially available instruments are quite limited in their ability to accurately measure the abundances of elements with $Z < 11$. XRF analyses also fail to distinguish variations among isotopes of an element.

A typical EDXRF spectra is shown in Figure 9.10.

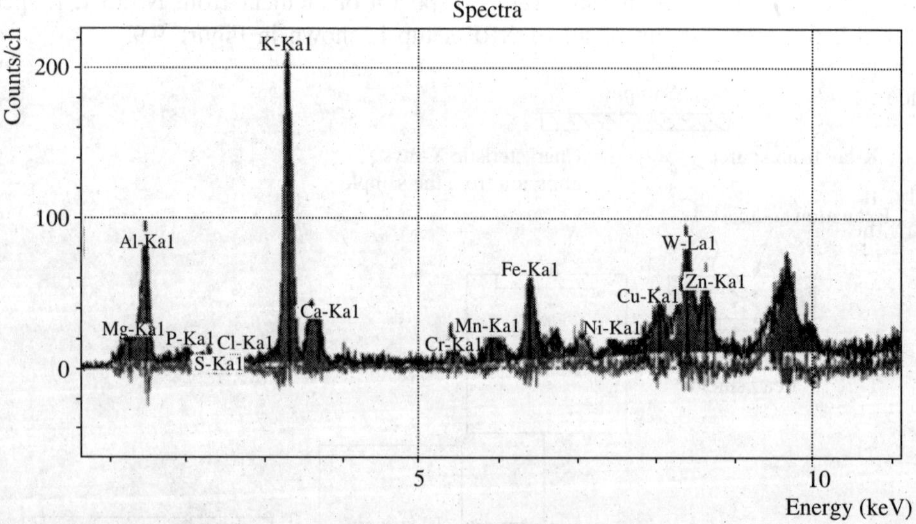

FIGURE 9.10 A typical EDXRF spectra of soil sample collected from 'Lalkuan' with Al target. [V. Singh, H.M. Agrawal, Analytical Spectroscopy for Environmental Assessment, LAP (2013)]

9.1.5 Mössbauer Spectroscopy

Mössbauer spectroscopy, also known by the name, *nuclear gamma resonance spectroscopy*, is a versatile method used to study nuclear structure with the absorption and the re-emission of γ-rays. It uses a combination of the Mössbauer effect and the Doppler shifts to probe the hyperfine transitions between the ground and the excited states of the nucleus. The Mössbauer effect is also known as *recoil-free nuclear resonance absorption*. The phenomenon was discovered in 1958 by the German physicist R.L. Mössbauer, who was awarded the Nobel Prize (Physics) for this work.

If a system absorbs a quantum of energy equal to the difference between the two of its energy states, the process is known as resonance absorption. As the energy states of the atomic nucleus are similarly quantized discrete values, it was expected that for γ-radiating radioactive source, resonant absorption will take place. However, such a resonance emission-absorption does not generally take place in the case of γ-radiation from nuclei, because of the relatively large momentum associated with the γ-photons. When a stationary nucleus of excitation energy E, emits a γ-photon in a transition to its ground state, the law of conservation of momentum dictates that the nucleus acquires an equal momentum, but in the opposite direction.

Thus, if m is the mass of the atomic nucleus, v is the velocity of the recoil nucleus resulting from the photon emission, and E_0 is the energy of the radiation emitted, the momentum conservation requires

$$\frac{E_0}{c} = mv \tag{9.14}$$

and, it follows that the nucleus acquires the recoil energy

$$R = \frac{1}{2}mv^2 = \frac{E_0^2}{2mc^2} \tag{9.15}$$

Hence, the energy of the photons of the radiation is R less as compared with the energy difference between the ground and excited states.

$$E_0 = E - R \tag{9.16}$$

Similarly, if a stationary nucleus is to be excited by an energy E due to the absorption of the radiation, the photon energy must be equal to $E_0 = E + R$ for the momentum to be conserved (Figure 9.11).

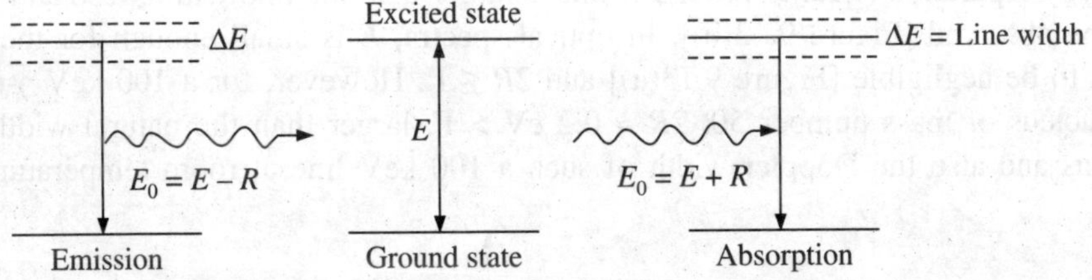

Figure 9.11 Effect of recoil on emission and absorption.

The line-width of the γ-radiation of energy E_0 can be calculated on the basis of the Heisenberg uncertainty principle. The natural width or FWHM of the energy distribution (Γ) depends on the lifetime (τ) of the excited state, i.e.,

$$\Gamma = \frac{\hbar}{\tau} \tag{9.17}$$

The energy distribution of the radiation is described by a Breit-Wigner Lorentzian type of function (Figure 9.12),

$$I(E) = \text{Constant}\left(\frac{\Gamma}{2\pi}\right)\frac{1}{(E - E_0)^2 + (\Gamma/2)^2} \tag{9.18}$$

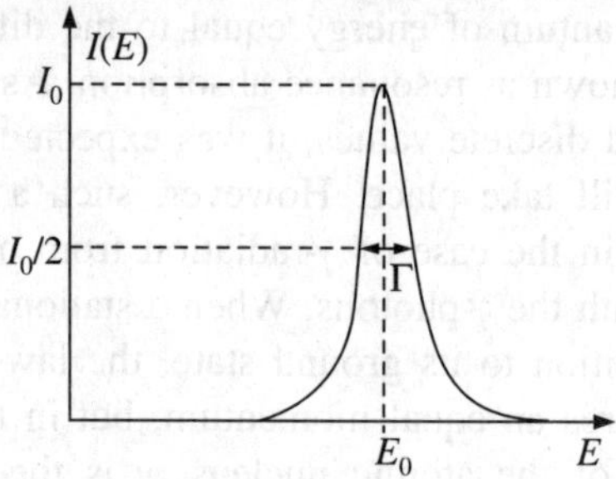

FIGURE 9.12 Energy distribution of the radiation.

Doppler Broadening

The measurable line-width is found to be considerably greater than the natural line-width defined by equation (9.17). Because of the thermal motion, the atoms of the absorbent and the source approach and move away from one another and the energy of the γ-quanta changes as a consequence of the Doppler effect. As a result, the emission and the absorption line-widths also increase accordingly. The higher temperature means higher kinetic energy and rate of motion of the atoms. Thus, the curve that describes the energy distribution is broader.

As we have already discussed, because of the recoil, the energy difference between the emitted and absorbed radiation is equal to

$$R + R = 2R = \frac{E_0^2}{mc^2} \tag{9.19}$$

Thus, the centres of the absorption and the emission lines are displaced relative to one another by an energy separation equal to $2R$ and if this is larger than the line-width, resonant absorption will not be observed [Figure 9.13(b)]. In optical spectra, E is small enough for the effects of the recoil to be negligible [Figure 9.13(a)] and $2R \leq \Gamma$. However, for a 100 keV γ-ray photon from a nucleus of mass number 50, $2R \approx 0.2$ eV > Γ, larger than the natural widths of most γ-radiations and also the Doppler width of such a 100 keV line at room temperature.

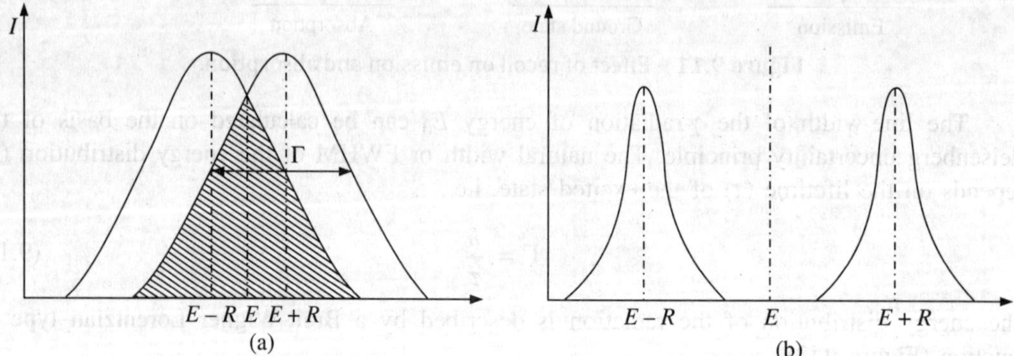

FIGURE 9.13 (a) Overlap of emission and absorption lines so that resonance absorption can occur like in case of optical transitions; (b) no overlap of the emission and absorption lines because of high recoil energy like in case of gamma-radiation.

Recoil-free Nuclear Resonance Absorption

In 1958, Mössbauer demonstrated that if the absorbing and the emitting nuclei are strongly bound in the crystalline solids, strong resonant emission and absorption of the γ-rays may take place, i.e., the energies of the source and the absorbent could be made to overlap. Thus, the recoil is suppressed by anchoring the emitter and the absorbent nuclei in large size crystal of mass >> mass of a nucleus. In this case, the momentum conservation condition is satisfied by the momentum of the crystal as a whole, so essentially no energy is lost. This fact is what makes Mössbauer spectroscopy possible, because it means that the γ-rays emitted by one nucleus can be resonantly absorbed by a sample containing nuclei of the same isotope, and this absorption can be measured.

In its most common form, a solid sample is exposed to a beam of γ-radiation and the transmitted beam intensity through the sample is measured by a detector. The atoms in the source emitting the γ-rays should be of the same isotope as the atoms in the sample absorbing them. Even after the recoil is suppressed by properly anchoring the source and the absorbent in large crystals, there will be a mismatch between the emission and the absorber frequencies. This is due to a difference in the chemical environments around the emitting and the absorbing nuclei. In order to observe the resonance absorption, the chemical shift has to be neutralized. This is done by externally induced Doppler motion by moving the source and the absorbent relative to each other. When either the source or the absorber is moving, the spectrum of the γ-radiation is Doppler-shifted by an energy ΔE, which is given by:

$$\Delta E = \left(\frac{v}{c}\right) E \tag{9.20}$$

where v is the relative velocity of the source and the absorber, c is the speed of light and E is the energy of the emitted γ-photon.

EXERCISE 9.5: There are several isotopes that exhibit Mössbauer characteristics. The most commonly studied isotope is ^{57}Fe.

(a) Compare the velocity of recoil for the free ^{57}Fe nucleus and a ^{57}Fe nucleus that is part of a rigid 100 mg crystal.

(b) Determine the Doppler shift in each case.

[Given: $^{57}_{26}\text{Fe}^*$ emits 14.4 keV γ-photon]

Solution:

(a) The momentum of the photon is

$$= \frac{E}{c} = \frac{14.4\times10^3\times1.6\times10^{-19}\ \text{J}}{3\times10^8\ \text{m/s}} = 7.685\times10^{-24}\ \text{kg m/s}$$

From conservation of momentum, the velocity of recoil of free ^{57}Fe of mass $57 \times 1.6 \times 10^{-27}$ kg

$$v = \frac{7.685\times10^{-24}\ \text{kg m/s}}{57\times1.6\times10^{-27}\ \text{kg}} = 84.267\ \text{m/s}$$

When the ^{57}Fe nucleus is tied to a 100 mg mass, from conservation of momentum, we have

$$v = \frac{7.685 \times 10^{-24} \text{ kg m/s}}{100 \times 10^{-6} \text{ kg}} = 7.7 \times 10^{-20} \text{ m/s}$$

As can be seen, the recoil is almost negligible in case the ^{57}Fe nucleus is tied to the crystal as compared to the case of free ^{57}Fe.

(b) From equation (9.20), the Doppler energy shift for the case of free ^{57}Fe

$$\Delta E = \left(\frac{v}{c}\right) E = \left(\frac{84.267 \text{ m/s}}{3 \times 10^8 \text{ m/s}}\right)(14.4 \times 10^3 \times 1.6 \times 10^{-19} \text{ J})$$

Therefore, the Doppler frequency

$$v = \frac{\Delta E}{h} = \frac{\left(\dfrac{84.267}{3 \times 10^8}\right)(14.4 \times 10^3 \times 1.6 \times 10^{-19})}{6.62 \times 10^{-34}} = 97.72 \times 10^{18} \text{ Hz}$$

Similarly, the Doppler frequency in the second case

$$v = \frac{\left(\dfrac{7.7 \times 10^{-20}}{3 \times 10^8}\right)(14.4 \times 10^3 \times 1.6 \times 10^{-19})}{6.62 \times 10^{-34}} = 8.93 \times 10^{-10} \text{ Hz}$$

Applications of Mössbauer Spectroscopy

There are typically three types of nuclear interactions that are observed in the Mössbauer spectrum, isomer shift (or chemical shift), quadrupole splitting and magnetic hyperfine splitting (or Zeeman splitting).

The isomer shift or chemical shift arises from the Coulomb interaction of the nuclear charge and the electron charge. The most effective part of this interaction is the result of the electron charge density at the nucleus. Quadrupole splitting is a reflection of the interaction between the nuclear energy levels and the surrounding electric field gradient. The charges distributed asymmetrically around the atomic nucleus give rise to an electric field gradient. Magnetic splitting is a result of any magnetic field surrounding the nucleus. The Zeeman splitting may be caused not just by an external magnetic field, but also by the internal magnetic fields (e.g., in the case of ferromagnetic and anti-ferromagnetic materials). The internal fields mainly originate from the atom's own electrons.

The three Mössbauer parameters—isomer shift, quadrupole splitting, and magnetic hyperfine splitting (as shown in Figure 9.14) are often used to identify a particular compound by comparison to spectra for standards. A typical Mössbauer spectrum for $NiGd_{0.04}Fe_{1.96}O_4$ nanoparticles is shown in Figure 9.15. Table 9.1 lists the Mössbauer Hyperfine Parameters for Gd-doped Nickel Ferrite.

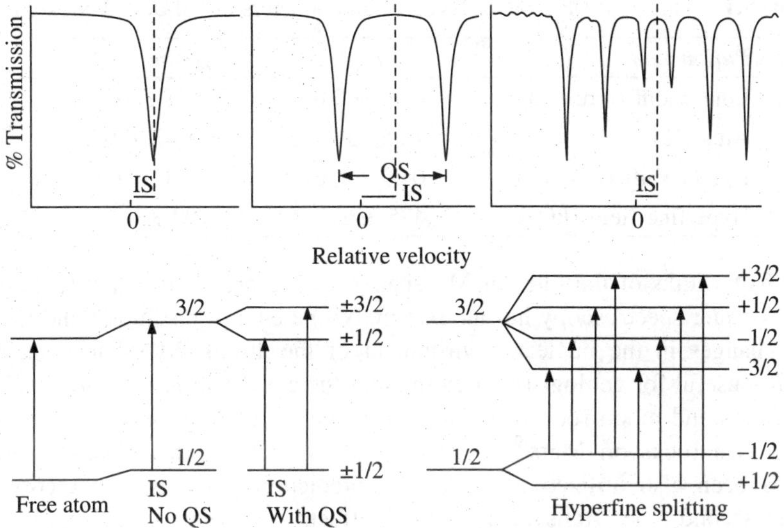

FIGURE 9.14 Isomer shift, quadrupole splitting and magnetic hyperfine splitting for the Mössbauer transition in ^{57}Fe.

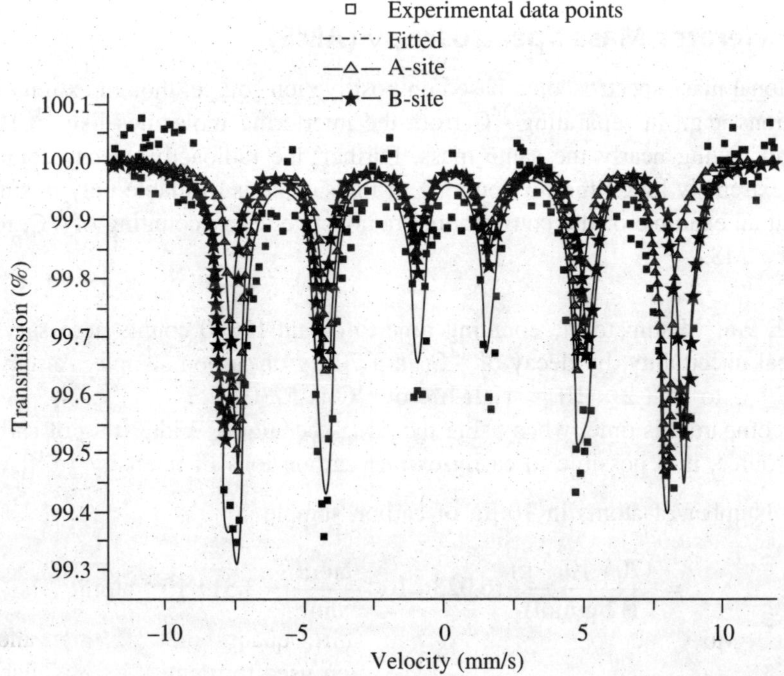

FIGURE 9.15 Mössbauer spectrum at room temperature for $NiGd_{0.04}Fe_{1.96}O_4$ nanoparticles. [J.P. Singh, G. Dixit, R.C. Srivastava, H.M. Agrawal and K. Asokan, 'Looking for the possibility of multiferroism in $NiGd_{0.04}Fe_{1.96}O_4$ nanoparticle system,' *J. Phys. D: Appl. Phys.*, 44, 435306 (6pp), 2011]

Table 9.1 Values of Mössbauer Hyperfine Parameters of Gd-doped Nickel Ferrite

Parameters	*A site*	*B site*
Line width (mm s^{-1})	0.46 ± 0.05	0.51 ± 0.04
Area (%)	48.57 ± 0.02	51.43 ± 0.02
Isomer shift (mm s^{-1})	0.29 ± 0.01	0.38 ± 0.01
Hyperfine field (kOe)	475 ± 1	509 ± 1

As the half-widths of the lines of Mössbauer spectra are fairly small (e.g., 4.9×10^{-9} eV for iron), Mössbauer spectroscopy has an extremely fine energy resolution and is able to detect even subtle changes in the nuclear environment of the atoms. Mössbauer spectroscopy has been especially useful for geologists for estimating the composition of iron-containing samples including meteors and moon rocks. *In situ* data collection of Mössbauer spectra has also been done on iron rich rocks on Mars.[2] Because of its very high energy resolution, Mössbauer technique has been also utilized to verify the prediction made by the theory of relativity regarding the second-order transverse Doppler effect.[3] The transverse Doppler effect is the small red-shift or blue-shift predicted by special relativity that happens when the emitter and the receiver are at the point of closest approach. Light emitted at this instant is red-shifted, while the light received is blue-shifted.

9.1.6 Accelerator Mass Spectroscopy (AMS)

The conventional mass spectroscopy, based on positive ion source, though extremely useful has some limitations, e.g., in separating ^{14}C from the interfering molecules like $^{12}CH_2$, ^{13}CH and isobar ^{14}N, all having nearly the same mass. Further, the radioactive decay counting method will take an extremely long time for counting, if the sample is available only in small amounts. Let us look at an example of the conventional radioactive decay counting of ^{14}C, to understand the power of AMS.

EXERCISE 9.6: Estimate the counting time to obtain 10000 counts necessary to achieve 1% statistical uncertainty, by decay of ^{14}C, in a 70 μg of carbon sample. Assume the usual ratio of $^{14}C/^{12}C$ to be 1.2×10^{-12}. Half-life of ^{14}C is 5730 y.

Now compare this time, when using the AMS technique. With 70 μg of carbon sample in the ion source, it is possible to easily extract carbon ions of intensity 1.6 μA.

Solution: Number of atoms in 70 μg of carbon sample

$$= \frac{(70 \times 10^{-6}\ \text{g})}{(12\ \text{g/mol})}(6.023 \times 10^{23}\ \frac{\text{atom}}{\text{mol}}) = 3.51 \times 10^{18}\ \text{atom}$$

2. Klingelhöfer, G. (2004), 'Mössbauer *in situ* studies of the surface of Mars'. Hyperfine Interactions 158 (1–4): 117–124.
3. Chen, Y.L.; Yang, D.P. (2007), 'Recoilless Fraction and Second-Order Doppler Effect'. Mössbauer Effect in Lattice Dynamics. John Wiley and Sons.

Therefore, the number of atoms of ^{14}C

$$(3.51 \times 10^{18} \text{ atom})(1.2 \times 10^{-12}) = 4.2 \times 10^6 \text{ atom}$$

For 10000 ^{14}C nuclei to decay, time taken

$$t = \frac{T_{1/2}}{0.693} \ln\left(\frac{N_0}{N}\right)$$

$$= \frac{5730 \text{ y}}{0.693} \ln\left(\frac{4.2 \times 10^6}{4.2 \times 10^6 - 10^4}\right) = 19.7 \text{ y}$$

Now, for AMS, a 1.6 μA current of carbon ions corresponds to a particle intensity:

$$\frac{1.6 \times 10^{-6} \text{ A}}{1.6 \times 10^{-19} \text{ C}} = 10^{13} \text{ s}^{-1}$$

Therefore, the particle intensity for ^{14}C

$$= 1.2 \times 10^{-12} \times 10^{13} \text{ s}^{-1} = 12 \text{ nuclei/s}$$

Operating the ion source at this intensity, time needed to count 10000 ^{14}C decays

$$= \frac{10000 \text{ nuclei}}{12 \text{ nuclei/s}} = 833 \text{ s} \approx 14 \text{ m}$$

The tremendous power of AMS can be judged by this example. In this technique, we do not wait for the atoms to decay to estimate the yield of the atoms of interest. In fact, the atoms, as they exist in the sample, are counted. This way, the significant enhancement in sensitivity at a fraction of the time taken for conventional decay counting, is achieved. As will be seen, in the AMS technique, the interference from the molecules and isobar is also completely eliminated.

Accelerator Mass Spectroscopy (AMS) has become a powerful technique for the analysis of stable and long-lived radioactive nuclei which are present in extremely small abundance. A schematic of an AMS setup is shown in Figure 9.16.

Negative ions from the ion source are accelerated first. While it is possible to produce negative ions of ^{14}C, it is not so for ^{14}N as $^{14}_{7}\text{N}^-$ ion is not stable. This feature is advantageous to get rid of interfering isobars in particular from the subsequent acceleration. The low-energy ions next enter an analyzing magnet, where they are deflected according to their radii of curvature. The radius of curvature of a particle of mass m, speed v, energy E and charge q in a uniform magnetic field B perpendicular to the plane of motion is given by the following equation:

$$r = \frac{mv}{Bq} = \frac{\sqrt{2mE}}{Bq} \tag{9.21}$$

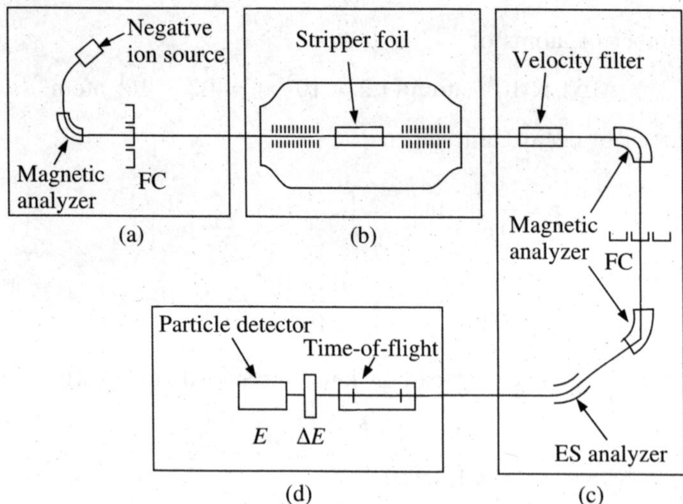

FIGURE 9.16 Schematic diagram of an AMS setup; (a) injector, (b) tandem accelerator, (c) positive ion analysis, (d) detector system. (FC = Faraday Cup, ES = Electrostatic).

Since, all the ions at this stage have the same energy and charge, this magnet is used to select ions of a given mass and direct them into the accelerator.

An offset Faraday cup is located after the injector magnet to monitor the source output. Then, the ions are made to pass through a stripper foil. The stripper is an integral part of AMS and is needed to remove unwanted molecules. In this process, the negative ions are converted to positive ions of different charge states. After passing through the stripper foil in the terminal, the ions of interest are free from interfering molecules. A combination of velocity filter, electrostatic analyzer and magnetic analyzers is needed to remove the scattered particles, molecular fragments and unwanted charged states. This step is essential to enhance the signal to noise ratio. The time of flight detector serves as an additional positive ion analyzer and is used to identify the mass. The $\Delta E - E$ counter telescope, which consists of a thin, passing detector ΔE, and a stopping detector E, is used for the identification and the quantification of the isotope or ion of interest. The ΔE signal is a measure of the stopping power, which is proportional to z^2, where z is the atomic number of the detected particle [see equation (7.36)]. Therefore, any particles, other than the ones of interest, will generate different signals in the passing detector and can be rejected. Starting from the ion source to the final detection, the interfering species are continuously reduced to obtain extremely low background and unique identification of the specific particle of interest. It is usual to calibrate an AMS system by comparing an unknown sample with one of the known isotope ratio.

Applications of AMS

AMS has become a powerful technique for the analysis of stable and long lived radioactive nuclei which are present in extremely small abundance and available in very small amount (mg-ng), respectively. AMS is widely used for the estimation of trace impurities in the semiconductors and for the measurement of low nuclear cross-sections. The small sample capability of AMS is fully exploited in the radiocarbon dating in archaeology, dating of ground water and exposure dating. As the AMS technique is very sensitive using which even small number of atoms in a large substrate can be detected, its use has been steadily growing in the biological and the environmental tracing programs. Study of human metabolism by using ^{14}C-labelled compounds is the most widely used application of biomedical AMS. In a similar manner, the bone loss can be monitored using ^{41}Ca as a tracer.

One of the most famous experiments carried out by AMS was that of the determination of the age of the Shroud of Turin. The Shroud of Turin, was popularly believed to be the cloth that was used to wrap Christ's body after crucifixion. To test its authenticity, its age had to be measured. This measurement was not a practical possibility at a time when radiocarbon dating required large samples to provide an accurate age. AMS provided the means of enabling radiocarbon dating with minimum amount of destructive testing. AMS measurements of radiocarbon have indicated that the linen of the Shroud of Turin is medieval (1260–1390 AD).

9.2 APPLICATIONS IN HEALTHCARE

Nuclear radiations find numerous applications in the field of healthcare by aiding diagnosis and therapy. In this section, we will touch upon very briefly, positron emission tomography (PET) and magnetic resonance imaging (MRI) used in the diagnostic process and gamma knife used in radiotherapy.

9.2.1 Positron Emission Tomography

In positron emission tomography, a compound labelled with a positron-emitting radioisotope is introduced into the patient, which accumulates at the target sites over time. The emitted positron traverses a short distance until it collides with an electron, annihilating into two γ-rays of 511 keV, close to 180° relative to each other. The coincident detection of the two γ-rays that are emitted 180° apart from each other can be recorded and used to reconstruct an image showing the location, which lies on the line of response connecting those two points, and the amount of the radionuclide in the body of the patient. Photons that do not arrive in pairs (i.e., within a few nanoseconds) are rejected. A ring of detectors, as shown in Figure 9.17, defines a slice through the patient and records coincidences from pairs of γ-rays along all possible directions within the plane. After reconstruction, the data obtained provides information on the 3-D tracer concentration within the body.

PET is both a medical and a research tool. PET scanning with the tracer 2-[^{18}F]-fluoro-2-deoxyglucose (FDG), called FDG-PET, is widely used in clinical oncology to detect and grade primary tumors before surgery and to evaluate their response to the surgery. FDG is a glucose analog that is selectively taken up by the glucose-using tissue, such as the malignant tissue, where the metabolic rate is abnormally high. PET is also used in cardiology and neurology.

More on PET can be found in the review article—Farwell, M.D., Pryma, D.A. and Mankoff, D.A., 'PET/CT imaging in cancer: Current applications and future directions,' Cancer, 120: 3433–3445, 2014, doi: 10.1002/cncr.28860.

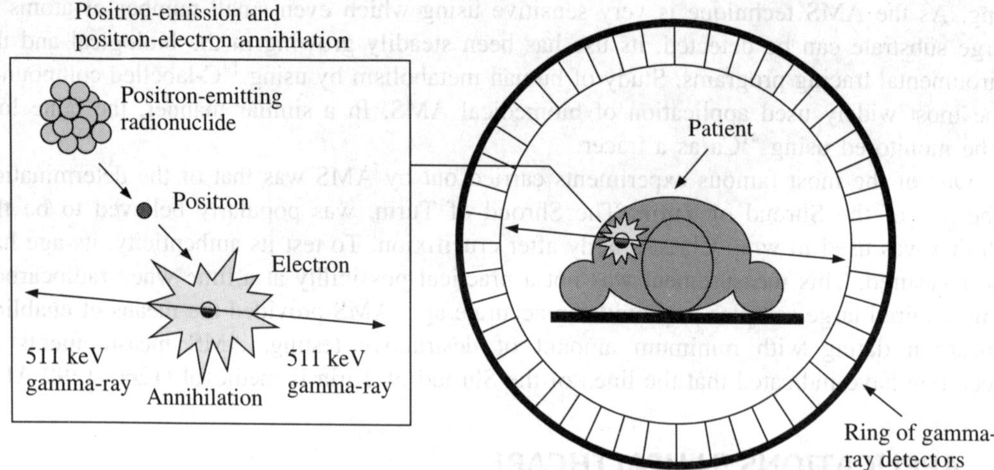

FIGURE 9.17 Schematic arrangement for performing PET. The positron-emitting radionuclide collected at the target sites inside the patient emit positrons which within a very short duration annihilate with an electron and pair of 511 keV gamma-rays are emitted in opposite directions, which are detected by the ring of gamma-ray detectors. [van der Veldt AAM, Smit EF and Lammertsma AA (2013), **Front. Oncol.** 3:208. doi: 10.3389/fonc.2013.00208]

9.2.2 Magnetic Resonance Imaging

Magnetic resonance imaging (MRI), previously known as magnetic resonance tomography (MRT), is a non-invasive technique used to obtain information about the interior of an object. MRI measures the magnetization due to nuclear magnetic moments in the object being scanned. MRI is based on the phenomenon of nuclear magnetic resonance (NMR). A strong static magnetic field is created by passing an electric current through the wire loops. As living tissue is dominantly water, the quantum spin states are mainly those of protons. While this is happening, oscillating magnetic field pulses at radio frequency are applied in a plane perpendicular to the magnetic field lines. This triggers protons in the body to change from their aligned positions. After each pulse, the nuclei relax back to their original configuration and in doing so they generate signals in the receiver. All the signals are then processed by a computer and it generates a 3D image representation of the area being examined. Also, by measuring the time variation of the recorded signal, it is possible to obtain information about the variations in the local chemical environment. Unlike PET or X-ray studies, no ionizing radiation is involved in MRI. In some instances, the likelihood of the MR images of the soft-tissue structures of the body, such as the heart and liver, to identify and accurately characterize diseases is more as compared to other imaging methods. This detail makes MRI an invaluable tool in the early diagnosis and the evaluation of many crucial lesions and tumors. A typical MR scan is shown in Figure 9.18.

Paul Lauterbur and Sir Peter Mansfield were awarded Nobel Prize (2003) in Medicine for their discoveries concerning magnetic resonance imaging. More on MRI can be found in the review article, Henderson, R.G. 'Nuclear Magnetic Resonance Imaging: A Review,' *Journal of the Royal Society of Medicine*, 76.3, 206–212, 1983.

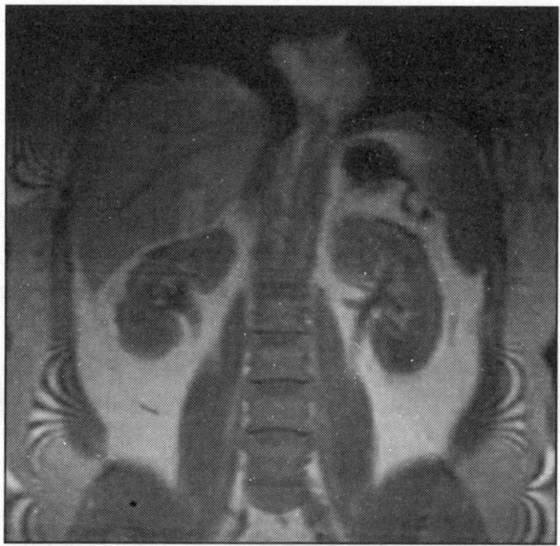

FIGURE 9.18 A typical MR scan of the abdomen showing liver and kidneys—frontal.

9.2.3 Gamma Knife

Gamma knife surgery represents one of the most sophisticated means available to manage brain tumors; pain or movement disorders, and malformations related to the arteries and the veins. Radiosurgery refers to the precise delivery of a high dose of radiation to a reconstructed 3D target site of the tumor. A head ring serves as a reference platform for reconstructing the patients head and the target lesion within (Figure 9.19). The radiation can be X-rays or γ-rays emitted from ^{60}Co source. The procedure is unique because, with the gamma knife, no surgical incision is performed to expose the target.

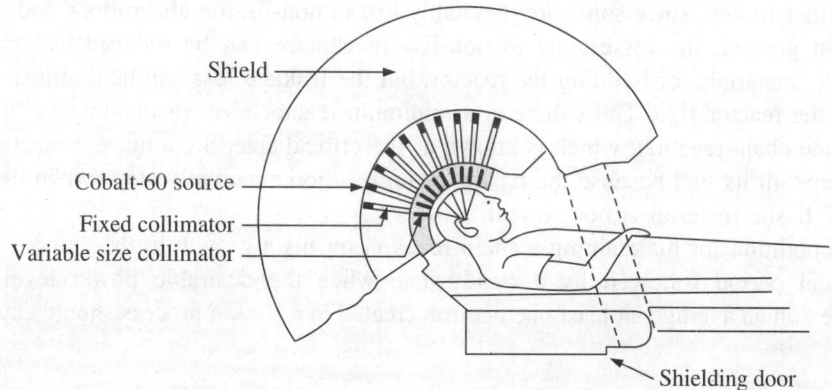

FIGURE 9.19 Schematic view of gamma knife.

Further details on the gamma knife can be found in the review article, Torrens, Michael, *Gamma Knife Neurosurgery–A Review*.

9.3 NUCLEAR REACTORS

The first nuclear reactor (5 MWe[4]) using uranium oxide as the fuel, graphite as the moderator and water as the coolant was commissioned in Russia. This heralded the beginning of electricity production by nuclear technology. Today, it is an industry, which contributes to about 18% of the world's total electricity. In view of the fast depleting fossil fuels and at the same time ever increasing demand for electric power in the country, even though the quest is on for alternate sources of energy, nuclear route looks like an inevitable option despite the prevailing apprehensions about safe operation, environmental issues, waste management and the cost intensive requirements.

EXERCISE 9.7: What is the number of fission per second in a 100 MW reactor? If the efficiency of the nuclear power plant is 33%, what is the electric output of the nuclear reactor?

Solution: Each fission of uranium nucleus releases about 200 MeV, so the number of fission per second in a 100 MW reactor

$$N = \frac{100 \times 10^6 \text{ W}}{200 \times 1.60219 \times 10^{-13} \text{ J}} = 3 \times 10^{18} \text{ s}^{-1}$$

The electric output of the nuclear reactor

$$100 \times 33\% = 33 \text{ MWe}$$

9.3.1 Reactor Physics Aspects

Energy by fission can be harnessed only when a controlled chain reaction is maintained at a specific rate in a nuclear reactor. Not all the neutrons emitted in fission can be used in bringing about a further fission, since some are inevitably lost in non-fission absorptions and some other leak out. In general, the losses due to non-fission capture can be reduced by appropriately choosing the materials for building the reactor, but the leakage loss can be diminished only by increasing the reactor size. Thus, there is a minimum reactor size (or quantity of fuel) needed to sustain the chain reaction, which is known as the critical size. So, a nuclear reactor does not ever consume all its fuel because the fission chain reaction eventually stops when the adequate quantity of fissile material is not present.

The condition for maintaining a chain reaction means a growth in the number of fissions in the initial period followed by a steady-state when the desirable power level has been reached, i.e., on an average, at least one neutron created in a fission process should cause another

4. Megawatt electric is the electrical output of a power plant in megawatt. The electrical output of a power plant is equal to the thermal power multiplied by the efficiency of the plant. The power plant efficiency of light water reactors is about 33%–35% compared to up to 40% for modern coal, oil or gas-fired power plants.
MWt is the unit for reactor thermal power; hence, for a nuclear reactor MWt is about three times MWe.

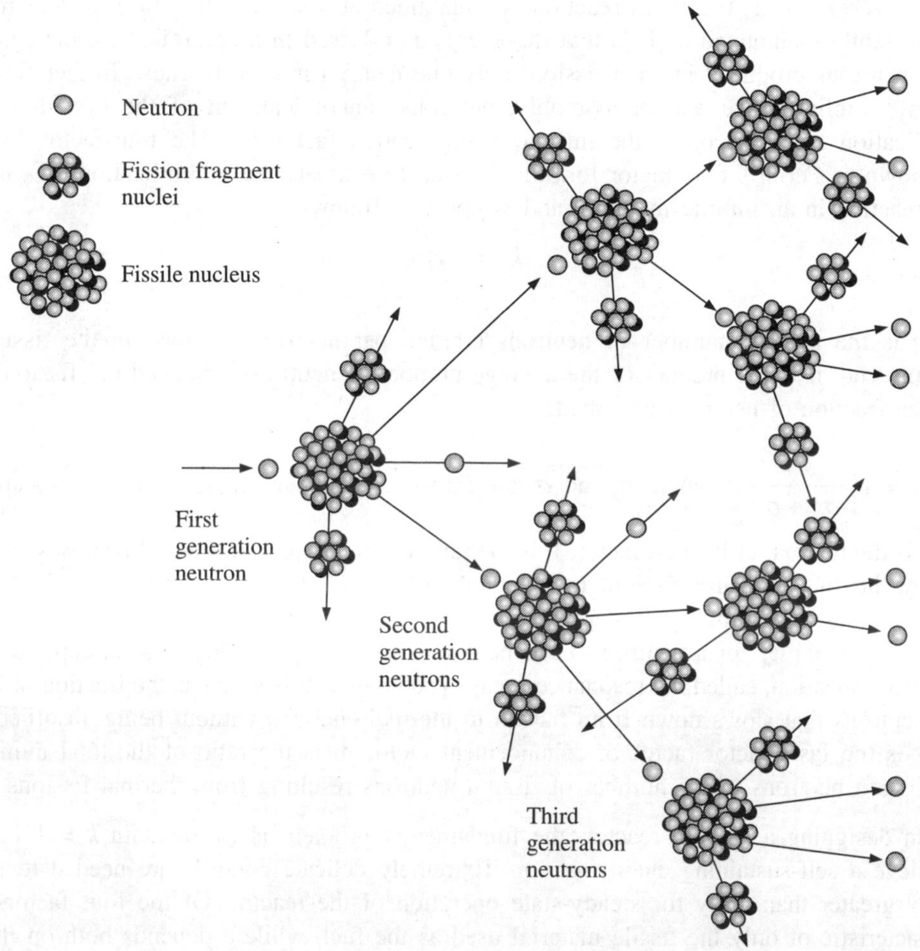

FIGURE 9.20 Neutron multiplication process.

fission. The factor which characterizes a fission chain reaction is called the neutron multiplication factor k, defined as the ratio of the number of fissions leading to the given generation of neutrons (Figure 9.20).

$$k = \frac{\text{Number of neutrons in one generation}}{\text{Number of neutrons in preceding generation}} \tag{9.22}$$

If $k > 1$, the chain reaction is said to be supercritical, and the neutron population will grow exponentially. If it continues to grow in an uncontrolled manner and the fuel is more than the critical mass, the result is a fission bomb or *atomic bomb*[5]. If $k < 1$, the chain reaction is subcritical, and the neutron population will exponentially decay. In such a scenario, the system is unable to sustain a chain reaction, and any beginning of a chain reaction eventually

5. A simple atomic bomb consists of two pieces of ^{235}U such that separately their masses are less than the critical mass. To detonate such a bomb, the two pieces of ^{235}U, initially apart at a safe distance from each other, are quickly brought closely together into a single supercritical mass.

dies out. When $k = 1$, the chain reaction is maintained at a steady-state. In a nuclear reactor, k is constantly maintained at 1 so that the energy is released in a controlled manner, i.e., out of 2–3 neutrons produced in each fission, only one neutron is used to cause further fission.

In an infinite size reactor assembly, neutrons cannot leak out of the system and the multiplication factor becomes the infinite multiplication factor k_∞. The four-factor formula, also known as Fermi's four factor formula, is used to estimate the multiplication of a nuclear chain reaction in an infinite medium, and is given as follows:

$$k_\infty = \eta f p \varepsilon \tag{9.23}$$

where

η is the average number of neutrons emitted per neutron absorbed in the fissionable material. It is the product of the average number of neutrons produced per fission ν and the fraction of neutrons absorbed, i.e.,

$\eta = \nu\left(\dfrac{\sigma_f}{\sigma_f + \sigma_c}\right)$; where σ_f and σ_c are fission and capture cross-sections, respectively.

f is the thermal utilization factor. It is the ratio of the number of thermal neutrons absorbed by the fuel to cause fission, to the total number of thermal neutrons absorbed by all processes (including by moderator, coolant, structural materials, etc.).

p is probability for a neutron to escape resonance absorption or any such capture during thermalization, called the resonance escape probability. It is given as the fraction of fission neutrons that slows down from fission to thermal energies without being absorbed.

ε is the fast reactor factor or enhancement factor. It is the ratio of the total number of fission neutrons to the number of fission neutrons resulting from thermal fissions.

In designing a nuclear reactor, the fundamental problem is to maintain $k = 1$ in order to achieve a self-sustaining chain reaction. Extremely delicate controls are needed to keep k slightly greater than unity for steady-state operation of the reactor. Of the four factors, η is a characteristic of only the fissile material used as the fuel, while ε depends both on the fuel element and size and shape of the assembly making up the reactor. The other two factors, f and p depend on the nature of the fuel element, the moderator substance and the other materials present, as well as on the geometry of the structure of the reactor, or on its lattice characteristics. The reactor is designed such that the product pf is made as large as possible by finding the best lattice structure of distributing the fuel and the moderator.

EXERCISE 9.8: If the multiplication factor for a nuclear reactor system is 1.005 and the average time between successive neutron generations τ is of the order of 10^{-8} s, estimate the increase in the number of neutrons after a lapse of 1 s.

Solution: Since, one neutron per fission is required to maintain the chain reaction, the number of neutrons increases by a factor of $(k - 1)$ in each generation. Thus, if the number of neutrons at any time t is N, the rate of increase of neutrons will be $N(k - 1)$ per generation. If τ is the average time between successive neutron generation, then

$$\frac{dN}{dt} = \frac{N(k-1)}{\tau}$$

Integrating the above equation, we get

$$N = N_0 \exp\left[\frac{(k-1)t}{\tau}\right] \quad (9.24)$$

where N_0 is the number of neutrons present at $t = 0$. Thus, after 1 s, the number of neutrons will increase by a factor of

$$\exp\left[\frac{(0.005)}{10^{-8}}\right] = \exp\,(5 \times 10^5)$$

9.3.2 Basic Components of a Nuclear Reactor

A reactor consists of four essential components namely, the fuel, the coolant, the moderator and control rods. The nuclear fuel is usually uranium or plutonium in metallic, oxide or carbide form, and sealed in a cladding material typically zirconium alloy, magnesium alloy or stainless steel, which ensures the retention of fission products in the fuel and prevents the coolant from chemically interacting with the fuel. The heat generated in the fuel rods is taken away by the coolant, which can be either light water, heavy water or a gas. The coolant is cooled in a suitable heat exchanger and returned back to the reactor; the heat energy exchanged is used to produce steam which turns the turbine to generate electricity. Some of the considerations that dictate the choice of coolant are it should have good thermal conductivity, high specific heat, compatibility with fuel and structural materials and low neutron absorption cross-section. Light water and heavy water are the most popular coolants. Carbon dioxide and helium are used in Magnox and high temperature gas cooled reactors. For fast reactors, sodium is the preferred choice for the coolant.

The neutrons generated in the fission have energy up to 10 MeV. These fast neutrons have to be slowed down, because slow moving neutrons cause more fissions than fast neutrons. Substances which are capable of slowing down the neutrons are called moderators. The moderator can be a liquid (e.g., light water or heavy water) or a solid (e.g., graphite). When light water is being used as a coolant, the same water stream acts as a moderator and no separate provision is made. If the moderator is different from the coolant, it must be separated from the coolant by a suitable intervening structure or should not react with the coolant.

In a reactor, the reactivity decreases with operation because of the consumption of fuel and the increase of the fission product concentration many of which, like xenon and samarium, are high neutron absorbers—the so-called *reactor poisons*. It is near to impossible to replace the fuel as it is consumed inside a reactor. It is, therefore, necessary to have a significant excess fuel in a reactor than required for criticality. The excess reactivity, which is the amount of reactivity in excess of that required for criticality, $\Delta k = k_{\text{eff}} - 1$, as a result of excess fuel is controlled using control rods. Reactivity is defined in terms of the effective multiplication factor k_{eff} as follows:

$$\rho = \frac{k_{\text{eff}} - 1}{k_{\text{eff}}} = \frac{\Delta k}{k_{\text{eff}}} \tag{9.25}$$

The control rods are made up of neutron absorbers like cadmium or boron. These rods are moved in and out of the reactor for controlling the fission chain reaction.

The entire reactor is housed in a leak tight containment to prevent the leakage of radioactivity into the environment. There may be an unusual leakage from the reactor. A secondary containment can also be added as a measure of added precaution. Further, the nuclear reactors are housed inside an exclusion zone and a buffer zone where any developmental activity is restricted.

Figure 9.21 gives a schematic illustration of the components of a nuclear fission reactor.

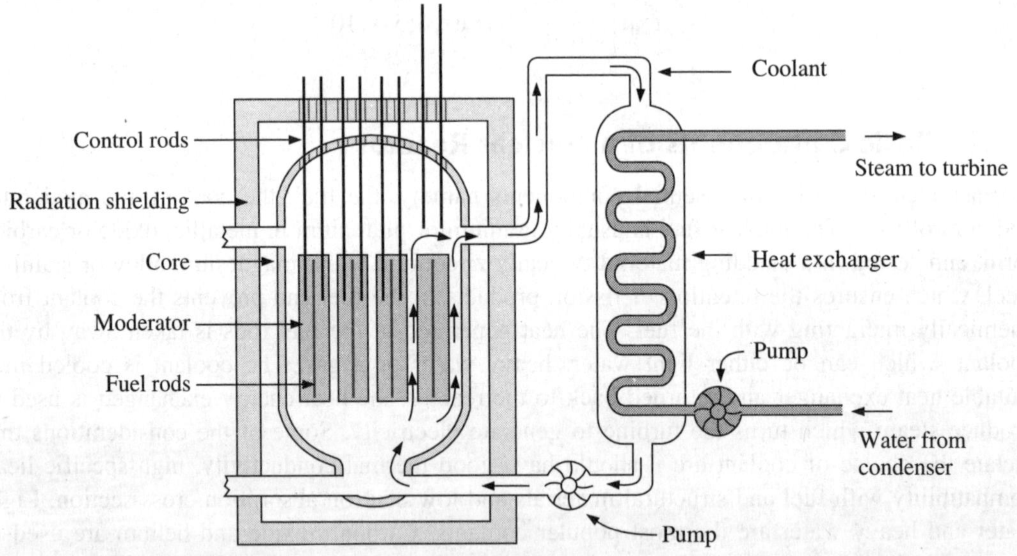

FIGURE 9.21 Schematic representation of a nuclear reactor system.

EXERCISE 9.9: What physical characteristics would make $^{24}_{12}$Mg a better moderator than $^{56}_{26}$Fe? In practice, why is neither of them used?

Solution: The largest energy transfer in an elastic collision between particles is when they are of the same mass. Since, the neutron mass is nearer to the $^{24}_{12}$Mg nucleus than $^{56}_{26}$Fe, $^{24}_{12}$Mg is a better moderator. Neither of them is used since both nuclei are heavier than neutrons and many materials exist with lighter nuclei.

9.3.3 Classification of Reactors

Nuclear reactors are classified depending on the various characteristics, such as:

Nuclear fuel: Natural uranium (0.7% ^{235}U), slightly enriched uranium (1–2% ^{235}U), highly enriched uranium (>90% ^{235}U), ^{233}U, ^{239}Pu, etc.

Energy of the neutrons that produce fission: Fast, intermediate (or epithermal) or thermal neutron reactors.

Materials used in the reactor components: Moderator, coolant, structure, shield, etc.

Method of heat removal by circulation of: Coolant only, fuel mixed with coolant, moderator-coolant, fuel-moderator-coolant.

Arrangement of fuel and moderator: Heterogeneous or homogenous, depending on whether the fuel and the moderator are in two different phases, or in one phase as a solution.

Purpose of the reactor: Power generation, plutonium extraction, isotope production, or purely research reactor used for material research, testing and training.

Often, more than one characteristic are combined to convey the full nature and scope of a reactor, e.g., thermal, natural uranium, heavy water, heterogeneous, water-cooled research reactor or pressurized heavy water reactor (PHWR), etc.

The job of choosing among the many possibilities is not so straightforward, because every system that has been proposed has some advantages and some disadvantages, e.g., water is cheap and good moderator and coolant, but it has a fairly high thermal neutron capture cross-section. It also has a relatively low boiling point, which means pressurization is required. Economic and other local factors also play an important part in choosing a reactor type.

9.3.4 Thermal Reactors

In thermal reactors, the fission is carried out by mainly using neutrons of thermal energy (0.025 eV range). This requires moderating the fast neutrons released in the fission, to the thermal energy range.

Light Water Reactors (LWRs)

These reactors use light water as the moderator and the coolant, e.g., Pressurized Water Reactors (PWR), Boiling Water Reactors (BWR) and Light Water Cooled Graphite Moderated Reactors (LWGR). This necessitates the enrichment of the fuel, usually uranium dioxide, to about 3% fissile content, since water has a fairly high thermal neutron absorption cross-section. In PWRs, the coolant is in the liquid condition under relatively high pressure, while in BWRs, as the name suggests, the coolant water boils and forms steam in the core.

A cutaway drawing of a typical PWR is shown in Figure 9.22.

Heavy Water Reactors (HWRs)

The enrichment requirement of uranium can be overcome by using heavy water as the moderator. The pressurized heavy water reactors (PHWRs) use heavy water as the moderator and the coolant. Natural uranium in the form of oxide is used as the fuel.

India has vast reserves of thorium, a fertile isotope, in contrast to the modest quantity of uranium fissile isotope. Thermal neutron cross-section of thorium is 7.4 b as against 2.7 b of natural uranium and is useful to capture extra neutrons. Thus, a lower fraction of neutrons will be lost to the structural materials in parasitic capture. India has embarked on utilization of

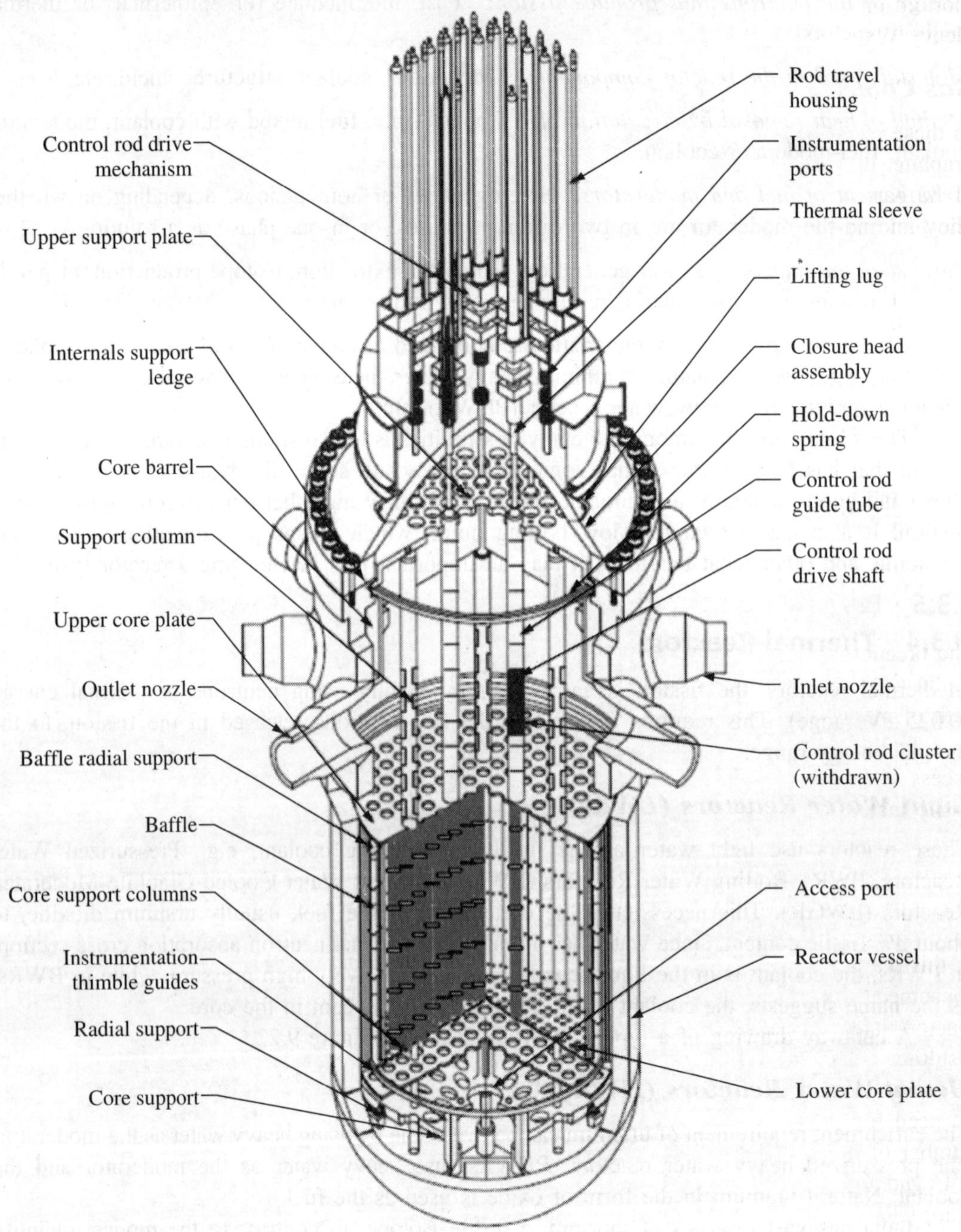

FIGURE 9.22 Cutaway drawing of a typical PWR. [Courtesy U.S. Nuclear Regulatory Commission (NRC)]

thorium on an industrial scale for power generation by designing a thorium fuelled Advanced Heavy Water Reactor (AHWR).

Gas Cooled Reactors (GCRs)

In these reactors, gas is used as the coolant. The moderator in a gas cooled reactor is usually graphite. In Magnox type gas cooled reactors, the coolant used is carbon dioxide at a pressure of 20 bar. The fuel elements consist of natural uranium metallic bars clad with magnesium alloy known by the trade name 'Magnox' from where the reactor derived its name.

One of the problems with CO_2-graphite gas cooled reactors is that the carbon dioxide coolant might extract carbon from the graphite moderator under high temperatures by the following reaction:

$$CO_2 + C \rightarrow 2CO$$

which not only causes corrosion and reduces the structural strength but also has serious explosion risk. Therefore, periodically the reactor is shut down and the graphite is annealed.

Helium is used as the coolant and graphite as the moderator in the high temperature gas cooled reactor (HTGR), which can be utilized for an efficient generation of electricity using direct cycle helium turbines.

9.3.5 Breeder Reactors

The breeder reactor, if it is successful, may become the most important source of heat in this century. A lot of effort and funding has gone into an attempt to solve many problems in the path of making commercial breeders a reality.

In the breeder reactor, neutrons from fissions in the fissile material, ^{235}U or ^{239}Pu, in excess of the number needed to maintain the chain reaction, are used to produce additional fissile material from the fertile material in the core, ^{238}U:

$$n + {}^{238}_{92}\text{U} + {}^{239}_{92}\text{U} \rightarrow {}^{239}_{93}\text{Np} + \beta^- + \bar{\nu}$$

$${}^{239}_{93}\text{Np} \rightarrow {}^{239}_{94}\text{Pu} + \beta^- + \bar{\nu}$$

In this way, the fuel is created at a rate equal to or greater than the rate at which it is being consumed. The far more abundant uranium isotope ^{238}U is used in this reaction, so breeder reactors have the potential of providing nearly unlimited supply of fissionable material. The resulting ^{239}Pu nucleus, fissions easily with both the thermal and the fast neutrons.

The breeder reactors are all fast reactors, i.e., they use the fast neutrons. Hence, there is no need for a moderator in the breeder reactors. In order for breeding to occur, a sufficient number of neutrons must be emitted per fission to provide for further fission, for capture in ^{238}U and for leakage and absorption in the structural and contaminant materials. Conversion ratio, CR, is defined as a measure of the breeding effectiveness of a reactor

$$\text{CR} = \varepsilon\eta - 1 - L \tag{9.26}$$

where L is a leakage factor. If CR = 1, the reactor just maintains its own fuel supply; for CR > 1, the reactor fuel actually grows, all the while producing heat which can be used in generating electricity.

EXERCISE 9.10: Explain in brief the operation of a breeder reactor. Which physical constant of the fission process is a prerequisite to the possibility of breeding? Could water be used as a moderator?

Solution: A breeder reactor contains a fissionable material and non-fissionable fertile material that can be made fissionable by absorbing a neutron. For example, uranium containing ^{235}U and ^{238}U isotopes. Suppose 3 neutrons are emitted per fission, then we already know that one is needed to keep the chain reaction going for the fissionable material, which in this case is ^{235}U. Now, we are left with 2 neutrons, which can be used to convert two non-fissionable fertile nuclei (of ^{238}U isotope) into fissionable ones (of ^{239}Pu), then two fuel nuclei are produced when one is consumed, and the reactor is said to be a breeder.

A prerequisite to breeding is that η, the average number of neutrons emitted per neutron absorbed in a fissionable material, should be greater than 2. As breeder reactors are fast reactors, i.e., they use fast neutrons, so no moderator is needed.

9.3.6 Three-stage Indian Nuclear Power Programme

The importance of nuclear energy, as sustainable energy resource for our country, was recognized at the very inception of our atomic energy programme more than five decades ago. A three-stage nuclear power programme, based on a closed nuclear fuel cycle, was then chalked out. This programme links the back-end of one stage to the next stage and maximizes the utilization of nuclear fuel. The three stages of India's Nuclear Power Programme are:

Stage I: 10000 MWe through natural uranium fuelled PHWRs.

Stage II: Fast breeder reactors (FBRs) utilizing plutonium-based fuel and thorium as blanket for breeding ^{233}U.

Stage III: Breeder reactors using ^{233}U as fuel and thorium as blanket.

For a large country like India, a major portion of energy ought to come from the domestic resources. From a long-term perspective, there are only limited options available. Solar and other renewable and non-conventional energy sources must be deployed to the maximum possible extent. Though, to meet the large concentrated energy needs for the industries and the urban centres, it may not be possible to overlook the nuclear energy option. Here too, India is in a somewhat unique place as regards to the availability of nuclear resources. India has meager reserves of uranium, the only naturally occurring fissile element, which can be directly used in a nuclear reactor to produce energy from fission. Nevertheless, it has almost a third of the earth's thorium, which is a fertile element, and needs to be first converted to a fissile material, ^{233}U, in a reactor. Therefore, any strategies for large scale deployment of nuclear energy should be focused towards utilization of thorium.

Recently, India's first indigenously designed 500 MW fast breeder reactor at Kalpakkam started operation.

9.4 ENVIRONMENTAL IMPACT

There is an old adage 'Little knowledge is a dangerous thing.' It goes without saying that wise decisions and judgements, which are beneficial in the long run, are founded on the bedrock of knowledge and understanding. This principle applies for nuclear power too. Therefore, it is pertinent that we also take keen interest in understanding the problems associated with nuclear power and our current limitations.

Some of these issues have been much highlighted such as its connection to the proliferation of nuclear weapons, risks related to severe nuclear accidents, meltdowns, natural disasters, terrorist and war attacks on nuclear sites, and radioactive waste which needs to be managed and heavily guarded for thousands of years. In 2011, an earthquake and tsunami led to explosions and partial meltdowns at the Fukushima I Nuclear Power Plant in Japan. The extent of the actual damage is still being debated; just to get the drift however, as of 2013, the Fukushima site continues to be highly radioactive, with some 160000 evacuees living in temporary housing; some portions of land will be unusable for centuries; and the complicated cleanup job will take forty years or more, which would cost tens of billions of dollar. With this kind of cost involved, even the economics of nuclear power has come under severe scrutiny. Numerous studies have been conducted on the possible effect of nuclear power in causing cancer. There is also an agreement that workers in other parts of the nuclear fuel cycle, particularly uranium mining have elevated risk of cancer. The waste generated from uranium mining operations and rainwater runoff can contaminate the groundwater and the surface water resources with heavy metals and traces of radioactive material. The structure of the reactor itself becomes radioactive and in turn requires decades of storage prior to it being economically dismantled and disposed of as waste.

But there is much less awareness about thermal pollution and waste heat management vis-à-vis nuclear power. Whenever the question of risks associated with nuclear radiation exposure becomes too hot to handle, the idea of using controlled nuclear fusion in future is floated as a panacea. Therefore, it is all the more important to acquaint ourselves with some of the facts, so that we could move towards a more holistic and sustainable solution.

Nuclear power plants use huge quantities of water for steam production and for cooling, and exchange about 60%–70% of their thermal energy by cycling with a water body or by evaporating water through a cooling tower (Figure 9.23). A single reactor can consume millions of litres of water each day; and this calculation does not include water consumption by uranium mines. Possibly, the most damaging environmental effect of a power plant is the many organisms that get sucked in through the water intake. Larger species, such as fish, seals and turtles are killed on the intake screens while smaller creatures that pass through the screens are subject to toxic stress. A U.S. Nuclear Information and Resource Service report details the destruction of delicate marine ecosystems and a great number of animals, including endangered species, by nuclear power plants. Most of the damage is caused by the water inflow pipes, whereas the expulsion of warm water causes further damage. The higher temperature of the water discharged from the power plant as well as water pollutants, diminishes the water quality, reduce dissolved oxygen and threaten aquatic life. Another documented problem is 'cold stunning'—fish adapt to the warm water but die when the reactor is taken offline and the warm water is no longer expelled. In the recent years, in several European countries, the nuclear reactors have been

FIGURE 9.23 Cooling towers of Nogent Nuclear Power Plant, France, which houses two reactors of 1300 MWe each, located on the bank of Seine River. [Courtesy Centrale Nucléaire de Nogent]

periodically operated at reduced output or have taken offline due to water shortages driven by climate change, drought and heat waves. The competition for water has been intensifying over the years among agriculture, power plants, industries, and environmental flows. The situation will only get worse in future.

EXERCISE 9.11: The waste heat from a 1000 MWe nuclear power plant having an efficiency of 32 per cent is rejected using once-through cooling method. Water is withdrawn from an estuary and after exchanging the latent heat of the steam condensing on the outer surface of the condenser tubes, it is discharged back to the estuary. If the maximum allowed limit for temperature change is 5 °C, determine the minimum water flow rate required to not exceed the limit.

Solution: The thermal power of the nuclear reactor

$$= \frac{1000}{32\%} = 3125 \text{ MWt}$$

Therefore, the amount of waste heat generation rate for this plant

$$= 68\% \ (3125) = 2125 \text{ MW}$$

Thus, the water flow rate through the condenser

$$= \frac{\text{Heat rejection rate/Specific heat of water}}{\Delta T}$$

$$= \frac{(2125 \times 10^6 \text{ W})/(4186 \text{ J/kg °C})}{5 \text{ °C}} = 1.015 \times 10^5 \text{ kg/s}$$

PROBLEMS

9.1 An alloy containing 5% copper was irradiated in a neutron flux of 10^6 n/cm^{2}s for 10 h. What weight of the sample should be taken if the desired activity after a cooling period of 5 h is to be 1500 cpm? Given: Half-life of ^{64}Cu is 12.7 h, σ = 4.5 b, r = 69.1% and the counting efficiency is 10%. [**Ans:** 0.525 g of alloy]

9.2 A sample containing an unknown amount of Ge metal is irradiated in a neutron flux of 10^{12} n/cm^{2}s for 1 h when the ^{76}Ge isotope forms ^{77}Ge radioisotope with half-life of 1 m. Suppose the activity measured 1 m after the 1 h irradiation is 2500 dps, find the amount of Ge in the same, given that the cross-section for the reaction is 3.28 mb, and the isotope abundance of ^{76}Ge is 7.8%. [**Ans:** 2.47 mg]

9.3 0.1 g of Mn powder was irradiated in a neutron flux of 10^6 n/cm^{2}s for 90 m. Find the activity of the sample in dpm after a cooling period of 5 h. Given: r = 100%, σ = 13.3 b, half-life of ^{56}Mn is 2.58 h. [**Ans:** 75,654 dpm]

9.4 A 5 MeV α-particle source is used in the Rutherford backscattering analysis. Calculate the energies of the backscattered α-particles from nuclei with A = 100. What detector energy resolution (in keV) for the α-particles would be needed to resolve ΔA = 1, for the above mass region? [**Ans:** 4260 keV, ΔE = 7 keV]

9.5 A thin target is bombarded with a 10 nA beam of 10 MeV α-particles in a backscattering experiment to determine the level of lead contamination. After 5 min, 30 counts corresponding to the scattering from lead in the target are recorded in a detector placed at 180°, which subtends a solid angle of 5×10^{-2} sr at the target. Calculate the level of lead contamination (mass per unit area). The atomic mass of lead is 207.2. [**Ans:** $\rho t = 6.3 \times 10^{-8}$ g/cm^2]

9.6 In a PIXE experiment, a thin target containing 10^{-11} g/cm^2 of an element (A = 120) is bombarded with a beam of protons of intensity 0.5 μA perpendicular to the target. The X-ray detection efficiency is 1%. If the count rate is 0.6 s^{-1} and the measurement lasted for 25 minute, what is the cross-section in barn for the production of X-rays and its accuracy (to one standard deviation)? The background may be neglected. [**Ans:** 900 ± 30]

9.7 A pulse of 10^{18} 100 keV X-ray photons per square metre is directly perpendicular to a 5 mm thick iron slab. Determine the rise in the temperature of the slab. Density of iron is 7870 kg/m^3, mass attenuation coefficient for 100 keV photons on iron is 0.04 m^2/kg and the specific heat capacity of iron is 106 J/kg K. [**Ans:** 3 °C]

9.8 In a $^{57}_{26}Fe^*$ Mössbauer experiment, it is found that to obtain resonance, the source, emitting 14.4 keV, had to be moved towards the absorber at 2.2 mm/s. What is the shift in the frequency between the source and the sample? [**Ans:** 25.5 MHz]

9.9 In an AMS measurement of a carbon sample, transmitted $^{14}_{6}C$ ions record 1000 counts in 5 m. A beam of 10 μA is measured when the system is set to transmit $^{12}_{6}C^{3+}$. Calculate the atomic ratio of $^{14}_{6}C/^{12}_{6}C$ in the sample assuming that the transmissions of ${}^{14}_{6}C$ and ${}^{12}_{6}C$ ions through the system are the same. Estimate the mass of ${}^{12}_{6}C$ was in the sample if it is completely consumed in half an hour. Assume a constant rate of consumption and a system efficiency of 2%? **[Ans:** 1.6×10^{-13}, 37 μg]

9.10 Assuming no loss of thermal or fast neutrons, calculate the multiplication factor for a reactor for which the fast fission factor is 1.03, the number of fast neutron generated per thermal neutron used up is 1.32, the resonance escape factor is 0.89 and the thermal utilization factor is 0.87. **[Ans:** 1.053]

9.11 Describe briefly the type of reaction on which a nuclear fission reactor operates. Why is the energy released, and roughly how much per reaction? Why are the reaction products radioactive? Why is a moderator necessary? Are light or heavy elements preferred for moderators, and why?

9.12 A particular cooling tower is designed to dissipate heat at a rate of 300 MW. If its cooling range is 8 °C, what is the water use rate? **[Ans:** 8960 kg/s]

BIBLIOGRAPHY

Accelerators for Frontiers in Science and Technology, IANCAS Bulletin, Vol. 5, No. 1, Mumbai, 2006.

Arnikar, H.J., *Essential of Nuclear Chemistry*, Wiley Eastern, New Delhi, 1990.

Devins, D.W., *Energy: Its Physical Impact on the Environment*, Wiley, New York, 1982.

Instrumentation for PIXE and RBS, IAEA-TECDOC-1190, Vienna, 2000.

Ionizing Radiation Sources, IANCAS Bulletin, Vol. 15, No. 4, Mumbai, 1999.

Lilley, J., *Nuclear Physics: Principles and Applications*, Wiley, New York, 2002.

Mani, H.S. and Mehta, G.K., *Introduction to Modern Physics*, EWP, New Delhi, 1988.

Mayer, M., *Rutherford Backscattering Spectroscopy,* Workshop on Nuclear Data for Science and Technology: Materials Analysis, Triesti, 2003.

Nuclear Analytical Techniques, IANCAS Bulletin, Vol. 2, No. 2, Mumbai, 2003.

Nuclear Reactors, IANCAS Bulletin, Vol. 1, No. 1, Mumbai, 2002.

Sood, D.D., Ramamoorthy, A.V.R., *Fundamentals of Radiochemistry*, IANCAS, Mumbai, 2000.

Use of Research Reactors for Neutron Activation Analysis, IAEA-TECDOC-1215, Vienna, 2001.

Vertes, A., Korecz, L. and Burger, K., *Mössbauer Spectroscopy*, Elsevier, Amsterdam, 1979.

10

Nuclear Astrophysics

"In some strange way, any new fact or insight that I may have found has not seemed to me as a 'discovery' of mine, but rather something that had always been there and that I had chanced to pick up."

—Subrahmanyan Chandrasekhar

Perhaps, the most important macroscopic manifestation of nuclear physics is its explanation of the stellar energy production and the nucleosynthesis. Astrophysicists now believe that they understand in some detail, the stellar histories from their origins as diffuse clouds through their successive stages of nuclear burning until their deaths as white dwarfs[1] or supernovae[2]. Through this process, concepts of nuclear physics have allowed us to understand how the initial mix of helium and hydrogen formed in the primordial universe has been transformed into the heavier elements that make terrestrial life possible.

In this brief chapter, we shall discuss about the evolution of the universe, primordial nucleosynthesis, stellar nucleosynthesis and heavy element production via *s* and *r*-processes.

10.1 EVOLUTION OF THE UNIVERSE

How do we determine the age of the universe, if it is assumed that the universe that we live in did have a beginning? In that case, how did all the matter in the universe come into being and how did it evolve? In 1929, an American astronomer by the name, Edwin Hubble, noticed that the galaxies outside our own Milky Way were all moving away from us at a speed proportional to their distance from us based on his observations of their absorption line spectra being red-shifted. Red-shift occurs when a light source moves away from its observer and because of the Doppler effect, the light's apparent wavelength stretches towards the red part (longer-wavelength end) of the spectrum.

1. A whitish star that is approximately the size of the earth, having low luminosity, has undergone gravitational collapse, thereby having great density, and is in the final stage of evolution for low-mass stars. They are formed when the stars use up their fuel and can no longer support nuclear reactions (Figure 10.1).
2. A large star in its death stages that suddenly explodes, increasing many thousands of times in brightness. Most heavy elements are created by nuclear reactions in supernovae and then returned to space (Figure 10.2).

FIGURE 10.1 NASA's Hubble Space Telescope captures a field of stellar husks. These ancient white dwarfs are 12 to 13 billion years old, only slightly younger than the universe itself. In theory, white dwarfs will eventually stop emitting light and heat and become black dwarfs. [Courtesy NASA/ESA and H. Richer (University of British Columbia)]

FIGURE 10.2 This Chandra X-ray photograph shows Cassiopeia A, the youngest supernova remnant in the Milky Way. [Courtesy NASA/CXC/SAO; Scale: 8 arcmin (arcmin is 1/60th of one degree) per side]

Hubble's discovery was the first observational support for the Big Bang hypothesis of the expanding universe. Big Bang hypothesis is the most popular explanation about how the universe began explosively from an extremely hot and dense state, and continues to expand today. There are no instruments to look back at the early universe, and needless to say that our time span of observation is so brief as compared to the evolution of the universe[3], which spans for billions of years according to the estimates, that we are left with no option but to draw our conclusions from more or less an instantaneous view. Thus, it is proper to point out that all the estimates based on the mathematical theory and models regarding the age of the universe, are humongous extrapolations and lounge on a web of inherent assumptions, the fact that should never be overlooked.

3. The historic detection of the feeble gravitational waves by Laser Interferometer Gravitational-Wave Observatory (LIGO) in September 2015, can pave the way to look back in time at the creation of the universe. More on the LIGO project can be read from <https://www.ligo.caltech.edu/>.

10.1.1 Age of the Universe

Hubble's law gives the correlation between the distance to a galaxy and its apparent recessional velocity as determined by the red shift. It can be stated as follows:

$$v = Hr \tag{10.1}$$

where v is the recessional velocity of the distant galaxies known from the red-shift, r is the distance to a galaxy and H is the Hubble constant. The value for the Hubble constant or more appropriately the Hubble parameter, since it has also been found to vary with time, is somewhat uncertain because of the difficulty of measuring the distances to remote galaxies. The best modern value for the Hubble constant as documented by the Particle Data Group (pdg.lbl.gov/) is

$$H = 72 \pm 10\% \frac{\text{km/s}}{\text{Mps}}$$

where Mps stands for million parsecs. Parsec is a unit of measurement for interstellar space that is equal to 3.26 light-year and is the distance to an object having a parallax of one second as seen from the points separated by one astronomical unit.

EXERCISE 10.1: Assuming that the universe is flat, homogenous (it looks roughly the same everywhere) and isotropic (same in every direction), and if the known physics laws are used to extrapolate into the past, what would be the estimate for the age of the universe?

Solution: The Hubble constant, which is a measure of the rate at which the universe is expanding uniformly, its reciprocal, should be the age of the universe, i.e.,

$$\begin{aligned} t_{\text{exp}} &= \frac{1}{H} \\ &= \left(\frac{1}{72\ \text{km s}^{-1}\ \text{Mpc}^{-1}}\right)\left(\frac{3.26\times 10^6\ \text{ly}}{1\ \text{Mpc}}\right)\left\{\frac{3600\times 24\times 365(3\times 10^5)\ \text{km}}{1\ \text{ly}}\right\}\left(\frac{1\ \text{y}}{3600\times 24\times 365\ \text{s}}\right) \\ &= 13.6\times 10^9\ \text{y} \end{aligned}$$

So, if there indeed was an instant in time when the entire universe was contained in a single point in space, as propounded by the Big Bang hypothesis, it must have been about 13.6 billion years ago!

There are several fundamental constants in physics. Two such fundamental constants involved in the general relativity are the gravitational constant $G = 6.6726 \times 10^{-11}$ Nm^2 kg^{-2} and the speed of light $c = 2.9979 \times 10^8$ ms^{-1}. Similarly, the Planck's constant $h = 6.6261 \times 10^{-34}$ J.s is the fundamental constant in the quantum theory. The three of them can be combined to arrive at a characteristic length called the Planck's length λ_p, which is given by:

$$\lambda_p = \sqrt{\frac{Gh}{c^3}} = 4.05\times 10^{-35}\ \text{m} \tag{10.2}$$

The Planck's length is significant, since at shorter distances, because of the uncertainty principle, the smooth geometry of the space is disrupted. Quantum theory and general relativity break down and fail to describe the physical reality at such small scales of length.

EXERCISE 10.2: After what instant of time, can the current physical laws say anything about the aftermath of the Big Bang?

Solution: The time that it takes for the light to travel across the Planck's length is called the Planck's time t_p, which is given by:

$$t_p = \frac{\lambda_p}{c} = 1.35 \times 10^{-43} \text{ s}$$

Thus, in absence of any unified quantum-mechanical theory of gravity, there is no way to understand the universe earlier than 10^{-43} s. So, our story of the how the universe might have evolved based on the existing physical laws, even though rudimentary at best, starts after 10^{-43} s. Of course, the laws of Physics do not shed any light on what happened just before the Big Bang, and why and how the space and the time came into existence in the first place!

10.1.2 The Future

The Hubble parameter can be related to the Big Bang model of expansion of the universe with the help of the Friedman equation:

$$H^2 = \frac{8\pi G\rho_m}{3} - \frac{k}{a^2} + \frac{\Lambda}{3} \tag{10.3}$$

where H is the Hubble parameter, G is the universal gravitational constant, a is the scale parameter that is proportional to the distance between the galaxies, ρ_m is the average mass density of the universe, k is the curvature parameter of the universe, and Λ is the cosmological constant.

EXERCISE 10.3:

(a) Determine the critical density of the universe.
(b) Based on the expression of critical density from part (a), show that the curvature parameter k in the Friedman equation is indicative of the future fate of the universe. For our purpose, the value of the cosmological constant Λ can be assumed to vanish.

Solution:

(a) If a galaxy of mass m were to escape permanently from the gravitational influence of the universe of mass M and radius R, the minimum escape velocity v required by the galaxy will be when its kinetic energy KE will just be equal to the magnitude of the gravitational potential energy U, i.e.,

$$\text{KE} = U \quad \text{or} \quad \frac{1}{2}mv^2 = \frac{GmM}{R} \tag{10.4}$$

Assuming spherical universe, the mass (M) of the universe in terms of its density ρ_m is given as follows:

$$M = \frac{4\pi R^3 \rho_m}{3}$$

Also, according to the Hubble's law, $v = HR$. Therefore, equation (10.4) can be written as follows:

$$\frac{1}{2}m(HR)^2 = \frac{Gm}{R}\left(\frac{4\pi R^3 \rho_c}{3}\right)$$

where ρ_c is the critical density of the universe.

or $$\rho_c = \frac{3H^2}{8\pi G}$$

(b) *Flat universe:* When we replace ρ_m with ρ_c in equation (10.3), we get $k = 0$. In such a scenario, the universe will expand perpetually at a decreasing rate. This is the case of flat universe and is often referred to as the *Einstein-de Sitter universe* in recognition of their effort in modelling it.

Closed universe: If on the other hand, ρ_m is high enough, the gravitational attraction will eventually stop the expansion and the universe will collapse backward to a 'Big Crunch[4].' In this case, the curvature parameter $k > 0$ and such a universe is described as being a closed universe.

Open universe: If $\rho_m < \rho_c$, i.e., $k < 0$, the universe is said to be open and the expansion will never cease.

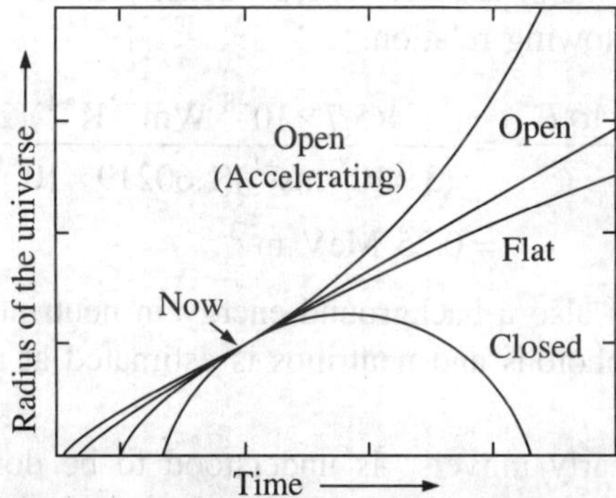

FIGURE 10.3 Four different cosmological models for the expansion of the universe. The open universe has low density, while the open accelerating universe is due to the negative attraction of the dark energy. The flat, critical density universe continues to expand, but more slowly, while high mass density universe eventually contracts and collapses under its own gravitational attraction.

4. It is the contraction of the universe to a state of extremely high density and temperature (a hypothetical opposite of the Big Bang).

Dark Matter

In 1998, observations based on supernovae data suggested that the expansion of the universe is actually speeding up or accelerating, which implies the existence of a strong negative force. Cosmologists have attributed it to a mysterious *dark energy*. In contrast to gravity which works to slow the expansion down, dark energy works to speed up the expansion.

Similarly, there is a growing agreement among cosmologists that the total density of matter is equal to the critical density, so that the universe is flat. However, the actual density of the luminous matter in the universe is only a small percentage of ρ_c. This is known as the *missing mass problem*. Much of the mass in the universe may be in the form of *dark matter*, which has yet not been observed.

10.1.3 Early Universe

In 1965, Arno Penzias and Robert Wilson[5], observed a faint cosmic microwave background (CMB) radiation that remained unchanged in intensity no matter in which direction in the sky they pointed their antennas. It is believed that this CMB was generated in the early universe and is filling all the space almost uniformly. The peak intensity of this radiation occurs at a wavelength of about 1.1 mm. Assuming the universe to be a blackbody radiator, from the Wien's displacement law, its temperature is given as follows:

$$\lambda_{\text{peak}} T = b$$

where b is the Wien's displacement constant whose value is equal to 2.898 mm K.

or
$$T = \frac{2.898}{1.1} \approx 2.7 \text{ K}$$

Now, from the Stefan–Boltzmann law, the energy density in this equilibrium CMB radiation can be calculated by the following relation:

$$u = \frac{4\sigma T^4}{c} = \frac{4(5.7 \times 10^{-8} \text{ Wm}^{-2}\text{K}^{-4})(2.7 \text{ K})^4}{(3 \times 10^8 \text{ ms}^{-1})(1.60219 \times 10^{-13} \text{ J/MeV})}$$
$$= 0.25 \text{ MeV m}^{-3}$$

Apart from photons, there is also a background energy in neutrinos. Taking that into account, the total energy density in photons and neutrinos is estimated at about 0.4 MeV m^{-3}.

EXERCISE 10.4: The early universe is understood to be dominated by radiation, with matter present in only small amounts. With this assumption, ρ_m in the Friedman equation can be taken as the energy density, related to the ratio of the temperature at that time and the current temperature of the CMB. Thus, assuming a flat universe, in the radiation-dominated early universe, determine a relationship between the age of the early universe (t_{exp} s) and its associated temperature (in K). The value of the cosmological constant Λ can be assumed to vanish.

5. Their observations supported Big Bang model and for their discovery of CMB, they were awarded the Nobel Prize (Physics) in 1978.

Solution: For a flat universe, $k = 0$. Therefore,

$$H^2 = \left(\frac{1}{t_{\text{exp}}}\right)^2 = \frac{8\pi G\rho_{\text{radiation}}}{3}$$

We have already estimated the value of $\rho_{\text{radiation}}$ for CMB as 0.4 MeV m^{-3}/c^2. At temperatures other than 2.7 K (the estimated CMB temperature), we can use the Stefan–Boltzmann law to estimate the value of $\rho_{\text{radiation}}$. Therefore, we get

$$\left(\frac{1}{t_{\text{exp}}}\right)^2 = \frac{8\pi G\left(\dfrac{0.4\ \text{MeV m}^{-3}}{c^2}\right)\left(\dfrac{T}{2.7\ \text{K}}\right)^4}{3}$$

or $$T^2 = \frac{1}{t_{\text{exp}}}\sqrt{\frac{(2.7\ \text{K})^4 \times 3 \times (3\times 10^8\ \text{ms}^{-1})}{8\pi(6.6726\times 10^{-11}\ \text{Nm}^2\ \text{kg}^{-2})(0.4\ \text{MeV m}^{-3})(1.60219\times 10^{-13}\ \text{J/MeV})}}$$

or $$T \approx \frac{2.1\times 10^{10}}{\sqrt{t_{\text{exp}}}} \tag{10.5}$$

If the temperatures are high enough, the matter and radiation should be in equilibrium. Now, we have a relationship between the age of the early universe and its temperature, so we can look at the various reactions to understand, which reactions would dominate at what period of time. This is how the timeline of the Big Bang cosmology can be generated for our universe, which consists of a mixture of the most fundamental particles and their antiparticles, and radiation.

$t \approx 10^{-43}$ **s:** Because of the restriction imposed by the Planck's time, we can only talk about the evolution of the universe from this point onwards. At this instant, the entire universe is smaller than a proton. Its temperature is close to 10^{32} K.

$t \approx 10^{-34}$ **s:** By this time, the universe had undergone an exponential inflation, increasing its size by a factor of about 10^{30}. This expansion of space–time itself was, in effect, at more than the speed of light, but since it was not the expansion of matter, there was no violation of the special theory of relativity[6]. The universe has become a hot soup of quarks, gluons and leptons, at a temperature of about 10^{27} K.

6. Our visible universe is effectively a 'bubble' in the larger Big Bang universe. In other words, the universe keeps on expanding, but the speed of light limits how much of it can be seen. The inflationary hypothesis was developed in 1981 to solve many cosmological problems with the Big Bang. One of such problems is the horizon problem, which points out that the different regions of the universe must not have contacted each other because of the great physical distances between them and the limit set by the speed of light for the transfer of any information such as energy temperature, etc. However, the cosmic microwave background radiation, which fills the universe, is nearly at the same temperature everywhere in the sky. The exponential expansion assumption means that the widely separated regions that we see at the horizon were closer together and could have been in contact by light signals before the inflation era.

$t \approx 10^{-4}$ s: After the inflation era, the universe continues to expand, but not nearly as quickly. As it expands, it becomes cools and less dense. Quarks could now combine to form protons and neutrons and their antiparticles. There are more protons than neutrons, as protons are slightly less massive. The temperature has dropped to about 10^{12} K and the universe is cool enough that photons lack the energy needed to break the protons and the neutrons back into the quarks. Particles and antiparticles collide and annihilate each other. The four forces of today have become distinct. Now, the universe is a soup of photons, electrons, neutrinos, protons, neutrons and their antiparticles.

$t \approx 3$ s: The temperature has now fallen to about 10^{10} K. This is the beginning of the nucleosynthesis—the protons and the neutrons collide and stick together to form the nuclei of light elements such as hydrogen, helium and lithium. Atoms are not able to form themselves because the intense radiation ionizes them as soon as they are formed. The radiation cannot travel far because of its interaction with the charged particles and nuclei. The universe is opaque.

$t \approx 300{,}000$ s: The universe continues to cool and as the temperature falls to 3000 K, the electrons get captured around the bare hydrogen and helium nuclei to form atoms. With the electrons now bound to the atoms, and with the radiation being free to travel great distances, the universe finally becomes transparent to light, making this the earliest observable period. This radiation forms the cosmic background radiation, which can still be detected today, as discussed earlier. The universe consists of a cloud of about 75% hydrogen and 25% helium, with traces of ${}^{7}_{3}\text{Li}$, which under the influence of gravity, begins the formation of the galaxies and the stars.

EXERCISE 10.5: At higher temperatures, the creation of the matter (i.e., particle-antiparticle) will dominate its annihilation (i.e., creation of radiation), however, as the temperatures lower, the latter reaction starts dominating.

For the reaction $e^+ + e^- \Leftrightarrow 2\gamma$, at what time after the Big Bang, the annihilation reaction starts dominating?

Solution:

$$\gamma \rightarrow e^+ + e^-$$

For the above reaction, the photons must possess the minimum energy equal to the rest mass energy of the electron–positron pair, i.e., 2(0.511 MeV) ≈ 1 MeV. If the average energy of the photons is k_BT, where k_B is the Boltzmann constant ($k_B = 8.62 \times 10^{-5}$ eV/K), the 1 MeV pair production threshold energy corresponds to a temperature

$$T = \frac{10^6 \text{ eV}}{8.62 \times 10^{-5} \text{ eV/K}} = 1.16 \times 10^{10} \text{ K}$$

From equation (10.5), this temperature corresponds to a time of about 3.3 s.

$$t_{\text{exp}} = \left(\frac{2.1 \times 10^{10}}{1.16 \times 10^{10}}\right)^2 \cong 3.3 \text{ s}$$

Thus, for $t > 3.3$ s, the annihilation reaction will continue to dominate over the pair production reaction.

10.2 PRIMORDIAL NUCLEOSYNTHESIS

If we go back to the formative years of the universe, at a time of 10^{-2} s ($T \sim 10^{11}$ K), the density of the universe dropped to ~10^6 kg/m^3 and this is known as the photon dominated era. At this time, the neutrons and the protons interconvert by the weak interactions.

$$\bar{\nu}_e + p \Leftrightarrow e^+ + n$$

$$\nu_e + n \Leftrightarrow e^- + p$$

At a certain point of time ($t \approx 3$ s), the universe became transparent to the neutrinos and they ceased to play a role in transforming the nucleons into each other. Shortly thereafter, the e^+e^- pair production ceased because $kT < 1$ MeV, and the positrons quickly disappeared by annihilation, leaving behind an excess of electrons equal to the number of protons. The net result was that the p/n ratio became frozen at ~5. The temperature was still too high for the fusion reaction to occur, however, after ~4 m had elapsed, the p/n ratio increased further to ~7 because during this extra time, some of the neutrons decayed into proton via β^- decay, and the first nucleosynthesis reactions occurred.

EXERCISE 10.6:

(a) Calculate the ratio of the protons to the neutrons, when the temperature of the universe was about 10^{10} K. There should be more protons than neutrons, because protons are slightly less massive.

(b) About 3 s after the onset of the Big Bang, the p/n ratio became frozen when the temperature was as high as about 10^{10} K. When ~ 4 m had elapsed, fusion reactions took place converting the neutrons and the protons into helium-4 nuclei. Show that the resulting ratio of the masses of hydrogen and helium in the universe at that time was close to 3. The neutron half-life is 10.24 m.

Solution:

(a) The ratio of the protons and the neutrons can be found from the Maxwell–Boltzmann statistical distribution based on the available energy of the protons ($E_p = m_pc^2 + K$) and the neutrons ($E_n = m_nc^2 + K$).

$$\frac{\text{Number of protons}}{\text{Number of neutrons}} = \frac{\exp\left(\dfrac{-E_p}{k_BT}\right)}{\exp\left(\dfrac{-E_n}{k_BT}\right)} = \exp\left(\frac{\Delta mc^2}{k_BT}\right)$$

$$= \exp\left\{\frac{(939.57 - 938.28)\times 10^6 \text{ eV}}{(8.62\times 10^{-5} \text{ eV/K})(10^{10} \text{ K})}\right\} \approx 4.5$$

As the temperature continued to decrease, the p/n ratio continues to increase. However, after a certain point, the p/n ratio stabilized due to the production of helium.

(b) At $t = 3$ s, the ratio of p/n ratio is 4.5. After time t, since some of the neutrons decayed and converted into protons, the p/n ratio changes, i.e.,

$$\frac{N_{p0} + N_{n0}(1 - e^{-\lambda t})}{N_{n0}e^{-\lambda t}}$$

where N_{n0} is the initial number of neutrons and N_{p0} is the initial number of protons.

$$= \frac{1 + 0.2\left\{1 - e^{-\left(\frac{0.693}{10.24\ \text{m}}\right)(4\ \text{m})}\right\}}{0.2e^{-\left(\frac{0.693}{10.24\ \text{m}}\right)(4\ \text{m})}} \approx 7$$

Therefore, for every 14 protons, there are two neutrons. Two protons and two neutrons combine to make one helium-4 nuclide. Thus, the ratio of the mass of proton (hydrogen nuclei) to helium-4 nuclei is equal to $\frac{(14-2)}{4} = 3$.

These primordial nucleosynthesis reactions began with the production of deuterium by the first fusion reaction:

$$n + p \rightarrow d + \gamma$$

At high temperatures, the reverse reaction occurs as quickly as the deuterium production and there is no accumulation of the deuterium nuclei. The photon energy required to dissociate the deuterium is 2.225 MeV (the binding energy of the deuterium). We know that the photons have a blackbody spectrum. When the number of photons in the high energy tail above the energy of 2.225 MeV is less than the number of nucleons participating in the deuterium formation, there will be too few photons to inhibit the deuterium production. This happens when the universe has cooled to a critical value and deuterium survived long enough to allow them to participate in the reactions leading to the $A = 3$ nuclei:

$$d + p \rightarrow {}^3_2\text{He} + \gamma$$
$$d + n \rightarrow t + \gamma$$

Both tritium and helium-3 are more stable than deuterium and produce the very strongly bound α-particles by allowing the following reactions to take place:

$$t + p \rightarrow {}^4_2\text{He} + \gamma$$
$$t + n \rightarrow {}^4_2\text{He} + n$$
$${}^3_2\text{He} + n \rightarrow {}^4_2\text{He} + \gamma$$
$$d + d \rightarrow {}^4_2\text{He} + \gamma$$

The α-particle is the most stable of these light nuclei, and helium-4 was the main end product of these early fusion processes.

Further reactions to produce $A = 5$ or $A = 8$ nuclei do not occur because there are no stable nuclei with $A = 5$ or $A = 8$. A small amount of ${}^7_3\text{Li}$ is produced in the reactions, which are Coulomb barrier restricted:

$$^4_2\text{He} + t \rightarrow {}^7_3\text{Li} + \gamma$$

$$^3_2\text{He} + {}^4_2\text{He} \rightarrow {}^7_4\text{Be} + \gamma \xrightarrow{\beta^-} {}^7_3\text{Li} + \gamma$$

But the $^7_3$Li nuclei are very weakly bound and are rapidly destroyed. Thus, the synthesis of larger nuclei was blocked. All the neutrons at the beginning of the process were consumed to form $^4_2$He nuclei. Since, the *n*/*p* ratio is close to 1/7 initially, this would have led to a mass ratio of $^4_2$He to $^1_1$H in the early universe to 1 : 3. This is in excellent agreement with the observed ratio of 0.32 ± 0.13, obtained from the observations of the various nebulae and stars.

10.3 STELLAR EVOLUTION

Nucleosynthesis occurred in two steps—the primordial nucleosynthesis (discussed in the previous section) forming only the lightest nuclei (hydrogen, helium and traces of $^7_3$Li) and stellar nucleosynthesis, which began ~10^6 year after the Big Bang explosion and continuing to the present in the stars. To learn about the nuclear reactions that make the stars shine and generate the bulk of the elements, one needs to understand how stars actually work. We will discuss this in brief in the following section.

After the onset of Big Bang, the matter in the universe was dispersed and its inhomogeneous distribution, under the influence of gravity, condensed to form galaxies. Within these galaxies, clouds of hydrogen and helium gas could collapse under the influence of gravity. As the gas becomes denser, the opacity increases and the gravitational energy associated with the collapse is stored in the interior rather than being radiated into the space. Eventually, a radiative equilibrium is established. This cloud of gas and dust that will eventually develop into a star, known as a *protostar*, continues to shrink due to gravity and the stellar interior continues to get heated. When the interior temperature reaches ~10^7 K, the thermonuclear reactions between the hydrogen nuclei begin as some of the protons have sufficient kinetic energy to overcome the Coulomb barrier between the nuclei.

The first generation of stars that are formed in this way is called *population III stars*. These were massive, with relatively short life-times, and consisted of hydrogen and helium. These are extinct now, and the debris from these stars has been dispersed and incorporated into the later generation stars.

The second generation of stars is called *population II stars*. These stars are made up of hydrogen, helium and about 1% of the heavier elements like carbon and oxygen. The third generation of stars is called *population I stars*. These are made up of hydrogen, helium and 2%–5% of the heavier elements. Our sun, which is 4.5×10^9 y old, is a typical population I star. It has a mass of about 2×10^{30} kg, a radius of about 7×10^6 m, a surface temperature of about 6000 K and a luminosity of about 3.83×10^{26} W.

Astronomers observed a well-defined correlation between the luminosity and the surface temperature of the stars. The correlation is shown in Figure 10.4, and is called a H–R diagram. Most stars, like our sun, fall in a narrow band called the main sequence, on the H–R diagram. For stars in the main sequence,

$$L \text{ (Luminosity)} \propto T_{\text{surface}}^{5.5} \tag{10.6}$$

or in terms of their mass:

$$L \propto M^{3.5} \tag{10.7}$$

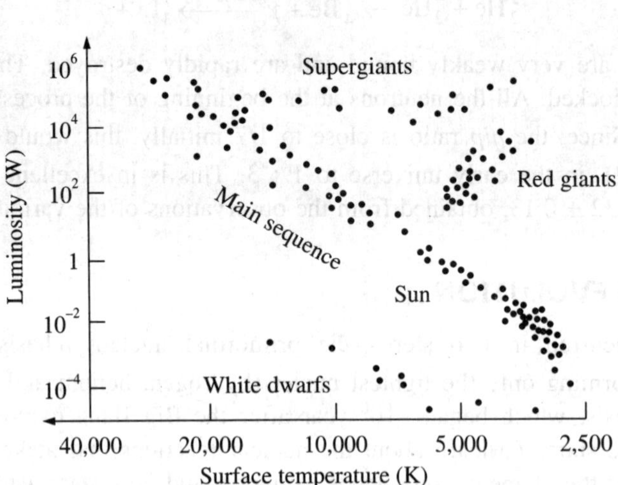

FIGURE 10.4 A schematic representation of an H–R diagram.

How long a star stays on the main sequence depends on its mass, which in turn, is related to the reactions rates in its interior. The rates of the reactions that take place in this thermal soup are known as *thermonuclear reaction rates*. They have been discussed in Chapter 5, in Section 5.10, on nuclear fusion.

10.3.1 Stellar Nucleosynthesis

After the primordial nucleosynthesis, the stellar nucleosynthesis continues the synthesis of the chemical elements in the interior of the stars up to the maximum in the nuclear binding energy curve at A ~ 60. An outline of the nuclear fusion reactions involved is given in Table 10.1.

Table 10.1 Outline of the Stellar Nuclear Fusion Reactions

Fuel	$k_B T$	*Products*
${}^{1}_{1}\text{H}$	0.002	${}^{4}_{2}\text{He}$
${}^{4}_{2}\text{He}$	0.02	${}^{12}_{6}\text{C}$, ${}^{16}_{8}\text{O}$, ${}^{20}_{10}\text{Ne}$
${}^{12}_{6}\text{C}$	0.07	${}^{16}_{8}\text{O}$, ${}^{20}_{10}\text{Ne}$, ${}^{24}_{12}\text{Mg}$
${}^{16}_{8}\text{O}$	0.2	${}^{12}_{6}\text{C}$, ${}^{28}_{14}\text{Si}$, ${}^{32}_{16}\text{S}$
${}^{20}_{10}\text{Ne}$	0.13	${}^{16}_{8}\text{O}$, ${}^{24}_{12}\text{Mg}$
${}^{28}_{14}\text{Si}$	0.3	$A < 60$

Hydrogen Burning in Stars

The first stage of the stellar nucleosynthesis, which is still occurring in the stars like our sun, is the hydrogen burning. When the interior of a star heats up to about 10^7 K, the first thermonuclear reactions begin to take place converting hydrogen into helium in a sequence of reactions known as the *proton–proton chain*:

$$p + p \rightarrow d + e^+ + \nu_e + 0.42 \text{ MeV} \tag{10.8}$$

where most of the released energy is shared between the two leptons. The *p–p* reaction is a weak interaction process and has, therefore, a very small cross-section (~10^{-47} cm^2) at these energies (k_BT ~ 1 keV). This is the calculated cross-section value because measurement of this so tiny cross-section is rather not possible. The resultant reaction rate is 5×10^{-18} reactions/s/proton.

EXERCISE 10.7: In 1939, Bethe pointed out that in the proton–proton fusion process, deuterium is produced by the weak interaction in a quark transformation that converts one of the protons to a neutron. Simple fusion up to two protons via strong interaction is impossible because of the di-proton being unstable.

Draw the Feynman diagram indicating the conversion of a bound proton into a neutron via weak interaction.

Solution:

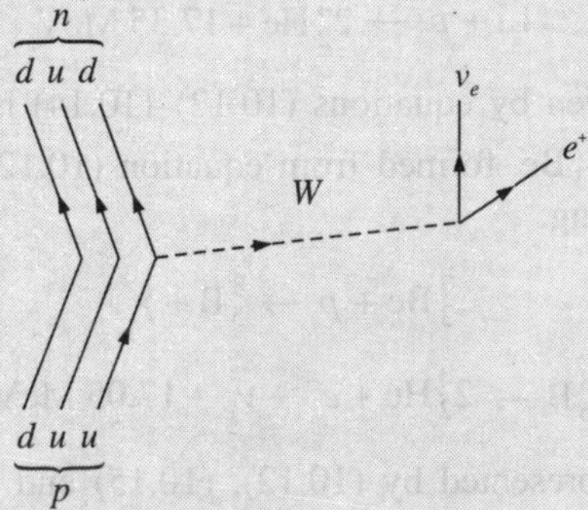

FIGURE 10.5 Feynman diagram depicting the conversion of a bound proton into a neutral via weak interaction.

There is an improbable (0.4%) variant of the *p–p* reaction, called the *pep reaction* that also leads to the deuteron production. This rare reaction is a source of energetic neutrinos from the sun.

$$p + e^- + p \rightarrow d + \nu_e + 1.42 \text{ MeV} \tag{10.9}$$

The next reaction in the sequence is

$$d + p \rightarrow {}^3_2\text{He} + \gamma + 5.49 \text{ MeV} \tag{10.10}$$

leads to the synthesis of ${}^3_2$He. The rate of this strong interaction is ~10^{16} times greater than the weak *p–p* reaction. If weak but steady supply of deuterium is there, the above reaction can proceed indefinitely, until no more hydrogen is left. The ${}^3_2$He that is formed in this way, can further react in three different ways to form ${}^4_2$He.

In the first way, two ${}^3_2$He nuclei react to form a ${}^4_2$He nucleus along with a proton.

$$ {}^3_2\text{He} + {}^3_2\text{He} \rightarrow {}^4_2\text{He} + p + 12.96 \text{ MeV} \tag{10.11} $$

The sequence of reactions given by (10.8)–(10.11) is known as the *ppI chain*. In a star consisting of pure hydrogen, ppI is the only option that is available. However, with the buildup of ${}^4_2$He, or it being already present from the beginning, the ppII and ppIII processes become possible.

In the ppII and ppIII processes, ${}^3_2$He available from the ppI chain, reacts with ${}^4_2$He in the following sequence to ultimately yield ${}^4_2$He:

$$ {}^3_2\text{He} + {}^4_2\text{He} \rightarrow {}^7_4\text{Be} + \gamma \tag{10.12} $$

${}^7_3$Be thus formed undergoes subsequent EC decay,[7]

$$ e^- + {}^7_3\text{Be} \rightarrow {}^7_3\text{Li} + \nu_e + 0.86 \text{ MeV} \tag{10.13} $$

The resulting ${}^7_3$Li nuclide undergoes proton capture to form two ${}^4_2$He nuclei,

$$ {}^7_3\text{Li} + p \rightarrow 2{}^4_2\text{He} + 17.35 \text{ MeV} \tag{10.14} $$

The sequence of reactions given by equations (10.12)–(10.14) is known as the *ppII chain*.

A small fraction of the ${}^7_3$Be formed from equation (10.12) reaction can undergo proton capture leading to the following:

$$ {}^7_3\text{Be} + p \rightarrow {}^8_5\text{B} + \gamma \tag{10.15} $$

$$ {}^8_5\text{B} \rightarrow 2{}^4_2\text{He} + e^+ + \nu_e + 17.05 \text{ MeV} \tag{10.16} $$

The sequence of reactions represented by (10.12), (10.15) and (10.16) is known as the *ppIII chain*.

These three chains of nuclear reactions constitute hydrogen burning and convert protons into ${}^4_2$He. The rate-limiting step in all these reactions is the first reaction to create the deuterium, which is referred to as the 'bottleneck.' In all the three processes, we see that the formation of one ${}^4_2$He nucleus involves two β-transitions, since in each case, we have in effect the reaction:

$$ 4p \rightarrow {}^4_2\text{He} + 2e^+ + 2\nu_e + 24.69 \text{ MeV} \tag{10.17} $$

This energy release of 24.69 MeV is, in fact, augmented by the energy (2 × 1.02 = 2.04 MeV)

7. This EC decay process does not involve the capture of the orbital electron of ${}^7_3$Be, since, it is fully ionized in a star, but rather involves the capture of a free continuum electron. As a consequence, the half-life of this decay is ~120 d, rather than the terrestrial half-life of 77d.

coming from the annihilation of the two positrons with the star's electrons. The ppI chain of reactions is responsible for ~91% of our sun's energy. The ppII process accounts for ~7% and the ppIII chain provides ~0.015% of the sun's energy. The rest 2% comes from the *CNO cycle*, which is discussed in the next section.

CNO Tricycle

In some stars, the presence of heavy elements like carbon, nitrogen and oxygen, which act as catalysts, leads to the occurrence of another set of reactions whose net effect is the conversion of four protons into ${}^4_2\text{He}$ nucleus (α-particle)

$$4p \rightarrow \alpha + 2e^+ + 2\nu_e$$

Figure 10.6 shows the CNO tricycle constituting of three distinct cycles that feed into each other, labelled by the two catalysts used in that particular cycle—carbon and nitrogen, oxygen-17 and nitrogen, or oxygen-18 and nitrogen.

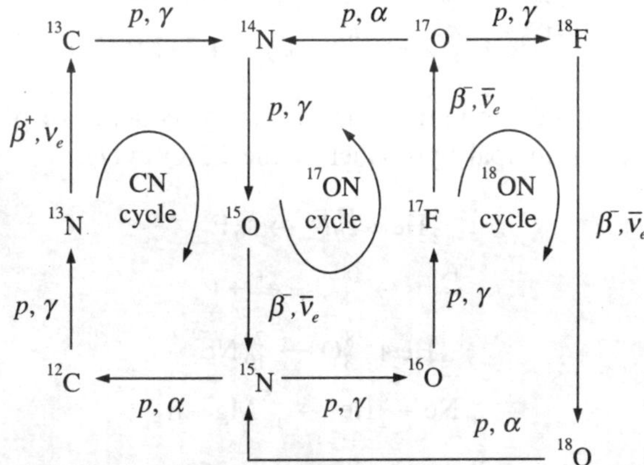

FIGURE 10.6 CNO tricycle constituting of three distinct cycles, labelled by the two catalysts used in that particular cycle.

The reactions in the CNO cycle are favoured at higher temperatures, where the Coulomb barrier for these reactions can be more easily overcome.

10.3.2 Later Stages of Stellar Burning

Having discussed how ${}^4_2\text{He}$ is synthesized out of hydrogen, we turn to the formation of heavier nuclides.

Helium Burning

As hydrogen is depleted at the centre of a star, the rate of energy generation through hydrogen burning will decrease and the core will begin to contract, so that with the rise in temperature, there will be onset of helium burning reactions.

Simple fusion of two ${}^4_2\text{He}$ nuclei according to

$$ {}^4_2\text{He} + {}^4_2\text{He} \rightarrow {}^8_4\text{Be} \quad (Q = -0.0191 \text{ MeV}) $$

seems to be unimportant because ${}^8_4\text{Be}$ is unstable ($T_{1/2} = 6.7 \times 10^{-17}$ s) and decays back into two α-particles. An exhaustive study shows that the only possibility of fusion is when three ${}^4_2\text{He}$ nuclei fuse together to form a ${}^{12}_6\text{C}$ nucleus,

$$ {}^4_2\text{He} + {}^4_2\text{He} + {}^4_2\text{He} \rightarrow {}^{12}_6\text{C} + \gamma $$

Three-body reactions are rare, but the aforementioned reaction proceeds through a resonance in ${}^{12}_6\text{C}$ at 7.65 MeV, corresponding to the second excited state of ${}^{12}_6\text{C}\,(I^+ = 0^+)$.

After a significant amount of ${}^{12}_6\text{C}$ is formed, the following α-capture reactions can take place:

$$ {}^4_2\text{He} + {}^{12}_6\text{C} \rightarrow {}^{16}_8\text{O} + \gamma + 7.16 \text{ MeV} $$

$$ {}^4_2\text{He} + {}^{16}_8\text{O} \rightarrow {}^{20}_{10}\text{Ne} + \gamma + 4.73 \text{ MeV} $$

At the same time, ${}^4_2\text{He}$ nuclide can initiate the following sequence of processes with any ${}^{14}_7\text{N}$ that may be present as the principal byproduct of the CNO cycle.

$$ {}^4_2\text{He} + {}^{14}_7\text{N} \rightarrow {}^{18}_9\text{F} $$

$$ {}^{18}_9\text{F} \rightarrow {}^{18}_8\text{O} + e^+ + \nu_e $$

$$ {}^4_2\text{He} + {}^{18}_8\text{O} \rightarrow {}^{22}_{10}\text{Ne} $$

$$ {}^{22}_{10}\text{Ne} + {}^4_2\text{He} \rightarrow {}^{25}_{12}\text{Mg} + n $$

The above sequence does not contribute significantly to the energy generation, but the neutrons produced in the last reaction of the sequence, can be radiatively captured to form new nuclides, especially the heavy nuclides with A > 56.

Synthesis of Nuclei with A < 60

In time, the helium of the star will be exhausted, leading to further gravitational collapse with a temperature increase equivalent to kT = 200 keV. At this temperature, the fusion reactions of the 'α-cluster' nuclei are possible. For example,

$$ {}^{12}_6\text{C} + {}^{12}_6\text{C} \rightarrow {}^{20}_{10}\text{Ne} + {}^4_2\text{He} $$

$$ {}^{12}_6\text{C} + {}^{12}_6\text{C} \rightarrow {}^{23}_{11}\text{Na} + p $$

$$ {}^{12}_6\text{C} + {}^{12}_6\text{C} \rightarrow {}^{23}_{12}\text{Mg} + n $$

$$ {}^{12}_6\text{C} + {}^{12}_6\text{C} \rightarrow {}^{24}_{12}\text{Mg} + \gamma $$

As ${}^{12}_6\text{C}$ is depleted, the temperature of the core will rise again and significant oxygen burning can occur.

$$^{16}_{8}O + ^{16}_{8}O \rightarrow ^{28}_{14}Si + ^{4}_{2}He$$

$$^{16}_{8}O + ^{16}_{8}O \rightarrow ^{31}_{31}P + p$$

$$^{16}_{8}O + ^{16}_{8}O \rightarrow ^{31}_{16}S + n$$

$$^{16}_{8}O + ^{16}_{8}O \rightarrow ^{32}_{16}S + \gamma$$

With further rise in the temperature (~5×10^9 K), a series of silicon burning reactions involving an equilibrium between the photo-disintegration and the radiative capture processes take place. For example,

$$^{28}_{14}Si + \gamma \rightarrow ^{24}_{12}Mg + ^{4}_{2}He$$

$$^{4}_{2}He + ^{28}_{14}Si \rightarrow ^{32}_{16}S + \gamma$$

One might also look for the fusion of two $^{28}_{14}Si$ nuclides to form $^{56}_{28}Ni$, which would then undergo two β^+ decays to form $^{56}_{26}Fe$, the most strongly bound nuclide. It must be the end product of all the processes in which the thermal equilibrium is maintained.

Synthesis of Nuclei with A > 60

The binding energy per nucleon curve peaks at $A \sim 60$ and decreases as A increases beyond 60. This indicates that the fusion reactions using charged particles are not energetically favoured for nuclei with mass $A \geq 60$. No more energy can be gained further from the fusion reactions. When the internal energy being produced is insufficient, the core cannot support the outer layers against gravitational collapse and the star begins its final contraction. A large amount of gravitational potential energy is converted into the kinetic energy and the star becomes a supernova when a large fraction of its mass is thrown out into space. For several days, the star's power output increases enormously by a factor of 10^{10} and many nuclear reactions take place, producing a flood of particles including neutrons. It is this supply of neutrons, which leads to the formation of the heaviest elements.

Neutron Capture Nucleosynthesis

The principal neutron reactions in a star during this stage are elastic/inelastic scattering and neutron capture (n, γ) reactions. Scattering leads back to the original nucleus and hence, only (n, γ) is mainly responsible for the formation of new elements. These neutron capture reactions have no Coulomb barriers to inhibit them. The reaction rates are governed by the Maxwell–Boltzmann distribution of velocities in a hot gas and the availability of free neutrons. Because of the $\sigma(n, \gamma) \propto \frac{1}{v}$ law [see equation (5.17) in Chapter 5], the reaction rate $N_n\langle\sigma v\rangle$ is largely governed by the neutron density N_n. If a particular nucleus is exposed to a flux of neutrons, it will be able to radiatively capture not just one neutron, but a whole succession of neutrons, in the process. However, a nucleus of fixed Z cannot absorb arbitrarily many neutrons without becoming unstable, since the neutron has a higher mass than the proton. At some point, a β^- decay must occur to restore the balance between the protons and the neutrons.

The distinction between r (rapid) and s (slow) processes has to do with the rate of the neutron capture vis-à-vis the β-decay rate. If the neutron flux is low, each capture that results in a β-unstable nucleus is followed by a β-decay before the next capture. This is the *s-process*. If the neutron flux is very high, then multiple neutron captures can occur in-between the β-decays. This is the *r-process*. Different sources of neutrons, one in hydrostatically stable stars, known as red giants, and the other in supernovae, give rise to these so-called s and r-processes, respectively.

The s-process

In the s-process, the mean time between neutron captures is long compared with most β-decay life-times. Consider the stable nucleus ${}^{56}_{26}\text{Fe}$. If it captures a neutron, the following reactions can occur:

$$ {}^{56}_{26}\text{Fe} + n \rightarrow {}^{57}_{26}\text{Fe (stable)} + \gamma $$

$$ {}^{57}_{26}\text{Fe} + n \rightarrow {}^{58}_{26}\text{Fe (stable)} + \gamma $$

$$ {}^{58}_{26}\text{Fe} + n \rightarrow {}^{59}_{26}\text{Fe } (T_{1/2} = 44.5\text{ d}) + \gamma $$

${}^{59}_{26}\text{Fe}$ will undergo β-decay before another neutron is captured, i.e.,

$$ {}^{59}_{26}\text{Fe} \rightarrow {}^{59}_{27}\text{Co (stable)} + \beta^- + \bar{\nu}_e $$

and further neutron captures will start with ${}^{59}_{27}\text{Co}$. The heaviest nucleus that can be formed in this way is ${}^{209}_{83}\text{Bi}$. The s-process terminates because of the following cyclic sequence:

$$ {}^{209}_{83}\text{Bi} + n \rightarrow {}^{210}_{83}\text{Bi} \xrightarrow{\beta^-} {}^{210}_{82}\text{Po} \xrightarrow{\alpha} {}^{206}_{84}\text{Pb} \xrightarrow{n} {}^{207}_{84}\text{Pb} \xrightarrow{n} {}^{208}_{84}\text{Pb} \xrightarrow{n} {}^{209}_{84}\text{Pb} \xrightarrow{\beta^-} {}^{209}_{83}\text{Bi} $$

In the production of s-process nuclei, which occurs over a very long time, the equilibrium between the production and the loss of adjacent nuclei is established. For such nuclei A with abundance N_A, produced through neutron capture by nuclei of mass number $A - 1$ and depleted by neutron capture leading to $A + 1$, its rate of change can be written as follows:

$$ \frac{dN_A}{dt} = (\sigma_{A-1}N_{A-1}) - (\sigma_A N_A) $$

where σ represents the neutron capture cross-section.

At equilibrium, $\frac{dN_A}{dt} = 0$, i.e.,

$$ \sigma_{A-1}N_{A-1} = \sigma_A N_A = \text{Constant} \tag{10.18} $$

The above relationship between the abundances of the neighbouring stable nuclei is a signature for the s-process. A similar equation relates the next set of nuclei giving $\sigma_A N_A = \sigma_{A+1}N_{A+1}$ and we thus, get the general result that $\sigma_A N_A$ should be independent of A within the entire s-process chain of neutron capture. Neutron capture cross-sections have been measured quite precisely on all stable nuclei, and Figure 10.7 shows $\sigma_A N_A$ values vs. A. After decreasing from a peak at the region of greatest stability (A ~ 60), the product approaches an approximately constant value beyond A ~ 100, consistent with the general expectation for the *s*-process nucleosynthesis.

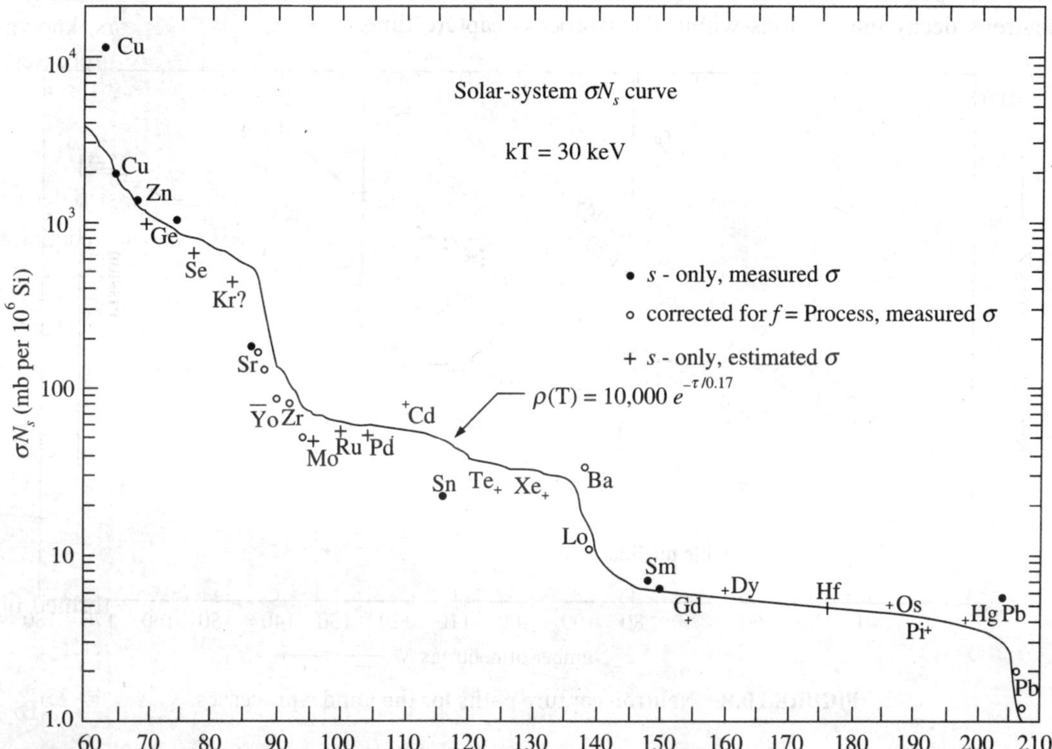

FIGURE 10.7 Solar system σN-curve. The product of the isotopic abundances (Si = 10^6) and neutron capture cross-sections at kT = 30 keV (in mb) is plotted versus mass number A for various nuclides. The solid line is a calculated curve corresponding to an exponential distribution of integrated neutron flux. [Seeger, P.A., W.A. Fowler, and D.D. Clayton, 'Nucleosynthesis of heavy elements by neutron capture,' *Astrophys. J. Suppl*, XI, 121-66, 1965]

The r-process

If the time scales of the neutron capture reactions are smaller than the β-decay life-times, then rapid neutron capture or the *r*-process occurs. For the *r*-process nucleosynthesis, large neutron densities are needed, e.g., in supernovae, neutron densities around $10^{28}/\text{m}^3$ are believed to be available. In the *r*-process, the sequence of the neutron captures by a nucleus of a given Z will not stop at the first β-unstable isotope that is encountered; rather, any such gap will be jumped and the neutron captures will continue until the nuclei become very neutron-rich, close to the neutron drip line. At this point, the β-decay times decrease to the point that the β-decay starts competing with the neutron capture reaction.

Figure 10.8 shows the neutron capture paths for the s and r-processes through the chart of nuclides. It can be seen, how the production of nuclei follows a zigzag path, with an increment in the mass number ($A = N + Z$) when a neutron is captured, and an increment in the atomic number Z when the β-decay precedes the next neutron capture. One can also notice that the r-process capture path climbs up in atomic number along the neutron magic numbers (N = 50, 82, 126). Near the magic number, the β-decay times become so short that added neutrons decay into protons within the r-process capture time.

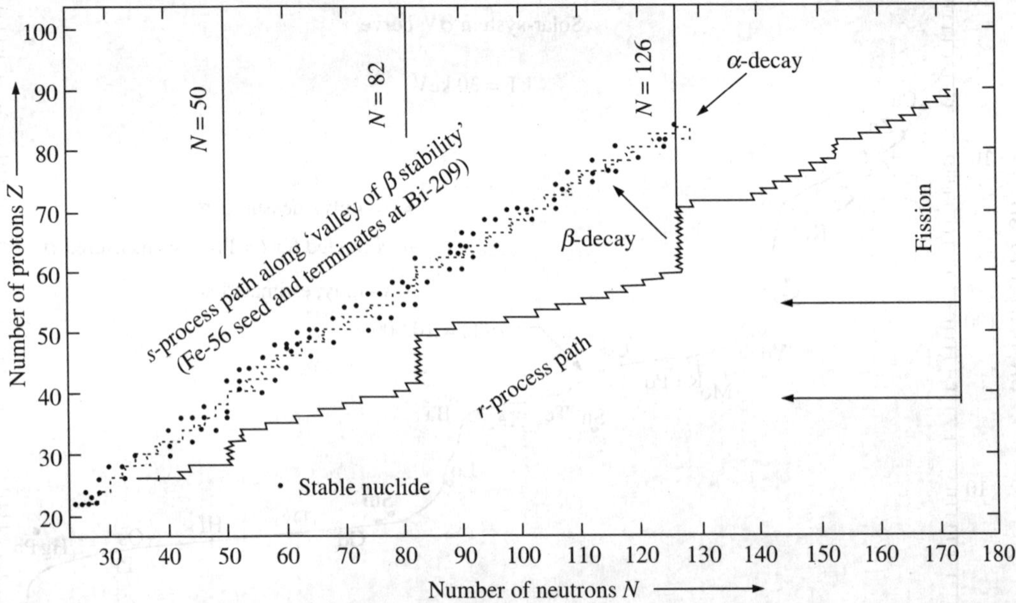

FIGURE 10.8 Neutron capture paths for the s and r-processes.

EXERCISE 10.8: There are seven stable isotopes of tellurium for $122 \le A \le 130$, viz. ${}^{122}_{52}\text{Te}$, ${}^{123}_{52}\text{Te}$, ${}^{124}_{52}\text{Te}$, ${}^{125}_{52}\text{Te}$, ${}^{126}_{52}\text{Te}$, ${}^{128}_{52}\text{Te}$, and ${}^{130}_{52}\text{Te}$. Indicate the main neutron capture processes (s or r) by which each one was created inside a star.

Solution: ${}^{122}_{52}\text{Te}$ is created by the s-process via ${}^{121}_{51}\text{Sb}\,(n,\gamma){}^{122}_{51}\text{Sb} \xrightarrow{\beta^-} {}^{122}_{52}\text{Te}$. (Since, ${}^{121}_{51}\text{Sb}$ is a stable isotope, after neutron capture and subsequent β-decay it will produce ${}^{122}_{52}\text{Te}$.)

${}^{123}_{52}\text{Te}$ is created by the s-process via ${}^{122}_{52}\text{Te}\,(n,\gamma)\;{}^{123}_{52}\text{Te}$. (Since, ${}^{122}_{52}\text{Te}$ is a stable isotope, after neutron capture it will produce ${}^{123}_{52}\text{Te}$.)

${}^{124}_{52}\text{Te}$ is created by the s-process via ${}^{123}_{51}\text{Sb}\,(n,\gamma){}^{124}_{51}\text{Sb} \xrightarrow{\beta^-} {}^{124}_{52}\text{Te}$. (Although relatively stable, ${}^{123}_{52}\text{Te}$ has an abundance of only 0.9% as compared to ${}^{123}_{51}\text{Sb}$, which is stable and has an abundance of 42.79%. Thus, the preferred channel for the reaction will be via ${}^{123}_{51}\text{Sb}$.)

$^{125}_{52}$Te is created by the *s*-process via $^{124}_{52}$Te (n, γ) $^{125}_{52}$Te. (Since, $^{124}_{52}$Te is stable)

$^{126}_{52}$Te is created by the *s*-process via $^{125}_{52}$Te (n, γ) $^{126}_{52}$Te. (Since, $^{125}_{52}$Te is stable)

$^{128}_{52}$Te and $^{130}_{52}$Te are populated by the *r*-process since, $^{127}_{52}$Te/$^{127}_{51}$Sb and $^{129}_{52}$Te/$^{129}_{51}$Sb are too unstable.

PROBLEMS

10.1 If Hubble's law holds, how far away a quasar (celestial objects that resemble stars but whose large red-shift and apparent brightness imply large distance and huge energy output) with an apparent recession of $0.93c$. **[Ans:** 11 billion light year]

10.2 The expression for the relativistic Doppler effect is as follows:

$$\lambda = \sqrt{\frac{1+\beta}{1-\beta}}\,\lambda_0$$

where λ is the light wavelength and $\beta = \frac{v}{c}$. Plot the red-shift $\left(\frac{\Delta\lambda}{\lambda}\right)$ versus the recession velocity based on the above equation. Note how the recession velocity approaches the speed of light with the increase in the red-shift.

10.3 Calculate the temperature of the universe when it has cooled enough that photons no longer dissociate deuterons. Use the mean value of the distribution. Binding energy of the deuteron $B_d = -2.2244$ MeV. **[Ans:** 2.58×10^{10} K]

10.4 Determine the nuclear mass difference between $^{7}_{3}$Li and $^{7}_{4}$Be. The half-life of an atom of $^{7}_{4}$Be is 53.3 days on earth. How do the terrestrial $^{7}_{4}$Be atom decay? Draw the Feynman diagram of the processes and show that each of the decay channels conserves the charge and the lepton number. Which of these processes is allowed energetically? **[Ans:** β^{+}, EC but energetically allowed is EC]

10.5 From being radiation-dominated, if 400000 years after the Big Bang, the universe changed to matter dominated, at what density $\rho_{\text{radiation}}$ did this occur? **[Ans:** 8.4×10^{-25} kg m^{-3}]

10.6 Explain why in the low-mass stars, helium cannot burn to create heavier mass nuclei?

10.7 Calculate the energy deposited in the sun by the ppI, ppII and ppIII cycles. Assume that all of the energy is deposited except for the energy carried off in neutrinos. For the neutrinos, neglect the $p + p + e^{-}$ reaction and assume that in β-decays the neutrino carries on an average half of the kinetic energy available to the outgoing neutrino and positron. **[Ans:** 26.3 MeV, 25.71 MeV, 19.02 MeV]

10.8 Which nucleosynthesis processes are responsible for the following nuclei:

$$^{7}_{3}\text{Li},\ ^{12}_{6}\text{C},\ ^{20}_{10}\text{Ne},\ ^{56}_{26}\text{Fe},\ ^{84}_{38}\text{Sr},\ ^{96}_{40}\text{Zr},\ ^{114}_{50}\text{Sn},\ ^{209}_{83}\text{Bi},\ ^{238}_{92}\text{U}$$

10.9 The first reaction in the proton-proton chain is $p + p \rightarrow d + e^{+} + \nu_e$. Calculate the Q-value of the reaction and determine the maximum neutrino energy.
[**Ans:** 0.42 MeV, maximum neutrino energy = 0.42 MeV]

10.10 Estimate the temperature of the universe at which the ratio of the neutrons to the protons in the universe is 6.7. Assume that the particles follow the Maxwell–Boltzmann distribution of velocities in a hot gas. [**Ans:** $T = 7.9 \times 10^{9}$ K]

BIBLIOGRAPHY

Basdevant, J., Rich, J. and Spiro, M., *Fundamentals in Nuclear Physics: From Nuclear Structure to Cosmology*, Springer-Verlag, New York, 2005.

Krane, K.S., *Introductory Nuclear Physics*, Wiley, New York, 2008.

Lilley, J., *Nuclear Physics: Principles and Applications*, Wiley, New York, 2002.

Loveland, W.D., Morrissey, D.J. and Seaborg, G.T., *Modern Nuclear Chemistry,* Wiley, New Jersey, 2005.

Pearson, J.M., *Nuclear Physics: Energy and Matter*, Adam Hilger, Bristol, 1986.

Rose, W.K., *Advanced Stellar Astrophysics,* Cambridge University Press, Cambridge, 1998.

Thornton, S.T. and Rex, A., *Modern Physics*, Brooks/Cole Cengage Learning, New Delhi, 2007.

Index